M000268658

Second Edition

PAVEMENT ENGINEERING

Principles and Practice

Second Edition

PAVEMENT ENGINEERING

Principles and Practice

Rajib B. Mallick and **Tahar El-Korchi**

CRC Press
Taylor & Francis Group
Boca Raton London New York

CRC Press is an imprint of the
Taylor & Francis Group, an **informa** business

CRC Press
Taylor & Francis Group
6000 Broken Sound Parkway NW, Suite 300
Boca Raton, FL 33487-2742

© 2013 by Taylor & Francis Group, LLC
CRC Press is an imprint of Taylor & Francis Group, an Informa business

No claim to original U.S. Government works

Printed in the United States of America on acid-free paper
Version Date: 20121107

International Standard Book Number: 978-1-4398-7035-8 (Hardback)

This book contains information obtained from authentic and highly regarded sources. Reasonable efforts have been made to publish reliable data and information, but the author and publisher cannot assume responsibility for the validity of all materials or the consequences of their use. The authors and publishers have attempted to trace the copyright holders of all material reproduced in this publication and apologize to copyright holders if permission to publish in this form has not been obtained. If any copyright material has not been acknowledged please write and let us know so we may rectify in any future reprint.

Except as permitted under U.S. Copyright Law, no part of this book may be reprinted, reproduced, transmitted, or utilized in any form by any electronic, mechanical, or other means, now known or hereafter invented, including photocopying, microfilming, and recording, or in any information storage or retrieval system, without written permission from the publishers.

For permission to photocopy or use material electronically from this work, please access www.copyright.com (http://www.copyright.com/) or contact the Copyright Clearance Center, Inc. (CCC), 222 Rosewood Drive, Danvers, MA 01923, 978-750-8400. CCC is a not-for-profit organization that provides licenses and registration for a variety of users. For organizations that have been granted a photocopy license by the CCC, a separate system of payment has been arranged.

Trademark Notice: Product or corporate names may be trademarks or registered trademarks, and are used only for identification and explanation without intent to infringe.

Library of Congress Cataloging-in-Publication Data

Mallick, Rajib Basu, 1966-
    Pavement engineering : principles and practice / Rajib B. Mallick and Tahar El-Korchi. -- 2nd ed.
      p. cm.
    Includes bibliographical references and index.
    ISBN 978-1-4398-7035-8 (hardback)
    1. Pavements--Design and construction. I. El-Korchi, Tahar. II. Title.

TE251.M3245 2013
625.8--dc23   2012043213

Visit the Taylor & Francis Web site at
http://www.taylorandfrancis.com

and the CRC Press Web site at
http://www.crcpress.com

*This book is dedicated to the past, present,
and future generations of civil engineers*

Contents

Preface to the Second Edition

Along with the encouragement of our publishers, the main motivations for the second edition of this book have been the desire to include emerging topics in the area of pavement engineering, to strengthen some of the areas that were not covered adequately in the first edition, and to correct the errors that had inadvertently crept into the first edition. All chapters have been updated with relevant material, and explanatory schematics and photos have been added.

The emerging areas that have been included in this new edition are sustainable pavement engineering and environmental mitigation in transportation projects. They have been included with the view that these areas are becoming universally important and that in many cases professionals, such as transportation planners, who do not deal directly with pavement materials, do need to become more knowledgeable about different options that are available for pavement design and construction.

The second edition consists of new figures that include more informative schematics and photographs to better explain construction details, materials processing, and relevant equipment and methods. This was done because of two reasons. First, for an applied field such as pavement engineering, a good understanding of the construction practices is essential for newly graduating engineers. Second, a good understanding of materials processing and construction is essential for solving numerous problems—it helps in problem formulation, development of solution approaches, and refinement of techniques.

Finally, more relevant issues and recently developed techniques and guidelines for practical problems (and web-based references) have been included, such as selection of type of pavements, effect of vehicle tires, use of smart sensors in rollers, image analysis, use of software for drainage analysis, pavement surface characterization and forensic investigation of pavement failures, and continuously reinforced concrete pavement (CRCP).

Despite all these updates, we are certain that the book can still be improved upon, and we request and encourage our readers to suggest improvements. However, we hope that the book, with the updated materials, will be an appropriate textbook or reference in any pavement-related course—it will be able to provide the basics and at the same time generate interest in the readers, which is so very essential in the learning process!

Preface to the First Edition

This book has been written for undergraduate and graduate students of civil engineering as well as practicing engineers. We have been greatly influenced by two classic books in this area: E. J. Yoder and M. W. Witczak, *Principles of Pavement Design*, 2nd edn. (John Wiley & Sons, 1975), and Yang Huang, *Pavement Analysis and Design* (Prentice Hall, 1993). These two books lay down most of the important principles of pavement engineering, and it is difficult, if not impossible, to find any alternative for these two books.

However, it is our belief that most students at the undergraduate level can spend only a few terms or semesters studying pavements, and for this reason they need a textbook that presents a broad but concise view of the more relevant topics in pavement engineering. The need is the same for many practicing engineers, who can use a reference textbook that is current and practical.

It is important for graduate students to get exposed to a broad overview of different topics in pavement engineering before starting in-depth research in any specific area. This can be helpful when they are focusing on a narrow topic of research to view the relevance of their work in the larger scope of pavement engineering.

For professionals who go back to university, it is often the case that they gain professional experience in either a specific area of pavement engineering or some other area (such as structures), and they need a broad overview of the subject before getting started with an in-depth study. This book can be used for providing such an overview on pavement engineering for graduate students.

The Internet has changed both teaching and learning forever. In most cases, the use of a single textbook is just not enough. A vast treasure-house of online documents is accessible on the Internet. When possible, this textbook has referenced numerous websites with a wealth of information and documentation relevant to pavement engineering. Students can supplement their textbook with the relevant online documents. This is important because students can now view, read, and understand "real-world" case studies, research reports, and specifications from pavement professionals and are no longer restricted to the knowledge confined within a single textbook. And since learning through multiple sources has become easy and accessible through the Internet, students are encouraged to tap into this. This book makes reference to Internet resources wherever they are appropriate and encourages students to download freely available design guides/charts and procedures to work out the practice problems provided at the end of each chapter.

The objective of this book is to explain the basic principles of the various design and construction procedures with relevant examples. The required resources in the form of charts, tables, and weblinks have been provided wherever appropriate for such examples. Note that different agencies have developed different types of pavement design and construction methods over the years and have published comprehensive manuals. We have purposefully not provided the full range of charts or tables from any specific organization in this textbook for two reasons. First, the manuals published by the different organizations are complete by themselves and are the best source of information, and they should be consulted for design or construction. Second, it is the objective of this book to assist the user in self-learning about design and construction and not to actually be a design and construction manual or guide. We firmly believe that while learning can be accomplished with a textbook, the actual application of knowledge requires the utilization of resources provided by the industry and pavement agencies. As such, students are encouraged to use appropriate guidelines from these sources.

Which unit system should be used in this book? We have debated this question at length, since both the English/U.S. and International System of Units (SI)/metric systems continue to be used in pavement engineering. While most people use the SI units outside the United States, many of the original methods/charts that are still used today were developed using English/U.S. units. One perfect example is the California Bearing Ratio (CBR). A good approach would be to utilize both systems by putting one or the other units in parentheses. However, we feel that there are two drawbacks to this approach. First, this system does not encourage students to learn conversions from˙one system to the other, and, second, equations, tables, numbers, and figures often become cluttered and clumsy with the use of a double-unit system. It is important for students in pavement engineering to be conversant with both systems of units. With these considerations in mind, we have used both systems, but in many cases, not together. The units corresponding to the original models/ equations have been retained. A conversion guide is provided in the appendix to assist the user.

This book is offered as a textbook. Obviously, it draws information from a number of sources— notes, research reports, specifications, and manuals. The importance of these documents cannot be overstated, and students are encouraged to review them, either online or in school libraries. Some more relevant sources are as follows:

ACI Committee 211. *Standard Practice for Selecting Proportions for Normal, Heavy-Weight and Mass Concrete, ACI 211. 1-91*. Farmington Hills, MI: American Concrete Institute, 1991.

ACI Committee 214. *Recommended Practice for Evaluation of Strength Test Results for Concrete, ACI 214-77*. Reapproved 1997. Farmington Hills, MI: American Concrete Institute, 1977.

American Association of State and Highway Transportation Officials. *AASHTO Guide for Design of Pavement Structures*. Washington, DC: AASHTO, 1993.

American Association of State and Highway Transportation Officials. *MDM-SI-2, Model Drainage Manual, 2000 Metric Edition*. Washington, DC: AASHTO, 2000.

ASCE. *Introduction to Mechanistic-Empirical Pavement Design: Reference Manual*. Seminar. Baltimore, MD: American Society of Civil Engineers, 2004.

Barksdale, R. D., Ed. *The Aggregate Handbook*. Washington, DC: National Stone Association, 1991.

Coduto, D. P. *Geotechnical Engineering: Principles and Practices*. Upper Saddle River, NJ: Prentice Hall, 1999.

Federal Aviation Administration. FAA Advisory Circular AC No. 150/5320-6D. Washington, DC: U.S. Department of Transportation, 1995.

Federal Highway Administration. *Pavement Deflection Analysis*, NH1 Course No. 13127, Participant workbook, FHWA Publication No. FHWA-HI-94-021, 1994. Ullidtz, Per. *Pavement Analysis*, New York: Elsevier, 1987.

Huang, Y. *Pavement Analysis and Design*. Englewood Cliffs, NJ: Prentice Hall, 1993.

Read, J. and D. Whiteoak. *Shell Bitumen: The Shell Bitumen Handbook*, 5th edn. London, U.K.: Thomas Telford, 2003.

Roberts, F. L., P. Kandhal, E. R. Brown, D. Y. Lee, and T. W. Kennedy. *Hot Mix Asphalt Materials, Mixture Design, and Construction*. Lanham, MD: NAPA Education Foundation, 1996.

Yoder, E. J. and M. W. Witczak. *Principles of Pavement Design*, 2nd edn. New York: John Wiley, 1974.

Acknowledgments

For the second edition, specifically, we would like to acknowledge the help of Professor Suzanne LePage, who has contributed to Chapters 24 and 25 on sustainable pavement engineering and environmental mitigation in transportation projects.

Several colleagues and friends have provided valuable feedback on the first edition, and the improvements would not have been possible without their help. In particular, we would like to thank the following: Dr. A. Veeraragavan, Dr. Sivakumar Babu, Dr. Soheil Nazarian, Dr. Jorge Prozzi, Dr. Linbing Wang, Dr. Okan Sirin, Dr. Kevin Hall, and Dr. Elie Y. Hajj.

This book is the collective result of many years of learning and experience. Several people have contributed to this book, and it is not possible to list all of them here. There are some people, though, without whom this book would not have been possible: Ray Brown, Prithvi Kandhal, Frazier Parker, and Dan Brown of Auburn University; Rick Bradbury, Dale Peabody, Brian Marquis, and Wade McClay of the Maine Department of Transportation; Sudhansu Bhattacharjee, Consultant; Robert Pelland of the Massachusetts Port Authority (Massport); Mike Marshall of Wirtgen GmbH; Ed Kearney of Wouter Gulden of the American Concrete Pavement Association (ACPA—SE Chapter); Robert Otto Rasmussen of The Transtec Group, Inc.; Maureen Kestler of the U.S. Forest Service; and Frances Wychorski, Agata Lajoie, Martins Zaumanis, Soomin Kim, Yeesock Kim, and Don Pellegrino of Worcester Polytechnic Institute. We thank them from the bottom of our hearts.

Writing a book requires support and encouragement from the family. Special thanks go to Sumita and Urmila, and to Marya, Miriam, and Youssef, for their patience, help, and encouragement. This book would not have been possible without the encouragement of our parents, Bimalangshu Kumar Basu Mallick and Monika Basu Mallick, and M'hamed El-Korchi and Bahija El-Bouchti El-Korchi.

Last but not least, we would like to thank the ever helpful staff of Taylor & Francis Group, especially Joe Clements, acquiring/contact editor; Andrea Dale, editorial assistant; Jill Jurgensen, senior project coordinator; Marc Gutierrez, marketing manager; Joette Lynch, project editor; and Vijay Bose, SPi project manager. It has been a pleasure to work with them.

Authors

Dr. Rajib Mallick has extensive experience in the study of asphalt materials, mixture design, construction, pavement design, recycling, and nondestructive testing. From 1992 to 1998, he worked as a research assistant and as a senior research associate at the National Center for Asphalt Technology (NCAT) and is currently the Ralph White Family Distinguished Professor of Civil and Environmental Engineering at Worcester Polytechnic Institute (WPI). He has completed numerous research projects for several departments of transportation, the Federal Highway Administration (FHWA), the Massachusetts Port Authority (Massport), the National Science Foundation (NSF), the Federal Aviation Administration (FAA), and several private practitioner organizations. Dr. Mallick has coauthored more than 100 papers for journals and conference proceedings, as well as several practical reports, manuals, and state-of-the-practice reports for federal, state, and local highway agencies. He has taught professional courses on asphalt technology, presented courses on recycling for the FHWA, as well as lectured in the national workshops at the Indian Institute of Technology (IIT) at Kanpur, Kharagpur, and Madras in India. He has presented a series of seminars on hot mix asphalt paving for the Baystate Roads Programs for the Massachusetts Highway Department. He has also introduced many innovations in his courses and has received several awards from the NSF. He is a member of several professional organizations and a registered professional engineer (PE) in Massachusetts. Dr. Mallick has served as a consultant in several projects for practitioner organizations and has served IIT Madras as a professor of civil engineering for one year.

Dr. Tahar El-Korchi is a professor and head of civil and environmental engineering at Worcester Polytechnic Institute. He has been at WPI since 1987. He does research and teaches courses on pavement analysis, design and management, construction materials testing and evaluation, cement and concrete materials and construction processes, structural analysis and design, and high-performance structural materials at the undergraduate and graduate level. He has published more than 80 publications and presentations and has served at numerous WPI global project centers, including Puerto Rico; Washington, District of Columbia; Morocco; Costa Rica; and Panama. Dr. El-Korchi was granted the prestigious Presidential Young Investigator Award by the National Science Foundation in 1991. His research funding has been generated from federal and state sources, including the National Science Foundation, the U.S. Army Waterways Experiment Station (WES), the U.S. Geological Survey's Columbia River Research Laboratory (CRRL), the Small Business Innovation Research (SBIR), the Federal Highway Administration, National Institute of Standards and Technology (NIST), the New England Transportation Consortium, the Massachusetts Highway Department, and numerous industries. He was an invited scientist at the Turner Fairbanks Highway Research Center (Federal Highway Administration), Washington, District of Columbia (1993–1994). He was also an advisor to the Mass Highway QC/QA Implementation Program, the LTPP Expert Technical Group, and the FHWA LSPEC. Dr. El-Korchi is also a consultant to several major corporations. He provides recommendations for the design and evaluation of pavements, materials, and structures.

1 Introduction and Description of Pavements

1.1 IMPORTANCE

Pavements are an essential part of our life. We use them as roads, runways, parking lots, and driveways. Pavements are engineered structures and are important for our everyday life, commerce and trade, and defense. Surface transportation is the most widely used mode of transportation in the world, and a country's development is often measured in terms of its total paved road mileage. The construction of roads is and will continue to be a major industry in developing countries, and as the infrastructure matures, it will be a major industry in developed countries as well.

Like any other engineered structure, pavements are expected to be adequately strong and durable for their design life. They are expected to function properly by providing a smooth traveling surface for the traffic under various conditions of the environment. In order to ensure this, pavements must be designed, constructed, maintained, and managed properly.

Pavements can be broadly classified into asphalt (or flexible) and concrete (or rigid) pavements (Figure 1.1). Pavements consist of different layers, more so in the case of asphalt pavements than concrete ones. From the bottom up, these layers are known as the subgrade, subbase, base, and binder and/or surface. There are certain pavements with asphalt surface layers on top of concrete layers.

In the United States, there are about 4 million miles of roads, of which approximately 2.5 million are paved. The federal government and state departments of transportation (DOTs) spend a significant portion of their budget on maintaining and managing existing pavements and rehabilitating old pavements. More than 90% of commodities are transported on highways in the United States. Roads in poor condition end up costing the DOTs a lot of money for repairs, as well as to the users for repairing damaged vehicles. These roads are also unsafe for travel—for example, more than 30% of traffic fatalities in the state of Massachusetts in the United States have been reported to be due to poor road conditions.

1.2 FUNCTIONS

The most important function of the pavement is to withstand the load applied from a vehicle such as a truck or an aircraft, without deforming excessively. The layered structure of the pavement is meant for ensuring that the load is spread out below the tire, such that the resultant stress at the bottom layer of the pavement, the subgrade, is low enough not to cause damage. The most significant load applied to a pavement surface comes from a truck or an aircraft tire. The approach in a flexible pavement is to spread the load in such a way that the stress at the subgrade soil level is small enough so that it can sustain the stress without any major deformation. When the existing soil is not stiff enough to support the relatively small stress, then there is a need to improve the soil. There is also a need to improve the soil if it is susceptible to moisture. Such a problem can be solved by treating the soil with an additive, such as lime and a Portland cement.

Since pavements are exposed to the environment, a very important factor in the design of pavements is the consideration of water, which could be coming from rain/snow (surface water) and/or from the ground (ground/subsurface water). Since water can be detrimental to a pavement, a basic necessity of designing a proper pavement is to provide adequate drainage for both surface and

(a)   (b)

FIGURE 1.1   (a) Example of asphalt pavement. (Courtesy of Mike Marshall, Wirtgen, GmbH, Windhagen, Germany.) (b) Example of concrete pavement. (Courtesy of Wouter Gulden, ACPA, SE Chapter, Duluth, GA.)

subsurface water. Standing water on a pavement can cause hydroplaning, skidding, and accidents. There is a need to make sure that water from precipitation is drained away quickly and effectively and that there is no depression on the roads to collect water. Water present in frost-susceptible soils in the subgrade can freeze, causing heaving and failure of the pavement. Therefore, frost-susceptible materials should be avoided. If this is not possible, then the pavement structure above the subgrade should be thick enough to prevent the freezing front from reaching the frost-susceptible soil. Similarly, as one expects some water to make its way through random cracks and joints, proper subsurface drainage must be provided, and the material within the pavement structure should be made resistant to the actions of water—otherwise the aggregates, for example, would be washed away due to repeated traffic-induced pressure or freeze–thaw pressures.

In most cases, it is not possible to completely avoid water flowing inside a pavement. Such ingress of water can physically remove materials from inside a pavement structure and also freeze and cause deformations in the pavement. To prevent this, the pavement material must be selected properly and, if needed, modified.

1.3   DESIGN AND CONSTRUCTION

Generally the layers in a pavement improve in quality as one goes up from the bottom to the surface layer. The surface layer, which can be asphalt or concrete, is the most expensive and stiff/durable layer in the entire pavement structure. Components of this layer are mostly naturally occurring materials, for example, asphalt binder is a by-product of the petroleum distillation process and aggregates are obtained from rock quarries or riverbeds. These materials are combined and used in different proportions to produce the final material that is used in the pavement. For example, asphalt binder is mixed with aggregates to produce hot mix asphalt (HMA) for asphalt pavements, while Portland cement is mixed with aggregates in Portland cement concrete (PCC) pavements. In both cases, the mixing must be conducted in the correct proportions to ensure adequate quality of the mixture. It is important to find out whether the resultant mix has the adequate strength and stiffness through testing. Such *testing* is generally conducted in the laboratory during the mix design process. During this process, loading and effect of environmental conditions can be simulated in the laboratory. If the responses to this testing do not meet our *expectations*, then the mix needs to be redesigned and/or the materials need to be reselected. These "expectations" are specifications that have been developed on the basis of experience and research. State DOTs or agencies can use their own specifications and/or use specifications from organizations such as the American Association of State Highway and Transportation Officials (AASHTO) or the Federal Aviation Administration (FAA). Similarly, testing is also conducted according to standards, which again can be from state

DOTs or the AASHTO. Whether or not the material can be placed and compacted is also determined during the mix design process.

In most cases, pavement engineers are restricted to using locally available materials, with or without some modifications, because of economic and practical reasons. With these available materials it is important to determine what thickness of each layer, and hence the entire pavement, is required to carry the loads under different environmental conditions without any problem. This step, known as the structural design, makes sure that the pavement structure as a whole can withstand traffic for its design life, even though the traffic might increase and the properties of the layer might change cyclically and/or progressively during its design life.

Generally several layers are present in an asphalt pavement. From the bottom up, the layers are known as the subgrade, subbase, base, and binder and/or surface. Generally, the bottommost layer is soil; the subbase and/or base layers can be granular soil, or aggregates or asphalt-aggregate mixtures (mixes); and the binder and surface are asphalt mixes. While designing, adequate thickness to each layer is assigned, so as to obtain the desirable properties in the most cost-effective way. Concrete pavements may not have as many layers, and in many cases the concrete slab rests on a stabilized subgrade, which consists of soils modified with some additives.

Once the pavement materials/mix and structure are designed, it must be constructed properly. To ensure this, the material must be laid down and cured (if needed) and compacted in the proper way so that it has the desirable qualities, such as density and/or stiffness. While selecting the materials and designing the mix and the structure, it is important to keep workability in consideration, since the best-designed mix would be worthless if it cannot be constructed properly. Furthermore, quality control must be carried out during construction to ensure strict adherence to specification and hence uniformly good quality over the entire project duration.

1.4 MAINTENANCE AND REHABILITATION

Starting from day one after construction, a pavement starts deteriorating in quality. Even though a properly designed and constructed pavement will not deteriorate so as to cause total failure within its design life, if no maintenance is performed, the entire pavement will become totally worthless at the end of its design life. Furthermore, ingress of water through random openings such as surface cracks, and well-defined openings such as joints, can lead to quick deterioration of the quality of the pavement. The best approach is to regularly perform maintenance operations, similar to any engineered structure or product, through actions such as joint filling, crack repair or filling, or pothole patching. Note that these operations do not specifically increase the design life but do prevent its rapid deterioration. For pavements such as those on highways and airport runways, proper maintenance is critical for the safety of the vehicles and their occupants.

In most cases, once built, a pavement can be "recycled" at the end of its design life almost infinite times, by reusing the existing materials solely or in combination with new materials. Off-course rehabilitation of pavements is a costly process. Hence, for any pavement network, a proper inventory of the condition of different pavements must be kept and utilized effectively to determine the time/order in which the different pavement sections should be rehabilitated. This is important because of two reasons. First, there is never enough funding for rehabilitation of all of the roads in a network at the same time, and secondly, different roads deteriorate at different rates and hence are in different conditions at a specific time. It is important to "catch" the pavement at the most "appropriate" condition such that the rehabilitation can be done economically—a totally damaged pavement will need too much money to rehabilitate/replace it. This process of keeping an inventory of condition and selecting pavements for rehabilitation/reconstruction is called *pavement management*. This step includes the use of tools for determining the condition of existing pavements. The fastest growing method of such detection is nondestructive testing and evaluation (NDT and NDE). One example of a widely used form of NDT is the falling weight deflectometer (FWD) that works on the principle of evaluating deflections of pavements under known loads and making an assessment of the structure and the condition of the pavement.

1.5 IMPORTANT ISSUES

Traffic keeps on increasing, while the costs of materials and methods keep on climbing, as budgets dwindle everywhere. The only way to keep up and still have good roads is through learning and applying good principles, implementing proven new concepts and technologies, and continually researching for better materials and methods.

Therefore, this can be summarized as follows:

1. Drainage is needed to drain water away from the pavement.
2. The materials must be evaluated and selected properly so that they can withstand the effects of the traffic and the environment.
3. The mix must be designed properly such that it can withstand traffic and environmental factors.
4. The structure should be designed properly such that it has adequate thickness to resist excessive deformation under traffic and under different environmental conditions.
5. The pavement must be constructed properly such that it has desirable qualities.
6. The pavement must be maintained/managed properly through periodic work, regular testing, and timely rehabilitation.
7. Generation of knowledge through research is critical for ensuring good pavements in the future.

This book aims at providing the relevant principles and practical concepts of pavement engineering for both students and currently practicing engineers.

1.6 FUNCTIONAL REQUIREMENTS

A pavement's primary purpose is to provide a functional surface for a specific transportation need. The basic function is to withstand load, under different seasonal environmental conditions, without deforming or cracking, since either of these distress conditions would reduce the functionality of the pavement. The function of the different layers in the pavement is to spread out the load on the surface and reduce its intensity with depth, such that the pressure on the subgrade is much less than the pressure on the surface and can be tolerated by the subgrade without undergoing excessive deformation.

A pavement consisting of asphalt mixes (and aggregate and soil layers) only is referred to as a flexible pavement, since the pavement layers deflect under a traffic load. The typical applied concept of a flexible (or asphalt) pavement is that a layered structure (Figure 1.2) with better materials near the top would distribute the load in such a way that the resulting stress in the bottommost layer will be small enough so as to cause no significant deformation of the layer. The bottommost layer is the existing layer or the existing layer modified with some materials. The materials and the thicknesses of the different layers will be such as to be able to withstand the different effects of temperature and moisture due to changes in season in a specific location. The subbase, in addition to providing structural support, may also serve as a platform for constructing the base and prevent the fine materials from the subgrade from contaminating the base layer. If the subgrade is of frost-susceptible material, then the subbase could be made up of non-frost-susceptible materials to prevent frost-related damage.

Rigid (or concrete) pavements, which deflect very little under traffic loads, behave differently than flexible pavements under loads. The wearing layer, which is in contact with the traffic, is a Portland cement concrete slab that ranges in thickness between 5 and 12 in. depending on traffic

FIGURE 1.2 Function of the pavement is to decrease the tire contact stress on the subgrade to a tolerable level; flexible pavement (a), rigid pavement (b).

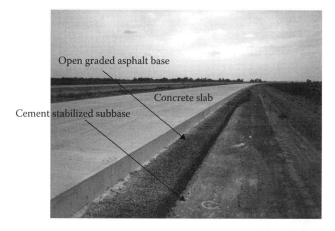

FIGURE 1.3 Different layers in a rigid pavement. (Courtesy of Wouter Gulden, ACPA, SE Section, Duluth, GA.)

loading. The thicker pavements are typical of heavier and more repetitive loads. The slab, due to its higher stiffness as characterized by elastic modulus, usually distributes the loading across a large pavement area. This in turn reduces the stresses experienced by the underlying base and subgrade layers. Rigid pavements may or may not have a base or subbase layer, and could be placed directly over the subgrade (Figure 1.3). However, in high-performance pavements, a base or subbase is typically included. Besides providing a wearing course and a contact surface with traffic, the slab provides friction, drainage, smoothness, noise control, and waterproofing for the underlying layers.

1.7 TYPES AND USES OF PAVEMENTS

Under the broad definition, there exist several different types of pavements with specific functions. Let us review the more important types.

1. *Pavement for roads*: There are different types of roads ranging from high-traffic-volume interstate highways to low-traffic-volume local roads. These roads have different types and volumes of traffic. Accordingly, they are designed and constructed in different ways. Thicker pavements are constructed for both heavier and high-volume traffic, of which trucks are the major load applicators. Thinner pavements are for low-volume roads, although in many cases low-volume roads may carry heavy trucks such as log trucks.

(a)   >300 mm thick   (b)   200 mm thick

(c)   130 mm thick low

FIGURE 1.4 Thicker pavements are provided for heavier loads. (a) Airport pavement, (b) highway pavement, and (c) low traffic volume pavement.

2. *Pavement for airports*: Pavements are required in airports in aircraft holding/terminal areas, taxiways, and runways. Just as road pavements are subjected to a wide range of vehicles, airport pavements are subjected to a wide range of aircrafts—a small, general aviation airport may have only light aircrafts (e.g., <30,000 lb), while a large hub/major airport would have major aircrafts. The different pavement areas are runways, taxiways, and aprons. Special considerations such as protection from fuel are required in many cases.
3. *Pavement for parking lots*: Parking lots are essential features in cities and towns, and are commonly found adjacent to business/office buildings, including those near hospitals, schools, and airports.
4. *Loading and unloading areas in ports and other areas*: Heavy-duty pavements are constructed to support equipment and materials unloaded from ships, rail, and trucks. These areas may also require special protections such as those from fuel droppings/spillage.

Figure 1.4 shows a comparison of thickness of different types of pavements. Note that in addition to thicker layers, pavements carrying heavier and more traffic also consist of better materials. For example, a >300 mm thick full depth asphalt pavement is found in an airport, whereas a 63 mm hot mix asphalt layer over a 75 mm aggregate base course is sufficient for a low-traffic-volume local road.

1.8   DIFFERENT FEATURES OF TYPICAL ASPHALT PAVEMENTS

The bottom layer on which the pavement is built is called the *subgrade*. The other layers, upward in order, are the subbase, base, binder, and surface layers. In many cases, the pavement can be a full depth asphalt pavement, in which case all of the layers above the subgrade are composed of hot mix asphalt constructed in several layers (or lifts). In other cases, the subbase and base may be combined to form one single layer, as far as materials are concerned, but constructed in multiple layers. And finally, above the surface layer of HMA, there may be a very thin wearing course of specialty material such as open graded friction course (OGFC), to provide better friction and drainage of water.

Generally, the materials become better and costlier as one moves up from the subgrade layer to surface layer. On a per unit weight cost basis, the surface layer is probably the most expensive layer. This is because, as layers at the surface and near it are subjected to the direct application of traffic loads as well as environmental effects, they need to be "fortified" to resist the traffic and environmental effects. These "fortifications" come mostly in the form of higher asphalt binder content, which is the more costly component. Furthermore, stricter specifications on size, shape, and quality of aggregates are also responsible for higher prices of the surface and near-surface layers.

A binder layer is provided because it would be difficult to compact the surface and the binder layers in one step, and also the binder layer part can be made up of larger aggregates and with lower asphalt content, thus costing less.

The base layer could be designed as a permeable base layer to let any water inside the pavement flow down to the sides for drainage. This water can be coming from the top—rain and snow through open voids or cracks in the surface—or from groundwater. The surface layer should be compacted to a high density, and cracks should be filled as soon as possible to prevent the flow of rain and snow water inside. On the other hand, groundwater level should be reduced before construction of the pavements by providing interceptor drains. However, there could still be some water coming inside the pavement which needs to be drained out.

The layers in the pavement provide different functions. While the wearing layer in an OGFC may provide friction and channels for quick draining of water, the binder and layers below provide the stiffness to prevent excessive deflection and hence cracking and structural rutting. Good performing surface and binder layers should have sufficient stiffness to prevent rutting related to poor materials. Furthermore, the layer (surface or wearing) exposed to the environment should not be too stiff at low temperatures (during the winter months), to avoid cracking caused by a drop in temperature. There could be a design in which the surface layer is stiff, followed by less stiff binder, base, and subbase layers. There could also be a design in which the bottom parts are less stiff but made with high asphalt contents to prevent cracking, whereas the top parts are stiffer with low asphalt content to prevent rutting, and the surface is sealed with a thin layer of mix with high asphalt content to provide a smooth and dense wearing layer. The point is that the materials and the structure can be engineered to provide a pavement with desirable qualities.

The density of the different layers is crucial for the proper functioning of an asphalt pavement. Soil/aggregate layers are compacted to maximum density using an optimum additive content. In the case of the HMA layers, they are compacted to a specific density (or voids) that has been found to be optimum for that specific mix (e.g., 6%–8% air voids for dense graded mixes), using a design asphalt content. There is a subtle difference between the two approaches that are used for the selection of the optimum additive content for the layers other than HMA and the one that is used for selecting the design asphalt content for HMA. For non-HMA layers, the optimum additive content is that which produces the maximum density, whereas for HMA layers, the design asphalt content is one that produces 4% void in compacted mixes in the laboratory.

The base course and subbase course are sometimes made up of different materials since the subbase can be made up of less costly material. If the base is open graded, then the subbase can serve as a filter with fine graded material between the base course and the subgrade. Even if no outside material is added before starting construction of the pavement, the top 6 in. of the existing subgrade are usually scarified and compacted to a high density using optimum moisture content.

One important requirement of an asphalt pavement is that there must be sufficient bonding between the different layers so as to prevent any slippage at the interface between the layers. Tack coatings of asphalt emulsions, for example, are applied between two HMA layers to achieve this. A tack coat is used between two asphalt layers or between an old PCC layer and a new asphalt layer to form a good bond. The spraying must be uniform and of the right amount per unit area, and if emulsion is used it must reach the desirable condition before the application of the upper layer. A prime coat is used between a granular material layer and an asphalt mix layer. The requirement is that the prime coat material must penetrate the voids and seal them up in the granular material.

In a full depth asphalt pavement, all layers starting from the base layers and up are made of asphalt mix. This type of pavement has a higher initial cost, but if designed and constructed properly, it provides several advantages over layered structures with different materials, such as little or no ingress of moisture and reduction in time of construction. Such pavements may deteriorate only at the surface with time, and hence rehabilitation work will be far less than that in pavements with different materials.

1.9 DIFFERENT FEATURES OF TYPICAL CONCRETE PAVEMENTS

There are three conventional types of concrete pavements: jointed plain concrete pavement (JPCP), Figure 1.5a and b, jointed reinforced concrete pavement (JRCP), Figure 1.5c, and continuously reinforced concrete pavement (CRCP), Figure 1.5d. For all conventional rigid pavement types, a concrete slab is usually poured directly on a subgrade, base, or subbase. The base or subbase could be a bonded or unbonded material that provides adequate support, drainage, and an even working surface during construction. All the three common rigid pavements carry traffic loading through flexural strength of the concrete. However, they differ in the slab lengths, joint details, and the type and amount of reinforcement they use.

1.9.1 JOINTED PLAIN CONCRETE PAVEMENTS

Jointed plain concrete pavements (JPCP) are the most common type of rigid pavements due to their cost and simplicity (Figure 1.6a). Contraction joints are typically constructed every 12–20 ft apart to control mid-slab cracking (Figure 1.6b and c). In JPCP, no slab reinforcement is used except for dowel bars placed at transverse joints or tie bars at longitudinal joints. Dowels are used for load transfer across transverse joints and allow the joints to move along the longitudinal axis of the dowel. Conversely, tie bars keep the longitudinal joints held tightly together.

FIGURE 1.5 Jointed plain concrete pavement (JPCP) with tiebars and (a) no dowels, (b) with dowels, (c) jointed reinforced concrete pavements with tiebars and dowels, and (d) continuously reinforced concrete pavements (CRCP). The slab has continuous reinforcement.

(a)

(b)

(c)

FIGURE 1.6 Jointed plain concrete pavements (JPCP) (a) showing transverse and longitudinal joints, (b) JPCP dowel bar and chair assembly set on an asphalt bonded base prior to slipform paving operations, and (c) JPCP under construction with slip form operation.

(Figure 1.7a through c shows dowel bar assembly for joint construction, tiebars at longitudinal joints, and a finished contraction joint.)

If dowels are not used, then load transfer across the joint can be achieved through aggregate shear interlock. Aggregate interlock is developed once a joint is cut in the young concrete pavement and a crack develops along the depth of the slab. Aggregate interlock is effective as long as the joint widths remain narrow. If a joint opening becomes too wide due to drying shrinkage and temperature effects, the aggregate shear interlock becomes ineffective. Repeated heavy loads across the transverse joint can cause the aggregate bond to wear down along the crack wall, rendering the aggregate shear bond ineffective. This will result in joint deformations that become excessive, resulting in faulting, a bumpy ride, and joint failure. According to a 1999 survey by ACPA, about 70% of the state highway agencies in the United States use JPCP (ACPA, 1999).

1.9.2 JOINTED REINFORCED CONCRETE PAVEMENTS

Jointed reinforced concrete pavements (JRCP) are similar to JPCP except for the longer slabs and the added light reinforcement in the slab. Joint spacing is typically 25–40 ft, although joint spacing of 100 ft has been used (Huang, 2004). For the longer slabs and joint spacing, dowels are highly recommended since joint openings will be wider and aggregate interlock will be ineffective for load transfer across the joint. The percent steel used in the longitudinal direction is typically 0.1%–0.25% of the concrete cross-sectional area, with less used in the transverse direction. The steel reinforcement in JRCP is not used for carrying structural loads but to hold the cracks tightly together to preserve shear load transfer across the cracks. The steel is placed near mid-slab depth or

(a)

(b)

(c)

FIGURE 1.7   (a) Jointed plain concrete pavements (JPCP) with tiebars and dowel bars set for construction of adjacent lane, (b) JPCP under construction with header and dowel with basket assembly, and (c) JPCP showing sawed joint and crack below cut.

close to the neutral axis, where bending stresses are close to zero. This reinforcement is also termed "temperature steel" since its purpose is to hold the cracks caused by temperature stresses tightly together. The steel reinforcement can be steel rebar, typically No. 5, or a wire fabric grid. Figure 1.8 shows JRCP and construction operations. According to a 1999 survey by ACPA, about 20% of the state highway agencies in the United States use JRCP (ACPA, 1999).

1.9.3   Continuously Reinforced Concrete Pavements

Continuously reinforced concrete pavements (CRCP) are heavily reinforced concrete slabs with no contraction joints. Hairline cracks do occur in CRCP but are not a concern to the performance of the pavement (Huang, 2004). This characteristic cracking pattern consists of cracks typically spaced every 2.0–8.0 ft. They are held tightly together by reinforcing steel, allowing for aggregate interlock and shear transfer. If this aggregate shear interlock is not maintained and compromised, then punchout failure at the pavement edge occurs, which is a typical distress in CRCP. The amount of reinforcing steel used in the longitudinal direction is typically 0.6%–0.8% of the cross-sectional area of the concrete, with less being used as temperature steel in the transverse direction. Experience has shown that if the amount of steel used in CRCP is less than 0.6%, the potential for punchout distress occurring is greater (Huang, 2004). CRCP require special anchors at both the beginning and the end of the pavement to keep the ends from contracting and to allow

FIGURE 1.8 Continuously reinforced concrete pavements (CRCP) (a) under construction, (b) rebar being set on an asphalt bonded base, and (c) CRCP under slipform construction.

the slab to develop the characteristic crack pattern (Dellate, 2008). Figure 1.8 shows CRCP and slipform paving construction operations.

The cost of CRCP is much higher than that of JPCP or JRCP because of the heavy reinforcing steel used. However, CRCPs may prove cost-effective in high-traffic-volume roadways due to their better long-term performance compared to the other types of concrete pavements. According to a study of 14 CRCP using the LTPP database, all but 3 CRCP had serviceability indices of 4 or better even though all the pavements carried heavy traffic and had exceeded their 20 year design life (Delatte et al., 2000). CRCP is not used widely, according to a 1999 survey by ACPA, only eight state highway agencies build CRCP (ACPA, 1999).

1.9.4 Composite Pavements

Although all pavements consist of different types of layers and hence can be called composite, generally the term refers to those pavements that consist of asphalt mix layer on top of a concrete pavement or a concrete layer on top of another concrete pavement. The concrete part could be PCC, CRCP, JRCP, or roller compacted concrete. For concrete over concrete pavements, generally the lower layers consists of locally available aggregates or recycled concrete while the upper layer (which could be a porous concrete layer) contains high-quality smaller size aggregates.

For pavements with asphalt mix layer over concrete, it is important that the asphalt layer bonds well and has full friction with the concrete surface. A good asphalt mix layer on top of a concrete layer can help in providing smoothness, reducing noise, and good friction. The common distresses in these pavements are de-bonding from the concrete layer and reflection cracking as a result of movement along a joint or a crack in the concrete slab. Sawing and sealing over existing joints in a concrete pavement can be used to reduce the potential of reflection cracks. The key concept in such a pavement is that the concrete slab will not have fatigue damage, and that the relatively thick layer of asphalt mix could be milled and replaced easily, to address the surface distresses. To make sure that the concrete slab does not experience fatigue damage, many factors must be taken into consideration, which include providing adequate base course, thickness, joint spacing, and load transfer at joints, in addition to the strength and thermal response of the concrete.

Generally, for concrete overlays over existing asphalt or concrete or composite pavements, a distinction is made between bonded and unbonded, or, correspondingly, thin and thick. Bonded, thin overlays (or ultrathin, thin/white topping) are generally 2–5 in. in thickness and are generally used when the existing pavement is structurally sound, with surface distress, such as friction loss. For design of such overlays, materials, bonding, and jointing (sawing after laydown) are critical factors for good performance. Generally the materials include high cement content and fiber, and produce a concrete that develops a high strength within a short period of time. The overlay can be placed directly on the existing asphalt pavement, unless there is rutting, in which case it needs to be milled and filled with concrete. Ultrathin-bonded PCC is used for areas such as parking lots or city streets, whereas thin-bonded overlays are used in secondary or state roads.

Thick concrete overlays (4–11 in.) are generally used for increasing the structural capacity of an existing pavement with stable underlying layers. For the design of such a layer, the structural support of the entire existing pavement (i.e., asphalt) is utilized in terms of an effective strength parameter. And if laid on an asphalt pavement, the asphalt mix layer is considered as a base course.

Comprehensive information on concrete overlays can be obtained from the American Concrete Pavement Association website at http://www.pavement.com/. The reader is also suggested to download the NCHRP Synthesis 338 on thin and ultrathin white topping from: http://onlinepubs.trb.org/onlinepubs/nchrp/nchrp_syn_338.pdf.

1.9.5 SELECTING THE TYPE OF PAVEMENT

The selection of the appropriate type of pavement for a specific project should be made on the basis of primary factors that include traffic, soil type, construction considerations, climate, recycling options, and cost and secondary factors, consisting of energy and material conservation and availability, performance history, and adjacent pavements. Traffic is affected by many factors such as fuel prices, competition among different modes of transportation in government policies, and proper margin of safety—comparison of alternative short- and long-term strategies to minimize traffic disruption should be made. Both loads and volumes of traffic are important, and the consideration of flexibility of timing of construction (for day-night) as well as stage construction should be considered. Concrete pavements can provide long life, whereas well-designed asphalt pavements, with periodic surface distress maintenance, can also provide good performing and long-lasting pavements. Subgrade soils under the pavement must be stiff and durable enough to resist both construction and traffic loading without excessive distress. For subgrade soils, the key factors are geographical and seasonal variability. The climate and the type of soil (with respect to gradation, drainage, and stiffness) will dictate the need for the use of additives, if required, during stabilization and compaction. Periodic rainfall and freeze thaw can have significant seasonal effects on the stiffness of subgrade. Furthermore, the temperature of the project region also needs to be considered for the proper selection of materials for the pavements. In this respect, the availability of the selected material is also an important issue. Lastly, for construction, the factors of the option of staged construction, the speed of construction, maintenance of traffic flow during construction, timing, experience and ease of construction, and the ability to accommodate consideration of future traffic growth should be considered.

1.10 RESEARCH ON PAVEMENTS

Continuous research is needed to produce better materials, processes, and designs to make pavements safer, durable, and more cost-effective. Such research can be basic or applied, for solving specific problems. In the United States, the benefits of the Strategic Highway Research Program (SHRP) have been estimated to be a savings of up to $785 million annually (for an investment of $50 million in asphalts), not considering the benefits to the motorist of a reduction in delays and vehicle wear and tear (Halladay, 1998). Benefits of investments in research can be quantified in terms of reduction of road agency costs also. The benefit-to-cost ratios are dependent on the inflation-free discount rate and have been reported to range from 4 to 5. In addition, unquantifiable benefits, such as boosting innovation and improved cooperation of researchers and practitioners, have also been noted (Rose and Bennett, 1994).

Since pavement engineering involves innumerable different disciplines, research potentials are abundant in this field. In a broad way, research on pavement engineering can be grouped into three major categories—materials, structures, and construction related. Material-related research work involves the development of innovative materials, mixtures, and additives, and chemical and material characterization. Structure-related research involves the development and use of advanced techniques, sensors, equipment, and models for understanding the pavement structure and its different components. Research related to construction includes activities such as the development of new procedures and equipment for faster and better construction, construction under difficult conditions, and/or the utilization of recycling/marginal materials, as well as those related to better quality control and specifications. Some of the emerging areas of research include those related to the environment, maintenance materials, microscale characterization, the use of nondestructive and accelerated testing, and recycling.

Research on pavements is conducted by industry and government organizations all over the world. In the United States, pavement research is conducted at local, state, and federal levels. The National Cooperative Highway Research Program (NCHRP), which brings together the AASHTO, FHWA, U.S. Department of Transportation (USDOT), and National Academies, is the primary sponsor of national-level pavement research in the United States. Similarly, the Federal Aviation Administration sponsors pavement-related research through the Asphalt Airfield Pavement Technology Programs (AAPTP) and the Innovative Pavement Research Foundation (IPRF). The FAA and FHWA also conduct research on material characterization as well as accelerated loading and testing, through both in-house as well as academia-based centers.

The longest road test in history to be implemented was the Long-Term Pavement Performance (LTPP) program. The motivation behind the LTPP was to create a pavement performance database on all different pavement types in different climatic and soil conditions throughout the United States and Canada. The LTPP program started in 1987 as part of the SHRP, a 5 year applied research program that was supported by Congress, and managed by TRB. The intent of the 20 year long LTPP program was to conduct a long enough pavement performance study that would cover the life cycle of typical pavements. Ultimately, the most valuable outcome of the LTPP program was the creation and development of a comprehensive "national" pavement performance database. Analysis studies using the database will enhance our knowledge of how pavements behave, and will also help us develop tools and engineering products to analyze, design, construct, and rehabilitate pavements in a sustainable cost-effective manner. http://www.fhwa.dot.gov/research/tfhrc/programs/infrastructure/pavements/ltpp/index.cfm

Research on pavements requires basic knowledge of principles, formulation of problems, development and application of appropriate theory, utilization of appropriate equipment and tools, and proper inference of results. A thorough literature review is the key to successful research. The Transportation Research Information System (TRIS; http://ntlsearch.bts.gov/tris/index.do) is a very powerful database that can be accessed and queried online to conduct a literature review on pavement engineering. Information is also available from state DOT websites (see the following Google site: http://www.google.com/cse/home?cx=006511338351663161139%3Acnklqdck0dc), as well as industry websites such as those of the National Asphalt Pavement Association (www.hotmix. org) and the Portland Cement Association (www.cement.org).

A number of general and specialty conferences are held every year or on alternate years all over the world on pavement-related research. One of the biggest assets in pavement research (as in any other case) is the teamwork/cooperation of industry/academic and government agencies, as well as the cooperation between researchers in different countries. Researchers and students should take opportunities to learn about new things as well as to network with fellow students and researchers in different schools/industries and countries. So wide is the area of pavement research that persons from a wide variety of backgrounds can find their niche areas of research. However, in general, knowledge on numerical techniques, mechanics, basic physics, and chemistry is essential.

Readers should also be aware of the currently (2012) ongoing (U.S. Congress mandated) Second Strategic Highway Research Program (SHRP 2). The focus of this program is to conduct applied research in the following four areas: Safety, renewal, reliability and capacity. The goals are to prevent or reduce severity of highway crashes by understanding driver behavior (safety), address the aging infrastructure through rapid design and construction methods that cause minimal disruption and produce long-lived facilities (renewal), reduce congestion through incident reduction, management, response, and mitigation (reliability), and integrate mobility, economic, environmental, and community needs in the planning and designing of transportation capacity (capacity). More information on SHRP2 is available at http://www.trb.org/StrategicHighwayResearchProgram2SHRP2/General.aspx.

QUESTIONS

1.1 What are the main functions of an interstate highway pavement and an airport taxiway pavement?

1.2 Why are these pavements generally built in layers?

1.3 Why are not all pavements very thick or very thin?

1.4 What happens if fuel such as kerosene or diesel drops in large amounts on asphalt pavements?

1.5 Why are roads and airport pavements important?

1.6 What are the steps in building a stable and long-lasting pavement?

1.7 Why is maintenance of pavements important?

1.8 How can roads be improved?

1.9 While designing a road pavement, are both weight and number of vehicles important?

1.10 How is the repetitive loading taken into account?

1.11 What are the steps in designing and constructing a road?

1.12 How can a road be rehabilitated in the most cost-effective way?

1.13 Is there a difference between airport and road pavement?

1.14 Are there special requirements for airport pavements?

1.15 How can roads be made safer for travel?

1.16 Can a pavement be designed and built in an environment-friendly way?

1.17 What is the best way to rehabilitate an old pavement?

1.18 What are the different types of rigid pavements?

1.19 What are the advantages and disadvantages of the different types of rigid pavements?

1.20 What are the considerations for the selection of the type of pavement—flexible or rigid?

1.21 Why is research on pavements important?

1.22 What are the general areas of research on pavements?

1.23 Conduct an Internet search and list the primary pavement-related research being carried out in the different parts of the world.

2 Principles of Mix and Structural Design and Construction of Asphalt Pavement

2.1 OVERVIEW

Flexible/asphalt pavements are generally made up of layers. Each layer is made up of a combination of materials—generally aggregate, binder, and any other additive. The process of designing the optimum combination of materials is called *mix design*. Once the structural properties of each layer are known (or estimated accurately), the optimum structure of the pavement is determined; that is, the thickness of each layer is determined. This process is called *structural design*, which also involves structural analysis—determination of stresses, strains, and deformations in different layers.

Note that mix design and structural design are dependent on each other; for example, a structurally better mix may only require a thin layer of the material to sustain a given load. However, the mix design is conducted in such a way that locally available (and hence relatively low-cost) materials are utilized for economic reasons. Hence, this can be viewed as one of the most important "constraints" in the overall design of an asphalt pavement. The next logical consideration is that of the locally available materials; one would utilize the relatively less costly materials more (that is, in thicker layers) than the more costly components. Hence, for a given condition of traffic and environment, and for a given period of time for which the pavement is expected to be usable (design period), it is the most judicious combination of materials and thicknesses that results in the optimum design of a pavement structure. In this process, the two factors, material quality *and* material (as well as construction) cost, are balanced. Note that a pavement with very cheap materials would probably be able to sustain traffic loads in the short term but would require frequent maintenance, which could be costly in terms of both money and user delay.

2.2 TRAFFIC AND LOAD DISTRIBUTION CONCEPT

One of the basic purposes of providing a good mix and structural design is that the pavement can withstand traffic loads without deforming or deteriorating to a degree that it becomes unusable within the design period. Heavy freight truck traffic is the main contributor of load on highways, since their weight is significantly higher than that of cars, small pickup trucks, or other passenger vehicles. Similarly, the load from big multiengine aircrafts is significantly higher than that coming from small single aircrafts on airport pavements.

The load (from a vehicle/aircraft) is transferred to the pavement through load-bearing axles and pressurized tires. The resulting pressure or stress on the pavement, at any depth, is dependent on many factors, such as total load, the number of axles and tires, and the condition of the tires. The stress on the surface of the pavement just below any tire is concentrated in the tire contact area, often assumed as a point for structural analysis. Theoretical concepts and validations have shown that this stress gets distributed in an inverted V form from the surface downward. In other words, the stress intensity decreases along the depth of the pavement. However, note that if the stress on the surface is coming from two tires placed, say 13 in. apart (or even axles placed close to each other), the stresses from the two tires might overlap at some point below the surface. This can result in a combined

FIGURE 2.1   Gear/axle layout of a Boeing 737 aircraft and a semi-tractor trailer.

stress that is higher than the stress resulting from each of the two tires. Obviously, whether or where (in the pavement) this would happen depends on the tire/axle configuration and the thickness of the pavement. Since the pavement needs to sustain the stresses, one needs to take into consideration not only the gross load but also relative positions of the axles and the gears during the structural analysis part. Figure 2.1 shows the gear/axle layout of a Boeing 737 aircraft and a semi-tractor trailer, and Figure 2.2 illustrates the concept of distribution of stress in the inverted V form.

2.3   MATERIALS AND LAYERS

The layer on the surface of the pavement has to withstand the maximum stress and bear the changing conditions of the environment. Therefore, this surface layer usually consists of the "best" and most costly materials. Also, this layer is always "bound"—that is, mixed with a "binder," in this case an asphalt binder—to prevent raveling of materials under traffic, as well as to provide a dense surface to prevent ingress of water, unless it is an open-graded friction course. Therefore, the surface layer has

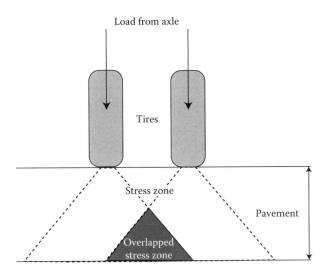

FIGURE 2.2 Concept of distribution of stress in the inverted V form.

two major components—asphalt binder and aggregates, with relatively small amounts of additive, if any. The mixture (referred to as *mix*) of this layer needs to be designed properly to give it adequate stiffness, strength, and durability. This mix is usually prepared by combining hot aggregates and asphalt binder (hot mix asphalt or HMA).

The next layer could be either binder or base. Generally, the binder layer is almost similar to the surface layer, except that it may consist of aggregates of larger size. The base may be made up of a bound layer or an unbound aggregate layer. The base layer has to be sufficiently strong in shear as well as bearing capacity, but need not be as good as the binder and surface course, since the stresses at this level are much lower. However, the base needs to be stiff enough to provide an overall stiffness to the pavement structure such that it does not deflect too much under load. How much is too much? That depends on the specific mode of failure we are concerned with, but it would suffice to say here that if the pavement can sustain repetitive recoverable deflection without cracking or having permanent deformation within the design period, then the deflection is acceptable. One of the most important criteria for designing the base course is to provide adequate density. If it is a bound layer, it may be with aggregates and asphalt binder or Portland cement, and if it is an unbound layer, it is most likely with aggregates only, with water used as an additive for aiding compaction during construction. The function of the subbase is almost similar to that of the base, except that it could be of lower-quality materials, since the stress is even less at this level. Generally, it is made up of aggregates only.

The subgrade is made up of the existing soil or the soil mixed with some additives to enhance its properties. It serves as the foundation of the pavement and should be of such quality as to resist excessive deflection under load. One important requirement is to make sure that the subgrade is capable of resisting the detrimental effects of water. Generally, the presence of excessive amounts of fine materials such as clays and silts is not desirable.

2.3.1 Soils

Soils or aggregates, which can be defined as processed soils, are present in every layer, either mainly by themselves, such as in unbound layers, or as a major component, such as in a bound layer. In the subgrade, the soil generally consists of unprocessed native soil, compacted to a maximum achievable density, using an optimum moisture content and roller compaction.

If the soil is incapable of resisting the detrimental effects of the environment or is too weak in shear strength and stiffness, it is generally "stabilized" with additives and/or adequate compaction.

FIGURE 2.3 Soil type map of the State of Maine, United States. (Courtesy of Brian Marquis, Maine Department of Transportation, Augusta, ME.)

Stabilization is the process of improving the quality of the soil (in most cases, reducing its susceptibility to the effect of moisture) with the addition of some other materials (known as *additives*).

Soils can vary widely over a region or a state (e.g., see Figure 2.3). To determine whether the native soil is adequate for the proposed pavement (and if not, what kind of stabilization process is needed), one needs to test and characterize the soil. These tests are mostly conducted to evaluate its physical properties and the effect of moisture on some of those properties. Strength/stiffness under different stress levels is also determined. Based on the results of these tests, one decides whether the soil can be compacted and used as it is or should be mixed with an additive and then compacted. Note that some compaction is almost always necessary to improve its bearing capacity to sustain the stress from the weight of the construction equipment and layers of the pavements. In general, since the subgrade is the lowest layer, the stresses from these two sources are much greater than the stress from a vehicle. One important factor is the consideration of the thickness of the subgrade layer. That is, to be adequate, the subgrade must have desirable properties throughout a specific depth.

2.3.2 Aggregates

Mostly naturally occurring aggregates are used in pavements. Aggregates are soil/rock processed to produce materials of a certain range of sizes (and other properties). *Aggregates* generally refer to soil particles whose major portion is of a size greater than 0.075 mm. These particles are tested for their physical properties and characterized to determine their suitability. Unlike existing soil in the subgrade, the aggregates used in bound courses are actually the result of a combination of different proportions of different sized particles—the combination is done as part of the mix design process.

Therefore, not only each aggregate particle but also the blend of the different size particles should be of desirable quality. Furthermore, such aggregates, when used in combination with asphalt, must also be of such (physical as well as chemical) quality as to resist cracking and retain the coating of asphalt binder under the action of traffic and harsh environmental conditions, such as repetitive pore pressure due to the presence of moisture. Thus the characterization of aggregate involves a host of testing and evaluation steps that range from those used for determination of size and shape to the action of moisture under different conditions.

2.3.3 Asphalt

Paving asphalt, also known as bitumen, is a product that is mostly obtained from crude petroleum through a series of refining steps. Asphalt can also be used in other forms, such as in emulsified form in asphalt emulsions. With reference to paving mixes, asphalt is commonly referred to as *asphalt binder*, since its basic purpose is to "bind" the aggregate particles together. In general, asphalt binders are semisolid or solid at room temperature and liquid at a relatively high temperature. Different "grades" of asphalt binders are produced by changing the source (of crude petroleum) as well as the refining conditions to meet different paving demands, mainly arising from differences in environmental and traffic conditions. Asphalt binders are classified into different "grades" with the help of characterization tests.

The properties of asphalt binders are affected significantly by the temperature and time of loading. Thus, characterization of asphalt binders is required to determine the effect of temperature and stress on the relevant engineering properties—mainly for two basic applications. The first application is the actual mix design process, where the asphalt should be of such grade as to produce a mix with aggregates that would be able to sustain the effects of traffic and the environment for the design life of the pavement. The second application of the characterization test results is for the construction process, where the asphalt binder must be transported from a refinery, stored in a mixing plant, pumped through pipes and mixed with aggregates, transported to the job site, and laid down and compacted at specific temperatures that would allow the completion of all those steps in the most convenient way. Therefore, the characterization tests of asphalt binders involve similar tests at multiple temperatures, using specific loading rates.

2.4 ENVIRONMENT

Since pavements are exposed to the environment, proper consideration of environmental factors must be made. There are three major factors—temperature, moisture, and the effect of temperature on moisture. These factors affect the properties of the pavement materials significantly. For example, a higher temperature decreases the stiffness of the asphalt materials, whereas a lower temperature increases its stiffness. A higher moisture content would reduce the strength/stiffness of unbound materials, whereas a frozen moisture would increase the overall stiffness of the pavement. In addition to one time effect, the environmental factors affect the properties of the pavement materials through their repetitive/cyclic nature. Frost-susceptible soils, with enough capacity to draw capillary water, can form ice lenses during the winter and result in frost heaves, and as the ice lenses melt in spring (spring thaw), voids form underneath the top layers of the pavement, weakening it and ultimately resulting in destruction of the pavement under traffic load. Figure 2.4 shows the different regions of the United States with respect to freeze–thaw.

To consider these factors in the appropriate way, three steps are taken. First, the effects of the different factors on the pavement materials are evaluated and those materials that are proven to be susceptible to the detrimental effects are eliminated from consideration. Second, to use available

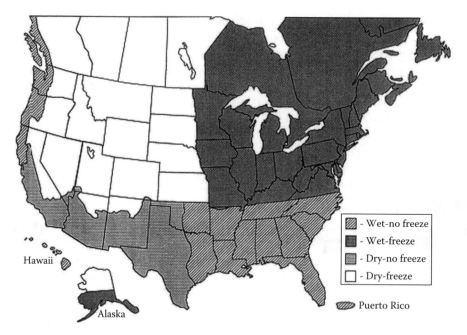

FIGURE 2.4 Freeze–thaw regions in the United States. (From Hanna, A.N., SHRP-LTPP Specific Pavement Studies: Five-Year Report, Strategic Highway Research Program, National Research Council, Washington, DC, 1994. Reprinted with kind permission of the Transportation Research Board.)

materials, and to avoid excessive costs, if it is not possible to eliminate all such materials, the available materials are modified with additives to make them resistant to the detrimental effects of the environment. Third, during the structural analysis and design, the effect of the environmental conditions on the relevant parameters (such as stiffness) is duly taken into consideration, and any pavement layer/material needed to eliminate the detrimental effect of the environment (such as frost action) is provided.

One of the important pavement-related issues directly related to the environment is providing an adequate surface and subsurface drainage system in the pavement. No matter what, some rain/snow water will always find its way into the pavement, and some groundwater will always seep inside the pavement. There should be a proper system to divert this water away from the pavement into side ditches/drains.

2.5 MIX DESIGN

The basic purpose of mix design is to select and combine the different components in such a way as to result in a mix that has the most optimum levels of all relevant properties so as to produce the best mix possible, within the constraints of available materials and funds available to produce the mix.

Therefore, the process starts with the selection of the most appropriate standard specification (or the development of one, if needed). The specification should clearly spell out the requirements for the materials in terms of properties to be determined by specific tests, details of the tests, and acceptable/not acceptable levels of results that are to be used. Then, according to the specification, aggregates, blends of aggregates, and asphalt binder are selected. Next, different blends of aggregates are tested for their properties. Then a selected blend is tested with different asphalt contents, and the optimum asphalt content is determined. Finally, the mix (design mix) with the selected gradation and the optimum asphalt content is tested at appropriate levels of controllable properties (such as air voids or density) to make sure that it is resistant to the detrimental effects of the environment. Furthermore, one needs to estimate the relevant structural design parameters for the designed mix.

Depending on the importance of the project, and the levels of sophistication employed in testing and evaluation of the materials, the mix design process can range from a simple process to a fairly complex one. In any case, it is always a trial-and-error experimental process, involving the testing of multiple samples for different test properties. Figure 2.5 shows the key steps in the mix design of an asphalt mixture.

2.6 STRUCTURAL DESIGN

Structural design of a pavement is conducted to determine the thickness of the different layers to prevent the occurrence of problems or distresses in the pavement due to traffic loading and the environment. Such distresses may include rutting or depressions in the wheel path and cracking. Figure 2.6 shows the effect of different stresses and strains on distress or problems in asphalt pavements.

The desirable thickness of the soil layers is determined from the consideration of their stiffness/resistance to deformation properties (such as with the use of the test property California bearing ratio or CBR), as well as the change in the properties that can be expected due to a change in the environmental conditions, primarily moisture and the effect of temperature on moisture (Figure 2.7). However, note that in a pavement structure, the stress in the subgrade or subbase/base is governed by the imposed stress (vehicle load) as well as the thickness of the layers above them. Therefore, if the subgrade/subbase or base is known to be relatively weak, it should be protected by a proper selection of binder/surface mix and layer thickness. It follows that the structural design process in pavements is an iterative process that balances the availability of materials and the cost of the total pavement structure.

The basic purpose of the structural design process is to combine the different layers in such a way as to result in the most cost-effective functional pavement structure. This can be achieved by primarily two different techniques: (1) by using empirical methods—that is, charts and equations developed from experimental studies carried out with a set of traffic, environment, and pavements or (2) by using a mechanistic method, in which concepts of mechanics are used to predict responses and performance of the pavement. Note that a purely mechanistic approach is not possible at this time—the responses can be predicted by employing concepts of mechanics, but the performance has to be predicted by empirical models. Hence, it is more appropriate to say that pavements can be designed either by using the empirical approach or by using the mechanistic-empirical approach (ME).

There are obviously differences in steps and levels of sophistication between the two processes. In the empirical approach, considering the traffic and environment, and the available subgrade, an appropriate "chart" may be used to "read off" the required base course and/or the asphalt mix layer thicknesses. Such methods can be simple or fairly sophisticated, utilizing different factors such as axle numbers, drainage parameters, concepts of reliability, and performance-related properties of the pavement. Figure 2.8 shows an example of an empirical procedure for designing airport pavements for the Boeing 747 aircraft. Note that in this case, the CBR of the soil is used, along with aircraft information, to determine the thickness of the pavement (subbase + base + surface) required on the subgrade.

The ME process is more of an iterative nature, and can be divided into three basic steps. It starts with the consideration of traffic, design period, projected total traffic (expected over the design period), and environmental conditions. In the second step, considering the load as well as environment-related properties of the available materials, a trial structure with different layers is designed, and stresses/strain/deflections at different critical levels are determined. Finally, the stresses and strains are utilized in the empirical models to predict the performance of the pavement over the design period. If the performance is not found to be satisfactory, the process loops back to the second step

Consider traffic and environmental factors

Select appropriate materials

Prepare mixes with different percentages of asphalt binders
compact samples using appropriate laboratory compactor
to simulate field compaction

Test samples for volumetric properties

Analyze: plot volumetric properties against asphalt content
and select optimum asphalt content on the basis
of desirable volumetric properties

Compact samples at optimum asphalt content and
check for mechanical properties such as resistance against
moisture damage after appropriate conditioning

Finalize design asphalt content and recommend job mix formula

    1. Aggregate information
    2. Asphalt amount and type
    3. Temperature information
    4. Limits and target values

FIGURE 2.5   Key steps in asphalt mix design.

A flexible pavement "flexes" under a load; repeated loadings
cause repeated compressive strains as well as tensile strains

FIGURE 2.6   Major distress-causing responses in a "flexible" or asphalt mix pavement.

and starts with a different trial structure. This process continues until a pavement structure with satisfactory performance over the design period is obtained.

Inherent in the design processes are the very important considerations of economics and the cost of the pavement over its entire "life" (referred to as *life cycle cost* or LCC). Considerations of both initial cost and regular maintenance and salvage costs must be made during the design process, particularly for large projects (e.g., for pavement projects costing more than $1 million).

2.7   LINK BETWEEN MIX AND STRUCTURAL DESIGN

Mix design is generally conducted on the basis of volumetric properties. This process produces the job mix formula (JMF), according to which the mix to be used for paving is produced. This mix is tested for its structural properties, and, using the structural properties, the structural performance of the pavement with the specific mix is predicted. Therefore, it is important to note that the structural performance is very much dependent on the mix design, even though structural properties may not be generally considered explicitly in the mix design process.

An example is shown in Figure 2.9. It shows the effect of two parameters on the rutting performance of a mix under medium-volume traffic. The two parameters are the type of asphalt and the modulus of the base. The type of binder and the asphalt content of the base are selected during mix design. The years-to-failure chart shows that a "stiffer" asphalt binder takes a longer time to fail by rutting. That is, it has a higher resistance to structural rutting. At the same time, a mix with a high modulus base shows more resistance to structural rutting than the one with a low modulus.

2.8   THEORETICAL CONSIDERATIONS FOR STRUCTURAL DESIGN

Before proceeding to structural design, it is necessary to review some of the basic mechanistic techniques that enable us to compute stresses and strains due to loads.

FIGURE 2.7   Basic steps in structural design.

2.8.1   HOOKE'S THEORY OF ELASTICITY

Ratio of stress over strain is a constant, Young's Modulus, E (modulus of elasticity). This constant relates stress to strain: $\sigma = E\varepsilon$. Poisson's ratio (μ) is the ratio of radial and longitudinal strains. Figure 2.10 explains the different parameters:

$$\varepsilon_1 = \frac{\Delta l}{l}; \quad \varepsilon_r = \frac{\Delta r}{r}; \quad \sigma = \frac{P}{A}; \quad E = \frac{\sigma}{\varepsilon_1}; \quad \mu = \frac{\varepsilon_r}{\varepsilon_1}$$

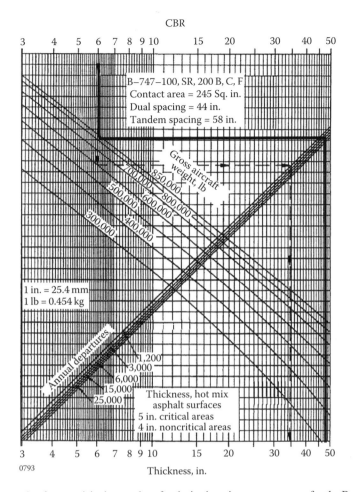

FIGURE 2.8 Example of an empirical procedure for designing airport pavements for the Boeing 747 aircraft. (Adapted from Federal Aviation Administration, FAA Advisory Circular No. AC 5320-6D, U.S. Department of Transportation, Washington, DC, 1995, Figures 3–7.)

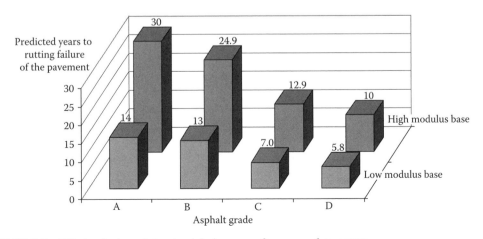

FIGURE 2.9 Effects of mix and structural design on performance of pavement.

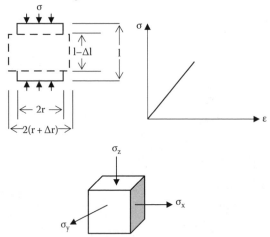

FIGURE 2.10   Schematics of stress and strains.

For a three-dimensional case,

$$E\varepsilon_x = \sigma_x - \mu\sigma_y - \mu\sigma_z$$

$$E\varepsilon_y = \sigma_y - \mu\sigma_x - \mu\sigma_z$$

$$E\varepsilon_z = \sigma_z - \mu\sigma_x - \mu\sigma_y$$

Note that for pavement materials, E and μ are not constants but are affected by different factors such as temperature, moisture content, and stress conditions.

Example 2.1

An elastic bar is subjected to the force shown next. Determine the total elongation of the bar. Area through which load is applied = 113 in.²; modulus = 22,000 psi.

Stress: σ = load divided by cross-sectional area = 4500/113 = 39.8 psi.
Strain: ε = elongation divided by original length = $\Delta/12$.
By definition, modulus of elasticity, E = stress/strain = σ/ε.
Therefore, 39.8/Δ/12 = 22,000.
Or Δ = 0.022 in.
What happens if E was exactly double, that is, 44,000 psi?
Δ = 0.011 in. (half of the original elongation).

2.8.2 BOUSSINESQ'S METHOD

This method provides a way of determination of stresses, strains, and deflections of homogeneous, isotropic, linear elastic, and semi-infinite space under a point load. Consider the schematic shown in Figure 2.11.

Vertical stress:

$$\sigma_z = \frac{-3Pz^3}{2\pi R^5}$$

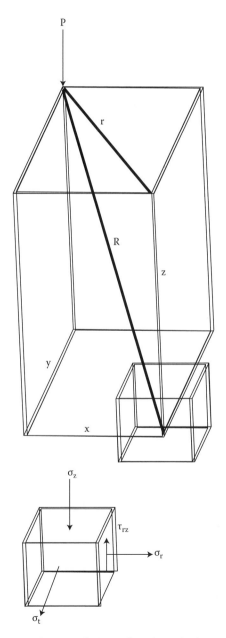

FIGURE 2.11 Coordinate system and stresses for Boussinesq's method.

Radial stress:

$$\sigma_r = -\frac{P}{2\pi} \left[\frac{3r^2z}{\left(r^2+z^2\right)^{5/2}} - \frac{1-2\mu}{r^2+z^2+z\sqrt{r^2+z^2}} \right]$$

Tangential stress:

$$\sigma_t = -\frac{P}{2\pi}(1-2\mu) \left[\frac{z}{\left(r^2+z^2\right)^{3/2}} - \frac{1}{r^2+z^2+z\sqrt{r^2+z^2}} \right]$$

Shear stress:

$$\tau_{rz} = \frac{P}{2\pi} \left[\frac{3rz^2}{\left(r^2+z^2\right)^{5/2}} \right]$$

Vertical deformation below the surface:

$$u_{zr} = \frac{P(1+\mu)}{2\pi E} \left[\frac{2(1-\mu)}{R} + \frac{z^2}{R^3} \right]$$

Surface (i.e., at z=0) vertical deflection:

$$u_r = \frac{(1-\mu^2)P}{\pi ER}$$

where
   r is the radial distance from the point load
   z is the depth

$$R^2 = r^2 + z^2 = x^2 + y^2 + z^2$$

As an example of the use of the aforementioned formulae, consider a case in which a 4500 lb point load is applied on the surface of a soil layer with modulus of 7.2 ksi and Poisson's ratio of 0.45. Consider the conditions that are applicable for the use of Boussinesq's formulae. Determine the vertical and radial stress and vertical deformation at a depth of 6 in. and radial distance of 3 in.

$$P=4500 \text{ lb}; \quad z=6 \text{ in.}; \quad r=3 \text{ in.}; \quad E=7.2 \text{ ksi}=7200 \text{ psi}; \quad \mu=0.45$$

$$R = \sqrt{(6^2+3^2)} = 6.71 \text{ in.}$$

$$\text{Vertical stress} = \sigma_z = \frac{-3Pz^3}{2\pi R^5} = -34.14 \text{ psi (compressive)}$$

$$\text{Radial stress} = \sigma_r = -\frac{P}{2\pi} \left[\frac{3r^2z}{\left(r^2+z^2\right)^{5/2}} - \frac{1-2\mu}{r^2+z^2+z\sqrt{r^2+z^2}} \right]$$

$$= 7.69 \text{ psi (compressive)}$$

Vertical deformation below the surface:

$$u_{zr} = \frac{P(1+\mu)}{2\pi E}\left[\frac{2(1-\mu)}{R} + \frac{z^2}{R^3}\right] = 0.04 \text{ in.}$$

For a distributed load under a circular area, the responses can be found out by numerical integration. Along the centerline of the load,

Vertical stress at depth z:

$$\sigma_z = \sigma_0\left[1 - \frac{1}{\left\{1+(a/z)^2\right\}^{3/2}}\right]$$

Radial and tangential stress at depth z:

$$\sigma_r = \sigma_t = \sigma_0\left[\frac{1+2\mu}{2} - \frac{1+\mu}{\sqrt{1+(a/z)^2}} + \frac{1}{2\left[\left\{1+(a/z)^2\right\}^{3/2}\right]}\right]$$

Vertical strain at depth z:

$$\varepsilon_z = \frac{(1+\mu)\sigma_0}{E}\left[\frac{z/a}{\{1+(z/a)^2\}^{3/2}} - (1-2\mu)\left\{\frac{z/a}{\sqrt{1+(z/a)^2}} - 1\right\}\right]$$

Deflection at depth z:

$$d_z = \frac{(1+\mu)\sigma_0 a}{E}\left[\frac{1}{\sqrt{1+(z/a)^2}} + (1-2\mu)\left\{\sqrt{1+\left(\frac{z}{a}\right)^2} - \frac{z}{a}\right\}\right]$$

where
  σ_0 is the stress on the surface
  E is the elastic modulus
  a is the radius of the circular area of the load
  z is the depth below pavement surface
  μ is the Poisson's ratio

Surface vertical deflection along the centerline of a rigid circular plate, with radius, a:

$$u_r = \frac{2(1-\mu^2)\sigma_0 a}{E}$$

As an example, consider the same problem as before, but now the load of 4500 lb is applied through a circular area of 45 in.[2] Determine the responses 6 in. below the surface along the centerline of the load.

$$\text{Area} = \pi a^2 = 45$$

$$a = 3.8 \text{ in.}$$

$$\text{Stress on the surface} = (4500/45) = 100 \text{ psi}$$

$$\sigma_z = \sigma_o \left[1 - \frac{1}{\left\{ 1 + (a/z)^2 \right\}^{3/2}} \right] = 39.7 \text{ psi}$$

$$\sigma_r = \sigma_t = \sigma_0 \left[\frac{1+2\mu}{2} - \frac{1+\mu}{\sqrt{1+\left(a/z\right)^2}} + \frac{1}{2\left[\left\{ 1+\left(a/z\right)^2 \right\}^{3/2} \right]} \right] = 2.24 \text{ psi}$$

$$d_z = \frac{(1+\mu)\sigma_0 a}{E} \left[\frac{1}{\sqrt{1+(z/a)^2}} + (1-2\mu)\left\{ \sqrt{1+\left(\frac{z}{a}\right)^2} - \frac{z}{a} \right\} \right] = 0.4 \text{ in.}$$

For flexible pavement design, layers are often simplified as homogeneous, isotropic, linear elastic materials. However, in reality the traffic load is not applied at a point but spread over an area; for practical purposes, the equations for a point load can be used for a distributed load at distances equal to or greater than 2a from the centerline of the load. From Boussinesq's equations, note that modulus and deflection are inversely related. When the modular ratio of the pavement (i.e., the combination of all layers above the subgrade) and the subgrade is close to unity, the equations for one-layer elastic solutions for subgrade stress, strain, and deflection are applicable. Tests show that in most cases stresses and deflections predicted by this method are larger than measured values.

2.8.3 Application, Extension, and Refinement of Boussinesq's Method

Foster and Ahlvin developed solutions for vertical and horizontal stress and vertical elastic strains due to circular loaded plate, for $\mu = 0.5$. Ahlvin and Ulery refined Foster and Ahlvin's charts to develop solutions for complete pattern of stress, strain, and deflection at any point in the homogeneous mass for any value of μ. The solutions are as follows. Examples of coefficients in equations are presented in Table 2.1.

$$\sigma_z = p[A + B]$$

$$\sigma_r = p[2\mu A + C + (1-2\mu)F]$$

$$\sigma_t = p[2\mu A + D + (1-2\mu)E] \qquad \Delta_z = \frac{p(1+\mu)a}{E_1}\left[\frac{z}{a}A + (1-\mu)H \right]$$

$$\tau_{rz} = \tau_{zr} = PG \qquad\qquad \theta = \sigma_z + \sigma_r + \sigma_t$$

$$\varepsilon_z = \frac{p(1+\mu)}{E_1}[(1-2\mu)A + B] \qquad \varepsilon_\theta = \varepsilon_z + \varepsilon_r + \varepsilon_t$$

$$\tau_{zt} = \tau_{tz} = 0$$

$$\varepsilon_r = \frac{p(1+\mu)}{E_1}[(1-2\mu)F + C] \qquad \sigma_{1,2,3} = \frac{(\sigma_z + \sigma_r) \pm \sqrt{(\sigma_z + \sigma_r)^2 + (2\tau_{rz})^2}}{2}$$

$$\varepsilon_t = \frac{p(1+\mu)}{E_1}[(1-2\mu)E - D] \qquad \tau_{max} = \frac{(\sigma_1 - \sigma_3)}{2}$$

TABLE 2.1
Example Values of Coefficients

r/a	z/a	A	B	C	D	E	F	G	H
0	0	1	0	0	0	0.5	0.5	0	2
	0.1	0.9005	0.09852	−0.04926	0.04296	0.45025	0.45025	0	1.80998
	0.5	0.55279	0.35777	−0.17889	0.17889	0.27639	0.27639	0	1.23607
	1	0.29289	0.35355	−0.17678	0.17678	0.14645	0.14645	0	0.82843
	2	0.10557	0.17889	−0.08944	0.08944	0.05279	0.05279	0	0.47214
	5	0.01942	0.03772	−0.01886	0.01886	0.00971	0.00971	0	0.19805
0.2	0	1	0	0	0	0.5	0.5	0	1.97987
	0.1	0.89748	0.10140	−0.05142	0.04998	0.44949	0.44794	0.00315	1.79018
	0.5	0.54403	0.35752	−0.17835	0.17917	0.27407	0.26997	0.04429	1.22176
	1	0.28763	0.34553	−0.17050	0.17503	0.14483	0.14280	0.05266	0.85005
	2	0.10453	0.18144	−0.08491	0.09080	0.05105	0.05348	0.02102	0.47022
	5	0.01938	0.03760	−0.01810	0.01950	0.00927	0.01011	0.00214	0.19785
1	0	0.5	0	0	0	0.5	0	0.31831	1.27319
	0.1	0.43015	0.05388	0.02247	0.07635	0.39198	0.03817	0.31405	1.18107
	0.5	0.28156	0.13591	0.00483	0.14074	0.21119	0.07037	0.26216	0.90298
	1	0.17868	0.15355	−0.02843	0.12513	0.11611	0.06256	0.18198	0.67769
	2	0.08269	0.11331	−0.04144	0.07187	0.04675	0.03593	0.07738	0.43202
	5	0.01835	0.03384	−0.01568	0.01816	0.00929	0.00905	0.00992	0.19455
2	0	0	0	0	0	0.12500	−0.12500	0	0.51671
	0.1	0.00856	−0.00845	0.01536	0.00691	0.11806	−0.10950	0.00159	0.51627
	0.5	0.03701	−0.02651	0.05690	0.03039	0.09180	−0.05479	0.03033	0.49728
	1	0.05185	−0.01005	0.05429	0.04456	0.06552	−0.01367	0.06434	0.45122
	2	0.04496	0.02836	0.01267	0.04103	0.03454	0.01043	0.06275	0.35054
	5	0.01573	0.02474	−0.00939	0.01535	0.00873	0.00700	0.01551	0.18450

2.8.4 BURMISTER'S METHOD FOR TWO-LAYER SYSTEMS

Burmister developed solutions for stresses and displacements in two-layer systems: $\mu = 0.5$ for each layer, in homogeneous, isotropic, and elastic materials, with the surface layer infinite in extent in the lateral direction but of finite depth, the underlying layer infinite in both horizontal and vertical directions, layers in continuous contact, and the surface layer free of shearing and normal stress outside the loaded area.

Total surface deflection, Δ_t, for a two-layer system:

Flexible plate:

$$\Delta_t = 1.5 \frac{pa}{E_2} F_2$$

Rigid plate:

$$\Delta_t = 1.18 \frac{pa}{E_2} F_2$$

where
p is the unit load on circular plate
a is the radius of plate
E_2 is the modulus of elasticity of lower layer
F_2 is the dimensionless factor depending on the ratio of moduli of elasticity of the pavement and subgrade (E_1/E_2) as well as the thickness of upper layer-to-radius of circular loaded ratio (h_1/a), as shown in Figure 2.12

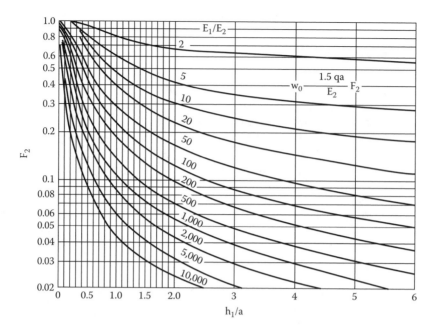

FIGURE 2.12 Values of dimensionless factor (F_2) in Burmister's method for two layer system. (From Huang, Y.H., *Pavement Analysis and Design*, 2nd edn., Upper Saddle River, NJ, p. 60, 2004. Reprinted by permission of Pearson Education, Inc.)

Note that stress and deflections are influenced by the ratio of the modulus of the pavement (everything above subgrade) and subgrade, there is significant effect of the layers above the subgrade, and there is significant difference in stress gradients obtained from Boussinesq's and Burmister's theory.

As an example of the application of Burmister's theory, consider the following problem. A 4500 lb load is being applied over a circular area with a stress (unit load) of 100 psi on a pavement with two layers—a hot mix asphalt (HMA) layer with a thickness of 6 in. and a modulus of 360 ksi and a subgrade (considered to be of infinite thickness) of 7.2 ksi. Consider a Poisson's ratio of 0.5 for both layers. What is the surface deflection?

$$\text{Area} = \pi a^2 = \frac{4500}{100} = 45\,\text{in.}^2$$

$$a = 3.8\,\text{in.}$$

$$p = 100\,\text{psi}$$

$$h_1 = 6\,\text{in.}$$

$$h_1/a = 1.6$$

$$E_1 = 360\,\text{ksi}$$

$$E_2 = 7.2\,\text{ksi} = 7200\,\text{psi}$$

$$E_1/E_2 = 50$$

From Figure 2.12, $F_2 = 0.2$

$$\Delta_t = 1.5\frac{pa}{E_2}F_2 = 0.016\,in.$$

2.8.5 ODEMARK'S METHOD OF EQUIVALENT LAYERS

Odemark developed a method whose principle is to transform a system consisting of layers with different moduli into an equivalent system where all the layers have the same modulus, and on which Boussinesq's equation may be used. This method is known as the method of equivalent thickness (MET).

$$Stiffness = \frac{IE}{(1-\mu^2)}$$

where I is the moment of inertia.

For stiffness to remain constant, this expression must remain constant, from which we can say the following:

$$\frac{h_e^3 E_2}{1-\mu_2^2} = \frac{h_1^3 E_1}{1-\mu_1^2}$$

$$h_e = h_1 \left[\frac{E_1}{E_2\left(\dfrac{1-\mu_2^2}{1-\mu_1^2}\right)} \right]^{1/3}$$

where h_e is the equivalent thickness. Note that this is an approximate method, and a correction factor is used to obtain a better agreement with elastic theory.

In many cases, Poisson's ratio may as well be assumed to be the same for all materials. Then, the aforementioned equation becomes

$$h_e = fh_1\left[\frac{E_1}{E_2}\right]^{1/3}$$

$$[\mu_1 = \mu_2]$$

where f is the correction factor that depends on layer thickness, modulus ratio, Poisson's ratio, and number of layers in the structure. Frequently used values of f are as follows: 0.9 for a two-layer system and 0.8 for a multilayer system, except for the first interface, where it is 1.0. A factor, f, of $1.1(a/h_1)^{0.3}$ can be used for the first interface when $a > h_1$.

For a multilayer system consisting of layers of thickness h_1 through h_n, with corresponding moduli of E_1 through E_n, and constant Poisson's ratio, the equivalent thickness of the upper $n-1$ layers with respect to the modulus of the layer n may be calculated as

$$h_{e,n} = f * \sum_{i=1}^{n-1}\left\{h_i * \left[\frac{E_i}{E_n}\right]\right\}^{1/3}$$

Note that to use the MET, moduli should be decreasing with depth, preferably by a factor of at least 2 between consecutive layers, and the equivalent thickness of a layer should preferably be larger than the radius of the loaded area.

To illustrate the application of the MET for reducing a two-layer pavement system to a one-layer system, consider the last example, where there are two layers—HMA of thickness 6 in. and modulus 360 ksi and a subgrade of 7.2 ksi modulus. If it is necessary to determine the stress at the bottom of the HMA layer (which is a critical factor for evaluating the fatigue cracking potential), and the only available equation is the one that is available for a one-layer system, that is,

$$\sigma_z = \sigma_o \left[1 - \frac{1}{\left\{ 1 + (a/z)^2 \right\}^{3/2}} \right]$$

How does one proceed?

First, convert the HMA layer thickness to an equivalent thickness of the subgrade layer, and then utilize the aforementioned equation (for finding stress) by considering z as that equivalent thickness:

$$h_e = fh_1 \left[\frac{E_1}{E_2} \right]^{1/3}$$

$$[\mu_1 = \mu_2]$$

$$f = 0.9$$

$$h_1 = 6 \text{ in.}$$

$$E_1/E_2 = 50$$

$$h_e = 19.6 \text{ in.}$$

using z = 19.6 in., a = 3.8 in., and σ_0 = 100 psi.
Vertical stress at the bottom of the HMA layer

$$\sigma_z = \sigma_o \left[1 - \frac{1}{\left\{ 1 + (a/z)^2 \right\}^{3/2}} \right] = 5.4 \text{ psi}$$

2.8.6 Fox and Acum and Fox's Solutions

Fox and Acum and Fox developed exact solutions for boundary stresses in the centerline of a circular, uniformly distributed load acting on the surface of a three-layer half-space, $\mu = 0.5$ for all layers. They provided a tabular summary of normal and radial stresses in three-layer systems at the intersection of the plate axis with the layer interfaces; Jones and Pattie expanded Fox and Acum and Fox's solutions to a wider range.

To find stresses, four parameters are needed:

$$K_1 = \frac{E_1}{E_2} \qquad A = \frac{a}{h_2}$$

$$K_2 = \frac{E_2}{E_3} \qquad H = \frac{h_1}{h_2}$$

The procedure to calculate the different stresses is as follows. For vertical stresses, obtain, from graphs, ZZ1 and ZZ2 (see Figures 2.13 and 2.14 for examples of values) and use the following:

$$\sigma_{z1} = p(ZZ1)$$

$$\sigma_{z2} = p(ZZ2)$$

For horizontal stresses, first obtain the following from a table (see Table 2.2 for an example of values):

$$[ZZ1 - RR1]$$

$$[ZZ2 - RR2]$$

$$[ZZ3 - RR3]$$

Then, use the following to calculate horizontal stresses from vertical stresses calculated earlier:

$$\sigma_{z1} - \sigma_{r1} = p[ZZ1 - RR1]$$

$$\sigma_{z2} - \sigma_{r2} = p[ZZ2 - RR2]$$

$$\sigma_{z2} - \sigma_{r3} = p[ZZ2 - RR3]$$

Knowing σ_{z1} and σ_{r1}, the horizontal strain at the bottom of layer 1, ε_{r1}, can be calculated as follows:

$$\varepsilon_{r1} = \frac{\sigma_{r1}}{E_1} - \mu_1 \frac{\sigma_{t1}}{E_1} - \mu_1 \frac{\sigma_{z1}}{E_1}$$

As an example, consider the following example. There is a three-layer pavement system—from the top they are, respectively (h denotes thickness and E denotes modulus),

1. HMA layer, $h_1 = 3.8$ in., $E_1 = 360$ ksi
2. Granular base course, $h_2 = 15$ in., $E_2 = 18$ ksi
3. Subgrade, $E_3 = 9$ ksi

The radius of the circular loaded area = 3.8 in.; the pressure at the surface is 100 psi.
    Determine the radial stress and strain at the bottom of the HMA layer (z = 3.8 in.), σ_{r1}, ε_{r1}.

$$K_1 = E_1/E_2 = 20; \quad K_2 = E_2/E_3 = 2; \quad a_1 = a/h_2 = 0.25; \quad H = h_1/h_2 = 0.25$$

From ZZ1 figure, ZZ1 = 0.2 (by interpolation)

$$\sigma_{z1} = pZZ1 = 100(0.2) = 20 \text{ psi}$$

$$\sigma_{z1} - \sigma_{r1} = p(ZZ1 - RR1)$$

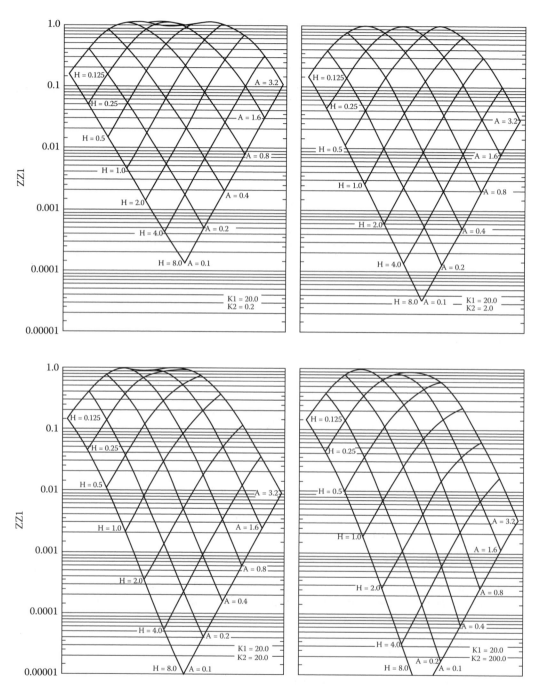

FIGURE 2.13 Examples of values of ZZ1. (From Yoder, E.J. and Witczak, M.W.: *Principles of Pavement Design*, 2nd edn. 1975. Copyright Wiley-VCH Verlag GmbH & Co. KGaA. Reprinted with permission.)

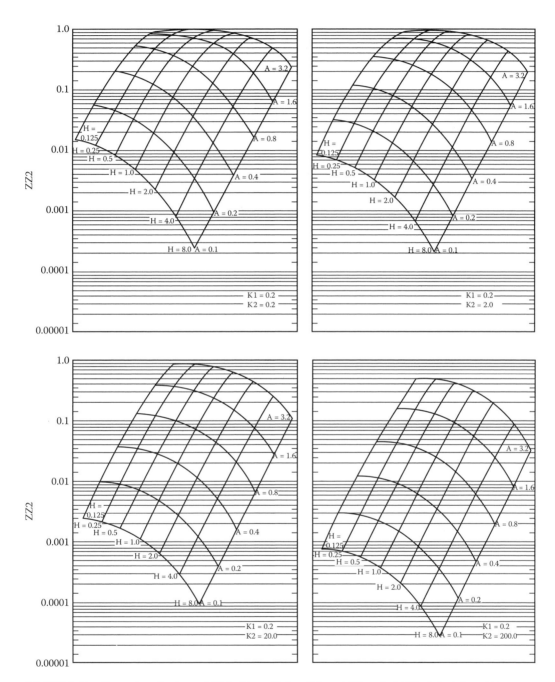

FIGURE 2.14 Examples of values of ZZ2. (From Yoder, E.J. and Witczak, M.W.: *Principles of Pavement Design*, 2nd edn. 1975. Copyright Wiley-VCH Verlag GmbH & Co. KGaA. Reprinted with permission.)

TABLE 2.2
Example Values of ZZ1 − RR1, ZZ2 − RR2, ZZ3 − RR3

a_1	H=0.125, k1=0.2			H=0.125, k1=2.0			H=0.125, k1=20.0			H=0.125, k1=200.0		
	(ZZ1−RR1)	(ZZ2−RR2)	(ZZ2−RR3)	(ZZ1−RR1)	(ZZ2−RR2)	(ZZ2−RR3)	(ZZ1−RR1)	(ZZ2−RR2)	(ZZ2−RR3)	(ZZ1−RR1)	(ZZ2−RR2)	(ZZ2−RR3)
			k2=0.2			k2=0.2			k2=0.2			k2=0.2
0.1	0.12438	0.00332	0.01659	0.71614	0.00350	0.01750	1.80805	0.00322	0.01611	2.87564	0.00201	0.01005
0.2	0.13546	0.01278	0.06391	1.01561	0.01348	0.06741	3.75440	0.01249	0.06244	7.44285	0.00788	0.03940
0.4	0.10428	0.04430	0.22150	0.83924	0.04669	0.23346	5.11847	0.04421	0.22105	15.41021	0.02913	0.14566
0.8	0.09011	0.10975	0.54877	0.63961	0.11484	0.57418	3.38600	0.11468	0.57342	9.70261	0.08714	0.43568
1.6	0.08777	0.13755	0.68777	0.65723	0.13726	0.68630	1.81603	0.13687	0.68436	7.02380	0.13705	0.68524
3.2	0.04129	0.10147	0.50736	0.38165	0.09467	0.47335	1.75101	0.07578	0.37890	2.35459	0.06594	0.32971
			k2=2.0			k2=2.0			k2=2.0			k2=2.0
0.1	0.12285	0.01693	0.00846	0.70622	0.01716	0.00858	1.81178	0.01542	0.00771	3.02259	0.00969	0.00485
0.2	0.12916	0.06558	0.03279	0.97956	0.06647	0.03324	3.76886	0.06003	0.03002	8.02452	0.03812	0.01906
0.4	0.08115	0.23257	0.11629	0.70970	0.23531	0.11766	5.16717	0.21640	0.10820	17.64175	0.14286	0.07143
0.8	0.01823	0.62863	0.31432	0.22319	0.63003	0.31501	3.43631	0.60493	0.30247	27.27701	0.45208	0.22604
1.6	−0.04136	0.98754	0.49377	−0.19982	0.97707	0.48853	1.15211	0.97146	0.48573	23.38638	0.90861	0.45430
3.2	−0.03804	0.82102	0.41051	−0.28916	0.84030	0.42015	−0.06894	0.88358	0.44179	11.87014	0.91469	0.45735
			k2=20.0			k2=20.0			k2=20.0			k2=20.0
0.1	0.12032	0.03667	0.00183	0.69332	0.03467	0.00173	1.80664	0.02985	0.00149	3.17763	0.01980	0.00099
0.2	0.11787	0.14336	0.00717	0.92086	0.13541	0.00677	3.74573	0.11697	0.00585	8.66097	0.07827	0.00391
0.4	0.03474	0.52691	0.02635	0.46583	0.49523	0.02476	5.05489	0.43263	0.02163	20.12259	0.29887	0.01494
0.8	−0.14872	1.61727	0.08086	−0.66535	1.49612	0.07481	2.92533	1.33736	0.06687	36.29943	1.01694	0.05085
1.6	−0.50533	3.58944	0.17947	−2.82859	3.28512	0.16426	−1.27093	2.99215	0.14961	49.40857	2.64313	0.13216
3.2	−0.80990	5.15409	0.25770	−5.27906	5.05952	0.25298	−7.35384	5.06489	0.25324	57.84369	4.89895	0.24495
			k2=200.0			k2=200.0			k2=200.0			k2=200.0
0.1	0.11720	0.05413	0.00027	0.67488	0.04848	0.00024	1.78941	0.04010	0.00020	3.26987	0.02809	0.00014
0.2	0.10495	0.21314	0.00170	0.85397	0.19043	0.00095	3.68097	0.15781	0.00079	9.02669	0.11136	0.00056
0.4	−0.01709	0.80400	0.00402	0.21165	0.71221	0.00356	4.80711	0.59391	0.00297	21.56482	0.43035	0.00215
0.8	−0.34427	2.67934	0.01340	−1.65954	2.32652	0.01163	1.90825	1.95709	0.00979	41.89873	1.53070	0.00765
1.6	−1.21139	7.35978	0.03680	−6.47707	6.26638	0.03133	−5.28803	5.25110	0.02626	69.63157	4.56707	0.02284
3.2	−2.89282	16.22830	0.08114	−16.67376	14.25621	0.07128	−21.52546	12.45068	0.06225	120.95931	11.42045	0.05710

a_1	$H=0.25$ $k_1-0.2$			$H=0.25$ $k_1-2.0$			$H=0.25$ $k_1-20.0$			$H=0.25$ $k_1-200.0$		
	(ZZ1 – RR1)	(ZZ2 – RR2)	(ZZ2 – RR3)	(ZZ1 – RR1)	(ZZ2 – RR2)	(ZZ2 – RR3)	(ZZ1 – RR1)	(ZZ2 – RR2)	(ZZ2 – RR3)	(ZZ1 – RR1)	(ZZ2 – RR2)	(ZZ2 – RR3)
			$k_2=0.2$			$k_2=0.2$			$k_2=0.2$			$k_2=0.2$
0.1	0.05598	0.00274	0.01370	0.28658	0.00277	0.01384	0.61450	0.00202	0.01011	0.86644	0.00090	00.0451
0.2	0.12628	0.01060	0.05302	0.72176	0.01075	0.05377	1.76675	0.00793	0.03964	2.71354	0.00357	0.01784
0.4	0.14219	0.03744	0.18722	1.03476	0.03842	0.19211	3.59650	0.02931	0.14653	6.83021	0.01365	0.06824
0.8	0.12300	0.09839	0.49196	0.88833	0.10337	0.51687	4.58845	0.08771	0.43854	13.19664	0.04624	0.23118
1.6	0.10534	0.13917	0.69586	0.66438	0.14102	0.70510	2.31165	0.14039	0.70194	13.79134	0.10591	0.52955
3.2	0.05063	0.11114	0.55569	0.41539	0.09804	0.49020	1.24415	0.07587	0.37934	2.72901	0.08608	0.43037
			$k_2=2.0$			$k_2=2.0$			$k_2=2.0$			$k_2=2.0$
0.1	0.05477	0.01409	0.00704	0.28362	0.01353	0.00677	0.63215	0.00962	0.00481	0.96553	0.00407	0.00203
0.2	0.12136	0.05484	0.02742	0.70225	0.05278	0.02639	1.83766	0.03781	0.01891	3.10763	0.01611	0.00806
0.4	0.12390	0.19780	0.09890	0.96634	0.19178	0.09589	3.86779	0.14159	0.07079	8.37852	0.06221	0.03110
0.8	0.06482	0.56039	0.28019	0.66885	0.55211	0.27605	5.50796	0.44710	0.22355	18.95534	0.21860	0.10930
1.6	-0.00519	0.96216	0.48108	0.17331	0.95080	0.47540	4.24281	0.90115	0.45058	31.18909	0.58553	0.29277
3.2	-0.02216	0.87221	0.43610	-0.05691	0.89390	0.44695	1.97494	0.93254	0.46627	28.98500	0.89191	0.44595
			$k_2=20.0$			$k_2=20.0$			$k_2=20.0$			$k_1=20.0$
0.1	0.05192	0.03116	0.00156	0.27580	0.02728	0.00136	0.65003	0.01930	0.00096	1.08738	0.00861	0.00043
0.2	0.11209	0.12227	0.00611	0.67115	0.10710	0.00536	1.90693	0.07623	0.00381	3.59448	0.03421	0.00171
0.4	0.08022	0.45504	0.02275	0.84462	0.39919	0.01996	4.13976	0.29072	0.01454	10.30923	0.13365	0.00668
0.8	-0.07351	1.44285	0.07214	0.21951	1.26565	0.06328	6.48948	0.98565	0.04928	26.41442	0.49135	0.02457
1.6	-0.40234	3.37001	0.16850	-1.22411	2.94860	0.14743	6.95639	2.55231	0.12762	57.46409	1.53833	0.07692
3.2	-0.71901	5.10060	0.25503	-3.04320	4.89878	0.24494	6.05854	4.76234	0.23812	99.29034	3.60964	0.18048
			$k_2=200.0$			$k_2=200.0$			$k_2=200.0$			$k_2=200.0$
0.1	0.04956	0.04704	0.00024	0.26776	0.03814	0.00019	0.65732	0.02711	0.00014	1.19099	0.01311	0.00007
0.2	0.10066	0.18557	0.00093	0.63873	0.15040	0.00075	1.93764	0.10741	0.00054	4.00968	0.05223	0.00026
0.4	0.04248	0.70524	0.00353	0.71620	0.57046	0.00285	4.26004	0.41459	0.00207	11.96405	0.20551	0.00103
0.8	-0.24071	2.40585	0.01203	-0.28250	1.92636	0.00963	6.94871	1.46947	0.00735	32.97364	0.77584	0.00388
1.6	-1.00743	6.82481	0.03412	-3.09856	5.35936	0.02680	8.55770	4.36521	0.02183	82.77997	2.63962	0.01320
3.2	-2.54264	15.45931	0.07730	-9.18214	12.64318	0.06322	10.63614	10.93570	0.05468	189.37439	7.60287	0.03801

Source: Yoder, E.J. and Witczak, M.W.: *Principles of Pavement Design*, 2nd edn. 1975. Copyright Wiley-VCH Verlag GmbH & Co. KGaA. Reprinted with permission.

From [ZZ1 − RR1] Table, ZZ1 − RR1 = 2.345 (by interpolation)

$$\sigma_{z1} - \sigma_{r1} = 100(2.345)$$

$$\sigma_{r1} = -214.5 \text{ psi}$$

$$\varepsilon_{r1} = \frac{1}{2E_1}(\sigma_{r1} - \sigma_{z1})$$

$$\varepsilon_{r1} = \frac{1}{2*360,000}(-214.5 - 20)$$

$$\varepsilon_{r1} = 325 \times 10^{-6}$$

2.8.7   COMPUTER PROGRAMS

Computer programs have been developed on the basis of the theory discussed earlier for calculating stresses, strains, and deflections of a layered elastic system. They have gradually become more sophisticated in capability to handle linear elastic materials, nonlinear elastic granular materials, vertical as well as horizontal loads, elastic multilayer systems under multiple wheel loads, and stress-dependent material properties, and in application of finite element linear and nonlinear analysis.

Examples include the following: BISTRO and BISAR (from Shell), ELSYM5 (from Chevron), ALIZEIII (LCPC) and CIRCLY (from MINCAD, Australia), DAMA (from the Asphalt Institute), SAPIV and ELSYM5 (from the University of California, Berkeley), PDMAP (Finn et al., 1986), the IILI-PAVE computer program (Raad and Figueroa, 1980), MICH-PAVE (Harichandran et al., 1989), and Everstress (from the Evercalc suite of software from the Washington State DOT [1995]).

The reader is advised to acquire any of the aforementioned software for the practice problems. One suggested software is Everstress (http://www.wsdot.wa.gov/Business/MaterialsLab/Pavements/PavementDesign.htm#PavementDesignRequirements; Washington State Department of Transportation, 2007). Note that a 3D-finite element software is also available from this website.

An example output (containing input information also) from EVERSTRESS with the last three-layer example problem is shown in Figure 2.15.

Moduli of soils are affected significantly by the stress that is imposed by loading as well as the self-weight of materials (overburden), and generally the moduli of coarse-grained soils is expressed as a function of the bulk stress whereas that of fine-grained soils is expressed as a function of deviator stress as follows:

$$E_{\text{coarse grained}} = \text{Multiplier} * (\text{Bulk stress})^{\text{Power}}$$

$$E_{\text{fine grained}} = \text{Multiplier} * (\text{Deviator stress})^{\text{Power}}$$

(Further discussion is in Chapter 7.)

In linear elastic analysis software, an iterative process is utilized to consider this nonlinear behavior of soils. Starting from an assumed (seed) modulus, the stresses are calculated, and the resultant modulus is checked against the seed modulus—this process is continued until a maximum tolerance (say 5%) of difference in modulus is reached.

A finite element software (3D-Move) that can account for viscoelastic behavior of asphalt mixes, and considers moving and multiple loads as well as nonuniform contact stress, has been recently made available by the University of Nevada, Reno, at: http://www.arc.unr.edu/Software.html.

CLayered Elastic Analysis by EverStress for Windows

No of Layers: 3	No of Loads: 1	No of X–Y Evaluation Points: 1

Layer *	Poisson's Ratio	Thickness (in.)	Moduli(1) (ksi)
1	0.5	3.8	360
2	0.5	15	18
3	0.5	*	S

Load No *	X-Position (in.)	Y-Position (in.)	Load (lbf)	Pressure (psi)	Radius (in.)
1	0	0	4500	100	3.785

Line
Line
Location No: 1   X-Position (in.): .000   Y-Position (in.): .000
Line

cNormal Stresses

Z-Position (in.)	Layer *	Sxx (psi)	Syy (psi)	Szz (psi)	Syz (psi)	Sxz (psi)	Sxy (psi)
3.799	1	221.02	221.02	−19.89	0	0	0
11.3	2	−0.39	−0.39	−7.44	0	0	0
18.8	2	2.61	2.61	−3.26	0	0	0

Line

cNormal Strains and Deflections

Z-Position (in.)	Layer *	Exx (10^{-6})	Eyy (10^{-6})	Ezz (10^{-6})	Ux (mils)	Uy (mils)	Uz (mils)
3.799	1	334.6	334.6	−669.2	0	0	15.828
11.3	2	195.82	195.82	−391.63	0	0	11.826
18.8	2	163.02	163.02	−326.02	0	0	9.343

FIGURE 2.15 Example output from EVERSTRESS analysis. Sxx, Normal stresses in X direction; Exx, Normal strain in x direction.

2.9 PRINCIPLES OF GOOD CONSTRUCTION

Pavement construction is a complex process involving many factors—each of which has the potential of affecting the quality of the constructed pavement in a very significant way. The process of good construction should start with the identification/adoption or creation of an appropriate specification. This specification should lay out details of steps, determine tests required to make sure that the quality of the product is acceptable, and monitor the key parameters during construction, such that errors can be identified and rectified quickly.

Paving mixes are produced in plants, which can be of different types. The basic processes are the same. Aggregates of different sizes are mixed in specific proportions, dried, mixed with asphalt at a high temperature, and then stored in silos or transported to the job site through insulated trucks. During the mixing process, the steps are monitored closely to identify any malfunction. Since the mixing process is conducted at a specific range of high temperature to make sure that the aggregates are dried and the asphalt is of a sufficiently fluid nature to allow coating and mixing of aggregates, and because the proportioning of different components is conducted in terms of weight, temperature and weight measurements need to be monitored very closely during the process.

Trucks take the mix to pavers at the job site. Pavers are self-propelled machines that lay down asphalt mix at a specific depth and provide the initial compaction. A paver is a complex piece of machinery and has a number of moving parts, including an auger and conveyors to mix the mix and send it to the back from the front hopper where it receives the mix from the trucks. The screed at the back of the paver maintains a specific depth of the mix and must be controlled carefully.

Right behind the paver come the rollers. Different rollers have different functions. Steel-wheel vibratory drum rollers help in initial compaction, and rubber-tired and static steel-wheel rollers provide intermediate and final compaction. The rollers must be of appropriate weight, and frequency and amplitude of vibration (for vibratory rollers) must be set at the appropriate level. The rollers should also move at the correct speed and in the correct pattern to ensure the best compaction possible.

Finally, all through the production and laydown operation, an appropriate quality control procedure must be adopted to monitor key properties and correct, if necessary, any errors as quickly as possible. There may also be a quality assurance process to make sure that the quality control process is actually working.

2.10 PUTTING EVERYTHING TOGETHER

The process of pavement construction actually begins during the mix design process and ends with compaction. While selecting materials for mix design, one needs to use experience and judgment, not only from the point of view of the optimum mix but also from the consideration of construction. In many cases, a mix that can be produced in small amounts in the laboratory during mix design may not be produced in large batches in the plant because of practical reasons—for example, due to the lack of locally available materials in sufficient amounts, the lack of appropriate fixtures in the plant machinery, or even the lack of sufficient paver and compaction equipment.

Hence, looking at the big picture is necessary from the very beginning. Consideration should be given to the type of project, environmental conditions expected during construction, and availability of plant and field equipment.

The mixing process in the plant involves batching or proportioning a variety of aggregates of different sizes as well as liquid asphalt binder. Any deviation in these parameters would be reflected in the constructed pavement—the key is to catch it through the use of appropriate tests in the plant and in the field, and to correct it quickly.

One important consideration is the availability of the fleet of the machinery required for transporting mix from the plant to the job site. The whole operation should be coordinated in such a way that the pavers and rollers keep moving smoothly at the appropriate speed. If required, a material transfer vehicle may be used to receive and store mix from the trucks, such that the paver can keep working even if the flow of mix through trucks is hampered for a specific time interval due to any reason.

One significant factor that affects the quality of the finished pavement is its air voids or density. Given that the mix design is accurate and the mixing process in the plant is working fine, four things have the most significant effect on the density of the pavement. These are the temperature of the mix, the ambient temperature (and wind), the availability of required rollers in appropriate numbers, and the thickness of the layer. It is crucial that appropriate considerations are given to these factors to ensure that for the given thickness and the existing environmental conditions at the site, within the allowable range of mixing temperature, there is sufficient time to compact the mix in the field with the use of the available equipment.

QUESTIONS

2.1 What is the purpose of mix design?

2.2 How does structural design differ from mix design?

2.3 What are the most important considerations for construction of an asphalt pavement?

2.4 Are mix design, structural design, and construction all related to a certain extent? Can you explain this schematically?

2.5 Determine the vertical and radial stresses at nine points for a 9000-lb point load on a homogeneous, isotropic, linear elastic, semi-infinite space. Consider a Poisson's ratio of 0.3.

Point	z (in.)	r (in.)	Point	z (in.)	r (in.)	Point	z (in.)	r (in.)
1	0	0	4	6	0	7	12	0
2	0	6	5	6	6	8	12	6
3	0	12	6	6	12	9	12	12

2.6 If the deflection at the center of a rigid plate is found out to be 0.03 in. from a load on a subgrade with Poisson's ratio of 0.35, what is the estimated modulus of the subgrade? Consider a load of 9000 lb and plate radius of 6 in.

2.7 Use any layered elastic analysis program to compute the vertical and radial stresses and strains directly below the load at a depth of 149 mm in a full depth 150 mm thick asphalt pavement with a modulus of 3500 MPa and Poisson's ratio of 0.35 for the following loading conditions. In each case, half of a standard 20 KM axle (only the main load-bearing axles, not including the steering axle) has been indicated. Can you sketch the axle/wheel configuration of the entire vehicles? For the subgrade consider a modulus and Poisson's ratio of 100 MPa and 0.4, respectively. Consider all layers to be stress insensitive:

A. Loads of 20 KN, with coordinates in cm (x,y): (0,0) (33,0); (0,122) (33,122); tire pressure of 690 kPa

B. Loads of 20 KN, with coordinates in cm (x,y): (0,0) (33,0); (0,122) (33,122); (0,244) (33,244); tire pressure of 690 kPa

2.8 Use a layered elastic analysis program to determine the radial stresses at the bottom of the surface layer directly under the load for a pavement with three layers as follows:

Layer	Modulus (psi)	Poisson's Ratio	Thickness (in.) Full Friction between All Layers
1	435,113	0.35	10.63
2	21,755.7	0.4	20.08
3	7,251.9	0.4	Infinite

Consider all layers to be stress insensitive.

Consider three different cases of loads as follows:

A. Single axle with dual tires

Tire #	X (in.)	Y (in.)	Load (lb)	Pressure (psi)
1	0	0	5000	100
2	13.5	0	5000	100

B. Tandem axle with dual tires

Tire #	X (in.)	Y (in.)	Load (lb)	Pressure (psi)
1	0	0	5000	100
2	13.5	0	5000	100
3	13.5	54	5000	100
4	0	54	5000	100

C. Tridem axle with dual tires

Tire #	X (in.)	Y (in.)	Load (lb)	Pressure (psi)
1	0	0	5000	100
2	13.5	0	5000	100
3	13.5	54	5000	100
4	0	54	5000	100
5	0	108	5000	100
6	13.5	108	5000	100

3 Principles of Structural Design, Mix Design, and Construction of Concrete Pavements

3.1 OVERVIEW

Producing high-performance and long-lasting rigid pavements is a complex process that integrates appropriate design and materials, quality construction practices, timely maintenance, and overall adequate pavement management. The design of pavements requires more than "thickness design" for carrying flexural stresses. The pavement is subject to a range of loads and stresses induced by traffic loading, climate, and environmental factors. The design process balances a large number of interacting factors and constraints to produce pavement systems that are functional, sustainable, cost effective, and long lasting.

Properly designed, constructed, and maintained rigid pavements can provide years of service life. However, service life could be greatly reduced if any part of this pavement process is poorly executed resulting in significant cost to the owner (State Highway Agency, County, Turnpike Authority, etc.). It is important for pavement engineers and students alike to understand that producing functional, cost-effective, and long-lasting pavements incorporates good science, good engineering, and adequate knowledge and experience about local conditions.

Similar to flexible pavements, the design process for rigid pavements requires broad knowledge about topics such as

- Traffic loads that vary in magnitude and frequency; incorporate truck loading and incremental damage
- In situ soils and layer bearing materials with respect to bearing capacity, compressibility, drainage issues, erosion, freezing environment, modeling nonlinear behavior, uniformity, and constructability
- Design requirements with respect to loads, environment, early age behavior, slab dimensions, reinforcement, joint design, and performance
- Concrete mix design with respect to quality materials, construction requirements, early age behavior, strength gain, structural performance, volume stability, durability, changing environmental conditions, local availability, and sustainability
- Design and construction details related to concrete placement, finishing, curing, steel reinforcement, joint design, and construction
- Life cycle cost analysis, sustainability, and energy considerations

The list of interactive variables and parameters is extensive. What is important is that pavement engineers must realize the complexity of this system and the interdependence of each component and subsystem in producing functional, cost-effective, and long-lasting pavements. Figure 3.1 shows a conceptual process to demonstrate the interaction between pavement structure, materials, and construction design.

Structural and mix design requirements based on considerations for traffic loads, environment, construction, short and long term pavement performance needs, pavement management strategies.

Develop a sustainable mix design with a low energy and carbon footprint, to satisfy fresh concrete properties for placement and finishing, early age properties, structural properties, dimensional and volume stability, environmental changes and long-term durability.

Assess concrete properties (i.e., compressive, flexural strength, stiffness); volume stability (shrinkage and CTE); durability (i.e., freeze-thaw, salt scaling, chloride permeability, etc.), maturity, air-void system, etc.

Use of computer software in design process.

FIGURE 3.1 Process of integrating mix design, structural design, and construction requirement for producing sustainable, long-lasting, and cost-effective rigid pavements.

3.2 STRUCTURAL DESIGN

Structural design is based on the concept of limiting stresses and deformations to prevent excessive damage and deterioration of pavements. Overstressed pavements due to traffic loads and environmental effects will result in pavement distress such as fatigue cracking, faulting, pumping, punchouts, and curling and warping as shown in Figure 3.2. The objective of pavement design is to recommend a pavement structure and configuration, including slab thickness, slab length, mix design, reinforcement requirements, joint details, and foundation support. And that can handle the required traffic loads, climate and environment conditions, and provide the expected level of performance at the desired cost. This is an ambitious requirement given the uncertainty and limits in predicting the behavior of all interacting components of such a complex system. Therefore, the approach is to study the subsystems that makeup rigid pavements and use the available tools to achieve and realize pavements that are functional, long lasting, and cost effective.

FIGURE 3.2   Major distress conditions considered for design in Portland cement concrete (PCC) pavements.

For analysis of stresses, a concrete pavement is idealized as a rigid slab resting on a spring-like foundation. Theories, as explained in the next section, are used to develop equations relating stresses and deformations to material and structural properties, and then analytical methods are used to predict these stresses and deformation on the basis of properties such as modulus, Poisson's ratio, and stiffness of the subgrade, as well as temperature fluctuations. The design process essentially consists of consideration of traffic and environmental factors and the use of the relevant properties (through material characterization tests) to ensure that the slab can sustain the stresses for the duration of its life, without failing.

3.3   THEORETICAL CONSIDERATIONS

A rigid pavement is basically a slab resting on a subgrade or base. The slab is much stiffer than the supporting base or foundation material, and therefore carries a significant portion of induced stresses. The load-carrying mechanism is similar to beam action, although a concrete slab is much wider than the beam and should be considered as a plate. Westergaard (1926b) developed stress equations for rigid pavement slabs supported on a Winkler or liquid foundation, which is a conceptual model that considers the foundation as a series of springs. When the slab is loaded vertically down, the springs tend to push back; when the environment-related loads are pulling up on the slab, the springs tend to pull down toward the foundation (Figure 3.3).

The induced stresses and their effects are controlled by a number of factors, such as restrained temperature and moisture deformation, externally applied loads, volume changes of the supporting material and frost action, continuity of subgrade support through plastic deformation, or materials loss due to pumping action.

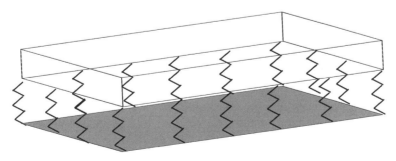

FIGURE 3.3   Slab on Winkler foundation.

Mathematical modeling of the true stress state of a PCC slab on a supporting foundation layer system is very complex due to variation in materials, nonlinear behavior, changing moisture and temperature conditions, changing support conditions, and complex interactions between components. However, simple models can be developed once assumptions are made.

3.3.1 STRESSES DUE TO CURVATURE AND BENDING IN SLABS

If we consider that the pavement slab as a beam on an elastic foundation, then the reactive pressure, p, can be related to the deformation, Δ, through the equation $p = k\Delta$, where k is the modulus of subgrade reaction, which is the ratio of the pressure applied to the subgrade using a loaded area divided by the displacement experienced by that loaded area. Figure 3.4a and b shows a schematic of the spring constant and a plate load test for determining k. It is assumed that the modulus of subgrade reaction k is a constant and that the behavior is elastic. This assumption is only true over very small deformations and is greatly affected by numerous conditions such as soil type, density, moisture content, and stress state.

A concrete pavement slab will deform under load. The resistance to deformation due to loading depends upon the characteristics of the foundation and the stiffness or modulus value of the slab. The moment due to bending in a beam is given by the following equation:

$$M = EI \frac{d^2y}{dx^2}$$

where
  E is the modulus of elasticity
  I is the moment of inertia of the beam

The stiffness term in the beam equation is expressed by the "EI" term. For an infinite thin plate, the moment Mx is given by the following equation:

$$Mx = \frac{Eh^3}{12(1-\mu^2)} \left(\frac{\partial^2 \omega}{\partial x^2} + \mu \frac{\partial^2 \omega}{\partial y^2} \right)$$

or

$$Mx = D \left(\frac{\partial^2 \omega}{\partial x^2} + \mu \frac{\partial^2 \omega}{\partial y^2} \right)$$

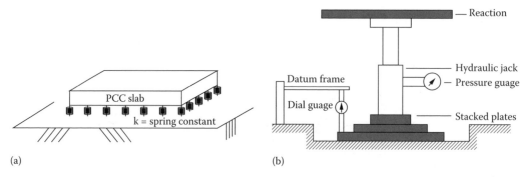

(a)   (b)

FIGURE 3.4 (a) Schematic showing k, the modulus of subgrade reaction, and (b) a plate load test for determining k.

where the stiffness term for the plate is given by the following equation and is represented by the letter D:

$$D = \frac{Eh^3}{12(1-\mu^2)}$$

Westergaard (1927) applied plate theory to a finite PCC slab and supporting foundation and developed the relative stiffness expression "ℓ," which is called the *radius of relative stiffness*. The radius of relative stiffness expression depends on both slab and foundation properties. The value for ℓ increases with a stiffer concrete, a thicker slab, and a weaker foundation. This term is used in many of the stress and deflection equations derived for rigid pavements by Westergaard and others.

$$\ell = \sqrt[4]{\frac{Eh^3}{12(1-\mu^2)k}}$$

where
   ℓ is the radius of relative stiffness, in.; mm
   E is the modulus of elasticity of the pavement, psi; MPa
   h is the thickness of the pavement, in.; mm
   μ is the Poisson's ratio of the PCC, in./in.; mm/mm
   k is the modulus of subgrade reaction, pound per cubic inch, pci or MN/m^3

3.3.2 STRESSES DUE TO TEMPERATURE CURLING

A concrete slab will undergo volume changes and develop stresses due to changes in temperature and moisture as shown in Figure 3.5. During the day, as the air temperature and sun increase the surface temperature of the concrete slab, the top of the slab will tend to expand relative to the neutral axis and the bottom of the slab will tend to contract as it is insulated by the soil in the base. However, the weight of the slab will prevent it from contraction or expansion, and compressive stresses will be induced in the top layer of the slab, while tensile stresses will be induced in the bottom layer. The opposite will occur at night where the air temperature will be cooler compared to the base of the slab since it is insulated by the base or subgrade. The top of the slab will be cooler compared to the bottom and will tend to contract. The slab weight will prevent the upward curling, and therefore tensile stresses will develop in the top of the slab while compressive stresses will be induced in the bottom of the slab. A similar effect is observed with moisture changes. As moisture is removed from the concrete slab, and specifically the hydrated cement paste, the concrete will shrink.

Another approach used to explain curling stress is by using the concept of plate theory and the concept of a "Liquid or Winkler Foundation" (Figure 3.3). When the surface temperature is greater than the bottom temperature, the top of the slab/plate tends to expand and the springs react to this movement. The outside springs are compressed and push the plate outward, and the inner springs are in tension and pull the plate inward. This condition induces compressive stresses in the top of the slab and tensile stresses in the bottom. When the surface temperature of the slab is lower than the bottom temperature, the top tends to contract and the bottom tends to expand. This produces an upward curvature of the slab (such that it can virtually hold water). The springs attached to the exterior edge of the slab are now in tension and pull the slab down, and the springs attached to the interior of the slab are in compression and push the slab outward, inducing tensile stresses in the top of the slab and compressive stresses in the bottom of the slab.

FIGURE 3.5 Slab curling and warping due to temperature and moisture variations. (a) Hot surface (sunny day), (b) cold surface (cool night), (c), wet surface and drier base, (d) dry surface and wetter base, (e) hot surface and wetter base (compounded curling and warping effect), and (f) cold surface and drier base (opposing effects of curling and warping).

The liquid foundation is analogous to the upward water force exerted on a boat. The weight of the boat displaces an amount of water that is equivalent to the total weight of the boat. A cubic foot of water displaced will create 62.4 pcf or 0.036 pci. The modulus of subgrade reaction for a soil could range from 50 to over 800 pci.

Curling stresses are attributed to temperature changes between the top and bottom of the concrete slab. Curling stresses occur once a day as the top of the slab is heated or cooled. For example, over 30 years of service, a pavement slab will experience approximately 10,950 fatigue cycles due to temperature effects. This built-in curvature can cause mid-slab cracking if joints and steel design is not appropriate or sufficient. Warping stresses, on the other hand, are defined as stresses due to moisture changes in the concrete. Warping stresses occur practically once during the life of the slab as the concrete is poured, cures and reaches some moisture equilibrium with the base. These stresses due to curling and warping are usually addressed by joint and steel design of the slab. Fatigue cycles due to traffic loading can range in tens of millions of cycles for highways and hundreds of thousands of cycles for airports. Therefore, thickness design for fatigue usually ignores curling and warping cycles, as is done in the PCA design method and which will be minimal according to Huang (Huang 2004).

Westergaard (1926a) developed equations for approximating the curling stresses in concrete pavements based on plate theory. These equations are very complex and will not be discussed in this book; they are discussed by Yoder and Witczak (1975) and Huang (2004). However, based on Westergaard's curling stress analysis, Bradbury (1938) developed correction coefficients C_x and C_y

to determine curling stresses due to temperature in a finite slab. The maximum total stresses at the interior of the slab, edge, and corner are given by the following equations:

$$\text{Interior stress, } \sigma_t = \frac{E\alpha\Delta t}{2}\left[\frac{C_x + \mu C_y}{1 - \mu^2}\right]$$

$$\text{Edge stress, } \sigma_t = \frac{CE\alpha\Delta t}{2}$$

An approximate equation for corner stress that is less critical than the interior or edge stresses was also developed later by Bradbury:

$$\text{Corner stress; } \sigma_t = \frac{E\alpha\Delta t}{3(1-\mu)}\left[\sqrt{\frac{a}{\ell}}\right]$$

where
  E is the modulus of elasticity of concrete
  μ is the Poisson's ratio of concrete
  α is the coefficient of thermal expansion
  $C = C_x$ and C_y are correction factors (after Bradbury 1938) for the finite slab and are dependent on L_x/ℓ and L_y/ℓ
  a is the radius of circular contact area applied at the corner

ℓ is defined as the radius of relative stiffness and is derived based on plate theory. The value of C increases with an increase in L/ℓ. When the ratio L/ℓ=6.45, C=1.0; when the ratio L/ℓ=8.5, C_{max}=1.084; for L/ℓ > 14, C=1.0 (see Figure 3.6a). L_x L_y are finite slab length along the x-axis (or longitudinal) and y-axis (or transverse), respectively (see Figure 3.6b). Δt is the change in temperature and a is the radius of wheel load distribution for corner load.

Example 3.1

Consider a concrete slab 30 ft (9.14 m) by 12 ft (3.66 m) and 8 in. (203 mm) thick, subjected to a temperature differential of 20°F (11.1°C). Assuming k=200 pci (54.2 MN/m³) and α=5 * 10⁻⁶ in./in./°F (9 * 10⁻⁶ mm/mm/°C), E=4 * 10⁶ psi, μ=0.15.
  Determine the maximum curling stress in the interior and at the edge of the slab (see Figure 3.7).

Solution

$$\ell = \sqrt[4]{\frac{Eh^3}{12(1-\mu^2)k}}$$

$$\ell = \sqrt[4]{\frac{4*10^6*9^3}{12(1-0.15^2)150}} = 35.88 \text{ in.} = 911.34 \text{ mm}$$

$L_x/\ell = 10.03$; $L_y/\ell = 4.01$
From Figure 3.6 the values for $C_x = 1.07$ and $C_y = 0.49$.

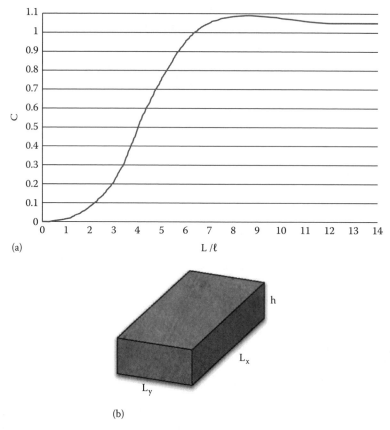

(a)

(b)

FIGURE 3.6 (a) Curling stress correction factors for a finite slab. (b) L_x, L_y are finite slab lengths along the x-axis or (longitudinal) and y-axis or (transverse) directions, and h is the slab thickness. (From Yoder, E.J. and Witczak, M.W.: *Principles of Pavement Design*. 2nd edn. 1975. Copyright Wiley-VCH Verlag GmbH & Co. KGaA. Reprinted with permission.)

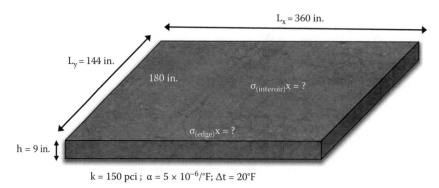

FIGURE 3.7 Example 3.1.

Interior stress x-direction; $\sigma_t = E\alpha\Delta t/2[(C_x + \mu C_y)/(1 - \mu^2)]$

$$\sigma_t = \frac{4*10^6 *5*10^{-6} *20}{2}\left[\frac{1.07+0.15*0.49}{1-0.15^2}\right] = 234\,\text{psi} = 1.61\,\text{MPa}$$

Edge stress x-direction; $\sigma_t = CE\alpha\Delta t/2$

$$\sigma_t = \frac{1.07 * 4*10^6 *5*10^{-6} *20}{2} = 214\,\text{psi} = 1.48\,\text{MPa}$$

Note that both the interior and edge critical stresses are in the x-direction.

3.3.3 STRESSES AND DEFLECTIONS DUE TO APPLIED LOADING

Loads from traffic will induce stresses in the concrete slab and will determine the fatigue life of the pavement. Loads at the corner will contribute to corner breaks, and loads at the interior and edge will contribute to mid-slab transverse cracking. The magnitude of the stress and deflection will depend on the magnitude and location of the applied load. Closed form solutions were developed by Westergaard et al. to determine stresses and deflections for three critical loading conditions applied to the corner, interior, and edge for idealized conditions. These equations are applicable to single wheel loads applied to very large slabs where the interior conditions are not influenced by edge effects and the edge effects are not influenced by corner effects.

3.3.3.1 Corner Slab Loading

The corner loading stresses for a slab were originally developed by Goldbeck (1919) and Older (1924) and discussed recently by Huang (2004). This solution is based on the simple bending stress equation

$$\sigma = \frac{Mc}{I}$$

It assumes a load "P" to be placed at the corner of the slab. It is also assumed that the slab has no subgrade support if the entire support layer is washed away or the slab is curling away from the base. The desired stress plane occurs at a distance "x" from the corner with a failure plane of width "2x" assuming the stress to be symmetrical with the bisector (Figure 3.8). Based on a slab thickness h, the moment of inertia for the cracked section becomes

$$I = \frac{bh^3}{12} = \frac{2x*h^3}{12}$$

The maximum moment produced at a distance x along the bisector from the corner is "Px," and thus

$$\sigma = \frac{Mc}{I} = \frac{Px*(h/2)}{2x*h^3/12} = \frac{3P}{h^2}$$

This solution, which shows that the stress produced decreases with the square of the slab thickness, is very conservative. It assumes no subgrade support under the slab, which is unlikely, and assumes a point load at the corner instead of the load being distributed over a circular tire contact area, which would produce a lesser moment and stress.

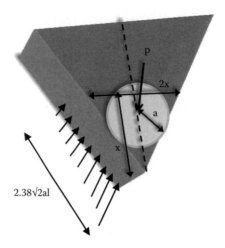

FIGURE 3.8   Slab cross section with an applied corner load for development of stress equation.

Westergaard (1926) developed an updated corner equation based on a method of successive iterations, which was later modified by Ioannides et al. (1985) and discussed by Huang (2004).

The corner stress σ_c and corner deflection Δ_c based on a circular applied load at the corner of the slab are given by

$$\sigma_c = \frac{3P}{h^2}\left[1-\left(\frac{a\sqrt{2}}{\ell}\right)^{0.6}\right]$$

$$\Delta_c = \frac{P}{k*\ell^2}\left[1.1-0.88\left(\frac{6\sqrt{2}}{\ell}\right)\right]$$

where
   P is the concentrate load applied over a circular contact area with a contact radius "a"
   $c = a\sqrt{2}$ is the side length of a square contact area
   "ℓ" is the radius of relative stiffness
   "k" is the modulus of subgrade reaction

The maximum moment occurs at a distance from the corner $= 2.38\sqrt{a\ell}$; note that the stress equation reduces to $\sigma = 3P/h^2$ when the contact radius a = 0. Therefore, Westergaard's updated stress equation will always produce a lower stress than the Goldbeck equation with the point load applied at the corner.

Slab deflections at the joints are important in contributing to pumping effects in rigid pavements. The larger the slab deflection, the greater the potential exists for pumping of fines and material supporting the slab at the joint to occur. Once enough material that supports the slab has been removed, the greater the potential for slab corner and edge breaks. However, it should be noted that corner deflections of slabs are more critical for pumping than edge deflections since they have a higher magnitude for deflection. Hence, controlling slab deflections by using thicker pavements or the use of dowels at joints is highly recommended. Some pavement designers believe the use of dowels, and a well-draining base course is more efficient and more cost effective than using a thicker pavement slab (Datelle, 2008).

Ioannides et al. (1985) proposed a different form of Westergaard's equations based on finite element analysis. Based on the FE method, they found that the maximum moment in the concrete slab occurs at a distance of $1.8c^{0.32}\ell^{0.59}$ from the corner. If a load is applied over a circular area, the value of c must be selected so that the contact area for the circle and the square are equal, and therefore $c = 1.772a$. The bending stress and deflection are provided as follows:

$$\sigma_c = \frac{3P}{h^2}\left[1-\left(\frac{c}{\ell}\right)^{0.72}\right]$$

$$\Delta_c = \frac{P}{k\ell^2}\left[1.205-0.69\left(\frac{c}{\ell}\right)\right]$$

Example 3.2

A concrete slab is subjected to a corner loading as shown in Figure 3.9. The slab thickness $h = 9$ in. (228.6 mm), $k = 150$ pci (27.2 MN/m³), $a = 6$ in. (152 mm), and $P = 9000$ lb (40.1 kN). Determine the maximum stress and deflection due to corner loading using both equations.

Solution

$$\ell = \left[\frac{4,000,000*9^3}{12\left(1-0.15^2\right)150}\right]^{0.25} = 35.88 \text{ in. (911.3 mm)}$$

Corner stress and deflection after Westergaard:

$$\sigma_c = \frac{3*9000}{9^2}\left[1-\left(\frac{6\sqrt{2}}{35.88}\right)^{0.6}\right] = 193.0 \text{ psi}$$

$$\Delta_c = \frac{9000}{150*35.88^2}\left[1.1-0.88\left(\frac{6\sqrt{2}}{35.88}\right)\right] = 0.0415 \text{ in.}$$

P = 9000 lb

a = 6 in.

σ = ?
Δ = ?

h = 9 in.
k = 150 pci

FIGURE 3.9 Example 3.2.

Corner stress and deflection after Ionnides et al.:

$$\sigma_c = \frac{3*9000}{9^2}\left[1-\left(\frac{1.772*6}{35.88}\right)^{0.72}\right]=194.5\,\text{psi}$$

$$\Delta_c = \frac{9000}{150*35.88^2}\left[1.205-0.69\left(\frac{1.772*6}{35.88}\right)\right]=0.0485\,\text{in.}$$

In this example, the difference in values between the two equations by Westergaard and Ionnides et al. yields a stress difference of less than 1% and a corner deflection difference of 16%. As the slab thickness increases, the difference in stress between the two equations increases while the deflection difference decreases.

3.3.3.2   Interior Slab Loading

Westergaard (1926b) developed an equation for the stress in the interior of a slab subjected to a circular loaded area with a contact radius "a." This equation is discussed by Huang (2004):

$$\sigma_i = \frac{3(1+v)P}{2\pi h^2}\left[\ln\frac{\ell}{b}+0.6159\right]$$

b = a when a ≥ 1.724 h.
   When a < 1.742 h,

$$b = \sqrt{1.6a^2+h^2}-0.675\,h$$

The equation can be simplified (see later) when using a value of Poisson's ratio of concrete of 0.15:

$$\sigma_i = \frac{0.316P}{h^2}\left[4\log\left(\frac{\ell}{b}\right)+1.069\right]$$

The deflection equation is based on Westergaard (1939):

$$\Delta_i = \frac{P}{8k\ell^2}\left\{1+\frac{1}{2\pi}\left[\ln\left(\frac{a}{2\ell}\right)-0.673\right]\left(\frac{a}{\ell}\right)^2\right\}$$

Example 3.3

Determine the interior stress and deflection for the concrete slab given in Example 3.2 and shown in Figure 3.10.

Solution
Solve for the value of 1.724 h = 15.5 in.; and a since <15.5

$$b = \sqrt{1.6a^2+h^2}-0.675h = 5.6978$$

FIGURE 3.10 Example 3.3.

$$\sigma_i = \frac{0.316*9000}{9^2}\left[4\log\left(\frac{35.88}{5.69}\right)+1.069\right]=149.76 \text{ psi}$$

$$\Delta_i = \frac{9000}{8*150*35.88^2}\left\{1+\frac{1}{2p}\left[\ln\left(\frac{6}{2*35.88}\right)-0.673\right]\left(\frac{6}{35.88}\right)^2\right\}=0.0068 \text{ in.}$$

Comparing the stresses produced at the corner and the interior of the slab, the interior stress is 23% less than the corner stress while the interior deflection is only 14% of the corner deflection. Note that the corner deflection using Westergaard's equation assumes no load transfer at the joint. The stresses will be much smaller at the corner with effective transfer load across the joint.

3.3.3.3 Edge Slab Loading

The equations for edge slab loading were presented by Westergaard (1948) and reviewed by Ioannides et al. (1985) and Huang (2004). The simplified equations for $\mu=0.15$ are presented here for a circular and semicircular loaded area:

$$\sigma_e = \frac{0.803P}{h^2}\left[4\log\left(\frac{\ell}{a}\right)+0.666\left(\frac{a}{\ell}\right)-0.034\right] \quad \text{(for a circle)}$$

$$\sigma_e = \frac{0.803P}{h^2}\left[4\log\left(\frac{\ell}{a}\right)+0.282\left(\frac{a}{\ell}\right)+0.650\right] \quad \text{(for a semicircle)}$$

$$\Delta e = \frac{0.431P}{k\ell^2}\left[1-0.82\left(\frac{a}{\ell}\right)\right] \quad \text{(for a circle)}$$

$$\Delta e = \frac{0.431P}{k\ell^2}\left[1-0.349\left(\frac{a}{\ell}\right)\right] \quad \text{(for a semicircle)}$$

Example 3.4

Determine the stresses and deflections for the edge loading for both circle and semicircle loaded areas as shown in Figure 3.11. Use the same slab properties as in Examples 3.2 and 3.3.

FIGURE 3.11   Example 3.4.

Solution

The edge stress due to a circular loaded area is given by

$$\sigma_e = \frac{0.803 * 9000}{9^2}\left[4\log\left(\frac{35.88}{6}\right)+0.666\left(\frac{6}{35.88}\right)-0.034\right]=284.1\,\text{psi}$$

The edge stress due to semicircle loading is given by

$$\sigma_e = \frac{0.803 * 9000}{9^2}\left[4\log\left(\frac{35.88}{6}\right)+0.282\left(\frac{6}{35.88}\right)+0.650\right]=339.4\,\text{psi}$$

The edge deflection due to a circular loaded area is given by

$$\Delta e = \frac{0.431 * 9000}{150 * 35.88^2}\left[1-0.82\left(\frac{6}{35.88}\right)\right]=0173\,\text{in.}$$

The edge deflection due to a semicircular loaded area is given by

$$\Delta e = \frac{0.431 * 9000}{150 * 35.88^2}\left[1-0.349\left(\frac{6}{35.88}\right)\right]=0189\,\text{in.}$$

Comparing all results for stress and deflection from Examples 3.2 through 3.4 as summarized in Table 3.1. It can be seen that the edge loading produces the largest stress, with almost twice the value for the interior stress and therefore more critical to mid-slab transverse cracking. The edge loading using the semicircle loaded area produces a higher stress than the circle loaded area. This is due to the fact that the centroid of the semicircle area is closer to the edge of the slab than for the circle and therefore produces higher stresses. The corner loading produces the largest deflection, but also assumes no load transfer at the joint. This analysis, even though simplified, still emphasizes the critical need for providing load transfer at the joints such as dowels and a well-draining base support for mitigating pumping distress.

TABLE 3.1

Stress and Deflection Comparison for Different Equations

	Corner	Interior	Edge Circle	Edge Semicircle
Stress (psi)	194	149.7	284	339.4
Deflection (in.)	0.0485	0.0068	0.0173	0.0189

The analytical solutions developed by Westergaard, Ioannides, and others are a good representation of idealized conditions. However, in practice, the assumptions are not realistic about the true pavement conditions. The corner and edge loads are very conservative. The edge loads must be right on the slab edge to create critical stresses calculated by the equations. These equations also assume very large slabs so that edge or corner effects are negligible. Tied and doweled joints and tied shoulders also reduce edge stresses considerably. Slab support, either full or partial, will also reduce corner stresses. Representing the foundation by a liquid foundation will also provide for an unrealistic k-value. The stress equations developed by Westergaard and verified by others can only be applied to a single wheel load with simple contact areas such as circular, semicircular, elliptical, and semielliptical.

In summary, the analytical solutions are good approximations for determining stresses and deflections, but fall short of truly characterizing the real conditions.

3.4 COMPUTER PROGRAMS FOR RIGID PAVEMENTS

Given the complexity of all the parameters that influence the state of stress in a rigid pavement, finite element methods are being used to determine the response of rigid pavements such as stresses, strains, and deflections using a mechanistic approach. The MEPDG method (NCHRP 1-37) uses FEM models for determining stresses, strains, and deformations from a particular load, assesses the damage of each particular load, and then uses empirical models to predict pavement distress for the loading condition. A summation of all cumulative distresses (cracking, faulting, and smoothness using IRI values) for the traffic spectra over the life of the pavement is predicted.

Computer programs have been developed over the years to determine the complex response behavior of computing stresses, strains, and deformations for various types of rigid pavements, applied loads, and environmental conditions. Some of the programs developed specifically for rigid pavements include ILLI-SLAB (Tabatabie and Barenberg, 1980), WESLIQID (Chou, 1981), J-SLAB (Tayabji and Colley, 1983), KENSLAB (Huang, 1993), KOLA (Kok, 1990), FEACONS-IV (Choubane and Tia, 1995), and EVERFE (Davids et al., 1998). Most of these programs can analyze multiwheel loading of one- or two-layered medium thick plates resting on a Winkler or liquid foundation or an elastic solid.

ISLAB2000 is a proprietary revision of ILSL2 (Khazanovich, 1994) and ILLI-SLAB and was developed by Applied Research Associates, Inc. (ERES Division). ISLAB2000 was used in the development of the MEPDG Guidelines. It was used for finite element model development for determining critical jointed plain concrete pavement (JPCP) bottom surface stresses, and finite element model development for determining critical continuously reinforced concrete pavement (CRCP) stresses.

ISLAB2000 is a more advanced software and can analyze many subgrade models (Winkler, Elastic Solid, Two Parametric, and others), partially bonded layers, the effects of linear and nonlinear temperature distribution throughout pavement layers, mismatched joints and partial depth cracks, and the effect of a widened base; can define many pavement layers and wheel loads; and includes an enhanced void analysis model. The software also has advanced graphical capabilities.

EverFE is an advanced 3D finite-element analysis software that can be used for simulating the response of JPCP systems subject to axle loads and environmental effects. EverFE can be used for linear or complex nonlinear analyses of JPCP. EverFE was developed jointly by the University of Maine and the University of Washington and is available for free download from http://www.civil.umaine.edu/everfe/. As a demonstration of the software, Example 3.1 is solved using the software EverFE.

Table 3.2 shows the input data, and Figure 3.12a through e shows the EverFE data input window for temperature gradient, materials properties, curling stresses for the top of the slab, curling stresses for the middle depth of the slab, and the results of the slab deflection due to curling stresses, respectively.

TABLE 3.2
EverFE Input and Output Results

Slab layout	1 row × 1 col
Slab geometry	L = 9140 mm
	W = 3660 mm
	T = 203 mm
Subgrade depth	300 mm
Slab materials	E = 27,600 MPs (4,000,000 psi)
	ν = 0.15
	α = 9 * 10^{-6} mm/mm/°C
	Concrete density = 2,400 kg/m³
Base material	E = 500 MPa
	ν = 0.15
	Density = 0
	k = 0.054 MPa (200 pci)
	Slab-base interface = unbound
Temperature differential	ΔT = 11.1°C
Results	Stress (interior) = 1.58 MPa
	With bounded base = 1.49 MPa
	Stress (edge) = 1.48 MPa
	With bounded base = 1.31 MPa

3.5   COMBINED STRESSES

The pavement community is divided on how to combine curling stresses with load stresses in the design process. Curling stresses due to temperature could be quite large and when coupled with loading stresses, they could be critical. For example, if the rigid pavement design is governed by edge stresses, then curling and loading stresses are additive during the day conditions and subtractive during night conditions as can be seen in Figure 3.5. Some believe that the joints and the steel reinforcement in the slab would mitigate temperature curling and stresses. The Portland Cement Association design method does not include curling stresses since the number of curling stress reversals is quite small during the life of the pavement compared to fatigue due to traffic loads. Warping stresses due to moisture changes in the pavement can also add significant stresses under the right conditions. The relative humidity in the concrete slab will vary from drier to wetter from the top to the bottom of the slab. However, some believe that rigid pavements basically go through one moisture cycle as the concrete is poured, cures and reaches some moisture equilibrium with the base. In addition, moisture changes are seasonal and moisture changes in the pavement are gradual.

3.6   STRESSES DUE TO FRICTION

Changes in temperature and moisture will create volume changes in concrete materials and will have significant effects on the concrete pavement slab. If a concrete slab, for example, is prevented from moving freely due to the friction between the slab and the supporting foundation, then stresses will develop in the concrete, in the reinforcing steel if it is included, and in the tiebars across longitudinal joints and shoulders. Large volume changes due to temperature and moisture changes can cause joints to open too wide, therefore reducing aggregate interlock and decreasing joint transfer efficiency.

The selection of joint spacing should be chosen appropriately so that mid-slab cracking in the concrete slab does not occur or should occur infrequently if it does. Otherwise, steel reinforcement should be included so that cracks, if they occur, are held tightly together. The design of tiebars is also controlled by frictional stresses developed in the concrete.

FIGURE 3.12 (a) EverFE input page for temperature gradient, (b) EverFE input page for materials properties, (c) EverFE output results for stress at the top of the slab, (d) EverFE output results for stress at the middle of the slab, and (e) EverFE output results for slab deflection due to temperature curling.

Frictional stresses between a concrete slab and the foundation can be determined using a simple mechanics approach. Consider a concrete slab subject to a decrease in temperature as shown in Figure 3.13. The concrete slab will tend to shrink toward the center of the slab from both ends. The frictional resistance between the concrete slab and the foundation will prevent the slab from moving and the stresses in the concrete will develop. The magnitude of the frictional forces developed will depend on the relative movement between the slab and the foundation. Frictional forces will range from zero at the center where no movement occurs to a maximum some distance away from the center where movement is fully mobilized (Huang, 2004). Figure 3.13a shows a slab with the frictional forces. An average coefficient of friction f_a is assumed. The maximum tensile stress in the concrete is achieved at the center of the slab and is determined by equating the frictional force per unit width of the slab, $\gamma_c L f_a/2$, to the tensile force $\sigma_c h$ in the concrete. Figure 3.13b shows the free body diagram of the frictional forces generated by the slab weight and frictional coefficient equated to the internal concrete tensile stresses. The concrete tensile strength is provided by the following equation, where σ_c is the concrete tensile strength, γ_c is the concrete unit weight, L is the length of the concrete slab, and f_a is the average coefficient of friction between the slab and the subgrade.

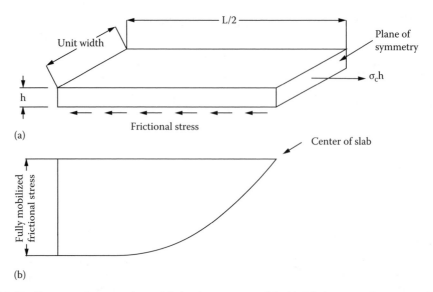

Frictional stress

(a)

(b)

FIGURE 3.13 Forces and stresses due to friction in a concrete slab. (a) Friction stress in concrete block and (b) friction stress distribution along concrete slab.

The value of f_a is typically selected at 1.5 but will vary with the type of subgrade or base material and could vary from 0.9 to 3 (ACPA):

$$\sigma_c = \frac{\gamma_c L f_a}{2}$$

Example 3.5

A concrete pavement slab as shown in Figure 3.14 has a joint spacing of 30 ft (6.1 m) and a coefficient of friction equal to 1.5, the concrete unit weight $\gamma_c = 150$ pcf, and the concrete compressive strength is 3500 psi. Determine the stress developed in the concrete due to frictional forces.

Solution
Given the concrete unit weight $\gamma_c = 150$ pcf $= 0.0868$ pci; $L = 30$ ft $= 360$ in.

$$\sigma_c = \frac{\gamma_c L f_a}{2} = \frac{0.0868 * 360 * 1.5}{2} = 23.4 \text{ psi } (0.16 \text{ MPa})$$

Given the compressive strength of the concrete of 3500 psi, the tensile strength based on the ACI approximate equations is $f_t = 5\sqrt{f_c'}$ to $7.5\sqrt{f_c'}$; since $f_c' = 3500$ psi, the tensile strength ranges between 295 psi (2.04 MPa) and 444 psi (3.06 MPa). Therefore, in general, it does not appear that the joint spacing in plain concrete pavements is governed by stress due to friction alone.

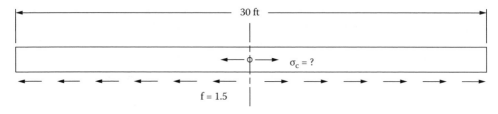

FIGURE 3.14 Frictional forces developed in a concrete slab for Example 3.5.

3.7 JOINT OPENING

The spacing of joints in JPCP will depend more on the shrinkage of the concrete than on other stresses generated in the concrete. Load transfer efficiency across the joint is greatly influenced by the extent of the width opening across the joint. If the joint opening is too wide (say 0.6 mm or greater), the aggregate interlock mechanism will not function effectively and cannot provide load transfer across the joint (Hansen et al., 2001).

An equation for approximating a joint opening was developed by Darter and Barenberg (1977) and discussed by Huang (2004) and FHWA (1989):

$$\Delta L = CL(\alpha \Delta t + \varepsilon)$$

where
ΔL is the joint opening caused by the change in temperature and the drying shrinkage of the concrete
α is the coefficient of thermal expansion (CTE) of concrete; the CTE value is typically selected between 5 and 6 * 10^{-6}/°F but varies greatly with aggregate type (FHWA, 1989)
ε is the drying shrinkage coefficient of concrete typically ranging between 0.5 and 2.5 * 10^{-4} in./in.; Figure 3.15 shows values of CTE for typical materials
L is the slab length or joint spacing
Δt is the temperature change, which is the temperature at placement minus the lowest mean monthly temperature
C is the adjustment factor due to slab-subbase friction, which is 0.65 for a stabilized base and 0.8 for a granular subbase

3.8 JOINTS

Joints are typically used in common pavements such as JPCP, JRCP, and CRCP to allow for controlled concrete slab movement and cracking. Transverse contraction joints are typically used in JPCP and JRCP, which are usually doweled for heavily trafficked pavements. Construction transverse or longitudinal joints are joints between slabs that result when concrete is placed

$\varepsilon = 2 \times 10^{-4}$ for concrete indirect tensile strength 700 psi
$\varepsilon = 4.5 \times 10^{-4}$ for concrete indirect tensile strength 500 psi
$\varepsilon = 8 \times 10^{-4}$ for concrete indirect tensile strength 300 psi (FHWA, ACI)

	Coefficient of Thermal Expansion	
	10^{-6}/°C	10^{-6}/°F
Aggregate		
Granite	7–9	4–5
Basalt	6–8	3.3–4.4
Limestone	6	3.3
Dolomite	7–10	4–5.5
Sandstone	11–12	6.1–6.7
Quartzite	11–13	6.1–7.2
Cement paste (saturated)		
w/c = 0.5	18–20	10–11
Concrete	7.4–13	4.1–7.3
Steel	11–12	6.1–6.7

FIGURE 3.15 Typical values of CTE for various materials.

FIGURE 3.16   Images of (a) transverse and longitudinal joints, (b) joint saw cutting, (c) dowel baskets set for transverse joint construction, and (d) construction joints.

at different times, for example, at the end of the daily pavement construction operation, due to equipment breakdown, or during long delays where concrete cold joints are not desirable. Transverse expansion (or isolation) joints are placed at specific locations to allow the pavement to expand without damaging adjacent structures such as bridges, drainage structures, or the pavement itself (FHWA, 1990a). Figure 3.16 shows images of transverse and longitudinal joints. Figure 3.17 shows various types of joint designs.

Well-performing joints are critical to good performing pavements. According to FHWA, most jointed pavement failures can be attributed to joint failure as opposed to a lack of structural capacity. If a joint failure occurs, distresses such as faulting, pumping, spalling, corner breaks, blowups, and mid-panel cracking may develop. Good practices that contribute to good performing joints should be included in the design, construction, and maintenance of pavements. These include adequate load transfer at the joint, good concrete consolidation, proper and timely joint cutting, appropriate sealant use, and good maintenance repair practices. Adopting these practices should produce pavement joints that are capable of performing well throughout the life of the pavement (FHWA, 1990a).

3.8.1   TRANSVERSE CONTRACTION JOINTS

A transverse contraction joint is a joint that is sawed, formed, or grooved in a concrete slab. The joint creates a weak vertical plane and a point of high stress where cracks initiate and propagate from. Contraction joints regulate the location of cracks formed due to dimensional changes caused by

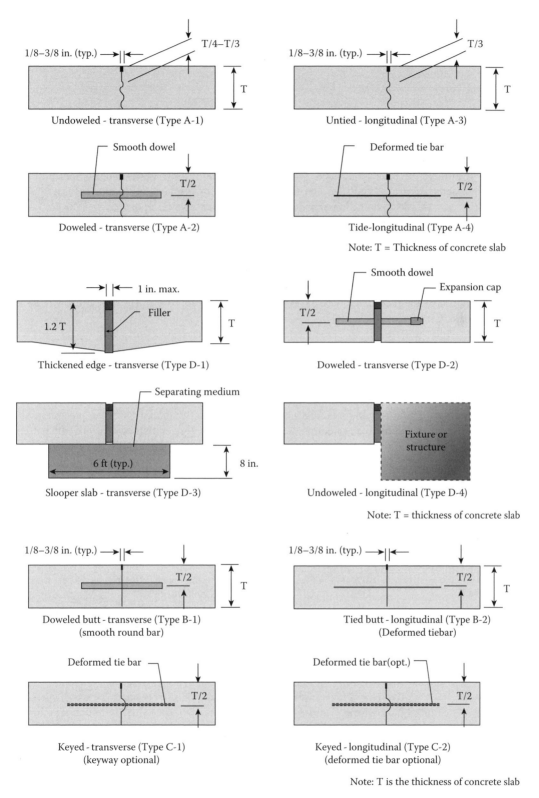

FIGURE 3.17 Examples of various transverse and longitudinal joints in PCC pavements. (Courtesy of ACPA.)

temperature and moisture changes in the concrete. Transverse joints in highways are used perpendicular to the direction of traffic, while longitudinal joints are placed parallel to the traffic and between traffic lanes. Airport pavements are much wider and use square concrete slabs that are doweled on all four sides.

3.8.2 Longitudinal Joints

Longitudinal joints are constructed between two slabs and allow the slabs to warp or deform without significant separation or cracking of the slabs. Although load transfer at longitudinal joints is achieved through aggregate interlock, tiebars are commonly used across longitudinal joints to prevent slab separation and faulting. The tiebars are usually thinner and longer deformed steel bars, unlike smooth and bigger diameter dowels. Tiebars can also be epoxy coated for corrosion protection. Longitudinal joints can be sawn or constructed similar to transverse joints. For sawn joints, the tiebars are preset in tiebar baskets similar to dowels. Otherwise they can be mechanically inserted at slab mid-depth to connect the old and new concrete slabs together. Longitudinal joints are highly recommended when slab widths exceed 15 ft, although some cases with slabs that wide have performed satisfactorily with minor longitudinal cracks occurring. It is also recommended that longitudinal joints should coincide with pavement lane lines to improve traffic operations whenever possible (FHWA 1990a).

Tiebar design guidelines are provided by *AASHTO Design Guidelines*, 1993. FHWA recommends that when using Grade 40 steel, a tiebar diameter of 5/8 and 30 in. length or 1/2 in. diameter and 24 in. length should be used. Similarly, when using Grade 60 steel, a tiebar with 5/8 in. diameter and 40 in. length or 1/2 in. diameter and 32 in. length should be used. Since tiebars are held tightly to the concrete by bond shear strength, the recommended lengths are necessary to develop adequate bond. Tiebar spacing varies directly with the thickness of the pavement and the distance from the joint to the nearest free edge. Tiebars longer than 48 in. are not recommended. A common tiebar design is the use of the No. 4 bar, 36 in. in length and spaced 30–40 in. at the center (Huang, 2004). Table 3.3 shows the recommended tiebar spacings guidelines from ACI 325.

TABLE 3.3
Tiebar Dimensions and Spacings

		Tiebar Spacing, mm (≈in.)			
		Distance to Nearest Free Edge or to Nearest Joint Where Movement Can Occur			
Slab Thickness mm (≈in.)	Tiebar Size × Length, mm (≈in.)	3.0 m (≈10 ft)	3.7 m (≈12 ft)	4.3 m (≈14 ft)	7.3 m (≈24 ft)
130 (5)	13 M × 600 (24)	760 (30)	760 (30)	760 (30)	700 (28)
150 (6)	13 M × 600 (24)	760 (30)	760 (30)	760 (30)	580 (23)
180 (7)	13 M × 600 (24)	760 (30)	760 (30)	760 (30)	500 (20)
200 (8)	13 M × 600 (24)	760 (30)	760 (30)	760 (30)	430 (17)
230 (9)	16 M × 760 (30)	900 (35)	900 (35)	900 (35)	600 (24)
250 (10)	16 M × 760 (30)	900 (35)	900 (35)	900 (35)	560 (22)
280 (11)	16 M × 760 (30)	900 (35)	900 (35)	860 (34)	500 (20)
310 (12)	16 M × 760 (30)	900 (35)	900 (35)	780 (31)	460 (18)

Source: Adapted from ACI 325.12R. American Concrete Institute, 2002.

(a)

(b)

FIGURE 3.18   Joint construction examples: (a) image of a header and (b) the construction of an expansion joint. (Courtesy of Wouter Gulden, ACPA-SE Chapter, Duluth, GA.)

3.8.3 CONSTRUCTION JOINTS

Construction joints are either transverse or longitudinal joints that result when concrete is placed at different times. A good practice is that transverse construction joints should be placed where a planned contraction joint should be located. Construction joints should not be skewed due to the difficulty in construction and concrete consolidation. Transverse construction joints should be doweled, butted, but not keyed. Experience with keyed construction joints show a tendency toward excessive spalling and are not recommended. However, it is recommended that transverse construction joints be sawed and sealed with similar sealant reservoir dimensions as transverse contraction joints (FHWA, 1990a).

Figure 3.18a shows a photo of a header and an expansion doweled joint and a basket assembly at a future transverse construction joint. During the paving operation, the concrete will be placed up to the header. The header will then be removed when the concrete has set. When paving operations resume, the new concrete will be placed butted against the old concrete.

3.8.4 EXPANSION JOINTS

An expansion (or isolation) joint is a joint placed at a specific location to allow the pavement to expand without damaging adjacent structures such as bridges, drainage, and utility structures or the pavement itself. Properly designed and maintained contraction joints have practically eliminated the need for expansion joints, except at fixed objects such as structures. As the pavement expands due to temperature and moisture changes, the expansion joints will tend to close over a period of several years. As this happens, adjoining contraction joints may open, which may destroy the joint seals and aggregate interlock. Typically, the width of an expansion joint is approximately 3/4 in. or more, while filler material is typically placed 3/4–1 in. below the slab surface to allow space for sealing material. A special type of dowel assembly is used to transfer load across expansion joints. The special joint dowel system is fabricated with a cap on one end of each dowel to create a void in the slab to accommodate the dowel as the adjacent slab closes the expansion joint. Figure 3.17 shows a diagram of the capped dowel system for expansion joints, and Figure 3.18b shows the construction of an expansion joint.

Keyed longitudinal construction joints have been used in the past. Figure 3.19 shows a typical keyed joint. Problems with keyed longitudinal construction joints in the past have prompted caution in using this type of joint detail. The top of the slab above the keyway usually fails in shear. When the slab thickness is less than 10 in., it is recommended that keys should not be used and tiebars

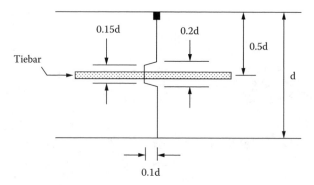

FIGURE 3.19 Schematic of a typical keyed joint.

should be designed to carry the load. When the slab thickness is greater than 10 in., keyways maybe used with great caution in addition to the tiebars. Keyways should be placed in the center of the concrete slab. Alternatively, keyways maybe eliminated all together and bigger and more tiebars can be used to carry the load across the longitudinal joint (FHWA, 1990a).

The effectiveness of tiebars to carry the load across the longitudinal joint and tie the adjacent slabs together is greatly dependent on the bond between the tiebar and the concrete. Tiebars could be mechanically inserted into the plastic concrete or installed as a two-part threaded tiebar and splice coupler system. To ensure that tiebars are securely anchored into the concrete, pullout tests should be conducted periodically for verification. Bending of tiebars is not recommended. However, if bending of tiebars is necessary (e.g., for construction purposes), it is recommended that a two-part threaded tiebar and splice coupler system be used in place of tiebars. If tiebars must be bent and later straightened during construction, Grade 40 steel is recommended since it is more ductile than the higher grades and would tolerate the bending better. It is also recommended that longitudinal construction joints be sawed and sealed and the sealant reservoir dimensions should be the same as those used for the longitudinal joints (FHWA, 1990a). Figure 3.20 shows the construction of bent tie bars.

3.8.5 JOINT DESIGN

Joints are provided in concrete pavements to reduce the curling and warping effects and premature slab cracking due to temperature and moisture changes. Joint spacing should be short enough so that curling stresses do not develop and transverse cracking and faulting do not occur.

The spacing of both transverse and longitudinal contraction joints depends greatly on local conditions such as the materials used and environmental conditions. The joint spacing required to reduce intermediate slab cracking will decrease with an increase in CTE, positive temperature gradient, base stiffness, or subbase frictional resistance; however, the spacing will increase with an increase in concrete tensile strength. Joint spacing is also related to slab thickness and joint sealant capabilities. As a rough guide, the joint spacing in feet for plain concrete pavements should not greatly exceed twice the thickness in inches. For example, the maximum spacing for an 8 in. slab is 16 ft. Also, the ratio of slab width to length should not exceed 1.25 (AASHTO, 1993).

According to FHWA (1990a), a maximum transverse joint spacing of 15 ft is recommended for plain concrete slabs. Performance however will be greatly dependent on local conditions, and slabs with longer slab lengths in some cases have also performed satisfactorily. Research studies have shown that pavement thickness, base stiffness, dimensional properties of the concrete, and climate have a great effect on the maximum joint spacing beyond which potential transverse cracking can be expected. In addition, it was observed that a relationship between the ratio of slab length (L) to

(a)   (b)

(c)

FIGURE 3.20   Tie bars applied during construction of (a) bent tie bars inserted in concrete, (b) bent tie bars ready for insertion into concrete, and (c) straight tie bars in concrete. (Courtesy Robert Rasmussen, Transtec Group Inc, Austin, TX.)

the radius of relative stiffness (ℓ) and the amount of transverse cracking exists (Smith et al., 1990). Research data indicate that there is an increase in transverse cracking when the ratio L/ℓ exceeds 5.0. Figure 3.21 was developed using the relationship $L/\ell = 5.0$, $E_c = 4 * 10^6$ psi and a poison's ratio $\mu = 0.15$. For example, if the pavement thickness is 10 in., and the slab foundation support, $k = 100$ pci, the slab length $L = 18$ ft. If the foundation support increases to $k = 500$ pci, then the joint spacing is shortened to $L = 12$ ft.

3.8.6   Joint Spacing for Airfields

Joint spacing recommendations for airfield pavements are provided by FAA-AC 150/5320-6D. For unstabilized bases, the recommended joint spacing for slab thickness greater than 12 in. is 25 ft, and for a slab thickness between 9–12 in., the joint spacing is 20 ft. For stabilized bases, which increase k values and curling stresses, the joint spacing is limited to "5ℓ." Airfield pavement joints are typically constructed as squares and are usually doweled. FAA recommendations for dowel bar diameter, length, and spacing for a given slab thickness are given in Table 3.4 (FAA-AC 150/5320-6D). Dowels are typically solid bars, but other types may also be used, such as high-strength pipe plugged on either end with a cap, a bituminous or mortar mixture. Tiebars should be deformed steel bars with 5/8 in. diameter and spaced 30 in. center-to-center (FAA-AC 150/5320-6D; FAA 2004).

FIGURE 3.21   Maximum joint spacing as a function of slab thickness and k for $L/\ell=5.0$, $E_c=4*10^6$ psi, $\mu=0.15$.

TABLE 3.4
Dowel Bar Diameter Recommendations

Slab Thickness (in.)	Dowel Diameter (in.)	Dowel Length (in.)	Dowel Spacing (in.)
<8	Not required		
8	1.25	18	12
10	1.25	18	12
11	1.5	18	12
12	1.5	18	12
14	1.75	20	12
16 and up	2	24	18

Source: ACPA 1998; ACI Committee 325, *Concrete Floor and Slab Construction*, American Concrete Institute, Farmington Hills, MI, 2002.

3.8.7   Variable Joint Spacing

Construction joints are usually placed at regular intervals and perpendicular to the centerline of the pavement. However, if evenly spaced joints are faulted, then this can introduce harmonic motion in vehicle suspensions at certain vehicle speeds. To mitigate this problem, some SHA have used variable joint spacing. A variable joint spacing of 13, 19, 18, 12 ft became a standard. However, the 19 and 18 ft panels were too long and were prone to mid-slab cracking. A new spacing of 12, 15, 13, 14 ft was then adopted as the new norm. According to FHWA-1990a, when using randomized joint spacing, the longest slab should not be longer than 15 ft (Delatte, 2008).

3.8.8   Skewed Joints

Skewed joints are cut at obtuse angles to the direction of traffic or the pavement centerline (Figure 3.22). The advantage of skewed joints is that the right and the left wheels do not arrive at the

FIGURE 3.22   Image of a typical skewed joint.

joint at the same time, which reduces the load transfer stresses on the joint. If a contraction joint is properly skewed, the left wheel of each axle will cross the joint onto the leave slab first. And only one wheel crosses the joint at a time.

3.8.9   Aggregate Interlock between Joints

Effective load transfer across joints is important to minimize deflections at the joint. Reduced joint deflections will decrease the potential for pumping of the subbase material and faulting. This can be achieved using load transfer devices such as dowels or aggregate interlock (Figure 3.1). Aggregate interlock is achieved through frictional shearing forces at the aggregate–paste interface that develops below a joint saw cut. If joints are typically saw-cut 1/4–1/3 of the way through the slab depth, then a crack develops naturally through the remaining of the slab depth. Hard, angular, and durable aggregates will contribute to an improved aggregate interlock and load transfer efficiency. A concrete mix with a larger size coarse aggregate that has a reduced cement paste content will shrink less and therefore keep the joint widths small. Any factors that increase crack width and coupled with heavy traffic loads will contribute to a decrease in aggregate interlock and joint load transfer efficiency. Therefore, the use of aggregate interlock for load transfer is recommended for use on local roads and streets, which carry a low volume of heavy traffic (FHWA, 1990a).

3.8.10   Design of Dowels

Dowels are used in transverse joints to enhance load transfer across the joint and prevent pumping and faulting distress (see Figure 3.16). Dowels are typically smooth metal rods that are lubricated to permit movement along the dowel's longitudinal axis and allow joints to open and close without stress buildup. The FHWA recommends that all highway pavements carrying more than a low volume of heavy trucks should use doweled joints (FHWA, 1990a).

Dowels are typically used for pavement slabs 8 in. or more in thickness. When dowels are used, they are typically 18 in. long and spaced 12 in. at the center. In a typical 12 ft lane, 11 dowels are commonly installed. However, during a pavement retrofit operation, only three dowels are usually placed in each wheelpath (ACPA, Dalette, 2008).

3.8.11 Dowel Diameter Design

Appropriate design of dowels for effective load transfer across joints requires the selection of an appropriate dowel diameter. This is required for effectively transmitting shear and bending forces between slabs and reducing the bearing stress between the steel dowel and the concrete material to an acceptable level. Usually, the concrete bearing stress will be the limiting design parameter. A sustained high stress level that is cyclic between the dowel and the concrete will erode the concrete material and loosen the dowel, reducing load transfer effectiveness (Delatte, 2008).

The dowel-bearing stress is theoretically dependent on the concrete strength (see next section). Using this relationship, dowel diameters can technically be much smaller for a very high strength concrete. In practice, this does not hold true. Recommendations for dowel diameter size should be based on the thickness of the pavement slab. FHWA recommends that the minimum dowel diameter should be at least one-eighth the pavement thickness. However, the dowel diameter should not be less than 1¼ in., the dowel length should be 18 in. with a dowel spacing of 12 in. The Portland Cement Association (PCA, 1991) recommended the use of 1.25 in. diameter dowels for highway pavements less than 10 in. thick and 1.5 in. diameter dowels for pavements 10 in. or thicker. A minimum dowel diameter of 1.25–1.5 in. is needed to control faulting by reducing the bearing stress in concrete. Corrosion of dowels should be avoided so that dowel joints do not seize-up and lock. To mitigate corrosion problems with dowels, epoxy-coated dowels and stainless steel dowels have been used with good results (FHWA, 1990a). Table 3.4 shows recommended dowel bar diameter, length, and spacing for different pavement thicknesses.

The Smith and Hall study (Huang, 2004) recommends dowel diameter size based on the traffic level. For a pavement designed to carry less than 30 million ESALs, 1.25 in. diameter dowels are recommended; for a pavement carrying 30–90 million ESALs, a 1.5 in. diameter dowel should be used; for traffic over 90 million ESALs, a 1.625 in. diameter dowel is recommended.

3.8.11.1 Allowable Bearing Stress

Because concrete is much weaker than steel, the size and spacing of dowels required are governed by the bearing stress between the dowel and concrete. The allowable bearing stress can be determined by the following ACI equation (ACI, 1956; Huang, 2004):

$$f_b = \frac{(4-d)}{3} f_c'$$

where
   f_b is the allowable bearing stress, psi
   d is the dowel diameter, in.
   f_c' is the ultimate compressive strength of the concrete

3.8.11.2 Bearing Stress on One Dowel

If the load applied to one dowel is known, the maximum bearing stress can be determined theoretically by assuming the dowel to be a beam and the concrete to be a Winkler foundation (Huang, 1993). Based on the original solution by Timoshenko (Timoshenko and Goodie, 1934/1951), Friberg (1938) developed a relationship for determining the maximum bearing stress (Figure 3.23).

FIGURE 3.23 Dowel deformation under wheel load.

These equations are discussed by Yoder and Witczak (1975) and Huang (2004). First, the deformation of the dowel at the interface of the concrete can be expressed by the following equation:

$$y_0 = \frac{P_t(2+\beta z)}{4\beta^3 E_d I_d}$$

where
  y_0 is the deformation of the dowel at the face of the joint
  P_t is the load on one dowel
  z is the joint width
  E_d is the Young's modulus of the dowel
  I_d is the moment of inertia of the dowel

$$I_d = \frac{1}{64}\pi d^4$$

β is the relative stiffness of a dowel embedded in concrete

$$\beta = \sqrt[4]{\frac{Kd}{4E_d I_d}}$$

  K is the modulus of the dowel support, which ranges from 300,000 to 1,500,000 pci
  d is the diameter of the dowel

The bearing stress in concrete σ_b is proportional to the deformation:

$$\sigma_b = Ky_0 = \frac{KP_t(2+\beta z)}{4\beta^3 E_d I_d}$$

The bearing stress obtained from this equation should compare with the allowable bearing stress in the equation given for f_b. If the actual bearing stress is greater than allowable, then dowel bars with a bigger diameter or a smaller dowel spacing should be used.

When a load is applied at the joint, the dowel bar immediately below the load carries a major portion of the load, while adjacent dowels will assume proportionally lesser amounts. Based on Westergaard's solutions, Friberg (1938) determined that the maximum negative moment for both interior and edge loadings occurs at a distance of 1.8ℓ from the load, where ℓ is defined earlier as the radius of relative stiffness. More recent research has shown the effective length to be at 1.0ℓ (Heinrichs et al., 1989).

Example 3.6

Figure 3.24 shows a concrete pavement, 9 in. thick, having a joint width of $z = 0.2$ in. (5.1 mm), a modulus of subgrade reaction of $k = 120$ pci (32 KN/m³), and a modulus of dowel support $K = 1.6 * 10^6$ pci (434 GN/m³). A wheel load of 9000 lb (40 KN) is applied to the outermost dowel at a distance of 6 in. (152 mm) from the edge. The dowels are 3/4 in. (19 mm) in diameter and 12 in. (305 mm) at the center. Determine the maximum bearing stress between the dowel and concrete. Assume for concrete: $E_c = 4 * 10^6$ psi, $\mu = 0.15$.

Solution
The radius of relative stiffness is determined by

$$\ell = \sqrt[4]{\frac{Eh^3}{12(1-\mu^2)k}} = \sqrt[4]{\frac{4*10^6*9^3}{12(1-0.15^2)120}} = 37.94 \text{ in.}$$

FIGURE 3.24   Example 3.6, dowel-bearing stress under single wheel loads.

Assume that the outermost dowel is directly under the wheel load and is subjected to a shear force P_t, then the forces on the dowels within a distance 1.8ℓ or 68.3 in. can be determined by assuming a linear variation as shown in Figure 3.22. The sum of forces on all dowels is 3.36 P_t. Assuming 100% load transfer efficiency at the joint, the affected dowel will carry half the applied load 9000/2 = 4500 lb (20 KN); therefore, P_t = 4500/3.36 = 1338 lb (5.95 KN).

Determine the moment of inertia for a circular dowel.

$$I_d = \frac{\pi d^4}{64} = \frac{\pi * 0.75^4}{64} = 0.0155 \text{ in.}^4 = 6450 \text{ mm}^4$$

$$\beta = \sqrt[4]{\frac{Kd}{4 E_d I_d}} = \sqrt[4]{\frac{1.6 * 10^6 * 0.75}{4 * 29,000,000 * 0.0155}} = 0.9034 \text{ in.} = 22.6 \text{ mm}$$

$$\sigma_b = K y_o = \frac{K P_t (2 + \beta z)}{4 \beta^3 E_d I_d} = \frac{1.6 * 10^6 * 1338(2 + 0.9034 * 0.2)}{4 * 0.9034^3 * 29,000,000 * 0.0155^3} = 3514 \text{ psi} = 24.23 \text{ MPa}$$

For a concrete compressive strength, $f_c' = 3000$ psi (20.7 MPa), the allowable stress is given as

$$f_b = \left(\frac{4-d}{3}\right) f_c' = \left(\frac{4-0.75}{3}\right) * 3000 = 3250 \text{ psi} = 22.4 \text{ MPa}$$

The actual bearing of 3514 psi > allowable stress of 3250 psi; therefore, the design is not adequate.

Now, let's modify some of the input parameters and observe the changes in the actual and allowable stresses.

- If the concrete compressive strength f_c' is increased to 3600 psi, the allowable stress increases to 3900 psi. This now provides an 11% margin and the designer should decide if this is an adequate margin.
- If the dowel diameter is increased to 1 in. (25 mm), then the actual bearing stress $\sigma_b = 2090$ psi (14.4 MPa) and the allowable stress $f_b = 3000$ psi (20.7 MPa). Using a dowel diameter of 7/8 in. (22 mm) is still acceptable, since the actual bearing stress $\sigma_b = 2660$ psi (14.4 MPa) and the allowable stress remains the same, $f_b = 3000$ psi (20.7 MPa).
- If the foundation becomes stiffer, and k = 200 pci (54 KN/m³), then the actual stress increases $\sigma_b = 3930$ psi (27 MPa) and the allowable stress stays the same $f_b = 3250$ psi (22.410 MPa).
- If the slab thickness h and the concrete stiffness E_c are increased, and the foundation becomes weaker k is lower, then the value for ℓ increases and now both truck wheels must be considered in determining the P_t on the most critical dowel.

Example 3.7

Figure 3.25 shows a 10 in. slab resting on a foundation with k = 50 pci (13.6 KN/m³). In the 12 ft lane, there are 12 dowels spaced 12 in. at the center. The two 9000 lb (40 KN) wheel loads are applied at points A and B, which are 6 ft (1.82 m) apart. Determine the maximum load on one dowel, and the maximum dowel-bearing stress.

Solution

$$\ell = \sqrt[4]{\frac{E h^3}{12(1-\mu^2)k}} = \sqrt[4]{\frac{4 * 10^6 * 10^3}{12(1-0.15^2)50}} = 51.10 \text{ in.} = 1275 \text{ mm}$$

$$1.8\ell = 92 \text{ in.}$$

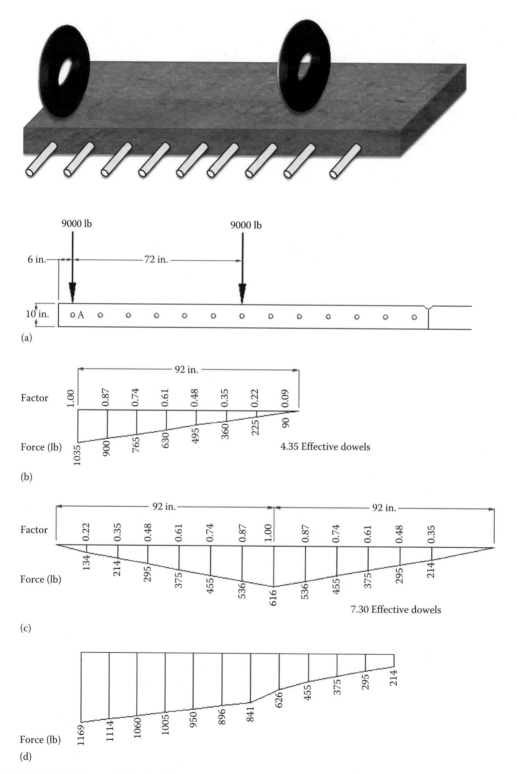

FIGURE 3.25 Example 3.7, dowel-bearing stress under duel wheel loads. (a) Location of loads and dowels, (b) dowel forces due to load at A, (c) dowel forces due to load at B, and (d) dowel forces due to both loads.

First, the left wheel load acting at point A will be considered. If the dowel at point A has a load factor of 1.0, assuming a linear relationship, then the values for the other dowels can be determined using similar triangles as shown in Figure 3.25b. The sum of these factors is 4.35 effective dowels, and the load carried by the dowel at point A is 4500/4.35 = 1035 lb (4.6 KN). The loads carried by the other dowels can be determined by linear proportion.

The next step is to consider the 9000 lb (40 KN) wheel load at point B. If the dowel at point B has a load factor of 1.0, then the load factor at the other dowels can also be determined from similar triangles as shown in Figure 3.25c. The sum of all load factors is 7.30 effective dowels. The dowels beyond the longitudinal joint are not considered effective in carrying the load and are not included in the summation. The load carried by the dowel at point B is given by 4500/7.30 = 616 lb (2.7 KN). The load carried by the other dowels is proportional to the load factor. Finally, the loads carried by the dowels due to the combined wheel loads at both points A and B are provided in Figure 3.25d. The most critical load is carried by the dowel located at the edge of the pavement. This load should be used for design purposes. Assuming similar values from the previous example, we then calculate the dowel-bearing stress $\sigma_b = 3070$ psi (21 MPa).

3.8.11.2.1 Tiebar Steel Design

Tiebars are placed along longitudinal joints to keep two adjacent concrete slabs tied together or across pavement slabs and concrete shoulders (see Figures 3.16 and 3.17). Tiebars are also used to maintain load transfer across the joints. However, load transfer for tiebars across longitudinal joints is less critical than for dowel bars across transverse joints since traffic load is applied only when traffic changes lanes. Tiebars also differ from dowel bar in that they are not expected to allow movement.

The amount of steel required to ensure adequate load transfer is also dependent on stresses generated due to friction. The design for tiebars is similar to the design for longitudinal and transverse temperature steel. The amount of steel required is given by Huang (2004):

$$A_s = \frac{\gamma_c h L' f_a}{f_s}$$

A_s is the area of steel required per unit length of slab
L' is the distance from the longitudinal joint to the free edge where no tiebars are present

For two- to three-lane highways, L' is the lane width. If tiebars are used in all three longitudinal joints of a four-lane highway, L' is equal to the lane width for the two outer joints and twice the lane width for the inner joint.

The length of tiebar design is based on the allowable bond stress. For deformed bars, an allowable bond stress of 350 psi is used and the length of the bar is determined using the following equation (Huang, 2004):

$$L_T = 2\left(\frac{f_s A}{\phi P}\right)$$

where
L_T is the length of the tiebar
f_s is the steel yield strength
A is the area of tiebar, $\pi d^2/4$
ϕ is the bond strength between steel tiebar and concrete
P is the tiebar perimeter, πd

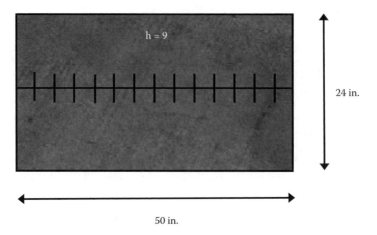

FIGURE 3.26 Example 3.8, tiebar design for longitudinal joint.

The equation can be reduced to

$$L_T = \frac{1}{2}\left(\frac{f_s d}{\phi}\right)$$

Example 3.8

A two-lane concrete pavement is 9 in. thick, 50 ft long, and 24 ft wide with a longitudinal joint in the center (see Figure 3.26).

    1. Determine the design of the tiebars.
        Use Billet steel with 40 ksi yield stress and an allowable stress of 27 ksi. The lane width
            L = 12 ft (144 in.; 3.66 m); unit wt. γ_c = 150 pcf = 0.0868 pci

$$A_s = \frac{\gamma_c hL'f_a}{f_s} = \frac{0.0868*9*144*1.5}{27,000} = 0.00624 \text{ in.}^2/\text{in.}$$

        Using No. 4 bars with (0.5 in. diameter; area = 0.2 in.^{2})
        The spacing of the tiebars = 0.2/0.00624 in.^{2}/in. = 32 in.
    2. Determine the length of the tiebar.
        Assume ϕ = bond strength between steel tiebar and concrete = 350 psi.

From the equation, $L_T = 1/2(f_s d/\phi) = 0.5 * 27,000 * 0.5/350 = 19.3$ in. + 3 in. for misalignment = 22.3 use 24 in. length. Therefore, use no 4. tiebars with 24 in. length and spaced 32 in. at the center.

3.8.11.2.2 *Traffic and Loads*

Rigid pavements, as well as most other highway and airport pavements, are subject to a mixed traffic stream. Slab thickness design is required to protect against various distresses including cracking, faulting, pumping, punchouts, joint spalling among other distresses. The AASHTO method relates the axle equivalency method to the pavement damage based on loss of pavement serviceability as related to the standard 18 kip single axle load (AASHTO, 1993). The PCA method (PCA, 1984) determines the damage caused by each vehicle type in the traffic stream and accumulates

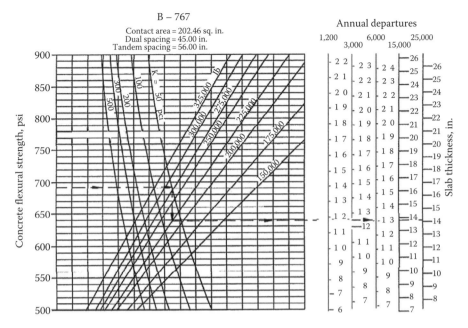

FIGURE 3.27　FAA pavement thickness design.

the total damage over the life span of the pavement. This method considers fatigue and erosion (subgrade pumping) as the failure criteria. Slab thickness design is determined based on either fatigue damage or erosion damage. It should be noted that pumping failures can be controlled with adequate base or subbase design. The MEPDG method also relates damage due to traffic loading but for a specific load group or traffic spectra within a specific season. For the MEPDG method, thickness design is more involved but traffic damage is compared to failure criteria for cracking, faulting, and smoothness. A different approach is used by the FAA empirical method. The FAA method uses the maximum gross weight of the aircraft for pavement thickness design. Figure 3.27 shows an example for slab thickness design for the Boeing 767, which requires additional input data such as the concrete flexural strength, subgrade bearing capacity, aircraft gear load, and gear spacing (FAA AC150/532-6D).

3.9　CONCRETE PROPERTIES AND MIX DESIGN

Portland cement concrete (PCC) consists of Portland cement, course and fine aggregates, water, chemical and mineral admixtures, and may or may not contain entrained air. Concrete mix design is the process of determining the proportions of the different components, and achieving the desirable mix characteristics based on the most sustainable, economical and practical combination of readily available materials that will sustain traffic loads and environmental conditions. Concrete properties that are most important for design include the strength and stiffness, dimensional stability to temperature and moisture, durability and constructability. Of course, these desirable qualities must be considered within economic and environmental sustainability considerations.

The flexural strength of concrete is a direct input into the design process. This could be determined directly using the bend beam test (ASTM C78; AASHTO T97) or indirectly using the compressive strength test (ASTM C39; AASHTO T22) values. When using the compressive strength, a correlation study for the selected mix design is highly recommended. The modulus of elasticity or stiffness of the concrete (ASTM C469) relates stress–strain behavior and predicts deformation under load. This is a good predictor for how the concrete slab will deflect and distribute traffic loading.

Dimensional stability of the concrete to expansion, contraction, warping, and curling stresses is more involved. Shrinkage properties in concrete are associated with the water content in the mix. But more specifically, it is related to the paste fraction and the resulting microstructure of the paste phase in the concrete. Temperature curling and volume changes are highly dependent on the type of aggregate. Generally, slabs made with a limestone coarse aggregate will have a lower thermal coefficient, while slabs made with quartz or sandstone have a higher thermal coefficient. Drying shrinkage and contraction will depend on the movement of moisture in and out of the paste. Dimensional changes have a large effect on rigid pavements and control curling and warping stresses, which can result in distresses such as corner breaks and mid-slab cracking, erosion and pumping, widened joint opening, and faulting.

Durability is a more complex desirable characteristic to attain over the life of the pavement. Producing durable pavements is a balancing act between mix design, available materials, strength needs, construction details, and dimensional and chemical stability with the surrounding environment. Freeze–thaw durability, for example, can be addressed by appropriate air-void structure (air bubble size and distribution within the cement paste), which in turn decreases the strength at a given cement content. Increasing the cement content will increase the water demand for a given workability. However, this in turn increases the potential for drying shrinkage, joint opening, a reduction of load transfer across joints, and the potential for joint failure.

The design of a quality concrete mixture should satisfy a number of competing desirable performance requirements for construction such as workability, ease of placement, consolidation and finishing; early age properties such as setting time; should not bleed excessively but just enough to avoid drying shrinkage cracking; appropriate strength gain; volume stability to minimize curling and warping; bond strength to reinforcement and tiebars; protection against steel corrosion; adequate aggregate shear friction for load transfer across cracks and joints; and sustainability, durability, appearance, and cost. A mix design leads to a concrete specification that imposes limits on used materials and properties. Ultimately, a desirable mix will be one that uses the maximum amount of aggregate volume, the least amount of cement materials that provide the required structural properties, and durability and construction requirements.

Prior to proportioning the ingredients, the desirable concrete properties must be determined based on the intended use of the concrete to handle load and environmental stresses and construction requirements. Strength requirements will dictate a minimum cement content and water–cement ratio. Workability needs require more fine aggregate and a higher water demand. Durability concerns for freeze–thaw durability, D-cracking, and corrosion of reinforcement may require more aggregate volume, less cement, and more fly ash (or other supplementary cementitious materials), which raises the water demand for more workability. Mix design is an interesting optimization process that requires elements of mix design, local experience, and good engineering decisions.

The selection of the appropriate water–cementitious ratio (W/CM) influences the quality of the cement paste mixture and ultimately the concrete mix. More importantly than strength, a concrete must be designed for durability, impermeability or water tightness, wear, and abrasion resistance. Once these characteristics are achieved, usually the required strength is satisfied too.

3.9.1 Hydration, Strength, and Materials

When water is added to cement, an exothermic reaction called *hydration* occurs, resulting in the formation of a number of compounds, the main one being calcium silicate hydrate. This process and the resulting chemical compounds provide the glue to bind the components of the mix together and provide the strength to hardened concrete.

The water–cementitious ratio is the mass of water divided by the mass of all cements, blended cements, and pozzolanic materials such as fly ash, slag, silica fume, and natural pozzolans. In most literature, *water–cement ratio* (W/C) is used synonymously with W/CM.

Assuming that a concrete is made with clean sound aggregates, and that the cement hydration has progressed normally, the strength gain is inversely proportional to the W/CM ratio by mass. The paste strength is proportional to the solid's volume or the cement density per unit volume. Differences in strength may also be influenced by aggregate gradation, shape, particle size, surface texture, strength, and stiffness or maybe due to factors associated with cement materials (i.e., different sources, chemical composition, and physical attributes), the amount of entrained air and air-void system, and effects of admixtures and curing. The strength of concrete is measured in terms of compressive strength of cylinders, and the strength gain is evaluated by checking strengths at different periods of time from the mixing of the concrete.

Aggregates have a significant effect on the workability of fresh concrete. The aggregate particle size and gradation, shape, absorption, and surface texture will influence the type of concrete that is produced with a given amount and quality of paste (cement plus water). The selection of the maximum size aggregate is governed by the thickness of the slab and by the closeness of the reinforcing steel. The larger size aggregates should not be obstructed and should flow easily during placement and consolidation. The CTE of the aggregate will have a large influence on the CTE of the resulting concrete and its dimensional stability to temperature. This in turn will influence concrete expansion and contraction and curling stresses.

The purposeful entrainment of air in concrete provides tremendous protection against freezing and thawing action and against deicing salts. Hydraulic pressure is generated when water in the paste pore structure freezes and pushes against the unfrozen water. The tiny entrained air bubbles act as relief valves for this developed hydraulic pressure.

Fresh concrete must have the appropriate workability, consistency, and plasticity suitable for construction conditions. Workability is a measure of the ease of placement, consolidation, and finishing of the concrete. Consistency is the ability of freshly mixed concrete to flow, and plasticity assesses the concrete's ease of molding. If a concrete is too dry and crumbly or too wet and soupy, then it lacks plasticity. The slump test is a measure of consistency and indicates when the characteristics of the fresh mix have been changed or altered. However, the slump is indicative of workability when assessing similar mixtures. Different slumps are needed for different construction projects. Slipform paving, for example, will require a much stiffer consistency than fixed-form paving.

The cementing materials content is usually determined based on the specified water to cementitious materials ratio. However, usually a minimum amount of cement is also specified to ensure satisfactory durability, finishability, and wear resistance of slabs even though strength needs may be satisfied at lower cement contents. For durability, the type of cement used is also important. And finally, admixtures are used in concrete to enhance the desirable characteristics in the fresh and hardened concrete.

Concrete admixtures are commonly used for producing high-quality concretes with low w/c ratios. Common concrete admixtures include air-entraining admixtures, water reducers, set modifiers. Other admixtures that are less commonly in use include lithium base chemicals for mitigating alkali–silica reaction, shrinkage reducing modifiers, and corrosion inhibitors.

Air-entraining agents are the most commonly used in concrete and are typically derived from pine wood resins, vinsol resins, and other synthetic detergents. They contribute in stabilizing tiny air bubbles that are uniformly distributed in the cement paste. In addition to protection against freezing and thawing and salt scaling, air entrainment can also enhance concrete workability at a reduced water content. Concretes with an average spacing factor of 0.008 in. (0.2 mm) will generally have adequate freeze–thaw durability.

Water reducers are the second most common admixtures used in concrete. Water reducers are specified by ASTM C 494/AASHTO M 194 and will reduce the amount of water needed for a given

FIGURE 3.28 Flowchart for concrete mix design.

consistency or slump. For a given cement content, air content, and slump, the concrete strength could be increased by 10% or more. High-range water reducers (HRWRs) also called superplasticizers, designated as ASTM Types F and G, can be used to reduce the required water content by 12%–30%. The rate of slump loss with HRWRs could be very high and may affect consistency for slipform operations. Water reducers work by altering the charge on cement particles and prevent cement particles from clumping, freeing up trapped water for more hydration.

Set-modifying admixtures are used to control (i.e., retard or accelerate) the rate of setting and strength gain of concretes. Altering setting functions for concrete are important for hot- and cold-weather concreting operations. These admixtures can also be useful for fast-track construction controlling production cycles and allow concretes to be hauled over longer haul times. They can also accelerate the slower strength gain of concretes made with some supplementary cementitious materials.

A flowchart showing the basic steps in mix design for concrete is shown in Figure 3.28.

3.9.1.1 Construction

The subgrade must be prepared to provide a uniform level surface. Protection against erosion, pumping, and frost action is important. If necessary, compaction and/or modification with additives such as lime treatment or cement treatment is conducted. The subgrade does not need to provide a large bearing capacity, since the concrete slab is rigid and carries most of the stresses. Uniformity is more important to avoid areas of soft material or hard material, which could cause local areas of high stress under traffic loading.

The base or subbase that is the layer supporting the concrete slab should provide adequate support and drainage. Uniformity is again more important than strength or bearing capacity. The free flowing of water is critically important to durability, pumping, and frost action. Bases may be bound or unbound. A granular base or a cement- or asphalt-treated base may be used. If bound material is used, an open-graded mixture should be used to provide adequate drainage and water flow.

Prior to construction, selection of the construction methods, materials, mix design, concrete production, quality control plan, and delivery should be studied and planned for. Trial mixes should be used to optimize the mix that will provide the required concrete for the needs of the project. Large batch testing and field testing is recommended, given the variability of materials and climatic conditions that could affect the concrete properties and construction process.

A pavement is commonly constructed by a slipform or fixed-form process. Proper placing, consolidating, finishing, curing, and jointing a pavement requires skilled and experienced people and is critical to producing a quality product. It is commonly said that "constructing a poorly designed pavement well, will outperform a well designed pavement that is constructed poorly" (Datelle, 2008).

QUESTIONS

3.1   What are the basic steps in PCC pavements' mix and structural design?

3.2   What are the primary distresses against which PCC pavements are designed?

3.3   A concrete slab that is 25 ft long, 12 ft wide, and 8 in. thick is subjected to a temperature differential $\Delta T = 24°F$. Assume $k = 100$ pci and $\alpha = 5.5 * 10^6$ in./in./°F.

   a.   Determine the maximum curling stress at the center edge and the center interior of the slab.

   b.   Determine the corner stress using the Bradbury equation, assuming $a = 6$ in. ($E_c = 4 * 10^6$ psi, $\mu = 0.15$).

3.4   A concrete slab 8 in. thick is placed on a subgrade with a modulus of subgrade reaction $k = 100$ pci. A 9000 lb single wheel load is applied at the corner of the slab having a circular contact area that produces a contact pressure of 90 psi ($E_c = 4 * 10^6$ psi, $\mu = 0.15$).

   a.   Determine the corner stress and deflection using the Westergaard equations.

   b.   Determine the corner stress and deflection using the Ioannides et al. equations.

3.5   Same as Problem 3.4 but the load is applied at the interior of the very large slab.

   a.   Determine the maximum stress and deflection due to the applied wheel load to the interior of the slab.

3.6   Same as Problem 3.4 but the load is applied at the edge of the very large slab.

   a.   Determine the maximum stress and deflection due to the applied wheel load to the edge of the slab.

3.7   A concrete slab, 50 ft long, 12 ft wide, and 10 in. thick, is placed on a subgrade with a modulus of subgrade reaction k = 100 pci. A 9000 lb single wheel load is applied at the center edge
       of the slab having a circular contact area and a contact pressure of 80 psi ($E_c = 4 * 10^6$ psi,
       $\mu = 0.15$, $\alpha = 5.5 * 10^6$ in./in./°F).

   a. Determine the curling tensile stress at the edge of the slab when the change in temperature differential between top and bottom of the slab is 3°F/in during the day and 1.5°F/in
       during the night.

   b. Determine the edge stress due to the 9000 lb wheel load applied at the center edge of
       the slab.

   c. Determine the combined maximum and minimum stress due to temperature curling and
       wheel load.

Contact pressure = 80 psi; $E_c = 4 \times 10^6$ psi; $\mu = 0.15$, $\alpha = 5.5 \times 10^6$ in./in./°F

3.8   A concrete pavement is subject to a temperature change of $\Delta T = 70$°F; a coefficient of thermal
       expansion $\alpha = 5.5 * 10^{-6}$ in./in./°F; a concrete shrinkage $\epsilon = 200 * 10^{-6}$; and a granular base
       with C = 0.8. Determine the joint spacing for an undoweled and a doweled joint having an
       allowable joint opening of 0.06 and 0.02 in., respectively.

3.9   A two-lane concrete pavement is 10 in. thick, 60 ft long, and 24 ft wide with a longitudinal
       joint in the center.

   a. Determine the diameter, length, and spacing of the tiebars (Assume $f_y = 40$ ksi billet steel).

60 in.

3.10   A concrete slab as shown in the following figure has a thickness of 11 in., a width of 12 ft, and
       a modulus of subgrade reaction k = 300 pci. A 32,000 lb axle load with a wheel spacing of 6
       ft is applied at the joint. The left wheel is located 6 in. from the outermost edge. Assume the
       maximum negative moment occurs at a distance 1.8ℓ from the load. Assume the load transfer
       across the joint is 100% efficient and a joint opening of 0.2 in. The dowel bar diameter is
       1.25 in. For concrete, $E_c = 4 * 10^6$ psi, $\mu = 0.15$.

   a. Determine the maximum bearing stress between the concrete and the steel dowel.

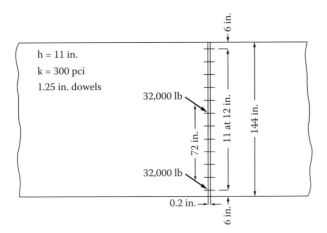

3.11 Same as Problem 3.10 except that the maximum negative moment occurs at 1.1ℓ from the load.

3.12 Same as Problem 3.10 except that the maximum negative moment occurs at 1.8ℓ from the load and the foundation support $k = 100$ pci.

3.13 Same as Problem 3.10 except that the maximum negative moment occurs at 1.1ℓ from the load and the foundation support $k = 100$ pci.

4 Standards

4.1 IMPORTANCE OF STANDARDS

Standards are documents provided by expert and recognized organizations, which provide guidelines or characteristics of common activities to ensure consistency, quality, and safety. Standards are developed on the basis of research and experience (and are subsequently revised, as needed) through consensus of groups of experts. Generally, standards used in pavement engineering are those that are developed by traditional formal organizations such as the American Society for Testing and Materials (ASTM International; www.astm.org), the American Association of State and Highway Transportation Officials (AASHTO; www.transportation.org), the Asphalt Institute (www.asphaltinstitute.org), the American Concrete Institute (ACT, http://www.aci-int.org), and the British Standards Institution (BSI; www.bsi-global.com). They are developed through a very open and transparent process, ensuring the representation of all stakeholders. Standards on safety can be obtained from the Occupational Safety and Health Administration (OSHA) website (http://www.osha.gov/pls/publications/publication.athruz?pType=Types&pID=10).

Each standard is an accumulation of years of experience in that specific topic. Use of appropriate standards can help the proper utilization of that combined experience; make the process (being used) rational, practical, and up-to-date; and, as required, ensure, to the maximum extent at that point of time, safety. Therefore, standards ensure uniformity and consistency, reliability, and safety in products and processes. Use of standards also ensures compatibility between different products, and the best use of money. This is important since it means, for pavement engineering, the best use of the taxpayers' dollars for road construction.

For pavement engineering, there are various types of standards—such as those for conducting tests, for interpreting results, as well as for writing specifications. Every test process is described in detail, with mentions of relevant other standards, steps in inference, and methods of reporting the results. If a new test is developed, standards provide guidance for conducting a round-robin study to determine the variability of results from that specific test. Once conducted, the results are considered and transferred into inter- and within-laboratory tolerance values. This ensures that in the future, if the tests are conducted, the variability in the results would point out inconsistencies or greater than expected errors, if any.

4.2 THE AMERICAN SOCIETY OF TESTING AND MATERIALS

A typical ASTM standard on a test consists of the following items:

1. *Scope*: The scope defines the purpose of the standard and provides relevant details related to specific applications of the test/procedure.
2. *Referenced documents*: This section lists the other standards that are used as part of this standard.
3. *Summary of test method*: The summary describes briefly the steps, ranging from sample preparation (for example) to reporting of results.
4. *Significance and use*: This section relates the results of the test/process that is described in this standard to practical problems/solutions and applications.

FIGURE 4.1   Example of American Society for Testing and Materials (ASTM) standard designation.

5. *Apparatus*: The apparatus section describes, in detail, each and every apparatus or piece of equipment that is used. Wherever applicable, appropriate dimensions/capacities of apparatus are provided.

6. *Sample*: This section describes how a sample, from which test specimens are to be prepared, is to be handled or modified, if required.

7. *Test specimen*: This part provides a detailed description of the process of preparing test specimens. If the test consists of determination of multiple parameters, it describes the steps needed for the preparation of samples for testing of the different parameters.

8. *Procedure*: This section describes the actual test procedure.

9. *Calculation*: This part presents the steps needed to calculate the different parameters and provides steps to analyze the data, make corrections, and plot the data, as required.

10. *Report*: The report section points out what needs to be indicated in the document that reports the results. Specific items include those such as other relevant properties of the test specimens, any special sample preparation procedures, results, and any other tests conducted as part of this.

11. *Precision and bias*: This part provides the maximum variation that can be expected between two tests conducted in the same laboratory and between tests conducted in different laboratories, as well as between the results obtained from this test and those from more accurate tests, considered as "true" values.

12. *Keywords*: The last part presents a list of keywords against which this standard has been indexed.

Other than tests, some of the other relevant ASTM standards include those defining different terminology, sampling materials, conducting ruggedness tests for detecting sources of variation in test methods, and establishing precision of test methods.

The ASTM standards are generally designated as a two-part number, the first number being the number of the standard, and the second number indicating the year in which it was first adopted (or, if revised, the year of the last revision). A letter (e.g., D) indicating the group that deals with the subject matter under which the specific topic presented in the standard falls is also mentioned in the designation. For example, see Figure 4.1.

With the advent of global business and economy, in more and more cases, organizations are working on globalization of standards such that materials and process become compatible regionally, nationally, and internationally.

4.3   THE AMERICAN SOCIETY OF STATE HIGHWAY AND TRANSPORTATION OFFICIALS

AASHTO standards are developed either separately or, if appropriate, on the basis of existing ASTM standards by the AASHTO Highway Subcommittee of Materials, with representatives of transportation department officials in the United States, associate members, and international members, such as those from the Canadian provinces and Korea. The set of standards is made available in two parts—Part 1 contains specifications for materials, and Part 2 contains methods of tests and specifications for testing equipment.

A typical standard test method from the AASHTO consists of the following:

1. *Scope*: This part states the objective of the test method.
2. *Referenced documents*: Relevant other AASHTO and ASTM standards are mentioned here.
3. *Terminology*: This part defines the different terms used in this standard.
4. *Significance and use*: In this section, the practical applications of the results of this test are indicated.
5. *Apparatus*: All of the required apparatus are listed and described briefly in this part.
6. *Sample*: The procedure of sampling or preparation of sample is described here. Required sample sizes are indicated.
7. *Procedure*: This section describes the test procedure along with required warnings and cautions.
8. *Calculation*: The equations and expressions for calculating the results are presented in this section.
9. *Precision and bias*: The precision section indicates the within- and between-laboratory standard deviations, and hence guidelines for maximum allowable differences in results obtained by tests conducted by the same operator and those conducted in different laboratories. The bias section presents data on differences between results obtained from this test and more accurate tests (considered as a "true" value).
10. *Keywords*: The index keywords are listed in this part.

4.4 USE OF STANDARDS IN MATERIALS SELECTION, MIX DESIGN, AND STRUCTURAL DESIGN

Standards are used in selecting materials such as aggregates and asphalt, conducting tests, and interpreting results in mix design and structural design of pavements. The selection part involves using appropriate project/application conditions to choose specific types of materials and avoid those that could be detrimental for such conditions. For example, for selecting aggregates, AASHTO Specification M 283 could be used. In the design step, standards are used for conducting tests, which involve sample preparation, fabricating test specimens, conducting tests, analyzing data, and reporting results. An example is AASHTO T 27 for sieve analysis of fine and coarse aggregates.

4.5 USE OF STANDARDS IN QUALITY CONTROL IN CONSTRUCTION

In quality control and testing of pavements, the different steps involve developing a plan for quality control, sampling materials from the plant and the field, conducting either field or laboratory tests, and interpreting the results as evidence against which a pavement section could be accepted or rejected by the owner. These steps could involve the use of a series of standards, for example, ASTM D3665 (Standard Practice for Random Sampling of Construction Materials) and ASTM D295 (Determination of Density of Pavement Using the Nuclear Method).

4.6 IMPORTANT SPECIFICATIONS

Two lists of important specifications are provided in Tables 4.1 and 4.2. The reader is encouraged to visit the ASTM (www.astm.org) and AASHTO (www.aashto.org, www.transportation.org) websites to learn more about the standards.

TABLE 4.1
List of Relevant ASTM and AASHTO Standards for Flexible Pavement

Item	AASHTO	ASTM
Specification: coarse aggregate	M 283	
Specification: fine aggregate	M 29	
Specification: mineral filler	M 17	
Specification: aggregate gradations		D 3515
Tests: sampling aggregates	T 2	D 75
Random sampling		D 3665
Reducing field samples to test size	T 248	C 702
Sieve analysis of fine and coarse aggregates	T 27	C 136
Amount of material finer than 0.075 (no. 200) sieve	T 11	C 117
Soundness of aggregates by freezing and thawing	T 103	
Soundness of aggregates by sodium or magnesium sulfate method	T 104	C 88
Resistance to abrasion of small-size coarse aggregate by LA abrasion	T 96	C 131
Resistance to abrasion of large-size coarse aggregate by LA abrasion		C 535
Insoluble residue in carbonate aggregate		D 3042
Specific gravity and absorption of fine aggregate	T 84	C 128
Specific gravity and absorption of coarse aggregate	T 85	C 127
Moisture in aggregate by drying	T 255	C 566
Coating and stripping of aggregate mixtures	T 182	C 1664
Mechanical analysis of extracted aggregate	T 30	C 136, C 117
Effect of water on cohesion of compacted mixtures	T 165	D 1075
Resistance of compacted mixtures to moisture-induced damage	T 283	
Effect of moisture on asphalt concrete paving mixtures		D 4867
Asphalt content by nuclear method	T 287	
Compressive strength	T 167	
Specification: mixing plants for hot mix asphalt	M 156	
Hot asphalt paving mixtures		D 3515
Plant inspection	T 172	D 290
Random sampling		D 3665
Sampling paving mixtures	T 168	D 979
Degree of particle coating	T 195	D 2489
Marshall tests	T 245	D 1559
Hveem tests	T 246	
Bulk-specific gravity and density of compacted mixtures	T 166	D 2726
Bulk-specific gravity and density of compacted mixtures, paraffin coated	T 275	D 1188
Density of pavement, nuclear method		D 2950
Degree of compaction	T 230	
Theoretical maximum specific gravity of mixtures	T 209	D 2041
Thickness of compacted mixtures		D 3549
Percentage air voids of compacted mixtures	T 269	D 3203
Moisture or volatiles in mixtures	T 110	D 1461
Extractions	T 164	D 2172

TABLE 4.2
List of Relevant ASTM and AASHTO Standards for Rigid Pavement

Item	AASHTO	ASTM
Resistance to freezing and thawing	T 161, T 103	C 666, C 682
Resistance to disintegration by sulfates	T 104	C 295, C 3398
Gradation	T 11, T 27	C 117, C 136
Fine aggregate degradation		C 1137
Uncompacted void content of fine aggregate	T 304	C 1252
Bulk density (unit weight)	T 19	C 29
Fine aggregate grading limits	M 6	C 33
Compressive and flexural strength	T 22, T 97	C 39, C 78
Definitions of constituents		C 125, C 294
Aggregate constituents—maximum allowed of deleterious and organic materials	T 21	C 40
Resistance to alkali reactivity and volume change	T 303	C 227
Sizes of aggregate for road and bridge construction	M 43	D 448
Bulk density ("unit weight") and voids in aggregate	T 19	C 29
Coarse aggregate for Portland cement concrete	M 80	—
Organic impurities in fine aggregate for concrete	T 21	C 40
Portland cement	M 85	C 150
Compressive strength of cylindrical concrete specimens	T 22	C 39
Making and curing concrete test specimens	T 23	C 31
Sodium chloride	M 143	D 632
Obtaining and testing drilled cores and sawed beams of concrete	T 24	C 42
Calcium chloride	M 144	D 98
Quality of water to be used in concrete	T 26	C 1602
Liquid membrane-forming compounds for curing concrete	M 148	C 309
Effect of organic impurities in fine aggregate on strength of mortar	T 71	C 87
Ready-mixed concrete	M 157	C 94
Air-entraining admixtures for concrete	M 154	C 233
Resistance to degradation of small-size coarse aggregate by abrasion and impact in the Los Angeles machine	T 96	C 131
Chemical admixtures for concrete	M 194	C 494
Flexural strength of concrete (simple beam with third point loading)	T 97	C 78
Lightweight aggregate for structural concrete	M 195	C 330
Fineness of Portland cement by the turbidimeter	T 98	C 115
Epoxy protective coatings	M 200	A 884
Soundness of aggregates by freezing and thawing	T 103	—
Soundness of aggregates by use of sodium sulfate or magnesium sulfate	T 104	C 88
Chemical analysis of hydraulic cement	T 105	C 114
Use of apparatus for the determination of length change of hardened cement paste, mortar, and concrete	M 210	C 490
Compressive strength of hydraulic cement mortar (using 50-mm or 2-in. cube specimens)	T 106	C 109
Clay lumps and friable particles in aggregate	T 112	C 142
Lightweight pieces in aggregate	T 113	C 123
Slump of hydraulic cement concrete	T 119	C 143
Mass per cubic meter (cubic foot), yield, and air content (gravimetric) of concrete	T 121	—
Blended hydraulic cements	M 240	C 595
Making and curing concrete test specimens in the laboratory	T 126	C 192
Concrete made by volumetric batching and continuous mixing	M 241	C 685

QUESTIONS

4.1 Why are standards important?

4.2 List five different standard-authoring organizations for pavements.

4.3 How are standards developed?

4.4 Contact any pavement industry in your area and ask them which standards they use regularly. List these standards; critically review the different parts of any commonly used standard.

5 Traffic

5.1 DIFFERENT TYPES OF HIGHWAY TRAFFIC

Different types of vehicles use roadways, and different types of aircrafts use airport pavements. For roads, the main destructive effect comes from trucks, since other vehicles such as passenger cars are significantly lighter than trucks. Again, actual truck traffic on pavements consists of a variety of loads and axles. Trucks can be of the single-unit type or of the tractor-semitrailer or trailer type.

In the United States, combination trucks with 53 ft trailers are the most common. The gross maximum permissible weight is dictated by road agencies (such as the states in the United States) and depends on the number of axles that the truck consists of. Gross weights can range from 70,000 to 164,000 lb. There are specific guidelines regarding maximum weights per axle. For example, for highways other than interstates in the United States, the single-axle maximum is 18,000 lb and the tandem-axle maximum is 32,000 lb. The nominal spacing between a pair of axles in a tandem is 4 ft. Axles can be spread apart to accommodate different weights. The federal bridge formula in the United States specifies the maximum weights as follows:

$$W = 500 \left[\frac{LN}{N-1} + 12N + 36 \right]$$

where

W is the maximum weight in pounds carried by any group of two or more axles
L is the distance in feet between the extremes of any group of two or more axles
N is the number of axles under consideration

Definitions of truck dimensions as well as some typical values are shown in Figures 5.1 and 5.2.

5.2 MEASUREMENT OF TRAFFIC LOADS

Automatic traffic recorders (ATR) with different technologies—inductive loop (change in magnetic field), piezo sensors (signal-voltage or current), road tube (air pulse), or infra-red/radar detectors—can constantly count the number of vehicles (traffic volumes) passing through a location. The average daily traffic (ADT), the annual average daily traffic (AADT), and monthly seasonal adjustment factors can be determined from the ATR data. The inductive loop is often used with piezo sensors for axle detection for obtaining comprehensive data about traffic number, length, and speed (automatic vehicle classification system, AVC). Data can be transmitted to a central location from the field by telemetry. A good example of state wide program of the use of ATRs and examples of traffic data analysis is available at http://www.sha.state.md.us/oppen/HISD%20Traffic%20Monitoring%20System%20Program.pdf for the Maryland State Highway Administration (SHA).

KCRA-Kingpin to Center of Rear Axle

KCRT-Kingpin to Center of Rear Tandem

Definition of truck dimensions

Different truck configurations

FIGURE 5.1 Truck dimensions and some truck configurations. (From Harwood, D.W. et al., Review of truck characteristics as factors in roadway design, NCHRP report 505, Transportation Research Board, Washington, DC, 2003.)

FIGURE 5.2 Typical truck dimensions. (From American Association of State Highway and Transportation Officials, A policy on geometric design of highways and streets, Washington, DC, 2001.)

For highway pavements, with the help of a weigh-in-motion (WIM) system, the load on each and every axle of each and every vehicle can be determined. In a mechanistic or mechanistic-empirical structural design process, such data can be used directly to determine the damage on the pavement by linking the stress/strain caused by such loads and the stress/strain versus performance relationships.

ASTM Standard E1318 describes WIM as a system to measure the dynamic tire forces of a moving vehicle and estimate the tire loads of the static vehicle. WIM is used to collect traffic data and classify them according to different time periods, such as days and weeks. The specific type of WIM that is used depends on the intended applications of the data, for example, collection of traffic data for design and/or weight enforcement. There are three primary types of WIM systems in use—load cell, bending plate, and piezoelectric sensor. In the load cell system, the load is directly measured with a load cell on scales placed on the pavement. In the bending plate system, strain gages attached at the bottom of a plate (i.e., inserted in the road) record the strain from a moving vehicle. The strain data are used to calculate the dynamic load and then the static load (using calibration constants). Piezoelectric sensors operate on the basis of generation of electricity in quartz-piezo sensors as a result of application of pressure on them. The quartz units in a sensor (placed in the pavement) are usually inserted in epoxy-filled aluminum channels. The electric charge generated from the sensor is detected, and the dynamic and hence the static load of the axles are calculated.

The data from the WIM system are processed by a data acquisition system and software at the site to generate time and traffic information (e.g., speed, weight of axle, and classification) and are transmitted to the office computer as data files (such as in ASCII format). These data files can then be processed, using separate software, to generate detailed reports, such as those required for checking and calibration of the WIM, and classification of the traffic according to days or weeks. Figure 5.3 shows the commonly used piezoelectric sensor, installation on the pavement, the on-site data acquisition system, the software output, and an example of traffic classification.

As noted in Figure 5.3, the traffic can be grouped into different types. For example, Table 5.1 shows the FHWA classification system.

In mechanistic-empirical processes, data from WIM and automatic vehicle classification (AVC) systems can be used to determine the number of axle applications for each axle type and axle load group (full axle-load spectrum) over the design period. Generally, vehicles in the range of classes 4–13 (FHWA classification system) are considered.

5.3   LOAD EQUIVALENCY FACTOR AND EQUIVALENT SINGLE-AXLE LOAD

The mixed stream of traffic is often considered in empirical design processes by converting the different axles with different loads to an equivalent number of standard axles, for example, the 18,000-lb-or 18-kip-load single axle. The equivalency is based on the assumption that the load equivalency factor (LEF; also known as the equivalent single-axle load factor or ESAL factor) of a specific load/axle combination is the ratio of damage caused by one pass of the load/axle to a single pass of a standard 18-kip single axle. The damage can be represented in different methods—each design method may use a different parameter. The AASHTO design process uses the loss in serviceability, which is represented in terms of the present serviceability index (PSI), as the selected parameter.

In the AASHTO design process, the terminal serviceability index, p_t, and the rigidity of the pavement structure (denoted by structural number, SN, for a flexible pavement and slab thickness, D, for a rigid pavement) are utilized, along with the specific axle/load information to determine a load equivalency factor. The equations that were used to generate the LEFs for different axle loads and configurations (single, tandem, and extended for tridem) are as follows (given in the 1986 *AASHTO Guide*, vol. 2, Appendix MM).

Sensors

Elastic material
Quartz element
Aluminum housing

WIM system   Peizo sensor

On-site data acquisition system   Real time visual information from WIM system

Summary report   09/28/2007 11:11:38

Site name   : ST000017
Date/Time   : From 03/01/2007   00:00   To 06/01/2007   00:00
Period   :
Report   : Gross weight distribution by classification
Unit   : Kip
Class   : 1, 2, 3, 4, 5, 6, 7, 8, 9, 10, 11, 12, 13
Directions : Toward capital   Away from capital
Lanes   : 0, 2, 4, 6   7, 5, 3, 1

Weight	Classification									
	1	2	3	4	5	6	7	8	9	10
under 5	1165	157384	13162	0	0	0	0	0	0	0
5 to 10	20	35556	69962	0	0	0	0	0	0	0
10 to 15	1	24	1611	199	2318	8	0	2	3	0
15 to 20	0	0	53	430	1026	170	0	22	4	7
20 to 25	0	0	0	498	800	359	1	70	5	20
25 to 30	0	0	0	103	539	244	1	210	84	85
30 to 35	0	0	0	68	340	195	4	243	368	827
35 to 40	0	0	0	45	169	78	2	76	437	2205
40 to 45	0	0	0	20	31	70	0	11	240	666
45 to 50	0	0	0	16	7	73	1	11	110	111
50 to 55	0	0	0	4	0	86	7	4	68	18
55 to 60	0	0	0	2	1	87	9	6	63	31
60 to 65	0	0	0	0	1	30	5	5	38	45
65 to 70	0	0	0	0	0	16	11	4	37	83
70 to 75	0	0	0	1	0	0	23	1	71	84
75 to 80	0	0	0	0	0	0	14	0	116	107
80 to 85	0	0	0	0	0	0	4	0	101	207
85 to 90	0	0	0	1	0	0	0	0	49	519
90 to 95	0	0	0	0	0	0	0	1	15	932
95 to 100	0	0	0	0	0	0	0	0	2	1258
100 to 105	0	0	0	0	0	0	0	1	8	1869
105 to 110	0	0	0	0	0	0	0	0	2	1975
110 to 115	0	0	0	0	0	0	0	0	1	844
115 to 120	0	0	0	0	0	0	0	0	0	248
above 120	0	0	0	0	0	0	0	0	0	69
TOTAL	1186	192964	84788	1387	5232	1416	82	667	1822	12210
%	0.4	64	28.1	0.5	1.7	0.5	0	0.2	0.6	4

Example of report on traffic classification

FIGURE 5.3 Commonly used piezoelectric sensor, installation on a pavement, on-site data acquisition system, software output and an example of traffic classification.

5.3.1 FLEXIBLE PAVEMENTS

$$\log_{10}\left[\frac{W_{t_x}}{W_{t_{18}}}\right] = 4.79 * \log_{10}(18+1) - 4.79 * \log_{10}(L_x + L_2) + 4.33 * \log_{10} L_2 + \frac{G_t}{\beta_x} - \frac{G_t}{\beta_{18}}$$

$$G_t = \log_{10}\left[\frac{4.2 - p_t}{4.2 - 1.5}\right]$$

TABLE 5.1
Vehicle Classification

Class	Description
0	Unclassified vehicles that do not fit into any other classification. Vehicles that do not activate the system sensors are also unclassified
1	Motorcycles: All two- or three-wheeled motorized vehicles. This category includes motorcycles, motor scooters, mopeds, and all three-wheel motorcycles
2	Passenger cars: All sedans, coupes, and station wagons manufactured primarily for the purpose of carrying passengers
3	Other two-axle, four-tire single units: Included in this classification are pickups, vans, campers, and ambulances
4	Buses: All vehicles manufactured as traditional passenger-carrying buses with two axles and six tires or three or more axles
5	Two-axle, single-unit trucks: All vehicles on a single frame, including trucks and camping and recreation vehicles
6	Three-axle, single-unit trucks: All vehicles on a single frame, including trucks and camping and recreational vehicles
7	Four or more axle, single-unit trucks: All vehicles on a single frame with four or more axles
8	Four or less axle, single-trailer trucks: All vehicles with four or less axles consisting of two units, one of which is a tractor or straight truck power unit
9	Five-axle, single-trailer trucks: All five-axle vehicles consisting of two units, one of which is a tractor or straight truck power unit
10	Six or more axle, single-trailer trucks: All vehicles with six or more axles consisting of two units, one of which is a tractor or straight truck power unit
11	Five or less axle, multitrailer trucks: All vehicles with five or less axles consisting of three or more units, one of which is a tractor or straight truck power unit
12	Six-axle, multitrailer trucks: All six-axle vehicles consisting of three or more units, one of which is a tractor or straight truck power unit
13	Seven or more axle, multitrailer trucks: All vehicles with seven or more axles consisting of three or more units, one of which is a tractor or straight truck power unit

$$\beta_x = 0.40 + \frac{0.081 * (L_x + L_2)^{3.23}}{(SN+1)^{5.19} * L_2^{3.23}}$$

where

L_x is the load on one single tire or one tandem-axle set (kips)
L_2 is the axle code (1 for single axle, and 2 for tandem axle)
SN is the structural number
p_t is the terminal serviceability
β_{18} is the value of β_x when L_x is equal to 18 and L_2 is equal to 1
W_{t_x} is the total applied load from a given traffic
$W_{t_{18}}$ is the total applied load from an 18-kip single axle

5.3.2 RIGID PAVEMENTS

$$\log_{10}\left[\frac{W_{t_x}}{W_{t_{18}}}\right] = 4.62 * \log_{10}(18+1) - 4.62 * \log_{10}(L_x + L_2) + 3.28 * \log_{10} L_2 + \frac{G_t}{\beta_x} - \frac{G_t}{\beta_{18}}$$

$$G_t = \log_{10}\left[\frac{4.2-p_t}{4.2-1.5}\right]$$

$$\beta_x = 1.00 + \frac{3.63*(L_x+L_2)^{5.20}}{(D+1)^{8.46}*L_2^{3.52}}$$

where D is the slab thickness, in.

The equivalency factor

$$e_x = \frac{w_{t_{18}}}{w_{t_x}}$$

The equation for converting total applications (w_{t_x}) of a given axle load and configuration into an equivalent number of applications of the standard 18-kip single-axle load is as follows:

$$w_{t_{18}} = w_{t_x}*e_x$$

Examples of load equivalence factors are shown in Table 5.2. Note that there are separate tables for single and tandem axles. The use of Table 5.2 can be illustrated with the following example. Consider a vehicle with two single axles, each with dual tires. The total load on each axle is 10,000 lb. Consider a p_t of 2.5 and SN of 4. The total ESAL for the vehicle is (from Table 5.2) 0.204 (2 * 0.102). Note that the effect (in terms of ESAL) of single tire (carrying the same load as dual tires) is 10% higher.

TABLE 5.2
Examples of Axle Load Equivalency Factors for Flexible and Rigid Pavements, Single Axles and $p_t = 2.5$[a]

Flexible Pavements (AASHTO, 1986, p. D-6, Table D.4)						
Axle Load	Pavement Structural Number (SN)					
(kips)	1	2	3	4	5	6
2	0.0004	0.0004	0.0003	0.0002	0.0002	0.0002
10	0.078	0.102	0.118	0.102	0.088	0.080
18	1.00	1.00	1.00	1.00	1.00	1.00
30	10.3	9.5	7.9	6.8	7.0	7.8

Rigid Pavements (AASHTO, 1986, p. D-15, Table D.13)									
Axle Load	Slab Thickness, D (in.)								
(kips)	5	7	8	9	10	11	12	13	14
2	0.0002	0.0002	0.0002	0.0002	0.0002	0.0002	0.0002	0.0002	0.0002
10	0.097	0.089	0.084	0.082	0.081	0.080	0.080	0.080	0.080
18	1.00	1.00	1.00	1.00	1.00	1.00	1.00	1.00	1.00
30	8.16	7.67	7.79	8.28	8.79	9.14	9.35	9.46	9.52

[a] p_t is the terminal serviceability.

TABLE 5.3
Growth Rate Factors

Analysis Period,	Annual Growth Rate,% (g)							
n (Years)	No Growth	2	4	5	6	7	8	10
10	10	10.95	12.01	12.58	13.18	13.82	14.49	15.94
15	15	17.29	20.02	21.58	23.28	25.13	27.15	31.77
20	20	24.30	29.78	33.06	36.79	41.00	45.76	57.28
25	25	32.03	41.65	47.73	54.86	63.25	73.11	98.35

For design of a pavement, it is important that the total ESALs expected over the design period of the pavement (that is being designed) are estimated. This is accomplished by determining the current ESALs and applying a growth factor (provided in *AASHTO Guide*, vol. 1, Table D-20, p. D-23; AASHTO, 1986 and p. D-24, AASHTO, 1993) for the expected growth rate (percentage) and design life (years). Examples of growth factors are shown in Table 5.3.

$$\text{Growth rate factor} = \frac{(1+g)^n - 1}{g}, \quad \text{for } g \neq 0$$

$$g = \frac{\text{Rate}}{100}$$

Finally, the estimated accumulated ESAL should be split into two directions, and split further considering the multiple lanes. The recommended lane-split factors are shown in Table 5.4. An example calculation is shown in Figure 5.4.

Quite often, since load station data are not available everywhere, many agencies determine an ESAL factor (also known as a *truck factor*) for each type of vehicle, considering the total number of axles for that vehicle. The vehicle types can be selected on the basis of the FHWA vehicle classification system (13 different types of vehicles, ranging from Class 1 to Class 13). Furthermore, an average truck factor can be determined by adding all of the products of number of axles times LEF and dividing the product by the total number of vehicles of that specific type. These data from weighing stations can then be used directly to estimate the expected total

TABLE 5.4
Lane Distribution Factors

Number of Lanes in Both Directions	% of 18-kip ESAL Traffic in Design Lane
1	100
2	80–100
3	60–80
4 or more	50–75

Source: American Association of State Highway and Transportation Officials, AASHTO, *AASHTO Guide for Design of Pavement Structures*, AASHTO, Washington, DC, © 1993. Used with permission.

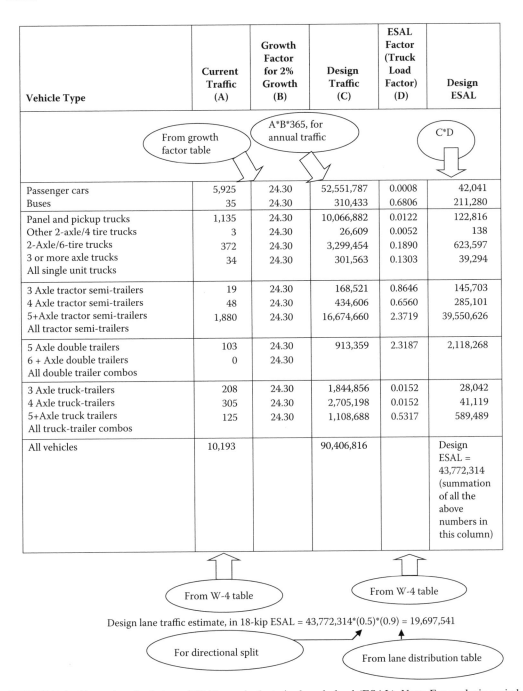

Vehicle Type	Current Traffic (A)	Growth Factor for 2% Growth (B)	Design Traffic (C)	ESAL Factor (Truck Load Factor) (D)	Design ESAL
	From growth factor table	A*B*365, for annual traffic		C*D	
Passenger cars	5,925	24.30	52,551,787	0.0008	42,041
Buses	35	24.30	310,433	0.6806	211,280
Panel and pickup trucks	1,135	24.30	10,066,882	0.0122	122,816
Other 2-axle/4 tire trucks	3	24.30	26,609	0.0052	138
2-Axle/6-tire trucks	372	24.30	3,299,454	0.1890	623,597
3 or more axle trucks	34	24.30	301,563	0.1303	39,294
All single unit trucks					
3 Axle tractor semi-trailers	19	24.30	168,521	0.8646	145,703
4 Axle tractor semi-trailers	48	24.30	434,606	0.6560	285,101
5+Axle tractor semi-trailers	1,880	24.30	16,674,660	2.3719	39,550,626
All tractor semi-trailers					
5 Axle double trailers	103	24.30	913,359	2.3187	2,118,268
6 + Axle double trailers	0	24.30			
All double trailer combos					
3 Axle truck-trailers	208	24.30	1,844,856	0.0152	28,042
4 Axle truck-trailers	305	24.30	2,705,198	0.0152	41,119
5+Axle truck trailers	125	24.30	1,108,688	0.5317	589,489
All truck-trailer combos					
All vehicles	10,193		90,406,816		Design ESAL = 43,772,314 (summation of all the above numbers in this column)

From W-4 table

From W-4 table

Design lane traffic estimate, in 18-kip ESAL = 43,772,314*(0.5)*(0.9) = 19,697,541

For directional split

From lane distribution table

FIGURE 5.4 Example calculation of 18-kip equivalent single axle load (ESAL). Note: For analysis period of 20 years and assumed thickness, D=9 in., for a four-lane rural interstate highway with equal traffic in both directions.

ESALs on the pavement (i.e., the design traffic, or design ESAL). AASHTO recommends the use of SN=5/D=9 in. and p_t=2.5 for picking the LEFs from the tables and designing the thickness, and if designed thickness is different from the assumed thickness by more than 1 in., repeat the design by assuming a different SN and hence picking a different LEF. An example calculation is shown in Tables 5.5 and 5.6.

TABLE 5.5
Grouping of LEF for Structural
Number = 5, $p_t = 2.5$[a]

Single-Axle Load (lb)	LEF
Under 3,000	0.0002
3,000–6,999	0.0050
7,000–7,999	0.0320
8,000–11,999	0.0870
12,000–15,999	0.3600
26,000–29,999	5.3890

Source: American Association of State Highway and Transportation Officials, AASHTO, *Guide for Design of Pavement Structures*, AASHTO, Washington, DC, © 1986. Used with permission.

[a] p_t is the terminal serviceability.

TABLE 5.6
Example of Calculation of Load Equivalency Factor for Structural Number = 5, $p_t = 2.5$[a]

Single-Axle Load (lb)	LEF	Number of Axles	18-kip Equivalent-Axle Load (EAL)
Under 3,000	0.0002	0	0
3,000–6,999	0.0050	1	0.005
7,000–7,999	0.0320	6	0.192
8,000–11,999	0.0870	144	12.528
12,000–15,999	0.3600	16	5.760
26,000–29,999	5.3890	1	5.3890
Tandem-Axle Load (lb)	**LEF**	**Number of Axles**	**18-kip EAL**
Under 6,000	0.0100	0	0
6,000–11,993	0.0100	14	0.140
12,000–17,999	0.0400	21	0.924
18,000–23,999	0.1480	44	6.512
24,000–29,999	0.4260	42	17.892
30,000–32,000	0.7530	44	33.132
32,001–32,500	0.8850	21	18.585
32,501–33,999	1.0020	101	101.202
34,000–35,999	1.2300	43	52.890
	18-kip EALs for all trucks		255.151

Source: American Association of State Highway and Transportation Officials, AASHTO, *Guide for Design of Pavement Structures*, AASHTO, Washington, DC, © 1986. Used with permission.

Note: The numbers in this table represent data obtained from a weigh station (W-4 form) for 165 trucks of 5-axle, tractor semitrailer type.

Therefore, for five-axle tracks, as weighed in this specific weigh station,

$$\text{Truck load factor} = \frac{18 \text{ kip EALs for all trucks}}{\text{Number of trucks}} = \frac{255.151}{165} = 1.5464$$

[a] p_t is the terminal serviceability.

5.4 ALTERNATIVE LOAD EQUIVALENT FACTOR CONCEPT

The concept of LEF can be approached in a different way also (as mentioned in Yoder and Witczak, 1975). Note that the ratio of damage by a single pass of the axle in question to a standard axle is the LEF. If N_a passes of an axle cause failure (note that the "failure" has to be defined) of a pavement, as do N_s passes of a standard axle, then the damage due to one pass of the axle and the standard axle can be denoted by $1/N_a$ and $1/N_s$, respectively. According to the definition of LEF,

$$LEF = \frac{1/N_a}{1/N_s} = \frac{N_s}{N_a}$$

Tests with fatigue cracking and analysis of cracking data with respect to tensile strain data show that the number of repetitions to failure of a pavement by fatigue cracking, due to repeated tensile strain, can be expressed as follows:

$$N_f = k_1 \left(\frac{1}{\varepsilon_t} \right)^{k_2}$$

where
N_f is the repetitions to failure
ε_t is the tensile strain in the asphalt mix layer
k_1 and k_2 are constants obtained by plotting the experimental data (log ε_t vs. log N_f)

It follows that LEF can also be then expressed as follows:

$$LEF = \left(\frac{\varepsilon_{ta}}{\varepsilon_{ts}} \right)^{k_2}$$

where ε_{ta} and ε_{ts} correspond to the strains for the axle in question and the standard axle. The value of k_2 has been reported to be between 3 and 6, with most common values being 4–5.

The most simple approach (and proven to be a good approximation) is to use the fourth power law to convert any axle load to an LEF as follows:

$$LEF = \left(\frac{W_a}{W_s} \right)^4$$

For example,

$$LEF \text{ for a } 40{,}000 \text{ lb load} = \left(\frac{40{,}000}{18{,}000} \right)^4 = 24.4$$

(In comparison, the AASHTO LEF for a 40-kip single-axle load for a flexible pavement with SN=5 and p_t=2.5 is 21.1 and that for SN=6 and p_t=2.5 is 23.0.)

5.5 EQUIVALENT SINGLE-WHEEL LOAD

For airport pavements, the concept of equivalent single-wheel load (ESWL, as opposed to ESAL for highways) has been used. For fixed levels of traffic (e.g., x number of departures), charts relating pavement thickness to CBR (California bearing ratio, a test property used to estimate strength of soil, Chapter 7) of the subgrade have been developed for specific wheel loads. This development has been done on the basis of conversion of multiple-wheel gear loads to ESWL. A critical aircraft is considered, and the ESWLs for all aircrafts are converted to those for the critical aircraft. The ESWL is defined as the load on a single tire that will cause the same response (such as stress, strain, or deflection) at a specific point in a given pavement system as that resulting from multiple-wheel loads at the same point in the same pavement system.

In determining the ESWL, either the tire pressure or the tire contact area of the ESWL is considered to be similar to those of the tires in the multiple-wheel gear. The choice of the response in developing the ESWLs for different gears depends on the agency using it—specifically, on its experience in correlating the response of choice to field performance. Derivations of ESWL can be made according to various procedures, such as the Asphalt Institute (AI), U.S. Navy, and U.S. Army Corps of Engineers methods. Examples are shown in Table 5.7.

In the Federal Aviation Administration (FAA) airport pavement design process, there are two steps in determining the equivalent load. First, each of the aircrafts (single-gear, dual-gear, dual-tandem-gear, and wide-body aircrafts) is converted into an equivalent gear, and then the number of departures is converted into the equivalent annual departures of the design aircraft. The sum of all of the equivalent annual departures is the design traffic. Furthermore, the departures are converted to "coverages" to take the lateral distribution of aircraft traffic into consideration. The following sections explain the process.

TABLE 5.7
Different Methods of Computing ESWL

Method	Equation	Considerations
U.S. Navy	For dual tires: $P_s = P_D/(1+(S_D/100))$ For dual tandem: $$P_s = \frac{P_{dt}}{\left[1+\left(\dfrac{S_d}{100}\right)\right]\left[1+\left(\dfrac{S_d}{100}\right)\right]}$$	Single-layer, equivalent stress concept is utilized at a depth of 30 in. for two tires separated center to center by a distance of S_d, tandem spacing of S_t, with duals contributing P_d load and dual tandem applying P_{dt} load
U.S. Army Corps of Engineers	$$P_e = \frac{P_k \sum_{i=1}^{n} F_{i\,max}}{F_e}$$ $$F_e = \frac{1.5}{\left[1+\left(\dfrac{Z}{a}\right)^2\right]^{0.5}}$$ F_e is determined from charts relating depth (in terms of radii) to deflection factor for various offsets (in terms of radii—see Figure 5.5	Single-layer, equal interface deflection criteria, n number of tires, each tire of radius a applying P_k load, separated center, considering the maximum deflection under the multiple gear; determination involves selecting multiple planes and determining the point where the deflection is maximum
Asphalt Institute	$$L = \frac{P_d}{P_e}$$ L is read off charts designed for specific subgrade CBRs, relating S_d/a to L, for different h/a ratios; h is the thickness	Two-layer interface deflection criteria for dual wheels, with asphalt mix modulus of 100,000 psi and subgrade soil modulus equal to 1500 times CBR (in psi)

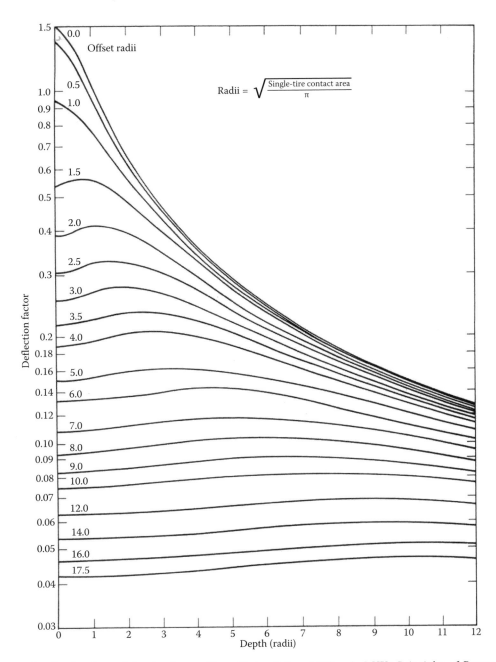

FIGURE 5.5 One-layer deflection factor. (From Yoder, E.J. and Witczak, M.W.: *Principles of Pavement Design*, 2nd edn. 1975. Copyright Wiley-VCH Verlag GmbH & Co. KGaA. Reproduced with permission.)

5.5.1 CONVERSION TO EQUIVALENT GEAR

The effect of a variety of aircrafts using an airport is considered by converting all aircrafts to the same landing gear type as the design aircraft, using conversion factors (same for flexible and rigid) that have been developed on the basis of relative fatigue effects. The conversion factors are as follows:

To Convert from	To	Multiply Departures by
Single wheel	Dual wheel	0.8
Single wheel	Dual tandem	0.5
Dual wheel	Dual tandem	0.6
Double dual tandem	Dual tandem	1.0
Dual tandem	Single wheel	2.0
Dual tandem	Dual wheel	1.7
Dual wheel	Single wheel	1.3
Double dual tandem	Dual wheel	1.7

5.5.2 CONVERSION TO EQUIVALENT ANNUAL DEPARTURE

The following formula is to be used:

$$\log R_1 = \log R_2 * \left(\frac{W_2}{W_1} \right)^{1/2}$$

where

R_1 is the equivalent annual departures by the design aircraft
R_2 is the annual departures expressed in design aircraft landing gear
W_1 is the wheel load of the design aircraft (95% of the gross weight is assumed to be carried by the main landing gear)
W_2 is the wheel load of the aircraft in question

Each wide-body aircraft is treated as a 300,000 lb dual tandem aircraft for the determination of equivalent annual departures. An example is shown next.

Assume an airport pavement is to be designed for the following forecast traffic.

Aircraft	Gear Type	Average Annual Departures	Maximum Take-Off Weight[a]	
			lb	kg
727-100	Dual	3760	160,000	(72,600)
727-200	Dual	9080	190,500	(86,500)
707-320B	Dual tandem	3050	327,000	(148,500)
DC-g-30	Dual	5800	108,000	(49,000)
cv-880	Dual tandem	400	184,500	(83,948)
737-200	Dual	2650	115,500	(52,440)
L-101 1-100	Dual tandem	1710	450,000	(204,120)
747-100	Double dual	85	700,000	(317,800)

Source: Federal Aviation Administration, FAA, Advisory Circular AC No. 150/5320-6D, 1995, 25; see also Federal Aviation Administration, FAA Advisory Circular AC 150/5300-13, 1989.

[a] Available from Federal Aviation Administration (1989).

Step 1: Determine design aircraft: Using charts provided in Section 2, Flexible Pavement Design, in FAA Circular 150-5320/6D, Federal Aviation Administration (1995), determine the thickness required for each aircraft. In this example, the thickest pavement is required for the 727-200 aircraft, which is therefore the design aircraft.

Step 2: Convert all aircrafts to landing gear of design aircraft.

Aircraft	Gear Type	Average Annual Departures	Multiplier	Equivalent Dual-Gear Departures
727-100	Dual	3760	—	3760
727-200	Dual	9080	—	9080
707-320B	Dual tandem	3050	1.7	5185
DC-g-30	Dual	5800	—	5800
cv-880	Dual tandem	400	1.7	680
737-200	Dual	2650	—	2650
L-101 1-100	Dual tandem	1710	1.7	2907
747-100	Double dual	85	1.7	145

Step 3: Convert equivalent dual-gear departures to equivalent annual departures for the design aircraft.

Aircraft	Equivalent Dual-Gear Departures	Maximum Take-Off Weight (lb)	Wheel Load of Design Aircraft (190,500 * 0.95 * 0.5 * 0.5)	Equivalent Annual Departure of Design Aircraft
727–100	3760	160,000	45,240	1,891 ($LOG R_1 = LOG R_2 * (W_2/W_1)^{0.5}$) $R_2 = 3,760$ $W_2 = 38,000$ $W_1 = 45,240$ $R_1 = 1891$
727–200	9080	190,500		9080
707–320B	5185	327,000		2763
DC-g-30	5800	108,000		682
cv-880	680	184,500		94
737–200	2650	115,500		463
L-101 1–100	2907	450,000		1184
747–100	145	700,000		83
Total design departures considering a 190,500 lb dual-wheel aircraft				16,241

5.6   TRUCK TIRE PRESSURE

The contact pressure at the interface of the tire and the pavement is important for determination of the structural response of the pavement. Quite often, the tire inflation pressure is assumed to be equal to the contact pressure. Either the tire inflation pressure is used, or, using an assumed elliptical or more commonly used circular contact area, the contact pressure is calculated from the load. The hot inflation pressure could be 10%–15% more than the cold pressure, and can range from 90 to 120 psi, with loads ranging from 3600 to 6600 lb. Assumed tire contact areas are shown in Figure 5.6. The tire inflation pressure controls the tire contact stress at the center of the tire, whereas the load actually dictates the contact stress at the tire edges. An increasing trend is the replacement of duals with super-single tires, which have significantly higher pressures and hence more damaging effects on the pavement.

Tire contact pressure could be nonuniform across the contact area, depending on the condition of the tire. If the tire is underinflated, the pressure will be higher at the edges, and if it is overinflated, then the pressure is higher at the center. The existence of high-shear stresses and/or tensile stresses due to anomalies in axle/loading/tire pressure can lead to top-down cracking in asphalt mix surface layers. Currently, this is not considered in the structural design of highway pavements.

FIGURE 5.6   Assumed tire contact areas.

5.7   TRUCK SPEED

For asphalt mix layers, the effect of loading time, which is dictated by the speed of the vehicle, is similar to the effect of temperature. In general, shorter loading times result in more elastic response and lower strains, whereas longer loading times result in more viscous response and larger strains. For structural design, the dynamic modulus can be determined at appropriate loading times to simulate the response under trucks moving at different speeds. For the same speed, any point at a greater depth from the surface will have a higher loading time compared to a point nearer to the surface. For an asphalt mix layer, one option is to assume that at the middle of the asphalt layer, the load is uniformly distributed over a circular area with a radius, r, where

$$r = a + \frac{h}{2}$$

in which a is the radius of the tire contact area and h is the thickness of the asphalt layer.

The loading time is expressed as t, where

$$t = \frac{2a + h}{s}$$

in which s is the speed of the truck.

The other option is to consider that the moving load varies as a haversine function with time. The stress at any point can be considered to be zero or negligible when the load is at a distance greater than 6a from that point, where a is the radius of the tire contact area. The loading time can be approximately denoted as t:

$$t = \frac{12a}{s}$$

where s is the speed of the truck. The concept is illustrated in Figure 5.7.

5.7.1   EFFECT OF LOAD AND TIRE PRESSURE

There have been numerous accelerated loading/testing studies related to the effect of tire pressure and loading/axle configuration on response and distress of pavements. It has been observed that the tensile strain at the bottom of asphalt layers and the compressive strain at base and subgrade are affected more by load, whereas the compressive stress within the asphalt layer is affected primarily by the tire pressure. The vertical stress in the center contact zone is primarily dictated by the tire pressure, whereas the edge zone stress level depends mainly on the wheel load. Rutting (primarily in the aggregate base) and fatigue cracking increase by up to 2.4 and 4 times, respectively, due to the

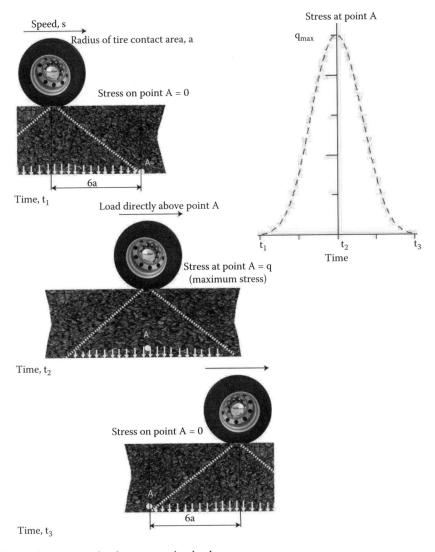

FIGURE 5.7 Stress at a point due to a moving load.

use of wide-based single tires instead of dual tires. For rutting in HMA, the increase has been found to be proportional to the increase in tire pressure. Transverse tensile contact stress has been found to develop across tread ribs, leading to a reduction in shear strength of the pavement material. Traffic speed affects the transient deflection (under wheel load) significantly—the deflection is reduced drastically from static (0 km/h) to a speed of 20 km/h, and the change is much less beyond that speed.

5.8 AIRCRAFT LOADING, GEAR CONFIGURATION, AND TIRE PRESSURE

As the amount of load that an aircraft can carry has increased over the years, so have the number of gears. At the same time, realizing that a higher number of gears means heavier weight and more fuel spent in carrying the aircraft itself, aircraft companies have also tried to reduce the number of gears, with a resulting increase in tire pressure and hence contact pressure on the pavement surface. It is expected that in future both range and payload of aircrafts will go up, resulting in increased loads on airport pavements. Figure 5.8 shows the load, gear configuration, and tire pressure of a few commonly used civilian aircrafts. The reader is advised to view the

Aircraft	Gross load on main landing gear group, lb	Gear layout	Dimensions, in	Tire pressure, psi
Airbus, A-300-B2	304,000		$x = 35$ $y = 55$	168
Airbus, A-330	460,000		$x = 55$ $y = 78$	200
Airbus, A380	942,700		$a = 53$ $b = 67$ $x_1 = 60$ $x_2 = 61$ $y = 67$	194
Boeing, B-737-100	100,000		$x = 55$	148
Boeing, B-747-200	833,000		$x = 44$ $y = 58$	200
Boeing, B-777-200B	634,500		$x = 55$ $y = 57$	215
Boeing, B-787-8	478,325		$x = 51$ $y = 58$	228

FIGURE 5.8 Gear layout, load, and tire pressure of a few commonly used civilian aircrafts.

aircraft characteristics of different aircrafts that are needed for airport planning at respective company websites, such as at http://www.boeing.com/commercial/airports and http://www.airbus.com/support/maintenance-engineering/technical-data/aircraft-characteristics/

Standard naming convention for aircraft landing gear configurations and pictures of different types of aircraft gear layout (for a wide variety of aircrafts) are available at http://www.faa.gov/airports/resources/publications/orders/media/Construction_5300_7.pdf.

QUESTIONS

5.1 What is a weigh-in-motion (WIM) sensor, and how is it used?

5.2 Determine the load equivalency factor for the following cases, for both flexible and rigid pavements, for $p_t = 2.5$: pavement structural number (SN) = 5, D for rigid pavement = 10 in., axle loads = 2, 18, and 30 kips.

5.3 Determine the design traffic for a 15 year design period for the following case, for a four-lane new rigid pavement, with traffic growth of 2%. The truck load factors are provided for the different vehicles.

Vehicle Type	Current Traffic (A)	ESAL Factor (Truck Load Factor; D)
Passenger cars	5925	0.0008
Buses	35	0.6806
Panel and pickup trucks	1135	0.0122
Other two-axle/four-tire trucks	3	0.0052
Two-axle/six-tire trucks	372	0.1890
Three or more axle trucks	34	0.1303
All single-unit trucks		
Three-axle tractor semitrailers	19	0.8646
Four-axle tractor semitrailers	48	0.6560
Five-axle-plus tractor semitrailers	1880	2.3719
All tractor semitrailers		
Five-axle double trailers	103	2.3187
Six-axle-plus double trailers	0	
All double-trailer combos		
Three-axle truck-trailers	208	0.0152
Four-axle truck-trailers	305	0.0152
Five-axle-plus truck trailers	125	0.5317
All truck-trailer combos		
All vehicles	10,193	

5.4 Compute the load equivalency factor for loads of 9,000, 18,000, 36,000, and 45,000 lb by the approximate method.

5.5 Determine the ESWL for dual ties of a Boeing 737 (see Figure 5.8 for dimensions) using the U.S. Navy method.

5.6 Determine the time of loading for truck speeds of 10, 20, 40, 60, and 70 mph, and plot speed versus time of loading data. Consider the radius of the tire contact area (circular) to be 6 in.

6 Drainage

6.1 SOURCE AND EFFECT OF WATER

Precipitation, in the form of rain and snow, may result in accumulation of water on the surface, and find its way through joints and cracks inside a pavement structure. Water can also enter the subgrade soils through capillary action from a nearby water table. Water puddles or films on the surface, especially in ruts, can cause unsafe driving conditions through hydroplaning and skidding of vehicles, whereas water inside the pavement structure can result in deterioration of the mixtures and materials. If frost-susceptible soils are present, ice lenses can form due to freezing temperatures in the winter, causing heaves. And then thawing in the spring causes loss of support and consequent cracking on the surface. The sources and effects of water are summarized in Figure 6.1.

The damage due to the presence of water includes the following:

1. The aggregate/soil layers lose shear strength due to increase in moisture content or saturation, resulting in an increase in deformation when subjected to traffic loads.
2. Repeated stress generation due to repeated freeze–thaw in asphalt as well as soil layers.
3. In the presence of water, vehicle loads cause hydrostatic pressures in soil layers, which cause movement of soil particles, especially near joints and cracks, leading to displacement of soil and hence loss of support. The end result is the failure of the pavement by cracking and deformation.

Of these three ways, the first two are taken care of in the drainage design, while the third one is considered in the materials selection and mixture design process.

Drainage refers to the system of providing ways to remove water from a pavement. In general, a drainage system consists of two separate subsystems—*surface* and *subsurface* drainage systems to avoid water coming from rain and/or snow and to remove water coming in through surface voids and/or cracks and from groundwater, respectively.

6.2 ESTIMATING FLOW

It is necessary to understand some hydrologic concepts before starting the design of drainage structures. The amount of water that needs to be drained away comes as runoff over areas and is dictated by the characteristics of the rainfall and the surface over which the water flows. This is illustrated with the rational method formula, which is the most commonly used equation to determine the runoff in drainage areas less than 80 ha. The individual parameters in the formula are explained in the subsequent sections.

The rational method equation is as follows:

$$Q = 0.00278 \, CIA$$

where
Q is the discharge, m^3/s
C is the runoff coefficient
I is the rainfall intensity, mm/h
A is the drainage area, hectares (ha) [1 ha = 10,000 m^2]

Precipitation: water and snow

Water in ruts causes unsafe driving conditions

Water enters through joints and cracks

Asphalt mix loses strength due to moisture

Soil layers lose strength due to moisture

Capillary moisture

FIGURE 6.1   Sources and effects of water.

Runoff coefficient, C, is the ratio of runoff to rainfall. It is affected by many characteristics, the more important of which are the soil group, land use, and average land slope. If the drainage area consists of different types of surfaces, then a weighted C value is used as follows:

$$\text{Weighted C, } C_w = \frac{A_1C_1 + A_2C_2 + A_3C_3 + \cdots}{A_1 + A_2 + A_3 + \cdots}$$

where

A_1, A_2, and A_3 are the subareas with different types of surfaces
C_1, C_2, and C_3 are the corresponding runoff coefficients

Typical values of runoff coefficients are shown in Figure 6.2.

Since the selected rainfall intensity depends on them, before discussing rainfall intensity, it is appropriate to discuss two other parameters, the *time of concentration* (T_c) and the *return period*. T_c is the time required for water to travel from the most hydraulically distant point of the drainage area to the drain system. It is used in determination of inlet spacing as well as pipe sizing. For inlet spacing, T_c is the time required for the water to flow from the hydraulically most distant point of the drainage area to the inlet (inlet time). This is the sum of the time required for water to flow across the pavement of overland back of the curb to the gutter plus the time required for flow to move through the length of the gutter to the inlet. If the total time of concentration to the upstream inlet is less than 7 min, then a minimum T_c of 7 min should be used. For pipe sizing, the T_c consists of the inlet time plus the time required for the water to flow through the storm drain to the point under consideration.

The time to flow overland can be determined from the chart shown in Figure 6.3 or by estimating the velocity from Figure 6.4, and dividing the length of flow by the velocity, whereas the time to flow within the storm drain can be determined simply by dividing the length of the pipe for runoff travel divided by the estimated normal velocity of water flow, which can be determined from Manning's equation, presented later in this chapter.

If there is more than one source of runoff to a given point in a drain system, the longest T_c is used. For municipal areas, a minimum T_c of 5 min is recommended.

6.2.1   RETURN PERIOD

It is not possible to design drainage structures for a maximum runoff that it could produce—say, in an interval of a long period of time. Therefore, for practical purposes, a realistic discharge, which could result from the biggest rainstorm that is *probable* (as opposed to possible) within the design life of

Recommended Coefficient of Runoff for Pervious Surfaces by Selected Hydrologic Soil Groupings and Slope Ranges

	Selected Hydrologic Soil Groupings and Slope Ranges			
Slope	A	B	C	D
Flat (0–1%)	0.04–0.09	0.07–0.12	0.11–0.16	0.15–0.20
Average (2–6%)	0.09–0.14	0.12–0.17	0.16–0.21	0.20–0.25
Steep (Over 6%)	0.13–0.18	0.18–0.24	0.23–0.31	0.28–0.38

Source: *Storm Drainage Design Manual,* Erie and Niagara Counties Regional Planning Board.

Recommended Coefficient of Runoff Values for Various Selected Land Uses

Description of Area		Runoff Coefficients
Business:	Downtown areas	0.70–0.95
Neighborhood areas		0.50–0.70
Residential:	Single-family areas	0.30–0.50
	Multi units, detached	0.40–0.60
	Multi units, attached	0.60–0.75
	Suburban	0.25–0.40
Residential (0.5 ha lots or more)		0.30–0.45
Apartment dwelling areas		0.50–0.70
Industrial:	Light areas	0.50–0.80
	Heavy areas	0.60–0.90
Parks, cemeteries		0.10–0.25
Playgrounds		0.20–0.40
Railroad yard areas		0.20–0.40
Unimproved areas		0.10–0.30

Source: Hydrology, Federal Highway Administration, HEC No. 19, 1984

Coefficients for Composite Runoff Analysis

Surface		Runoff Coefficients
Street:	Asphalt	0.70–0.95
	Concrete	0.80–0.95
Drives and walks		0.75–0.85
Roofs		0.75–0.95

Source: Hydrology, Federal Highway Administration, HEC No. 19, 1984

Characteristics	Soil Group			
	A	B	C	D
Runoff Potential	Low	Moderately Low	Moderately High	High
Primary type of soil	Deep well drained sands and gravel	Moderately deep to deep, moderately well to well drained soils with moderately fine to moderately coarse texture	Layer exists near the surface that impedes downward movement of water/soils with moderately fine to fine texture	Clays with high swelling potential, permanently high water tables, with claypan or clay layer at or near surface, shallow soils over nearly impervious parent material

FIGURE 6.2 Runoff coefficients. (From American Association of State Highway and Transportation Officials, *MDM-SI-2, Model Drainage Manual,* 2000 Metric Edition, AASHTO, Washington, DC, © 2000. Used with permission.)

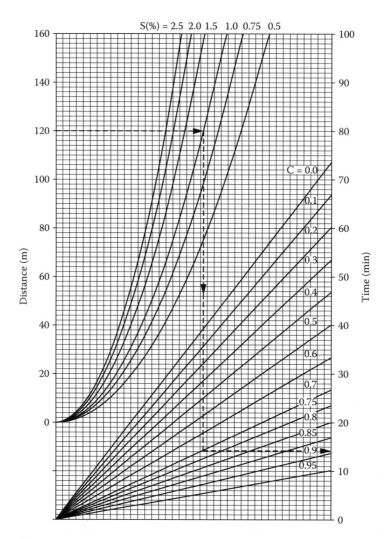

FIGURE 6.3 Direct estimation of overland time of flow. (From American Association of State Highway and Transportation Officials, *MDM-SI-2, Model Drainage Manual*, 2000 Metric Edition, AASHTO, Washington, DC, © 2000. Used with permission.)

the drainage structure, is considered. Of course, the probability and hence the size of the biggest storm/flood would be different for different types of pavements—a lower probability and a bigger storm/flood would be considered for a more important highway than a less important one. For design purposes, all of these considerations are made through two parameters, design frequency and return period (RP) or recurrence interval (RI). The frequency with which a given flood can be expected to occur is the reciprocal of the probability that the flood will be equaled or exceeded in a given year. For example, if a flood has a probability of 2% of being equaled or exceeded each year, over a long period of time, then the frequency is $1/(2/100) = 50$, which means that the flood will be equaled or exceeded on an average of once in every 50 years. The RI or RP is also 50 years. Looking at this in another way, the probability of being exceeded $= 100/RP$. Suggested values of RP are provided in Table 6.1.

6.2.2 RAINFALL INTENSITY

Rainfall intensity, I, is the intensity of rainfall in mm/h for a duration equal to the time of concentration. Intensity multiplied by the duration gives the amount of rain for that duration. The rainfall

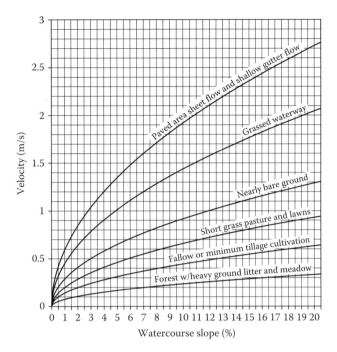

FIGURE 6.4 Estimation of velocity of flow to compute overland time of flow. (From American Association of State Highway and Transportation Officials, *MDM-SI-2, Model Drainage Manual*, 2000 Metric Edition, AASHTO, Washington, DC, © 2000. Used with permission.)

TABLE 6.1
Suggested Values of Return Period

Roadway Classification	Exceedance Probability (%)	Return Period (Years)
Rural principal arterial system	2	50
Rural minor arterial system	4–2	25–50
Rural collector system, major	4	25
Rural collector system, minor	10	10
Rural local road system	20–10	5–10
Urban principal arterial system	4–2	25–50
Urban minor arterial street system	4	25
Urban collector street system	10	10
Urban local street system	20–10	5–10

Source: American Association of State Highway and Transportation Officials, *MDM-SI-2, Model Drainage Manual*, 2000 Metric Edition, AASHTO, Washington, DC, © 2000. Used with permission.

Note: Federal law requires interstate highways to be provided with protection from the 2% flood event, and facilities such as underpasses, depressed roadways, and the like, where no overflow relief is available, should be designed for the 2% event.

FIGURE 6.5 Example of rainfall intensity curve. (From American Association of State Highway and Transportation Officials, *MDM-SI-2, Model Drainage Manual*, 2000 Metric Edition, AASHTO, Washington, DC, © 2000. Used with permission.)

intensity–duration–frequency plots for any weather station could be generated from the weather database at the NOAA Hydrometeorological Design Studies Center, Precipitation Frequency Data Server (http://www.nws.noaa.gov/oh/hdsc/). An example is shown in Figure 6.5.

6.3   HYDROPLANING AND SURFACE DRAINAGE SYSTEM

If the water from rainfall exceeds the depth that is required for hydroplaning for a specific pavement width and rainfall intensity, hydroplaning can be expected. In a broad sense, the potential of hydroplaning depends on vehicle speed, tire condition (pressure and tire tread), pavement micro- and macrotexture, cross-slope and grade, and pavement conditions such as rutting, depression, and roughness. Apart from the driver's responsibility to control speed, the following steps could be taken to reduce the potential of hydroplaning and/or prevent accidents:

1. Maximize transverse slope and roughness and use porous mixes (such as open graded friction course).
2. Provide gutter inlet spacing at sufficiently close spacing to minimize spread of water, and maximize interception of gutter flow above superelevation transitions.
3. Provide adequate slopes to reduce pond duration and depth in sag areas.
4. Limit depth and duration of overtopping flow.
5. Provide warning signs in sections identified as problem areas.

The basic principle of surface drainage is to design the pavement surface and the adjacent areas in such a way as to facilitate the quick flow of water falling in the form of rain and/or snow on the surface to the sides, and then drain it away to a nearest point of collection. This involves two major components—providing the adequate slope to the pavement surface and drainage channels with sufficient flow capacity along the pavement on either side.

The minimum grade for a gutter is 0.2% for curbed pavements. For sag vertical curves, a minimum of a 0.3% slope should be maintained within 15 m of the level point in the curve. In a very flat terrain, the slopes can be maintained by a rolling profile of the pavement. For one lane, the minimum pavement cross-slope is 0.015 m/m, with an increase of 0.005 m/m for additional lanes. Slopes of up to 2% can be maintained without causing driver discomfort, whereas it may need to be increased above 2% for areas with intense rainfall.

In a multilane highway, if there are three or more lanes inclined at the same direction, it is recommended that each successive pair of lanes from the first two lanes from the crown line have an increased slope compared to the first two lanes (by about 0.5%–1%), with a maximum pavement cross-slope of 4%. Note that the allowable water depths on inside lanes are lower because of high-speed traffic on those lanes, and hence sloping of inside lanes toward the median should be done with caution. Pavements also have longitudinal slopes—with a minimum of 0.5% and a maximum of 3%–6%, depending on the topography of the region.

A high level of macrotexture on the pavement is desirable for allowing the rainwater to escape from the tire–pavement interface and reduce the potential of hydroplaning. Tining of PCC pavement, while it is still in plastic state, could be done to achieve this. For existing concrete pavements, macrotexture can be improved by grooving and milling. Such grooving in both transverse and longitudinal directions is very effective in enhancing drainage. Use of porous asphalt mix layers such as open graded friction course allows rapid drainage of water. It is important that such layers are daylighted at the sides, and are constructed over layers that have been compacted adequately to prevent the ingress of water downward.

The coefficient of permeability of the surface mixture is dependent upon many factors, the more important of which are aggregate gradation and density. If the gradation consists of a relatively higher amount of coarse aggregate particles (compared to fine and filler materials), then the permeability is relatively high. On the other hand, permeability decreases with an increase in density.

Generally, the gradation is decided upon from other considerations, and the only available way for the pavement engineer to lower the permeability of the surface layer is through compaction and hence providing adequate density. Note that because of the effect of the gradation, the desirable density will be different for mixtures with different gradations. Also note that no matter how dense the surface layer is, some water will find its way through it, and hence there must be a way to get rid of this water—through the use of the subsurface drainage system (apart from the fact that the materials must be resistant to the action of water to a certain extent). The permeability of new rigid and asphalt mix pavement can be assumed to be 0.2 and 0.5 in./h, respectively.

There can be different types of drainage channels such as ditches, gutters, and culverts. The design of these channels means the design of the cross section. This design is accomplished with the help of Manning's formula (Daugherty and Ingersoll, 1954, equation):

$$Q = \frac{K}{n} S^{1/2} R^{2/3} A$$

where

Q is the pipe flow capacity, m^3/s

S is the slope of the pipe invert, m/m

n is the pipe coefficient of roughness (0.012 for smooth pipe, and 0.024 for corrugated pipe; FHWA, 1992)

A is the pipe cross-sectional area, m^2

$K = 1$

$R = A/P = D/4$

P is the wetted perimeter of pipe, m

D is the pipe diameter, m

The practical considerations for design include the elevation with respect to the subgrade (same or lower level than the subgrade), low construction and maintenance costs, and safeguard against slope failure (slope of 2:1 or less).

Curbs are concrete or asphalt mix structures provided along the side of the low-speed urban highways as well as bridge decks to facilitate collection of drained surface water from the pavement surface and protect pavement sides from erosion. Curbs are generally placed with gutters. In rural areas, roadside and median channels are provided instead of curbs and gutters.

For pavement drainage systems, two parameters are selected (Table 6.2), depending on the type of the pavement, design frequency, and spread (accumulated flow in and next to the roadway gutter, which can cause interruption to traffic flow, splash, or hydroplaning problems). The spread (T) is constant for a specific design frequency—for higher magnitude storms, the spread can be allowed to utilize most of the pavement as an open channel.

The nomograph shown in Figure 6.6 can be used to design gutter sections. An example is shown on the plot. Manning's coefficients are provided in Table 6.3.

TABLE 6.2

Parameters for Drainage Systems

Road Classification		Design Frequency (Years)	Design Spread
High volume	<72 km/h	10	Shoulder + 0.9 m
	>72 km/h	10	Shoulder
	Sag point	50	Shoulder + 0.9 m
Collector	<72 km/h	10	1/2 driving lane
	>72 km/h	10	Shoulder
	Sag point	10	1/2 driving lane
Local streets	Low ADT	5	1/2 driving lane
	High ADT	10	1/2 driving lane
	Sag point	10	1/2 driving lane

Source: American Association of State and Highway Transportation Officials (AASHTO), *MDM-SI-2, Model Drainage Manual*, 2000 Metric Edition, AASHTO, Washington, DC, 2000.

Note: These criteria apply to shoulder widths of 1.8 m or greater. Where shoulder widths are less than 1.8 m, a minimum design spread of 1.8 m should be considered.

FIGURE 6.6 Nomograph for designing gutter section. (Note: Chart 4 mentioned in step 2 refers to Figure 6.7.) (From American Association of State Highway and Transportation Officials, *MDM-SI-2, Model Drainage Manual*, 2000 Metric Edition, AASHTO, Washington, DC, © 2000. Used with permission.)

The next nomograph, Figure 6.7 (ratio of frontal to gutter flow), can be used to calculate the frontal flow for grate inlets and flow in a composite gutter section with width (W) less than the total spread (T), which can also be determined from Figure 6.8.

Example 6.1

Find spread, given flow

Step 1. Given: $S=0.01$, $S_x=0.02$, $S_w=0.06$, $W=0.6$ m, $n=0.016$, and $Q=0.057$ m³/s trial value of $Q_s=0.020$ m³/s.

Step 2. Calculate gutter flow: $Q_w=Q-Q_s=0.057-0.020=0.037$.

Step 3. Calculate ratios of $(E_o=)$ $Q_w/Q=0.037/0.057=0.65$, and $S_w/S_x=0.06/0.02=3$. And, using Figure 6.7, find an appropriate value of W/T; W/T=0.27.

TABLE 6.3
Manning's Coefficients

Type of Gutter or Pavement	Manning's n
Concrete gutter, troweled finish	0.012
Asphalt pavement	
Smooth texture	0.013
Rough texture	0.016
Concrete gutter-asphalt pavement	
Smooth	0.013
Rough	0.015
Concrete pavement	
Float finish	0.014
Broom finish	0.016

Source: Federal Highway Administration (FHWA), *Design Charts for Open-Channel Flow*, Hydraulic design series No. 3 (HDS 3), U.S. Government Printing Office, Washington, DC, 1961; American Association of State and Highway Transportation Officials (AASHTO), *MDM-SI-2, Model Drainage Manual*, 2000 Metric Edition, AASHTO, Washington, DC, 2000.

Note: For gutters with a small slope, where sediment may accumulate, increase above n values by 0.002.

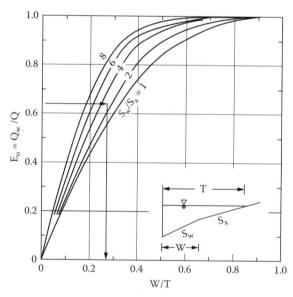

FIGURE 6.7 Ratio of frontal flow to gutter flow. (From American Association of State Highway and Transportation Officials, *MDM-SI-2, Model Drainage Manual*, 2000 Metric Edition, AASHTO, Washington, DC, © 2000. Used with permission.)

Step 4. Calculate T: $T = W/(W/T) = 0.6/0.27 = 2.22$ m.
Step 5. Find spread above the depressed section: $T_s = 2.22 - 0.6 = 1.62$ m.
Step 6. Use Figure 6.6 to determine Q_s: $Q_s = 0.014$ m³/s.
Step 7. Compare Q_s (0.014) from Step 6 to the assumed value of Q_s (0.020); since they are not close, try another value of Q_s (e.g., 0.023) and repeat until the calculated and assumed values are close.

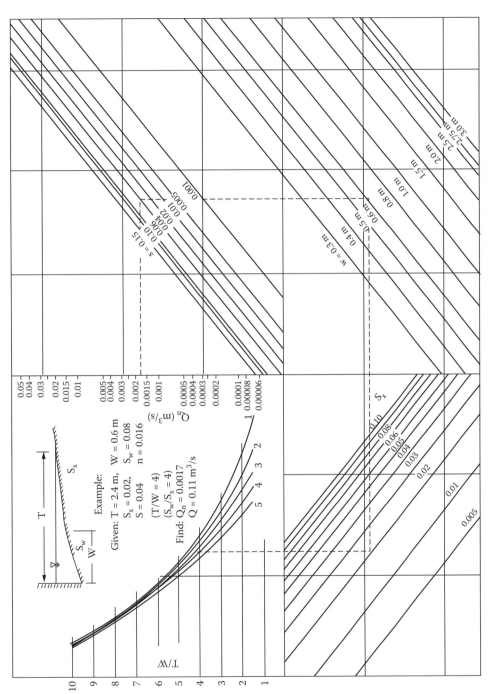

FIGURE 6.8 Flow in composite gutter section. (From American Association of State Highway and Transportation Officials, *MDM-SI-2, Model Drainage Manual,* 2000 Metric Edition, AASHTO, Washington, DC, © 2000. Used with permission.)

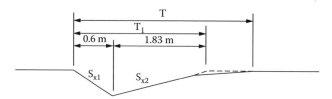

FIGURE 6.9 V-type gutter. (From American Association of State Highway and Transportation Officials, *MDM-SI-2, Model Drainage Manual*, 2000 Metric Edition, AASHTO, Washington, DC, © 2000. Used with permission.)

The use of Figure 6.8 is illustrated with an example on the figure.

Figure 6.6 can be used to solve the flow for a V-gutter section (utilizing Figure 6.9), such as those in triangular channel sections adjacent to concrete median barriers, using the following steps:

1. Determine S, S_x, n, and Q.
2. Calculate $S_x = (S_{x1} + S_{x2})/(S_{x1} + S_{x2})$.
3. Solve for Q (flow) using Figure 6.6.

6.4   INLETS

Inlets are provided at regular intervals to collect surface water and convey them to the storm drains. The inlets could be of the grate or curb-opening type or a combination of both. Grate inlets are suitable for continuous grades and should be made bike safe where bike traffic is expected. Curb openings are more suitable for sag points, since they can let in large quantities of water, and could be used where grates are hazardous for bikes or pedestrians. Inlets are spaced at regular intervals, as explained later, and also at points such as sag points in the gutter grade, upstream of median breaks, entrance and exit ramps, crosswalks and intersections, immediately upstream and downstream of bridges, side streets at intersections, the end of channels in cut sections, behind curbs, shoulders or sidewalks to drain low areas, and where necessary to collect snowmelt, and *not* in areas where pedestrian traffic is expected.

The inlet spacing should be calculated on the basis of collection of runoff. Inlets should be first located from the crest working downgrade to the sag point, by first calculating the distance of the first inlet from the crest and then computing the distances of the other successive inlets. The distance of the first inlet from the crest is calculated as follows:

$$L = \frac{10,000\,Q_t}{0.0028\,CiW}$$

where
    L is the distance of the first inlet from the crest, m
    Q_t is the maximum allowable flow, m³/s
    C is the composite runoff coefficient for contributing drainage area
    W is the width of contributing drainage area, m
    i is the rainfall intensity for design frequency, mm/h

6.5   SUBSURFACE DRAINAGE SYSTEM

Subsurface drainage involves providing a subsurface system to remove subsurface water. This water can be from any of the following:

1. Water coming from nearby groundwater
2. Water infiltrating the surface layers through voids and/or cracks
3. Water seeping up from artesian aquifers
4. Water resulting from the thawing of ice lenses in soil layers

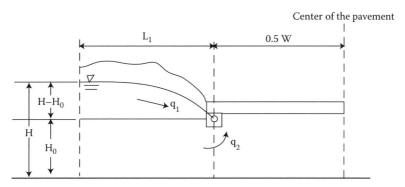

FIGURE 6.10   Flow from groundwater.

6.5.1   GROUNDWATER

Groundwater fluctuations can cause ingress of water, but generally the water table is low or lowered enough so as to avoid its effect. For cases where the groundwater table does need to be considered for seepage of water into the pavement, the following method can be used to estimate the inflow.

Refer to the general case illustrated in Figure 6.10.

Step 1. Determine the parameter, L_i, radius of influence: $L_i = 3.8(H - H_0)$.
Step 2. Estimate q_1, inflow from above the bottom of the drainage layer:

$$q_1 = \frac{k(H - H_0)}{2L_i}$$

where k is the permeability of the soil in the cut slope.
Step 3. Determine the parameter, $(L_i + 0.5\ W)/H_0$.
Step 4. Determine the parameter, W/H_0.
Step 5. Use the chart in Figure 6.11 to estimate the parameter: $k(H - H_0)/2q_2$.
Step 6. Knowing k, permeability of the soil in the subgrade, H, and H_0, determine q_2, the seepage inflow from the bottom of the drainage layer.
Step 7. Determine the total lateral inflow into the drainage pipe: $q_d = q_1 + q_2$, for collector pipes on both sides, and $q_d = 2(q_1 + q_2)$, for a collector pipe on one side of the pavement only.
Step 8. Determine the groundwater inflow into the drainage layer per unit area: $q_g = 2q_2/W$, for pavement sloped on both sides and collector pipes on both sides, and $q_g = (q_1 + 2q_2)/W$, for pavement sloped on one side with a collector pipe on one side.

6.5.2   WATER ENTERING THROUGH CRACKS

Water enters a pavement system through cracks and joints in the pavement surface or between the pavement and the shoulder, as well as in the shoulder and side ditches. The common method to determine the infiltration rate in an uncracked pavement is the use of the following equation:

$$q_i = I_c \left[\frac{N_c}{W} + \frac{W_c}{WC_s} \right] + k_p$$

where
q_i is the rate of pavement infiltration, m³/day/m² (ft³/day/ft²)
I_c is the crack infiltration rate, m³/day/m (ft³/day/ft); suggested value, $(I_c) = 0.223$ m³/day/m, 2.4 ft³/day/ft of crack (Ridgeway, 1976)

FIGURE 6.11 Estimation of K. (From Garber, N.J. and Hoel, L.A., *Traffic and Highway Engineering*, 3rd edn., Thomson Learning, Clifton Park, NY. © 2002, Nelson Education Ltd., Reproduced by permission. www.cengage.com/permissions.)

N_c is the number of longitudinal cracks

W_c is the length of contributing transverse joints or cracks, m (ft)

W is the width of permeable base, m (ft)

C_s is the spacing of contributing transverse joints or cracks, m (ft)

k_p is the pavement permeability, m/day (ft/day); suggested permeability of uncracked specimens of asphalt pavement (AC) after being subjected to traffic and Portland Cement Concrete pavement (PCC) are on the order of 1×10^{-9} cm/s (15×10^{-5} ft/day) (Barber and Sawyer, 1952)

6.5.3 ARTESIAN AQUIFERS

For seepage of water from artesian aquifers, the following equation is used to estimate the inflow:

$$q_a = K \frac{\Delta H}{H_0}$$

where
    q_a is the inflow from artesian source, ft³/day/ft² of drainage area
    ΔH is the excess hydraulic head, ft
    H_0 is the thickness of the subgrade soil between the drainage layer and the artesian aquifer, ft
    K is the coefficient of permeability, ft/day

6.5.4 MELTING SNOW

Melting of snow and ice also contributes to water flow inside the pavement. Moulton's chart (Figure 6.12) can be used to estimate the inflow.

The required parameters are overburdened pressure that can be determined from the density of the materials in the layer above this layer, the permeability of the subgrade material, as well as the frost susceptibility of the soil. The average rate of heave can be determined experimentally or from Moulton's table (Table 6.4).

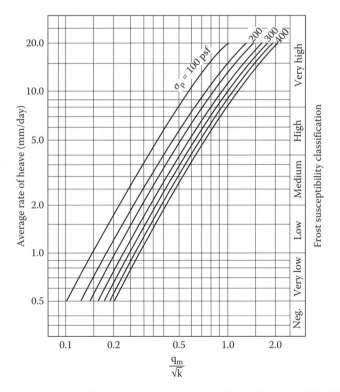

FIGURE 6.12 Chart for estimating inflow due to melting snow. (From Moulton, L.K., Highway subsurface design, FHWA-TS-80-224, U.S. Department of Transportation, Washington, DC, 1980.)

TABLE 6.4
Heave Rates for Various Soil Types

Unified Classification	Symbol	% <0.02 mm	Heave Rate (mm/day)	Frost Susceptibility
Gravels and sand gravels	GP	0.4	3.0	Medium
Gravels and sand gravels	GW	0.7–1.0	0.3–1.0	Low
Gravels and sand gravels	GW	1.0–1.5	1.0–3.5	Low to medium
Gravels and sand gravels	GW	1.5–4.0	3.5–2.0	Medium
Silty and sandy gravels	GP—GM	2.0–3.0	1.0–3.0	Low to medium
Silty and sandy gravels	GW—GM and GM	3.0–7.0	3.0–4.5	Medium to high
Clayey and silty gravels	GW—GC	4.2	2.5	Medium
Clayey and silty gravels	GM—GC	15.0	5.0	High
Clayey and silty gravels	GC	15.0–30.0	2.5–5.0	Medium to high
Sands and gravelly sands	SP	1.0–2.0	0.8	Very low
Silty and gravelly sands	SW	2.0	3.0	Medium
Silty and gravelly sands	SP—SM	1.5–2.0	0.2–1.5	Low
Silty and gravelly sands	SW— SM	2.0–5.0	1.5–6.0	Low to high
Silty and gravelly sands	SM	5.0–9.0	6.0–9.0	High
Clayey and silty sands	SM—SC and SC	9.5–35.0	5.0–7.0	High
Silts and organic silts	ML—OL	23.0–33.0	1.1–14.0	Low to high
Silts and organic silts	ML	33.0–15.0	14.0–25.0	Very high
Clayey silts	ML—CL	60.0–75.0	13.0	Very high
Gravelly and sandy clays	CL	38.0–65.0	7.0–10.0	High
Lean clays	CL	65.0	5.0	High
Lean clays	CL—OL	30.0–70.0	4.0	High
Fat clays	CH	60.0	0.8	Very low

Source: Moulton, L.K., Highway subsurface design, FHWA-TS-80-224, U.S. Department of Transportation, Washington, DC, 1980.

6.6   DESIGN OF SUBSURFACE DRAINAGE STRUCTURES

In general, the subsurface drainage system consists of a drainage layer in the subsurface part of the pavement and side drainage channels—the drainage layer of high-permeability material slopes away on both sides to intercept subsurface water and direct it sideways to drainage channels. It is important to provide filters in the drainage layer and the channels such that they remove water and water only, and do not let finer soil particles wash out with the water. The various components are shown in Figure 6.13.

For rigid pavements, there is a separator layer between the subgrade (to prevent migration of fines) and the permeable base over which the PCC layer is placed. In asphalt pavements, the permeable base layer is placed under adequate thickness of the asphalt mix layer.

The resultant slope of the permeable base is given by the following:

$$S_R = (S^2 + S_x^2)^{0.5}$$

where
   S_R is the resultant slope, ft/ft
   S is the longitudinal slope, ft/ft
   S_x is the cross-slope, ft/ft

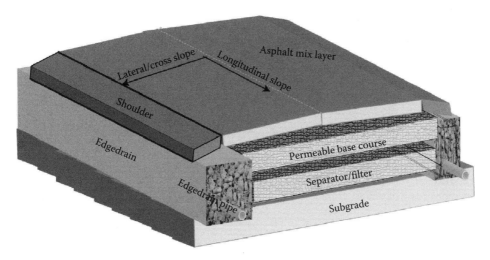

FIGURE 6.13 Various components of drainage system.

The resultant length of the flow path is the following:

$$L_R = W \left[1 + \left(\frac{S}{S_x} \right)^2 \right]^{0.5}$$

where
   L_R is the resultant length of flow path through permeable base, ft
   W is the width of permeable base, ft

The steps in designing a drainage system for pavements consist of determining the amount of water, designing a permeable base and the separation layer, determining the flow to edge drains and spacing of outflows, and checking the outlet flow.

The coefficient of permeability, k, is an important factor and is primarily dictated by effective grain size, D_{10}, porosity, n, and percentage passing the 0.075 mm sieve. The addition of stabilizer such as asphalt or Portland cement can counterbalance the effect of replacing the fine aggregate portion and hence the loss in stability.

Although k can be determined in the laboratory, quite commonly k is estimated from empirical equations such as the following:

$$k = C_k D_{10}^2$$

where
   k is the permeability, cm/s
   D_{10} is the effective grain size corresponding to size passing 10%
   C_k is the Hazen's coefficient, 0.8–1.2

Another equation proposed by Moulton (1980) is as follows:

$$k = \frac{(6.214 * 10^5) D_{10}^{1.478} n^{6.654}}{P_{200}^{0.597}}$$

where
   n is the porosity
   P_{200} is the percentage passing No. 200 sieve

The void ratio or porosity has a significant effect on k and the amount of water that can stay within the soil, and this is important since all of the water within a soil cannot be removed, since some of the water would remain as thin film. The porosity that is effective in determining how much water can be removed is called the *effective porosity*.

The total porosity, or porosity, is defined as follows:

$$n = \frac{V_v}{V_T}$$

where
   V_T is the total volume
   V_v is the volume of voids
   $V_v = V_T - V_S$
   V_S is the volume of solid

$$V_v = V_T - \frac{\gamma_d V_T}{\gamma_w G_S}$$

   G_S is the specific gravity of soil
   γ_d is the dry unit weight of soil
   γ_w is the unit weight of water

Effective porosity:

$$n_e = \frac{V_v - V_R}{V_T} = n - \frac{V_R}{V_T}$$

where V_R is the volume of the water retained in the soil.
   The volume of water retained in a soil:

$$V_R = \frac{\gamma_d w_c}{\gamma_w}$$

where w_c is the water content of the soil after draining.
   If $V_T = 1$,

$$n_e = n - \frac{\gamma_d w_c}{\gamma_w}$$

If a test is conducted to determine the volume of water draining under gravity from a known volume of material, then the effective porosity can be computed as follows:

$$n_e = \frac{V_e}{V_T}$$

where V_e is the volume of water draining under gravity.
   If W_L is the *water loss* percentage—that is, water drained from the sample—then the effective porosity can be expressed as follows (FHWA, 1992):

$$n_e = \frac{n W_L}{100}$$

Reported values of W_L expressed as a percentage are as follows:

	<2.5% Fines			5% Fines			>5% Fines		
	Filler	Silt	Clay	Filler	Silt	Clay	Filler	Silt	Clay
Gravel	70	60	40	60	40	20	40	30	10
Sand	57	50	35	50	35	15	25	18	8

Source: Federal Highway Administration (FHWA), Drainage pavement system participant notebook, FHWA-SA-92-008, Demonstration Project No. 87, U.S. Department of Transportation, Washington, DC, 1992.

6.6.1 DESIGN OF PERMEABLE BASE

The permeable base can be designed according to one of the two available methods. The Moulton (1979) method is based on the idea that the thickness of the base should be equal to or greater than the depth of the flow, which means that the steady flow capacity of the base should be equal to or greater than the rate of inflow. The design equations are as follows:

k is the permeability
S is the slope
L_R is the length of drainage
q_i is the rate of uniform inflow
H_1 is the depth of water at the upper end of the flow path

$$\text{Case 1}: \left(S^2 - \frac{4q_i}{k}\right) < 0$$

$$H_1 = \sqrt{\frac{q_i}{k}} L_R \left[\left(\frac{S}{\sqrt{\frac{4q_i}{k-S^2}}}\right)\left(\tan^{-1}\frac{S}{\sqrt{\frac{4q_i}{k-S^2}}} - \frac{\pi}{2}\right)\right]$$

$$\text{Case 2}: \left(S^2 - \frac{4q_i}{k}\right) = 0$$

$$H_1 = \sqrt{\frac{q_i}{k}} L_R^{-1}$$

$$\text{Case 3}: \left(S^2 - \frac{4q_i}{k}\right) > 0$$

$$H_1 = \sqrt{\frac{q_i}{k}} L_R \left[\frac{S - \sqrt{S^2 - \frac{4q_i}{k}}}{S + \sqrt{S^2 - \frac{4q_i}{k}}}\right]^{\frac{S}{2\sqrt{S^2 - \frac{4q_i}{k}}}}$$

The equations can be solved with the use of a chart (Figure 6.14). The chart can be used to determine the maximum depth of flow, or the required k of the material, when the other parameters are known.

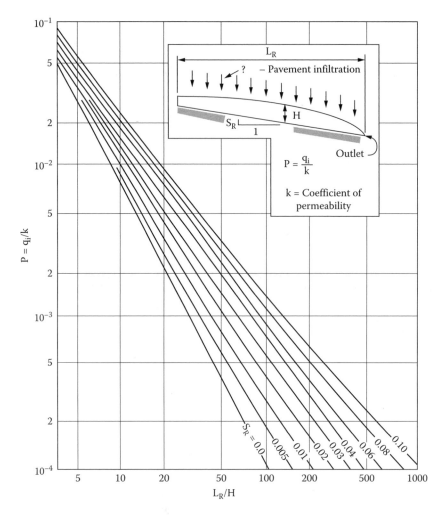

FIGURE 6.14 Nomograph for solving Moulton's (1980) equation. (From Moulton, L.K., Highway subsurface design, FHWA-TS-80-224, U.S. Department of Transportation, Washington, DC, 1980.)

The use of the chart is based on the assumption that H_1 equals H_{max}, which is the maximum depth of the flow.

The second approach is based on the concept of time to drain, specifically 50% drainage in 10 days, developed on the basis of a study with freeze–thaw-susceptible base courses by Casagrande and Shannon (1952).

For the following equations:

U is the percentage drainage (expressed as a fraction, e.g., 1% = 0.01)
S_1 is the slope factor = $H/(LS)$
H is the thickness of granular layer
L is the width of granular layer being drained
S is the slope of granular layer
T is the time factor = tkH/n_eL^2
t is the time for drainage, U, to be reached
k is the permeability of granular layer
n_e is the effective porosity of granular material

If U > 0.5,

$$T = \left(1.2 - \frac{0.4}{S_1^{1/3}}\right)\left[S_1 - S_1^2 \ln\left(\frac{S_1 + 1}{S_1}\right) + S_1 \ln\left(\frac{2S_1 - 2US_1 + 1}{(2 - 2U)(S_1 + 1)}\right)\right]$$

If U ≤ 0.5,

$$T = \left(1.2 - \frac{0.4}{S_1^{1/3}}\right)\left[2US_1 - S_1^2 \ln\left(\frac{S_1 + 2U}{S_1}\right)\right]$$

The equations can be solved by charts as shown in Figure 6.15.

Barber and Sawyer (1952) equations are as follows:

U is the percentage drainage (expressed as a fraction, e.g., 1% = 0.01)
S_1 is the slope factor = H/DS
H is the thickness of granular layer
D is the width of granular layer being drained
S is the slope of granular layer
T is the time factor = $(tkH)/(n_e L^2)$
t is the time for drainage, U, to be reached
k is the permeability of granular layer
n_e is the effective porosity of granular material

For 0.5 ≤ U ≤ 1.0,

$$T = 0.5S_1 - 0.48S_1^2 \log\left(1 + \frac{2.4}{S_1}\right) + 1.15S_1 \log\left[\frac{S_1 - US_1 + 1.2}{(1 - U)(S_1 + 2.4)}\right]$$

FIGURE 6.15 Nomograph for solving Casagrande and Shannon's (1952) equations. (From Casagrande, A. and Shannon, W.L., *Proc. Am. Soc. Civil Eng.*, 77, 792, 1952. With permission from ASCE.)

For $0 \leq U \leq 0.5$,

$$T = US - 0.48S_1^2 \log\left(1 + \frac{4.8U}{S_1}\right)$$

6.6.1.1 Materials for Permeable Base

The time to drain of a soil depends primarily on its coefficient of permeability, which is primarily dependent on the aggregate gradation. Gradations of typical materials (as suggested by FHWA, 1999) are given in Tables 6.5 through 6.7. Generally materials with permeability exceeding 1000 ft/day are used for highways.

6.6.2 Design of Separator or Filter Layer

The separator layer can be made up of aggregates or a geotextile layer. The functions of this layer include preventing pumping of fines from the subgrade to the permeable base, providing a stable

TABLE 6.5
Typical Unstabilized Permeable Base Gradations

State	2 in.	1 1/2 in.	1 in.	3/4 in.	1/2 in.	3/8 in.	No. 4	No. 8	No. 16	No. 40	No. 50	No. 200
AASHTO #57		100	95–100		25–60		0–10	0–5				
AAHSTO #67			100	90–100		20–55	0–10	0–5				
Iowa			100					10–35			0–15	0–6
Minnesota			100	65–100		35–70	20–45			2–10		0–3
New Jersey		100	95–100		60–80		40–55	5–25	0–12		0–5	
Pennsylvania	100			52–100		33–65	8–40		0–12			0–5

Source: Federal Highway Administration (FHWA), Pavement subsurface drainage design, FHWA-HI-99-028, NHI Course No. 131026, U.S. Department of Transportation, Washington, DC, 1999.

TABLE 6.6
Typical Asphalt-Stabilized Permeable Base Gradations

State	1 in.	1/2 in.	3/8 in.	No. 4	No. 8	No. 200
California	100	90–100	20–45	0–10		0–2
Florida	100	90–100	20–45	0–10	0–5	0–2
Illinois	90–100	84–100	40–60	0–12		
Kansas	100	90–100	20–45	0–10	0–5	0–2
Ohio	95–100			25–60	0–10	
Texas	100	95–100	20–45	0–15	0–5	2–4
Wisconsin	95–100	80–95	25–50	35–60	20–45	3–10
Wyoming	90–100	20–50		20–50	10–30	0–4

Source: Federal Highway Administration (FHWA), Pavement subsurface drainage design, FHWA-HI-99-028, NHI Course No. 131026, U.S. Department of Transportation, Washington, DC, 1999.

TABLE 6.7
Typical Cement-Stabilized Permeable Base Gradations

			Percent Passing Sieve Size				
State	1 1/2 in.	1 in.	3/4 in.	1/2 in.	3/8 in.	No. 4	No. 8
California	100	88–100	X ± 15		X ± 15	0–16	0–6
Virginia		100		25–60		0–10	0–5
Wisconsin		100	90–100		20–55	0–10	0–5

Source: Federal Highway Administration (FHWA), Pavement subsurface drainage design, FHWA-HI-99-028, NHI Course No. 131026, U.S. Department of Transportation, Washington, DC, 1999.

Note: X is the percentage submitted by the contractor.

platform to facilitate the construction of the overlying layers, directing water infiltrating from above to the side drains or edge drains and preventing it from entering the subgrade, and distributing the loads over the subgrade without deflecting excessively. Separator layers with aggregates range in thickness from 4 to 12 in. and can provide the stable construction platform as well as distribute loads over the subgrade without deflecting. Generally, geotextiles are used over stabilized subgrades. The separator layer should be such that subgrade fines do not move up to the separator layer and the fines from the separator layer do not move into the permeable base. The following requirements must be met:

D_{15} (separator layer) $\leq 5\ D_{85}$ (subgrade)
D_{50} (separator layer) $\leq 25\ D_{50}$ (subgrade)
D_{15} (base) $\leq 5\ D_{85}$ (separator layer)
D_{50} (base) $\leq 25\ D_{50}$ (separator layer)

The requirements of the aggregate separator layer are as follows:

1. Should have durable, crushed, angular aggregate
2. Maximum Los Angeles abrasion loss of 50%
3. Maximum soundness of 12% or 18% loss as determined by the sodium sulfate or magnesium sulfate tests (see Chapter 8)
4. Density should be at least 95% of the maximum density
5. Maximum percentage of materials passing the No. 200 sieve is 12%
6. Coefficient of uniformity is between 20 and 40

Examples of typical gradations used by state highway agencies are provided in Table 6.8.

6.6.2.1 Geotextile Separator Layer

The requirements for geotextile separators have been set as follows (FHWA, 1998):

$$AOS\ or\ O_{95\ (geotextile)} \leq B\ D_{85\ (soil)}$$

where
AOS is the apparent opening size, mm
O_{95} is the opening size in the geotextile for which 95% are smaller, mm
AOS $\approx O_{95}$
B is the dimensionless coefficient
D_{85} is the soil particle size for which 85% are smaller, mm

TABLE 6.8

**Typical Gradation Requirements
for Separator Layer**

Sieve Size	% Passing
1½ in.	100
¾ in.	95–100
No. 4	50–80
No. 40	20–35
No. 200	5–12

Source: Federal Highway Administration (FHWA), Drainage
requirements in pavements (DRIP), Developed by
Applied Research Associates for the Federal Highway
Administration, Contract No. DTFH61-00-F-00199,
U.S. Department of Transportation, Washington, DC,
2001.

For sands, gravelly sands, silty sands, and clayey sands (less than 50% passing 0.075 mm), B is
a function of C_u.

For

$$C_u \leq 2 \text{ or } \geq 8, \quad B = 1$$
$$2 \leq C_u \leq 4, \qquad B = 0.5 C_u$$
$$4 < C_u < 8, \qquad B = \frac{8}{C_u}$$

where

$$C_u = \frac{D_{60}}{D_{10}}$$

For silts and clays, B is a function of the type of geotextile:

For woven geotextiles, $B = 1$; $O_{95} \leq D_{85}$
For nonwoven geotextiles, $B = 1.8$; $O_{95} \leq 1.8 D_{85}$
And for both, AOS or $O_{95} \leq 0.3$ mm

6.6.3 DESIGN OF EDGE DRAINS

Guidelines from FHWA (1992) can be used for the design of edge drains. Generally, the edge drains
are designed to handle the peak flow coming from the permeable base:

$$Q = Q_p * L_o = (k S_x H) L_o$$

where

Q is the pipe flow capacity, m^3/day (ft^3/day)
Q_p is the design pavement discharge rate, m^3/day/m (ft^3/day/ft)
L_o is the outlet spacing, m (ft)
k is the permeability of granular layer, m^3/day (ft^3/day)
S_x is the transverse slope, m/m (ft/ft)
H is the thickness of granular layer, m (ft)

For design based on the pavement infiltration flow rate, the design flow capacity of the edge drain is as follows:

$$Q = Q_p * L_o = (q_i W) L_o$$

where

q_i is the pavement infiltration, m³/day/m² (ft³/day/ft²)
W is the width of the granular layer, m (ft)

For design based on average flow to drain the permeable base, the design flow capacity is as follows:

$$Q = Q_p * L_o = \left(WHn_e U \left(\frac{24}{t} \right) \right) L_o$$

where

n_e is the effective porosity of granular material
U is the percentage drainage (as 1% = 0.01)
t is the time for drainage, U, to be reached, days

The pipe for the edge drain is designed according to the Manning equation (Daugherty and Ingersoll, 1954):

$$Q = \frac{K}{n} S^{1/2} R^{2/3} A$$

where

Q is the pipe flow capacity, m³/s
S is the slope of the pipe invert, m/m
n is the pipe coefficient of roughness (0.012 for smooth pipe, and 0.024 for corrugated pipe; FHWA, 1992)
A is the pipe cross-sectional area, m²
K = 1
R = A/P = D/4, m
P is the wetted perimeter of pipe, m
D is the pipe diameter, m

$$Q = \frac{K}{n} S^{1/2} \left(\frac{D}{4} \right)^{2/3} \pi \left(\frac{D}{2} \right)^2$$

where

Q is in m³/s
D is in m
K is in (m^{1/3})/s

In English units:

$$Q = \frac{53.01}{n} S^{1/2} D^{8/3}$$

where

Q is in ft³/day
D is in inches (3–4 in., generally)

The equations are generally used for determining the spacing of the pipes, by fixing a specific type of pipe and with a specific diameter. This is done by setting the pipe capacity equal to the discharge from the unit length of the pavement times the distance between the pipe outlets (spacing).

For infiltration flow:

$$L_o = \frac{Q}{(q_i W)}$$

For peak flow:

$$L_o = \frac{Q}{kS_x H}$$

For average flow:

$$L_o = \frac{(Q t)}{(24 W H n_e U)}$$

The spacing is generally used as 250–300 ft. Pipe diameters and spacing are usually determined also in consideration of maintenance requirements.

A filter layer in the form of geotextiles (or, less commonly, aggregates of different gradations) could be used wrapped around the pipe or as an envelope to the edge drains to prevent the inflow of adjacent soil into the pipes but allow the free flow of water into them, and if slotted pipes are used, the filter material must be such that it does not enter the pipes.

Koerner and Hwu (1991) recommend requirements of prefabricated edge drain filters as follows:

Requirement	Method	Value
Core strength	GRI GG4	≥ 9600 lbf/in.^{2}
Core flow rate	ASTM D4716	≥ 15 gal/min—ft
Geotextile permeability	ASTM D4491	≥ 0.001 cm/s
Geotextile apparent opening size (AOS)	ASTM D4751	$\geq$ No. 100 sieve
Geotextile puncture	ASTM D3787	≥ 75 lb
Geotextile grab strength	ASTM D4632	≥ 180 lb
Geotextile tear strength	ASTM D4533	≥ 75 lb

Recommendations (NCHRP, 1994) are that geocomposite edge drains should be placed on the shoulder side, and the pavement side should be backfilled with suitable sand. The outlet drain pipe is selected so as to make sure that the capacity is equal to or greater than the capacity of the edge drain.

Recommendations (FHWA, 1989) regarding edge drains and outlets are as follows:

1. The preferable location of the edge drain is under the shoulder just adjacent to the pavement/shoulder joint, with the top of the pipe at the bottom of the layer to be drained, in a 12-in. trench.
2. The filter fabric could be avoided at the subbase–edge drain interface to prevent clogging of the filter by fines, and the trench backfill material should have adequate permeability.
3. Outlet spacing should be less than 500 ft, with additional outlets at the bottom of vertical sags, with rigid PVC outlets being desirable along with headwalls to protect it, prevent erosion, and help in locating the outlets.

6.7 CONSIDERATION OF DRAINAGE IN PAVEMENT DESIGN

Note that the effect of the quality of the drainage system provided in a pavement structure should and can be considered in its structural design. The basic idea is that if there is a good drainage system that removes water quickly from the pavement structure, then the total pavement structure will be

Step 1
Determine Quality of Drainage:

Water Removal Time	Quality of Drainage
2 h	Excellent
1 Day	Good
1 Week	Fair
1 Month	Poor
Water will not drain	Very poor

Step 2
Determine Drainage Coefficients

FIGURE 6.16 Drainage coefficients for AASHTO method. (a) Flexible pavements. (From American Association of State Highway and Transportation Officials (AASHTO), *AASHTO Guide for Design of Pavement Structures*, AASHTO, Washington, DC, 1993.)

(continued)

thinner than in the case in which a good drainage system is not provided. For example, in the current AASHTO design procedure, "drainage coefficients" are recommended to be used according to the quality of drainage provided. Figure 6.16 explains the steps, which include determination of the quality of drainage based on the time required for removal of water and selection of an appropriate drainage coefficient, based on the quality of drainage and the percentage of time the pavement is exposed to moisture levels close to the saturation level.

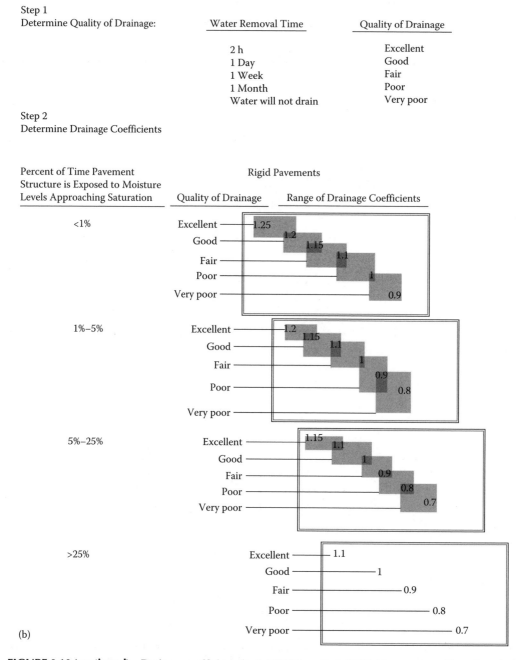

FIGURE 6.16 (continued)   Drainage coefficients for AASHTO method. (b) Rigid pavements. (From American Association of State Highway and Transportation Officials (AASHTO), *AASHTO Guide for Design of Pavement Structures*, AASHTO, Washington, DC, 1993.)

Problem

A pavement section consists of two 12-ft (3.66-m) lanes of 9-in.- (225-mm-) thick PCC pavement with 10-ft (3.05-m) AC shoulders on each side with a uniform cross slope (not crowned), and the width of the permeable base is the same as the PCC pavement. The transverse joint spacing is 20 ft (6.1 m). The slope in both the longitudinal (S) and transverse (S_x) directions is 2%. The permeable base is made up of AASHTO #57 material and has a unit weight of 100 pcf (1600 kg/m³), specific gravity of 2.65, and a minimum permeability of 3000 ft/day (914 m/day). The thickness of the permeable base is 4 in. (100 mm), based on construction considerations. Assume a unit weight of 162 pcf (2595 kg/m³) for the PCC material. The subgrade is Georgia Red Clay, which is actually a well-graded clayey-silt. Laboratory tests indicate the particle gradation for the subgrade material as follows: 92%, 67%, 55%, 42% and 31% passing the No. 4, 10, 20, 50 and 200 sieve, respectively, and a permeability of 0.0033 ft/day (0.001 m/day). Corrugated pipe edgedrains having 4-in. (100-mm) diameter are used on the project.

Solution

Step 1: Select unit (English), and Mode (Normal), Sensitivity (Length, Slope, Permeability, Inflow, Drain, Thickness, porosity), Plot Scale (Power 0.45) (Figure 6.18a)

Step 2: Enter information for roadway geometry (Figure 6.18b):
Enter the value of b as 24 ft (2 lanes of 12 ft each). Since the problem statement assumes that the width of the permeable base is same as that of the PCC layer, enter the value of c as 0. Clicking the calculator icon for width of the permeable base W results in a computed value of 24 ft. Enter the longitudinal and transverse slopes in the S and S_X edit boxes, respectively. Click the calculator icons for both S_R (resultant slope) and L_R (resultant length) to compute these parameters. The calculations should yield the following values: $S_R = 0.0283$ ft/ft and $L_R = 33.94$ ft/ft. Select a uniformly cross-sloped pavement section by clicking on the Geometry B radio button.

Step 3: Enter information for inflow:
Click on the Crack Infiltration Method radio button (since this method was required to be chosen to estimate inflow). Retain the default values of 2.4 ft³/day/feet for Ic and 0 for pavement permeability kp. Also retain the values for W_c and W that have been automatically carried forward from the calculations performed on the Roadway Geometry page. Enter the number of longitudinal cracks, Nc, as 3 (Nc = the number of contributing lanes + 1 = 2 + 1 = 3). Enter the given transverse spacing of contributing transverse joints Cs = 20 ft. Click the calculator icon to compute inflow q_i. This should yield a value of 0.42ft³/day/ft²

Step 4: (Continuing information for Inflow): meltwater computation
Activate the *Meltwater* sub-screen in the *Inflow* property page by clicking on the "Include Meltwater in the inflow calculations" checkbox. The heave rate (*Heave*) for the clayey-silt subgrade soil can be determined by clicking on the calculator button located to the left of the variable. This action calls up the heave rate table (Figure 6.18c). The heave rate table shows that the value for clayey silt soil is 0.51 in./day. This value can either be entered manually in the edit box on the *Meltwater* sub-screen or entered automatically by selecting the row corresponding to clayey-silt soil type. For materials with large heave rates, the mid-point values may be entered. Enter the given values for the subgrade permeability ($k_{sub} = 1$ ft/day), unit weight of the pavement surface ($\gamma_p = 162$ lb/ft³), the unit weight of the base material above the subgrade ($\gamma_b = 100$ lb/ft³), the thickness of pavement surface ($H_s = 9$ in.), and the thickness of the base material (H = 4 in.) in the appropriate boxes. Click on the calculator icon next to the variable σ to compute the stress imposed by the pavement on the subgrade. This will yield a value of 154.8 lb/ft². Click the calculator icon for determining the quantity of meltwater (q_m).

FIGURE 6.17 Flowchart for solving example problem with DRIP. (Adapted from Federal Highway Administration (FHWA), Office of Asset Management, *Life-Cycle Cost Analysis Primer*, U.S. Department of Transportation, Washington, DC, 2002.)

(*continued*)

Moulton's meltwater chart appears on the screen. (Figure 6.18d) On this screen a horizontal redline appears marking the heave rate of the subgrade (0.51 in./day). Use the mouse to slide the vertical tick line along the horizontal red line (0.51 in./day line) to an approximate location for a stress of 154.8 psf. The value for $q_m/(k_{sub})$, 0.5 will appear in the appropriate box. When satisfied that the mouse is positioned at the proper stress value, click the left mouse button to produce a computed value of q_m. For this example, the mouse is positioned to indicate a value of qm/(ksub), 0.5 of 0.91 ft³/day/ft². A click of the left mouse button produces a value for q_m of 0.05ft³/day/ft². Press OK to return to the *Inflow* property page (Figure 6.18e).

Step 5: Enter information for Permeable Base (design using depth of Flow method):
Select the *Permeable Base* tab to access the corresponding property page. Click on the Depth of Flow Method radio button. Most of the variables such as q_i+q_m, S_R, L_R, and H already have values carried forward from pervious screens. The term q_i+q_m is a summation of the inflow from rainfall and meltwater. Enter the coefficient of permeability of the base (k=3000 ft/day).Compute the H_{min} by clicking on the calculator icon. A H_{min} value of 0.1437 ft is reported, which is lower than the selected permeable base thickness (H) of 0.3333 ft. Therefore, the design is satisfied. If H_{min} were much greater than H, the designer would make adjustments to the design by changing design variables such as base thickness or permeability. If the base thickness H can be revised, this parameter must be changed to be at least equal to the H_{min}. Compute flow capacity of the permeable base (q_d) by clicking on the appropriate calculator button. This yields a value of 44.811 ft3/day/ft. The flow capacity is estimated from Moulton's chart (Figure 6.18f).

Additional discussion on depth of flow design
If the user wishes to perform a sensitivity analysis to see the influence of different parameters on the required thickness, the graph icon to the right of H_{min} variable in Figure 6.17 f can be clicked. DRIP internally performs the calculations over a pre-selected range of the independent parameter (in this case the required base thickness) and computes and plots the dependent variables. The charts are plotted using the DripPlot. The DripPlot window appears as soon as the graph icon next to the H_{min} variable is clicked and the following sensitivity plots are displayed: (Figure 6.18g)

- Required Base Thickness (H_{min}) vs. Pavement Inflow (q_i or q_i+q_m).
- Required Base Thickness (H_{min}) vs. Base Course Permeability (k).
- Required Base Thickness (H_{min}) vs. Resultant Slope (S_R).
- Required Base Thickness (H_{min}) vs. Resultant Length (L_R).

The user can control the type of plots to be displayed in the DripPlot window using the *Options | Sensitivity* menu command prior to clicking the graph icon.

Step 6: Enter information for separator layer (Design):
Select the *Separator Layer* tab. Select the No Separator radio button to evaluate the need for a separator layer. Click on the calculator icon next to the *Base Course* variable. This shifts the user to the *Sieve Analysis* property page. (Figure 6.18h.) The permeable base is made up of the AASHTO #57 material. The gradation for this material is already in the sieve analysis library. To initialize this gradation to the program, click on the dropdown material library list box and select AASHTO #57. Click on the calculator icon to determine the particle sizes for D_{15} and D_{50}. Select the *Separator Layer* tab again. The values of $D_{15}=0.2529$ in. and $D_{50}=0.5484$ in. appear for the *Base Course*. Click on the calculator icon next the *Subgrade* variable to compute the subgrade particle sizes. The user is returned to the *Sieve Analysis* property page again and the Subgrade radio button is automatically activated. (Figure 6.18i) Select the Value radio button and enter the subgrade gradation in the grid on the left-hand side of the property page. Compute the D_{50} and D_{85} by clicking on the particle size calculator icon. Return to the separator layer property page by selecting the *Separator* tab. The values of $D_{50}=0.0253$ in. and $D_{85}=0.1498$ in. are returned for the subgrade (Figure 6.18j). On the *Sieve Analysis* property page, check the filtration and uniformity criteria at the subgrade/base interface by clicking on the balance icon on the right-hand side. Both the criteria generate a *Pass* rating, which implies that no separator layer is required. (Figure 6.18k)

FIGURE 6.17 (continued)   Flowchart for solving example problem with DRIP. (Adapted from Federal Highway Administration (FHWA), Office of Asset Management, *Life-Cycle Cost Analysis Primer*, U.S. Department of Transportation, Washington, DC, 2002.)

Step 7: Enter information for Edgedrain (design):
Select the *Edgedrain* tab. Select the Pipe Edgedrain radio button. Click the "Corrugated" checkbox and a value of $n_{manning} = 0.024$ is displayed. Type in a value of D=4 in. The calculator icon is enabled for calculating the pipe capacity (Q). The computed pipe capacity should read 12,594 ft³/day. Click on the L_o button to compute the outlet spacing. This gives an L_o value of 391.6 ft. However, this value is based on the default selection of the *Pavement Infiltration* method for the pavement discharge rate. Two other methods are given to compute the maximum outlet spacing: *Permeable Base* and *Time-to-Drain*. Since the time-to-drain method was not used for designing the permeable base we will not examine this approach. For computing the outlet spacing L_o based on the *Permeable Base* approach, select the appropriate radio button from the "Discharge Rate Approach" options. Our previously computed base discharge q_d of 44.81 ft³/day/ft appears automatically. This results in a maximum outlet spacing of 281 ft. Thus, the permeable base discharge result is the critical value, so the user should specify an outlet spacing of 281 ft (Figure 6.18l)

↓

Step 8: Review input and output by clicking on File → Export Summary (Figure 6.18m)

FIGURE 6.17 (continued)   Flowchart for solving example problem with DRIP. (Adapted from Federal Highway Administration (FHWA), Office of Asset Management, *Life-Cycle Cost Analysis Primer*, U.S. Department of Transportation, Washington, DC, 2002.)

6.8   PUMPING IN RIGID PAVEMENTS

Pumping is defined as the ejection of material from underneath the PCC slab as a result of water pressure. Water accumulated underneath a PCC slab will pressurize when the slab deflects under traffic load. The source of the ejected material can be from the base or subgrade. To initiate the pumping process, a small void space needs to be created beneath the slab. This could occur by compression of plastic soil or due to warping of the PCC slab at the joint area. A free-draining soil will not accumulate the necessary water. A fine soil–water suspension underneath the slab is formed and aids in the ejection of the soil. After repeated cycles of slab deflection and soil–water suspension is ejected, a larger void space is created underneath the slab. At this stage faulting is observed, and eventually cracking and corner breaks could occur if significantly higher stresses are developed in the PCC slab. To minimize pumping, pump-susceptible soils such as clays and high-plasticity fine-grained soils should be avoided. Free-draining bases and erosion-resistant bases are highly recommended. A source of water is essential for pumping to occur. Water could be provided by the presence of a high water table or poorly functioning joint seals or cracks in the PCC slab. In any event, a well-draining base could greatly decrease the potential for pumping distress.

6.9   USE OF SOFTWARE FOR DESIGN OF DRAINAGE STRUCTURES

Figure 6.17 explains the step-by-step procedures for the use of Drainage Requirement in pavements (DRIP) software for the design of drainage system for an example pavement (DRIP User's Guide, FHWA, 2002); Figure 6.18 shows the screenshots for the example. DRIP, Version 2 can be downloaded from: http://mctrans.ce.ufl.edu/store/description.asp?itemID=221

(a)

(b)

FIGURE 6.18 Screenshots from DRIP for example problem. (a) Selection of unit and (b) roadway geometry.

(c)

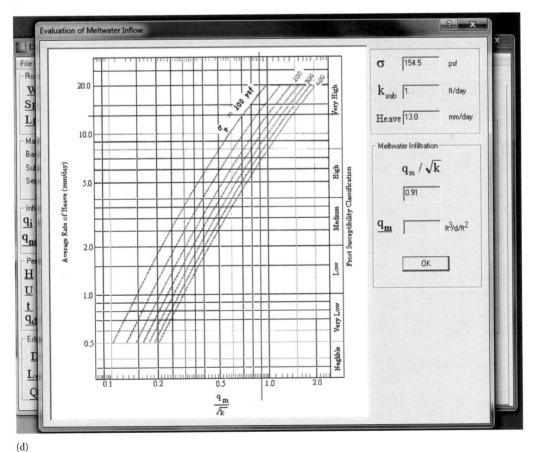

(d)

FIGURE 6.18 (continued) Screenshots from DRIP for example problem. (c) Heave and (d) meltwater inflow determination.

(continued)

(e)

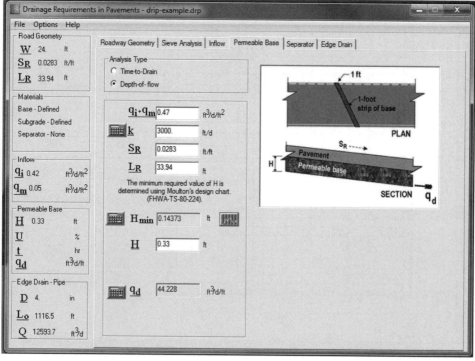

(f)

FIGURE 6.18 (continued)   Screenshots from DRIP for example problem. (e) Inflow method and (f) analysis type selection for permeable base.

(g)

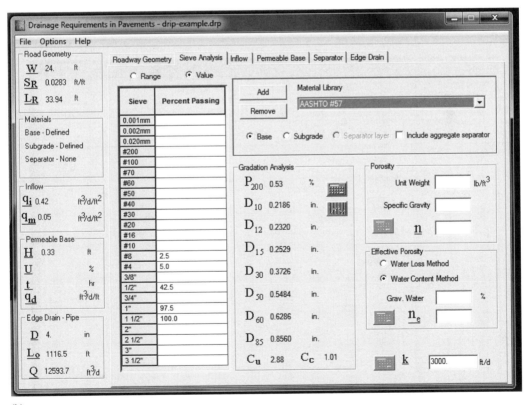

(h)

FIGURE 6.18 (continued)   Screenshots from DRIP for example problem. (g) Required base thickness and (h) sieve analysis results.

(continued)

(i)

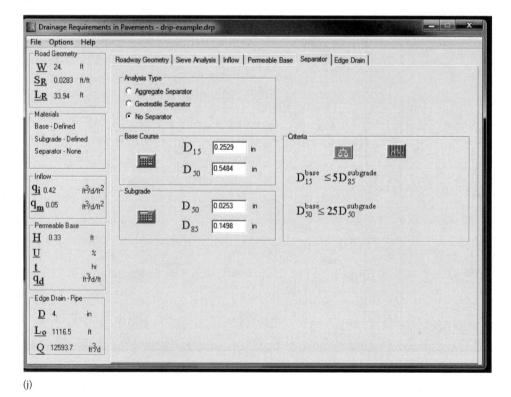

(j)

FIGURE 6.18 (continued) Screenshots from DRIP for example problem. (i) Sieve analysis results for subgrade and (j) separator selection.

(k)

(l)

FIGURE 6.18 (continued) Screenshots from DRIP for example problem. (k) Separator selection not required and (l) calculation of pipe capacity.

(*continued*)

(m)

FIGURE 6.18 (continued)   Screenshots from DRIP for example problem. (m) Summary of results—part showing input parameters.

QUESTIONS

6.1 What are the primary sources of water in a pavement?

6.2 What are the components of surface and subsurface drainage systems in asphalt and concrete pavements?

6.3 Estimate the maximum rate of runoff for a 10 year return period using the rational method for the following case:

Area: 25 ha, with two types of use: 60% unimproved and 40% residential single-family area; length of overland flow: 400 m; average overland slope: 1.5%

6.4 For a trapezoidal channel with Manning's coefficient of n=0.012 with a longitudinal slope of 0.0015, side slope of 2:1, and depth of flow 2.5 ft (bottom width of 6 ft), what is the rate of flow?

6.5 For an open channel to carry 650 ft^3/s of water, determine the most efficient cross section for a rectangular channel. Assume n=0.012 and S=0.0012.

6.6 For the following conditions, determine the groundwater infiltration.

New two-lane highway, lane width: 12 ft; shoulder width: 10 ft; length of cracks: 20 ft; rate of infiltration: 0.5 ft^3/day/ft^2; spacing of cracks: 25 ft; kp=15 * 10^{-5} ft/day

6.7 Compute the location of the first gutter inlet from the crest, considering the following information:

Maximum allowable flow=0.25 m^3/s
Composite runoff coefficient=0.45
Intensity of rainfall=120 mm/h
Width of drainage area=70 m

7 Soil

7.1 OVERVIEW

Soil can be defined as unconsolidated earth material composed of discrete particles with gas or liquids between, a relatively loose agglomeration of mineral and organic materials and sediments found above bedrock, or any earth material except embedded shale and rock.

The classification on the basis of particle size and gradation gives three groups of soils: coarse-grained or granular soils, consisting of sands and gravels; fine-grained or cohesive soils, consisting of clay; and silts, which lie in between.

The dominant factors for the coarse-grained soils are mass or gravitational forces between particles, and water has very little or no influence; whereas for fine-grained soils, electrical and chemical forces dominate particle interaction, and the amount and nature of pore fluid (such as water) affect the particle interaction significantly.

Particle size and gradation are important factors for coarse-grained soils. Uniform gradation means the soils consist of predominantly one size and should have a high permeability. *Well graded* means the presence of different sizes in sufficient proportions, which would result in higher density and strength. *Gap graded* would mean the absence of certain sizes of particles in the soil.

Mineralogy also has significant influence on the properties of the soil, more so in the case of the clay particles than in the case of coarse-grained soil. Coarse-grained soils mostly consist of siliceous materials with quartz and feldspar particles.

The physical properties that characterize clay particles are that they are very small, platy, and negatively charged. There are primarily three clay minerals (Figure 7.1). Kaolinite ranges in size from 0.5 to 5 μm. The specific surface area (particle surface area/particle mass) ranges from 10 to 20 m^2/g. In comparison, the specific surface of silt-size particles is less than 1 m^2/g. Kaolinite has a stable structure with strong bonds and is electrically almost neutral. The electrical activity is expressed by its cation exchange capacity, which for Kaolinite is 3–10 meq/100 g. Illite ranges in size from 0.1 to 2 μm, with a specific surface of 80–100 m^2/g. It has a moderately stable structure, with a cation exchange capacity of 10–40 meq/100 g. Montmorillonite ranges in size from 0.1 to 1 μm, with a specific surface of approximately 800 m^2/g. It has an unstable structure, with a cation exchange capacity of 80–120 meq/g.

Water is bound to negatively charged clay particles (hydrated), and the absorbed cations (Li^+, Na^{++}, Mg^{++}, Ca^{++}) influence the nature of the bound water layer. The bound water influences particle interaction.

7.2 SOILS IN SUBGRADE

Tests performed for the characterization of subgrade soil consist of the following:

1. Grain size distribution (through sieve analysis)
2. Specific gravity
3. Atterberg limits
4. Organic content
5. Moisture retention and hydraulic conductivity
6. Compaction

FIGURE 7.1 Different clay minerals. (From Holtz, R.D. and Kovacs, W.D., *Introduction to Geotechnical Engineering*, 1st edn., Prentice Hall, Upper Saddle River, NJ, p. 86, 1982. Reprinted with permission of Pearson Education, Inc.)

7. Frost susceptibility
8. Unfrozen moisture content
9. Resilient modulus or California bearing ratio (CBR)

Classification of the subgrade soil type and approximate judgment regarding the usability of the material can be made according to the American Association of Highway and Transportation Officials (AASHTO) and the American Society of Testing and Materials (ASTM). Subgrade soils (henceforth called *soils*) that are used for pavements can be primarily of two types—coarse grained (sands and gravel) and fine grained (silts and clays). Organic soils such as peat are generally not used in pavement subgrade. Tests are carried out on subgrade soils for characterization to select the proper materials, to develop specifications and hence quality control guidelines, and for estimation of properties such as modulus, which can be used in the structural design procedure for determination of responses under load. The starting point in characterizing the subgrade soil is the development of a boring/coring location plan based on existing soil maps. The soil samples are then tested in the laboratory for the different properties.

7.3 MASS–VOLUME RELATIONSHIPS

Any soil mass typically consists of three materials or "phases"—solid soil particles (or solid), water, and air. The relative proportions of these components greatly influence the other physical properties and load-carrying capacity of soils. Therefore, several parameters have been defined that help us determine one or two important parameters if a set of other parameters is known. Generally, a phase or block diagram is used to solve for such parameters. The parameters and different formulas are shown in Figure 7.2. It is important to note that these parameters help us to convert volume to mass and mass to volume, which is required for the estimation of materials during construction.

$$\text{Density: } \rho = \frac{M}{V} \text{ kg/m}^3; \quad \text{dry density: } \rho_d = \frac{M_s}{V} \text{ kg/m}^3$$

$$\text{Specific gravity of solids: } G_s = \frac{M_s}{V_s \rho_w} \quad \text{void ratio: } e = \frac{V_v}{V_s} \quad \text{porosity: } n = \frac{V_v}{V} * 100\%$$

$$\text{Water content: } w = \frac{W_w}{W_s} * 100\%; \quad \text{degree of saturation: } S = \frac{V_w}{V_v} * 100\%$$

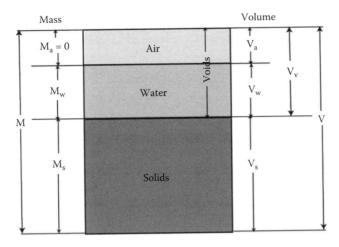

FIGURE 7.2 Soil volumetric properties.

7.4 GRAIN SIZE DISTRIBUTION: GRADATION

The grain size distribution or gradation of soil is one of its most important properties. A handful of soil from the ground would show us that the mass consists of particles of different sizes. The relative proportion of the particles of different sizes is very important in determining the soil's load-carrying capacity. The method to determine this proportion is called *particle size analysis*. In this procedure, particles of different sizes are mechanically separated by passing the soil through a stack of sieves, with progressively decreasing size openings from top to bottom. The dry weight of the soil in each sieve is then determined, and through a few steps the percentage passing each sieve is determined and plotted against the logarithm of sieve size opening. Generally the data are plotted (as a gradation curve) with the percentage passing in arithmetic scale in the y-axis and the sieve sizes in logarithmic scale in the x-axis. Note that the gradation of the soil is used in classification as well as indicators of different engineering properties, as mentioned later in this chapter.

For testing, either standard AASHTO 88 or ASTM D-422 can be used. The sieve sizes are selected on the basis of samples or specifications, and the particle size analysis of soil particles passing the 2.0 mm sieve can be conducted using a hydrometer. Particles retained on a 0.075 mm sieve after hydrometer tests should be subjected to sieve analysis also, and then all of the results should be combined to produce one combined gradation plot for the soil. The data are analyzed as shown in Figure 7.3.

The hydrometer analysis is conducted on the principle of Stokes' law, which relates the terminal velocity of a particle falling freely in a liquid to its diameter, as follows:

$$v = \frac{D^2 \gamma_w (G_s - G_L)}{18\eta}$$

where
    v is the velocity of settling soil particles
    D is the diameter of particle
    γ_w is the unit weight of water
    G_s is the specific gravity of solid particles
    G_L is the specific gravity of soil–water mixture
    η is the dynamic viscosity of soil–water mixture

Particle Size Analysis Using Sieves

Seive Size, mm	From Test Weight of Soil Retained, gm		Cumulative % Retained		% Passing
50	0		0		100
.25	300	100*(300/690) →	43 → 100−43 →		57
4.75	140	100*(300 + 140)/690 →	64 → 100−64 →		36
0.3	160		87		13
0.075	60		96		4.3
	30		100		0
Total	690				

Particle Size Analysis Using Hydrometer

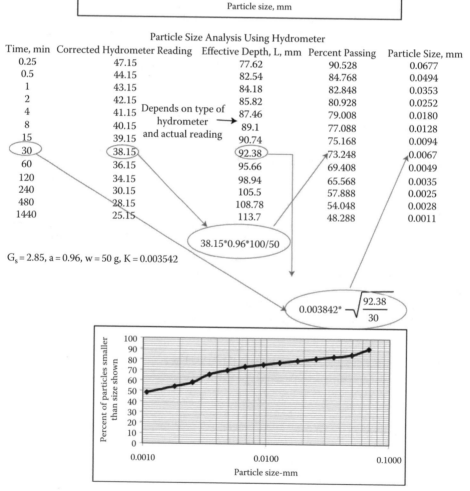

Time, min	Corrected Hydrometer Reading	Effective Depth, L, mm	Percent Passing	Particle Size, mm
0.25	47.15	77.62	90.528	0.0677
0.5	44.15	82.54	84.768	0.0494
1	43.15	84.18	82.848	0.0353
2	42.15	85.82	80.928	0.0252
4	41.15	87.46	79.008	0.0180
8	40.15	89.1	77.088	0.0128
15	39.15	90.74	75.168	0.0094
30	38.15	92.38	73.248	0.0067
60	36.15	95.66	69.408	0.0049
120	34.15	98.94	65.568	0.0035
240	30.15	105.5	57.888	0.0025
480	28.15	108.78	54.048	0.0028
1440	25.15	113.7	48.288	0.0011

Depends on type of hydrometer and actual reading →

$38.15*0.96*100/50$

$G_s = 2.85$, $a = 0.96$, $w = 50$ g, $K = 0.003542$

$0.003842* \sqrt{\dfrac{92.38}{30}}$

FIGURE 7.3 Particle size analysis.

The soil sample is suspended in water in a cylinder, using a dispersing agent such as sodium hexametaphosphate. The hydrometer, with a graduated scale on it, is then placed inside the liquid, and hydrometer readings are taken at the end of 120 s, and then at time intervals such as 5, 15, 30, 60, 250, and 1440 min, as particles continues to settle (sedimentation) down in the liquid. The two equations used for the determination of percentage finer and diameters of the particle are as follows:

$$P = \frac{Ra}{w} * 100$$

where
P is the percentage of originally dispersed soil remaining in suspension
R is the corrected hydrometer reading
a is the constant depending on the density of the soil specimen
w is the mass of soil, g

and

$$d = \sqrt{\frac{30 \, nL}{980(G - G_1)T}}$$

where
d is the maximum grain diameter, mm
n is the coefficient of viscosity of the suspending medium, Pa.s
L is the distance from the surface of the suspension to the level at which the density of the suspension is being measured, mm, known as the effective depth
T is the interval of time from beginning of sedimentation to the taking of the reading, min
G is the specific gravity of soil particles
G_1 is the specific gravity of the suspending medium

The aforementioned equation is simplified as follows, with values of K and L provided in standards:

$$d = K\sqrt{\frac{L}{T}}$$

An example is shown in Figure 7.3.
   Depending on the gradation, the soil can be classified as poorly graded (short range of sizes), well graded (good range of sizes), or gap graded (certain sizes missing). This is done by characterizing the gradation curves with the help of certain "coefficients" as follows.

Coefficient of uniformity:

$$C_u = \frac{D_{60}}{D_{10}} \quad \begin{array}{l} \text{poorly graded } \downarrow \\ \text{well graded } \uparrow \end{array}$$

Coefficient of curvature:

$$C_c = \frac{(D_{30})^2}{D_{60}D_{10}} \quad \begin{array}{l} \text{well graded} \approx 1-3 \\ \text{gap graded} < \text{or} > 3 \end{array}$$

Note that D_x refers to grain size corresponding to x% passing.
   Examples of different gradations are shown in Figure 7.4.

FIGURE 7.4   Different types of gradation.

7.5   EFFECT OF WATER

The presence of certain types of soil materials, such as clay, can cause the effect of water to be significant, and hence may make the soil not suitable for use in pavement subgrade. As the water content of the soil increases from very low to high, the soil properties change. The Atterberg limit parameters—the liquid limit, plastic limit, and plasticity index (PI) of a soil—tell us how sensitive the soil is to the effect of water.

Shrinkage limit: Shrinkage limit is defined as the maximum calculated water content, at which a reduction in water content will not cause a decrease in the volume of the soil mass. AAS-HTO T-92 can be used for determination of shrinkage limit (SL).

Plastic limit: Plastic limit (PL) is the moisture content below which the soil is nonplastic.

Experimental definition: Plastic limit is defined as the water content at which a soil thread just crumbles when it is rolled down to a diameter of 3 mm (approximately):

$$\text{Plastic limit} = \left[\frac{(\text{mass of water})}{(\text{mass of oven dry soil})} \right] * 100\%$$

The AASHTO T-90 procedure is used for the determination of the plastic limit of the soils.

Liquid limit: Liquid limit (LL) is the moisture content below which the soil behaves as a plastic material. Liquid limit is defined as the water content at which a part of soil placed in a brass cup, cut with a standard groove, and then dropped from a height of 1 cm will undergo a groove closure of 12.7 mm when dropped 25 times. A nearly linear plot of the water content (%) versus log of number of blows (N) is prepared. The moisture content corresponding to 25 blows is the liquid limit. The AASHTO T-89 procedure is used for the determination of liquid limit.

$$\text{Plasticity Index (PI)} : PI = LL - PL$$

$$\text{Liquidity Index (LI)} : LL = \frac{(w\% - PL)}{PI}$$

The different limits are shown in Figure 7.5.

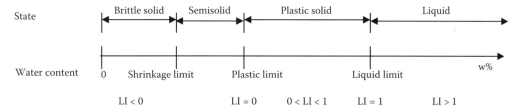

FIGURE 7.5 Atterberg limits and state of the soil. (Redrawn from Holtz, R.D. and Kovacs, W.D., *An Introduction to Geotechnical Engineering*, Prentice Hall, Upper Saddle River, NJ, 1981.)

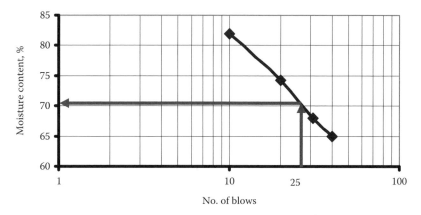

FIGURE 7.6 Plot of number of blows versus moisture content.

Example 7.1

The following observations were obtained from a liquid limit test of a soil:
Number of blows: 10, 20, 31, 40.
Moisture content: 82.0%, 74.3%, 68.0%, 65.0%.
Two tests for plastic limit were done that gave values of 28.2 and 29.4, respectively.
Determine liquid limit and PI of the soil.

Plot the moisture content versus number of blows data as shown in Figure 7.6, with number of blows on the x-axis in log scale and moisture content on the y-axis in arithmetic scale.

From Figure 7.6, LL = 71%

$$\text{Average of two plastic limit tests, plastic limit} = (28.2 + 29.4)/2 = 28.8$$

$$PI = LL - PL = 71 - 28.8 = 42.2$$

7.6 SOIL CLASSIFICATION

7.6.1 AASHTO Method

Taking particle size analysis and Atterberg limits into consideration, the AASHTO soil classification system has been developed, as shown in Table 7.1.

One can use this system to classify soil and make a conclusion regarding its suitability as subgrade soil. The criteria for the AASHTO classification system are based on performance as highway construction material. There are two general groups—coarse grained and fine grained (or granular and cohesive). The distinction between coarse- and fine-grained soil is 35% passing the No. 200 sieve. There is also organic soil.

TABLE 7.1
AASHTO M-145 Standard Specifications for Classification of Soils and Soil-Aggregate Mixtures for Highway Construction Purposes

| General Classification | Granular Materials (35% or Less Passing No. 200 Sieve) | | | | | | | Silt-Clay Materials (>35% Passing No. 200 Sieve) | | | | | Highly Organic |
| | A-1 | | A-3 | A-2 | | | | A-4 | A-5 | A-6 | A-7 | | A-8 |
	A-1-a	A-1-b		A-2-4	A-2-5	A-2-6	A-2-7				A-7-5	A-7-6	
Sieve analysis, % passing													
No. 10 sieve	≤50												
No. 40 sieve	≤30	≤50	≥51										
No. 200 sieve	≤15	≤25	≤10	≤35	≤35	≤35	≤35	≥36	≥36	≥36	≥36	≥36	
Characteristics of fraction passing no. 40 sieve													
Liquid limit				≤40	≥41	≤40	≥41	≤40	≥41	≤40	≥41	≥41	
Plasticity index	Max: 6		NP[a]	≤10	≤10	≥11	≥11	≤10	≤10	≥11	≥11	≥11	
Usual types of significant constituent materials	Stone fragments; gravel and sands		Fine sand	Silty or clayey grave and sand				Silty soils		Clayey soils			Peat or muck
General rating as subgrade	Excellent to good							Fair to poor					Unsuitable

Source: American Association of State Highway and Transportation Officials (AASHTO), *Standard Specifications for Transportation Materials and Methods of Sampling and Testing*, 27th edn. AASHTO, Washington, DC, 2007, M 145–91 (2004). Used with permission.

[a] NP, nonplastic; for A-7-5 soils, PI ≤ (LL-30); for A-7-6 soils, PI > (LL-30).

The groups range from A1 (best), which is the most coarse grained, to A8 (worst), which is the most fine grained. The parameter used for grouping is called the group index (GI), which rates soil as a pavement subgrade material, as shown in the following:

Group Index (GI)	Subgrade Rating
0	Excellent
0–1	Good
2–4	Fair
5–9	Poor
10–20	Very poor

The steps in classification consist of the following:

1. Use grain size distribution and Atterberg limits data to assign a group classification system and a GI.
2. Compute the GI using the following equation:

$$GI = (F - 35)[0.2 + 0.005(LL - 40)] + 0.01(F - 15)(PI - 10)$$

where
    F is the portion passing No. 200 sieve
    LL is the liquid limit
    PI is the plasticity index
    Express the GI as a whole number.
    A GI value of less than zero should be reported as zero.
    For A-2-6 or A-2-7 soils, use only the second term for the calculation of the GI.

3. Express the AASHTO soil classification as the group classification followed by the GI in parentheses.

Example 7.2

The following data were obtained for a soil sample.

Sieve No.	% Passing
4	98
10	93
40	87
100	77
200	69

Plasticity tests: LL = 48%, PL = 26%.
    Determine the classification of the soil, and state whether this material is suitable in its natural state for use as a subbase material.
    Refer to Table 7.1. Since the percentage passing the No. 200 sieve is 69, the soil must be a silt-clay material. The LL is 48%; therefore, it should fall in either the A-5 or A-7 category. The PI = LL − PL = 22 > 10, so it must be an A-7 soil. In this case, LL − 30 = 18; PI > (LL − 30); therefore, the soil can be classified as an A-7-6 material.

$$GI = (69 - 35)[0.2 + 0.005(48 - 40)] + 0.01(69 - 5)(22 - 10) = 14.6 \text{ (very poor)}$$

Conclusion: unsuitable as a subbase material in its natural state.

7.6.2 UNIFIED SOIL CLASSIFICATION SYSTEM (ASTM)

The first step in this system is to determine if the soil is highly organic (which has dark brown, gray, or black color; has an odor; is soft; and has fibrous materials) or not. If yes, it is classified as peat, and if not, the soil is subjected to particle size analysis and the gradation curve is plotted. Based on the percentage of weight passing the 3-in., No. 4, and No. 200 sieves, the percentages of gravel, sand, and fines are computed on the basis of Table 7.2. If less than 100% of the material passes through a 3-in. sieve, then base the classification on the basis of the percentage of the material that passes through the 3-in. sieve and note the percentage of cobbles and/or boulders. If 100% passes the 3-in. sieve, then check the percentage passing the No. 200 sieve—if 5% or more passes the No. 200 sieve, then determine liquid and plastic limits. Check whether more than 50% passes through the No. 200 sieve—if it does (fine-grained soil), then follow subsequent instructions for fine-grained soils; otherwise (coarse-grained soil), follow those for coarse-grained soils.

Fine-grained soils are primarily silt and/or clay, and Figure 7.7 is to be used for classification. The soils falling above the A line are classified as clays, and those falling below are classified as silts. The group symbols for fine-grained soils are as follows:

First Order	Second Order
M: predominantly silt	L: low plasticity
C: predominantly clay	H: high plasticity
O: organic	

The tests needed to determine whether a soil is organic or not are those testing the liquid limit on an unmodified and an oven-dried sample. The soil is considered to be organic if the liquid limit after oven drying is <75% of the original. For classification of inorganic soils, use Figures 7.7 and 7.8, and for classification of organic soils, use Figures 7.7 and 7.9.

TABLE 7.2
ASTM Particle Size Classification (ASTM D-2487)

Sieve Size		Particle Diameter		Soil Classification	
Passes	Retained on	in.	mm	Boulder	Rock Fragments
	12 in.	>12	>350		
12 in.	3 in.	3–12	75.0–350	Cobble	
3 in.	¾ in.	0.75–3.0	19.0–75.0	Coarse gravel	
¾ in.	#4	0.19–0.75	4.75–19.0	Fine gravel	
#4	#10	0.079–0.19	2.00–4.75	Coarse sand	
#10	#40	0.016–0.079	0.425–2.00	Medium sand	
#40	#200	0.0029–0.016	0.075–0.425	Fine sand	
#200		<0.0029	<0.075	Fine silt and clay	

Source: Coduto, D.P., *Geotechnical Engineering: Principles and Practices*, 1st edn., Prentice Hall, Upper Saddle River, NJ, p. 115, © 1999. Reprinted with permission from Pearson Education, Inc.

FIGURE 7.7   Plasticity chart, ASTM D-2487. (From Coduto, D.P., *Geotechnical Engineering: Principles and Practices*, 1st edn., Prentice Hall, Upper Saddle River, NJ, p. 143, © 1999. Reprinted with permission of Pearson Education, Inc.)

Coarse soils are primarily sand and/or gravel. The symbols for coarse-grained soils are as follows:

First Letter	Second Letter
S: predominantly sand	P: poorly graded
G: predominantly gravel	W: well graded
	M: silty
	C: clayey

Use Figure 7.10 to classify coarse-grained soils, according to the following guidelines:

1. Use two-letter groups to describe soils with <5% passing the No. 200 sieve.
2. Use four-letter groups to describe gradation and type of fines for soils with 5%–12% passing the No. 200 sieve.
3. Use two-letter symbols to describe the fines for soils with >12% passing the No. 200 sieve.

7.7   DENSITY AND OPTIMUM MOISTURE CONTENT

Compacting loose soil is the simplest way to improve its load-bearing capacity. Water is added to the soil, which is compacted in lifts, to lubricate the soil particles and air in the compaction process. As water is added to the soil, the density of the soil increases due to compaction. However, beyond certain water content, even though the soil becomes more workable, the unit weight of the soil would decrease. This is shown in Figure 7.11.

This can be explained by the fact that beyond the "optimum" water content, the water cannot force its way into trapped air pockets, and hence takes up the space originally taken by the soil solids—that is, it separates the soil particles and hence causes a corresponding decrease in density.

Generally, the density of the soil solids (generally referred to as *dry density*) is taken as an indicator of the load-carrying capacity of the soil and as an indicator of the degree of compaction of the soil—a higher density relates to higher stiffness and/or strength. Moreover, a high density also ensures low permeability, and minimizes settlement and volume change due to frost action, swelling, and shrinkage. A high density of the solids can be achieved in

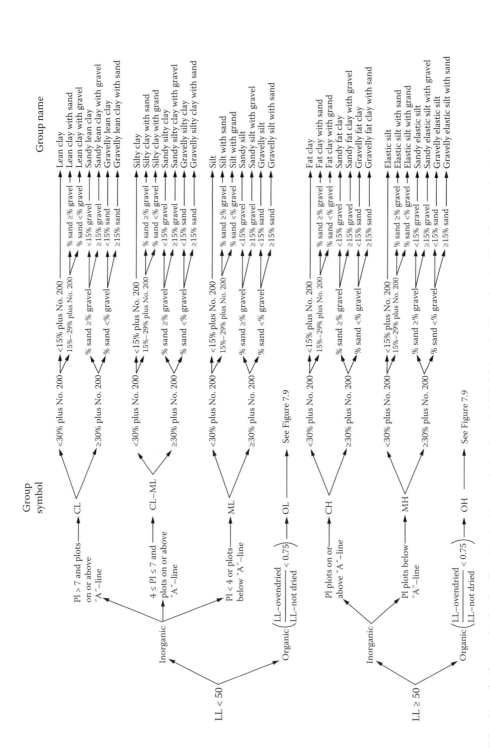

FIGURE 7.8 Chart for fine-grained soils. (Reprinted with permission from D2487-06 Standard Practice for Classification of Soils for Engineering Purposes [Unified Soil Classification System], © ASTM International, 100 Barr Harbor Drive, West Conshohocken, PA, 19428.)

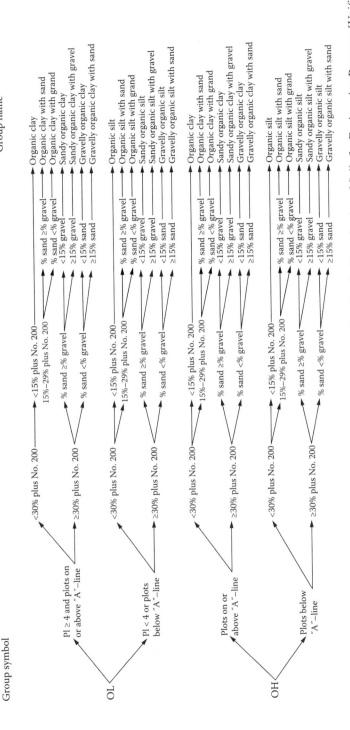

FIGURE 7.9 Chart for organic soil. (Reprinted with permission from D2487-06 Standard Practice for Classification of Soils for Engineering Purposes [Unified Soil Classification System], © ASTM International, 100 Barr Harbor Drive, West Conshohocken, PA, 19428.)

FIGURE 7.10 Chart for coarse-grained soils. (Reprinted with permission from D2487-06 Standard Practice for Classification of Soils for Engineering Purposes [Unified Soil Classification System], © ASTM International, 100 Barr Harbor Drive, West Conshohocken, PA, 19428.)

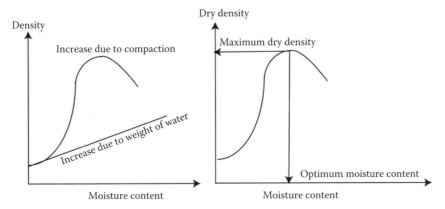

FIGURE 7.11 Moisture content versus density.

the field only by applying compactive effort (through rollers) and moisture. To achieve this, it is important to determine in the laboratory the "optimum" moisture content of the soil—the word *optimum* here refers to that moisture content which produces the highest dry density, for a given compactive effort. In soils with 10% or more passing the No. 200 sieve, the change in dry density versus change in moisture content can be expressed in the form of the curve shown in Figure 7.11. The moisture content corresponding to the maximum dry unit weight is called the *optimum moisture content.*

Note that the moisture content versus dry density curve is different for different compactive efforts for the same soil. The test that is carried out to determine the optimum moisture content is commonly referred to as the *Proctor compaction test.* The compactive effort to be applied for compaction of soils during the Proctor test is dependent on the level of compactive effort expected in the field—the higher the expected effort, the higher is the effort in the laboratory.

Accordingly, there are two standards—AASHTO T-99 (Moisture-Density Relations of Soils Using a 2.5 kg [5.5 lb] Rammer and a 305 mm [12 in.] Drop; energy level of 12,400 ft lb per ft³ of soil) and AASHTO T-180 (Moisture-Density Relations of Soils Using a 4.54 kg [10 lb] Rammer and a 457 mm [18 in.] Drop; energy level of 56,200 ft lb of energy per ft³ of soil).

Once the optimum moisture content and the maximum dry density are determined from the laboratory, these values are used for specifying moisture levels (generally, a little dry of the optimum) and density (generally, a percentage of the maximum). One practice is to use a fill in the site to determine the soil lift thickness and number of passes of compaction equipment that is required for the available equipment (or, if necessary, with other equipment) to achieve the required compaction level, and then prepare the method specification accordingly.

An example of determination of optimum moisture content is shown in Figure 7.12.

7.8 HYDRAULIC CONDUCTIVITY

Based on results of experiments conducted by D'Arcy (1856), the flow of water through soils is characterized by the following expression:

$$v = ki$$

where
v is the velocity of flow
i is the hydraulic gradient (head loss per unit length of flow)
k is the coefficient

Sample no.	Volume, ft³	Weight, lb	Bulk Density, lb/ft³	Moisture Content, %	Dry Density, lb/ft³
1	0.03333	4.16	124.8	4	120.0
2		4.39	131.7	6.1	124.1
3		4.6	138	7.8	128.0
4		4.68	140.4	10.1	127.5
5		4.57	137.1	12	122.4
6		4.47	134.1	14	117.6

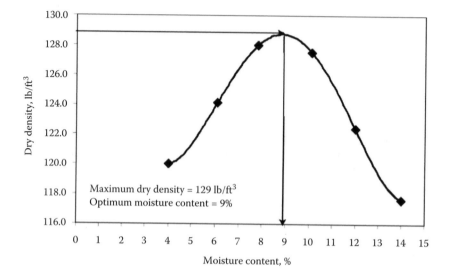

Example calculation: bulk density = (4.16)/(0.03333) = 124.8 lb/ft³;
Dry density = bulk density/(1+moisture content) = (124.8)/(1+0.04) = 120.0 lb/ft³

FIGURE 7.12 Example problem.

Generally k is referred to as a coefficient of permeability or permeability or hydraulic conductivity. The permeability of different types of soils is shown in Table 7.3.

Permeability for soils can be measured in the laboratory by using either a constant head (ASTM D-2434, AASHTO T-215) or a falling head method (ASTM D-4630). In the constant head test, a constant head of water is maintained in a soil sample, and the flow rate (volume divided by time) is used along with the sample cross-sectional area to determine the permeability. In a falling head test, the level of water in a standpipe is noted at regular time intervals as the water is made to flow through a soil sample, and the permeability is calculated on the basis of the total length and area of the soil sample, the area of the standpipe, and the initial and final levels of water in the standpipe in the specified time.

In the absence of test data, correlations with gradation properties can be used to estimate the permeability of different types of soils. For example,

$$k = C_k D_{10}^2$$

where

k is the permeability, cm/s
D_{10} is the effective grain size (cm) corresponding to size passing 10%
C_k is the Hazen's coefficient, 0.8–1.2 (applicable for 0.1 mm < D_{10} < 3 mm, C_u < 5)

TABLE 7.3
Hydraulic Conductivity of Soils

	Hydraulic Conductivity, k	
Soil	cm/s	ft/s
Clean gravel	1–100	$3*10^{-2}$–3
Sand-gravel mixtures	10^{-2}–10	$3*10^{-4}$–0.3
Clean coarse sand	10^{-2}–1	$3*10^{-4}$–$3*10^{-2}$
Fine sand	10^{-3}–10^{-1}	$3*10^{-5}$–$3*10^{-3}$
Silty sand	10^{-3}–10^{-2}	$3*10^{-5}$–$3*10^{-4}$
Clayey sand	10^{-4}–10^{-2}	$3*10^{-6}$–$3*10^{-4}$
Silt	10^{-8}–10^{-3}	$3*10^{-10}$–$3*10^{-5}$
Clay	10^{-10}–10^{-6}	$3*10^{-12}$–$3*10^{-8}$

Source: Coduto, D.P., *Geotechnical Engineering: Principles and Practices*, 1st edn., Prentice Hall, Upper Saddle River, NJ, p. 222, © 1999. Reprinted with permission from Pearson Education, Inc.

In general, coarse-grained soils have higher permeability compared to fine-grained soils. As such, coarse-grained soils have good drainage capabilities and are preferred for pavement applications compared to fine-grained soils.

7.9 FROST SUSCEPTIBILITY

In certain areas, where freezing cold weather is common during the winter months, frost heaves are common sights. In spring these heaves break apart under traffic, leading to potholes. The main causes of this distress are the environmental conditions and the existing soil in the lower pavement layers.

Soils that have low permeability and retain water, and change in volume in freezing temperatures, are known as being frost susceptible. The degree of heaving in such soils cannot be explained only on the basis of the expanding of water into ice.

Portions of a pavement can experience a rise or "heave" (frost heaves) due to the formation of ice lenses in these frost-susceptible materials, which typically contain excessive amounts of fine materials (i.e., those less than 0.02 mm). Formation of ice lenses leads to pressure exerted by the ice and soil. There is differential movement in the pavement structure, and as a result there is roughness and cracking. Most of the water that forms the ice lenses is drawn up by capillary action from water tables. When the soil thaws in warm weather in spring (from the surface downward), this new water cannot drain quickly, leading to a significant increase in the moisture content of the subgrade and hence a significant reduction in shear strength. This is illustrated in Figure 7.13. The effects are relatively more significant in pavements with untreated layers.

The process starts with the soil and a water table close to the soil. A soil such as silt, which has relatively high ability to draw water through small pores (high capillarity, 0.9–9 m rise compared to 0.15 m rise in coarse sand) as well as ability to move water in the pores relatively rapidly (moderate permeability, 0.9 m/day compared to 0.0009 m/day for clay), is considered to be frost susceptible. Such soil would draw water from a relatively close water table, and this water would freeze in the

Temperature below freezing point

Resulting heave at the surface

Asphalt mix layer

Frost susceptible soil

Soil is pushed up
due to expansion

Ice lenses form
from water under
freezing temperatures

Capillary action feeds
water into soil

Water table

Heaves caused by freezing in winter

Temperature above freezing point

Loss of support

New water from thawed soil
cannot drain quickly and reduce
strength of soil

Water table

Loss of support due to spring thaw

FIGURE 7.13   Freeze–thaw in soils.

winter time as the freezing front moves downward. The expansion of water upon freezing into ice lenses exerts pressure on the layers above them, resulting in surface heaves. With the coming of spring, as the temperature rises, the ice lenses start thawing from top downward. This water in the subgrade and the void created (by the difference in volume of ice and water) result in a weaker saturated subgrade and a weaker pavement structure as a whole.

Subsequently, when traffic moves over the heaved surface, which has no or little support underneath, it breaks down into smaller pieces, forming potholes. Frost-susceptibility tests can be conducted by subjecting soils to freeze–thaw cycles, noting the heave rate during the first 8 h and then determining the CBR, a strength-indicating test property, on the thawed soil after allowing time for drainage of the water from the soil. The criteria shown in Table 7.4 can be used for classification of the soil.

Plots of frost susceptibility and heaves, shown in Figure 7.14, can be used for the frost susceptibility classification of a soil. The use of dielectric constant devices, both in place and the laboratory, has been developed to evaluate and monitor freezing and thawing processes in the soil.

The frost susceptibility of the soil is affected primarily by the size and distribution of voids (where ice lenses can form), and is determined through empirical correlations from the soil classification

TABLE 7.4
Frost Susceptibility Classification

Frost Susceptibility Classification	8 h Heave Rate (mm/day)	Thaw CBR (%)
Negligible	<1	>20
Very low	1–2	20–15
Low	2–4	15–10
Medium	4–8	10–5
High	8–16	5–2
Very high	>16	<2

Source: Chamberlain, E.J., A freeze thaw test to determine the frost susceptibility of soils, Special Report 87-1, USA Cold Regions Research and Engineering Laboratory, Hanover, NH, 1987.

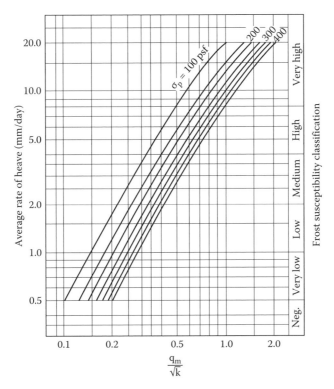

FIGURE 7.14 Frost-susceptibility classification chart. (From Moulton, L.K., Highway subsurface design, FHWA-TS-80-224, U.S. Department of Transportation, Washington, DC, 1980.)

(Unified Soil Classification System) and the percentage finer than 0.02 mm. Table 7.5 relates these parameters to the frost codes.

Frost heaves are nonuniform and tend to make pavements rougher, and expressions have been developed to relate the variability of soil frost susceptibility and the winter roughness. Affected pavement surface profiles can be characterized in terms of dominant wavelength, which has been found to range from 10 to 80 m.

TABLE 7.5

Frost Codes and Soil Classification by the Federal Aviation Administration

Frost Group	Type of Soil	Percentage Finer Than 0.02 mm by Weight	Soil Classification
FG-1	Gravelly soils	3–10	GW, GP, GW-GM, GP-GM
FG-2	Gravelly soils sands	10 to 3 to 15	GM, GW-GM, GP-GM, SW, SP, SM, SW-SM SP-SM
FG-3	Gravelly soils	Over 20	GM, GC
	Sands, except very fine silty sands	Over 15	SM, SC
	Clays, PI above 12		CL, CH
FG-4	Very fine silty sands	Over 15	SM
	All silts		ML, MH
	Clays, PI = 12 or less		CL, CL-ML
	Varied clays and other fine-grained banded sediments		CL, CH, ML, SM

Source: Federal Aviation Administration (FAA), *LEDFAA User's Manual*, AC 150/5320-16, U.S. Department of Transportation, Washington, DC, October 1995.

Note: The higher the frost group number, the more susceptible the soil, i.e., soils in frost group 4 (FG-4) are more frost susceptible than soils in frost groups 1, 2, or 3.

Seasonal frosts can cause nonuniform heave and loss of soil strength during melting, resulting in possible loss of density, development of pavement roughness, restriction of drainage, and cracking and deterioration of the pavement surface. Whether such effects are possible or not depend on whether the subgrade soil is frost susceptible, there is free moisture to form ice lenses, and the freezing front is deep enough to penetrate the subgrade soil.

During the design of pavements, the effect of the frost susceptibility needs to be checked by determining the depth of frost penetration—measures should be taken if the depth is found to penetrate into the frost-susceptible soil layer. The frost penetration depth can be estimated from Table 7.6. The required inputs are the air freezing index (degree days °F) and the unit weight of the soil (in pcf). The frost penetration depth is calculated from the interpolation of the data in Table 7.6, developed on the assumption of a 12-in. (300-mm) thick rigid pavement or a 20-in. (510-mm) thick flexible pavement. Note that the air freezing index is a measure of the combined duration and magnitude of below-freezing temperatures occurring during any given freezing season. The average daily temperature is used in the calculation of the freezing index. For example, if the average daily temperature is 12° below freezing for 10 days, the freezing index would = 12° * 10 days = 120 degree days. It is recommended that the design air freezing index be based on the average of the three coldest winters in a 30 year period, if available, or the coldest winter observed in a 10-year period. Air freezing indices are available at http://www.ncdc.noaa.gov/oa/fpsf/AFI-seasonal.pdf.

The steps that could be taken to avoid or reduce the potential of the freeze–thaw problem are replacing the frost-susceptible soil, lowering the water table, increasing the depth of the pavement with nonfrost-susceptible soil over the frost-susceptible soil, and providing means of draining away excess water, such as that from thawing.

Frost protection can be provided by completely or partially removing the soil to the depth of frost penetration, or by assuming a reduced modulus of the soil during design (so as to result in a thicker

TABLE 7.6
Frost Penetration Depths

Frost Penetration Depth, in Degree Days (°F)	Soil Unit Weight (pcf)			
	100	115	125	150
200	20.5	21.5	23.8	25.5
400	27.5	30.5	35	38.5
600	34	38	44.5	49
800	40	44.5	54	59
1000	45	51	62	69
2000	69.5	79	102	113
3000	92	105	140	156
4000	115	130	177	205
4500	125	145	197	225

Source: Federal Aviation Administration (FAA), *LEDFAA User's Manual*, AC 150/5320-16, U.S. Department of Transportation, Washington, DC, October 1995.

pavement structure, which in turn would provide a cover of nonfrost-susceptible materials on the frost-susceptible soil).

7.10   SWELL POTENTIAL

Soils consisting of minerals such as montmorillonite would expand significantly in contact with water. The presence of such soils in the subgrade would lead to differential movement, surface roughness, and cracking in pavements. Dominant wavelengths of affected pavements have been reported to range from 3.0 to 10.0 m.

The swell potential of a soil is primarily dictated by its percentage of clay-size particles and the PI, as illustrated in the following by one of the many equations developed (Gisi and Bandy, 1980):

$$S(\%) = aA^b C^{3.44}$$

where $S(\%)$ is the swell potential.

$$A = \text{activity} = \frac{PI}{C - 5}$$

where
   PI is the plasticity index
   C is the percentage of clay size particles (finer than 0.002 mm)
   $a = 3.28 * 10^{-5}$, $b = 2.259$, for $PI \leq 20$
   $a = 2.40 * 10^{-5}$, $b = 2.573$, for $21 \leq PI \leq 30$
   $a = 1.14 * 10^{-5}$, $b = 2.559$, for $31 \leq PI \leq 40$
   $a = 0.72 * 10^{-5}$, $b = 2.669$, for $PI > 40$

The swelling potential can be determined as the percentage change in volume due to absorption of water when tested for the CBR or in a Proctor mold. In the latter test, a soil is placed in a standard

TABLE 7.7
Federal Aviation Administration Recommendations for Treating Soils with Swelling Potential

Swell Potential (Based on Experience)	% Swell Measured (ASTM D-1883)	Potential for Moisture Fluctuation	Treatment
Low	3–5	Low	Compact soil on wet side of optimum (+2% to +3%) to not greater than 90% of appropriate maximum density[a]
		High	Stabilize soil to a depth of at least 6 in.
Medium	6–10	Low	Stabilize soil to a depth of at least 12 in.
		High	Stabilize soil to a depth of at least 12 in.
High	Over 10	Low	Stabilize soil to a depth of at least 12 in.
		High	For uniform soils (i.e., redeposited clays), stabilize soil to a depth of at least 36 in. below pavement section, or remove and replace with nonswelling soil. For variable soil deposits, depth of treatment should be increased to 60 in.

Source: Federal Aviation Administration (FAA), *LEDFAA User's Manual*, AC 150/5320-16, U.S. Department of Transportation, Washington, DC, October 1995.

Note: Potential for moisture fluctuation is a judgmental determination and should consider proximity of water table, likelihood of variations in water table, other sources of moisture, and thickness of the swelling soil layer.

[a] When control of swelling is attempted by compacting on the wet side of optimum and reduced density, the design subgrade strength should be based on the higher moisture content and reduced density.

Proctor mold at its optimum moisture content, with a surcharge of 1 psi, and the mold with a perforated base in a pan full of water. The test is actually conducted on two specimens at optimum moisture content (OMC) +3% and OMC −3% moisture contents, and the results are used to determine the percentage swell at the OMC.

The use of total suction characteristic curves has also been developed for the identification of swelling potential of soils.

The presence of swelling soil can affect the performance of the pavement and deteriorate its serviceability. The loss in serviceability with time due to the effect of swelling soils can be modeled as an exponential equation. Generally, soils exhibiting a swell percentage of greater than 3 (as determined with the CBR test, ASTM D-1883) are required to be treated. The Federal Aviation Administration (in AC 150/5320-6D; FAA, 1995) recommends modified compaction efforts and careful control of compaction moisture, as well as removal of such soils and replacement by stabilized materials, as shown in Table 7.7.

7.11   STIFFNESS AND STRENGTH OF SOILS

The most important characteristic feature that needs to be determined for a soil is its response, such as strain and deformation to load (or, more specifically, stress) from the traffic. In this respect, both "stiffness" and "strength" are important parameters that need to be measured in the laboratory. However, note that some design methods may use stiffness only and hence do not require the use of strength tests, while others may require it.

For a strength test, a triaxial testing system is used in general, to simulate the triaxial state of stress in the field. Shear failure in soils is caused by slippage of soil particles past each other, and the resistance against such failure (shear strength) is provided by two components, friction and cohesion.

Frictional strength, commonly denoted as Normal Force * Coefficient of Friction, is expressed in soils in terms of the effective friction angle (or effective angle of internal friction), φ', which is defined as follows:

$$\varphi' = \tan^{-1}\mu$$

where μ is the coefficient of friction.

Shear strength, S, due to friction is defined as $S = \sigma' \tan \varphi'$, where σ' is the effective normal stress.

The frictional strength is dictated by many factors, including density, mineralogy, shape of the soil participles, gradation, void ratio, and the presence or absence of organic materials. The other component of soil shear resistance, cohesion (C), present in some soils (cohesive), is defined as the strength that exists in the soil even if the effective stress (σ') appears to be zero. C is a result of different factors such as cementation, electrostatic attraction, adhesion, and negative pore water pressures and is significantly affected by factors such as density, moisture, and drainage conditions. Therefore, the combined shear resistance of the soil can be expressed as follows:

$$S = C' + \sigma' \tan \varphi'$$

Shear stress–strain curves for sand and gravel soils generally show ductile curves, whereas those for clays show brittle behaviors. A difference in test condition would result in a difference in stress–strain behavior, and as such, tests for the determination of shear strength are conducted under standard conditions or expected field conditions.

For evaluation of shear strength of soils, tests are conducted at different normal stress levels, and the failure shear stress corresponding to each normal stress level is determined. A series of such data points are connected to obtain the Mohr–Coulomb failure envelope and determine the cohesion and frictional components of the shear strength equation. This is explained in Figure 7.15.

Laboratory shear strength measurements can be conducted using the Direct Shear Test (ASTM D-3080), the Unconfined Compression Test (ASTM D-2166), or the Triaxial Test (ASTM D-2850 and ASTM D-4767).

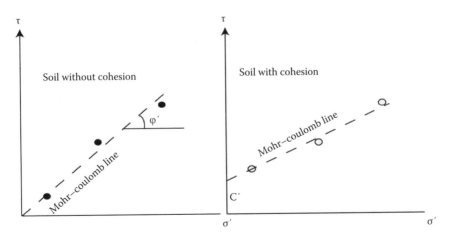

FIGURE 7.15 Mohr–Coulomb failure envelope. (Note: The τ-*intercept* of this line is the effective cohesion, C'. The *slope* of this line is the effective frictional angle, φ'. The Mohr–Coulomb line is often referred to as the *failure envelope*—points in the soil that have (σ', τ) values that fall *below* this line *will not* fail in shear, whereas the points with (σ', τ) values that fall *above* this line will fail in shear.)

In the direct shear test, a cylindrical soil sample is subjected to a vertical and a shear force, and the shear stress and shear displacements data are obtained. The test is repeated with different samples for different values of the normal force, and the peak shear stress in each case is obtained. A plot of peak shear stress versus the normal stress provides the information for the components of the shear strength equation. Although simple and inexpensive, and suitable for sandy soils, the drawbacks include nonuniform strain, failure along a specific plane, and inability to control drainage conditions (especially important for clay soils).

The unconfined compression test uses an unconfined cylindrical soil sample on which normal stress is applied and the failure stress is noted. The shear strength is obtained from the unconfined compression strength. It is a simple, inexpensive, and quick test but lacks the ability to apply confining stress, which is usually present in the field, and hence underestimates the shear strength (since confinement increases the shear strength).

The triaxial compression test is the most sophisticated of the three tests; in it, a soil sample is subjected to confining stress (by pressurized water) and loaded to failure. This test enables the determination of pore water pressure, if any, and hence the calculation of the effective stress (effective stress = total stress − pore water pressure). Depending on the soil and the specific project, triaxial testing can be conducted under unconsolidated undrained, consolidated drained, or consolidated undrained conditions.

7.11.1 California Bearing Ratio Test (AASHTO T-193)

This procedure was originally developed by the California Division of Highways in the 1920s, was later modified by the U.S. Army Corps of Engineers, and finally adopted by ASTM and AASHTO.

The CBR test is actually a penetration test that uses a standard piston (3 sq in.) which penetrates the soil at a standard rate of 0.05 in./min. A unit load is recorded at several penetration depths, typically 0.1 and 0.2 in. The CBR value is computed by dividing the recorded unit load by a standard unit load that is required to penetration for a high-quality crushed stone material (1000 and 1500 psi for 0.1 and 0.2 in., respectively). The CBR test is conducted on a soaked sample of soil—the soaking in water (for 96 h) is conducted to simulate the worst (saturated) condition under which the pavement would perform. Figure 7.16 shows a mold with compacted soil (soaking) in a water bath.

FIGURE 7.16   California bearing ratio (CBR) mold with soil in water bath.

7.11.2 Resilient Modulus Test (AASHTO T-307)

The main disadvantage of all of the strength tests presented is that the loading is static as opposed to dynamic, which is the case due to traffic loading in the field, and that the soil sample is subjected to failure, which is not quite the case in the field, where stress levels at the subgrade level are in most cases too small to cause failure. These two deviations from actual field conditions make the strength tests relatively less useful for predicting deformation and strains under real-world traffic applications.

The more appropriate test to characterize subgrade soil is the stiffness test—one which applies dynamic loads on a soil in a way in which the traffic loads are applied in the field, and also at levels that are more representative of those coming from the traffic. The resilient modulus test is one such widely recognized and increasingly used test procedure. Resilient modulus is the elastic modulus to be used with the elastic theory. The elastic modulus based on the recoverable strain under repeated loads is called the *resilient modulus*, or M_R. It is defined as the ratio of the amplitude of the repeated axial stress to the amplitude of the resultant recoverable axial strain:

$$M_R = \frac{\sigma_d}{\varepsilon_r}$$

where
  σ_d is the deviator stress, which is the axial stress in an unconfined compression test or the axial stress in excess of the confining pressure in the triaxial compression test
  ε_r is the recoverable strain (Figures 7.17 and 7.18)

The basic principle of this test is to apply a series of stresses (with rest periods in between) under triaxial conditions and measure the resulting strain. A haversine or a triangular stress pulse is applied in order to simulate the traffic loading on pavements.

As the magnitude of the applied load is very small, the resilient modulus test is a nondestructive test, and the same sample can be used for many tests under different loading and environmental conditions. This modulus is input in the mechanistic part of the mechanistic-empirical design procedure to estimate the response (such as strain) due to a traffic load. Generally, pulse loading and rest periods of 100 and 900 ms are used, respectively.

The resilient modulus of the soil is determined at different stress levels (to simulate different levels of traffic load). These differences happen in terms of the deviator stress as well as the confining stress. Generally, designers try to model the modulus (y) with respect to one or more relevant stresses (such as deviator stress, as x, and/or confining stress, as x). Note that fine-grained soils generally show stress-softening behavior, which means modulus decreases with an increase

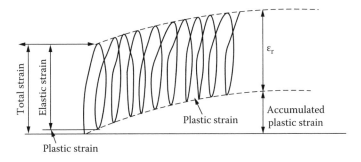

FIGURE 7.17 Plastic and elastic strain during resilient modulus testing.

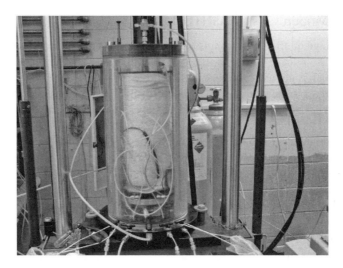

FIGURE 7.18 Soil inside cell during resilient modulus testing. (Courtesy of Mingjiang, T., Worcester Polytechnic Institute, Worcester, MA.)

in deviator stress, whereas coarse-grained soils show stress-hardening behavior, which means an increase in modulus with an increase in the sum of the principal stresses. These important observations indicate the significant difference that can be expected in the response from the same soil materials at different depths (and hence difference stress levels) of a pavement. In the resilient modulus test, the stress levels are kept low enough to ensure an elastic response of the soil. However, researchers have used experiments to correlate elastic strain with permanent strain of soils.

Confining stress is applied in this stress using a membrane and water pressure, whereas axial stress is applied through an actuator mounted on a reaction-loading frame. In the standard resilient modulus test for soils, a soil specimen is subjected to multiple (such as 16) stress stages, each consisting of the application of a specific confining stress and a fixed count of pulsed axial deviator stress. The software, which controls the test, begins testing the soil specimen for the next stress state as one state ends, until all of the stress states are completed. The first stress state is a conditioning state. Using sensors and software for the vertical and confining stress, the vertical and lateral specimen displacements are measured, and for each stress state the mean and standard deviation of the different parameters for the last five pulses are reported. An ASCII file is created with the data, which can then be imported to a spreadsheet for further analysis.

Any outlier, such as negative deviator stress values, very high or low values of resilient modulus, and deviator stress values, should be discarded before statistical analysis of the data is conducted. Next, the resilient modulus and the various stress data are evaluated to develop models relating the resilient modulus to one or more of the stress variables, according to the following equation:

$$M_r = K_1[f(\sigma)]^{K_2}$$

where
  M_r is the resilient modulus, kPa
  K_1, K_2 are the regression constants which represent the nonlinear elastic coefficients and exponents. They should be determined for each specimen to ensure that the coefficient of correlation exceeds 0.90
  σ is the stress, kPa
  $f(\sigma)$ is the stress parameter

Three values of stress parameter can be chosen: first stress invariant or bulk stress, J_1; second stress invariant/octahedral stress, J_2/τ_{oct}; and octahedral stress, τ_{oct}.

$$J_1 = \sigma_1 + \sigma_2 + \sigma_3$$
$$= 3\sigma_c + \sigma_d$$

$$J_2 = \sigma_1\sigma_2 + \sigma_2\sigma_3 + \sigma_3\sigma_1$$
$$= \sigma_c^2 + 2\left(\sigma_c\left(\sigma_c + \sigma_d\right)\right)$$

$$\tau_{oct} = \frac{1}{3}\sqrt{\left(\sigma_1 - \sigma_2\right)^2 + \left(\sigma_2 - \sigma_3\right)^2 + \left(\sigma_3 - \sigma_4\right)^2}$$
$$= \frac{1}{3}\sqrt{\left(2\left(-\sigma_d\right)\right)^2}$$

where
   σ_1, σ_2, and σ_3 are the principal stresses
   σ_c is the confining stress
   σ_d is the deviator stress

The equation can be simplified by applying log on both sides as follows:

$$\log M_r = \log K_1 + K_2 \log f(\sigma)$$

An example of raw data and analysis is shown in Figure 7.19.

The resilient modulus also needs to be conducted at different moisture and density levels to determine their effect, such that the response of a soil layer under varying environmental/construction conditions can be determined during design. One way of considering the effect of differing moisture- and stress-level conditions on the resilient modulus is to determine an effective roadbed soil resilient modulus, by considering the reduction of the resilient modulus of the soils in the different months.

The AASHTO (1993) standard guide for design of pavement provides relationships between resilient modulus and layer coefficients, whereas the relationship between CBR and resilient modulus is provided by the U.K. Transportation Research Laboratories (Powell et al., 1984). The relationships are as follows:

$$M_r = 2555(CBR)^{0.64}$$

where M_r, resilient modulus, is in psi.

$$M_r = 30,000\left(a_i/0.14\right)$$

where
   M_r is in psi
   a_i is the AASHTO layer coefficient

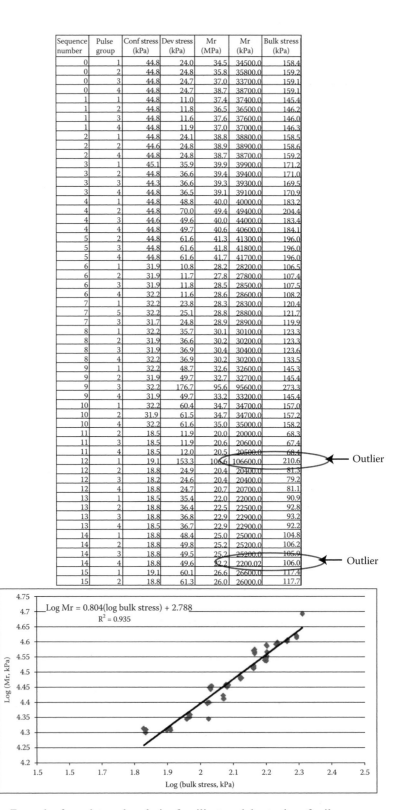

Sequence number	Pulse group	Conf stress (kPa)	Dev stress (kPa)	Mr (MPa)	Mr (kPa)	Bulk stress (kPa)	
0	1	44.8	24.0	34.5	34500.0	158.4	
0	2	44.8	24.8	35.8	35800.0	159.2	
0	3	44.8	24.7	37.0	33700.0	159.1	
0	4	44.8	24.7	38.7	38700.0	159.1	
1	1	44.8	11.0	37.4	37400.0	145.4	
1	2	44.8	11.8	36.5	36500.0	146.2	
1	3	44.8	11.6	37.6	37600.0	146.0	
1	4	44.8	11.9	37.0	37000.0	146.3	
2	1	44.8	24.1	38.8	38800.0	158.5	
2	2	44.6	24.8	38.9	38900.0	158.6	
2	4	44.8	24.8	38.7	38700.0	159.2	
3	1	45.1	35.9	39.9	39900.0	171.2	
3	2	44.8	36.6	39.4	39400.0	171.0	
3	3	44.3	36.6	39.3	39300.0	169.5	
3	4	44.8	36.5	39.1	39100.0	170.9	
4	1	44.8	48.8	40.0	40000.0	183.2	
4	2	44.8	70.0	49.4	49400.0	204.4	
4	3	44.6	49.6	40.0	44000.0	183.4	
4	4	44.8	49.7	40.6	40600.0	184.1	
5	2	44.8	61.6	41.3	41300.0	196.0	
5	3	44.8	61.6	41.8	41800.0	196.0	
5	4	44.8	61.6	41.7	41700.0	196.0	
6	1	31.9	10.8	28.2	28200.0	106.5	
6	2	31.9	11.7	27.8	27800.0	107.4	
6	3	31.9	11.8	28.5	28500.0	107.5	
6	4	32.2	11.6	28.6	28600.0	108.2	
7	1	32.2	23.8	28.3	28300.0	120.4	
7	5	32.2	25.1	28.8	28800.0	121.7	
7	3	31.7	24.8	28.9	28900.0	119.9	
8	1	32.2	35.7	30.1	30100.0	123.3	
8	2	31.9	36.6	30.2	30200.0	123.3	
8	3	31.9	36.9	30.4	30400.0	123.6	
8	4	32.2	36.9	30.2	30200.0	133.5	
9	1	32.2	48.7	32.6	32600.0	145.3	
9	2	31.9	49.7	32.7	32700.0	145.4	
9	3	32.2	176.7	95.6	95600.0	273.3	
9	4	31.9	49.7	33.2	33200.0	145.4	
10	1	32.2	60.4	34.7	34700.0	157.0	
10	2	31.9	61.5	34.7	34700.0	157.2	
10	4	32.2	61.6	35.0	35000.0	158.2	
11	2	18.5	11.9	20.0	20000.0	68.3	
11	3	18.5	11.9	20.6	20600.0	67.4	
11	4	18.5	12.0	20.5	20500.0	68.4	
12	1	19.1	153.3	106.6	106600.0	210.6	← Outlier
12	2	18.8	24.9	20.4	20400.0	81.3	
12	3	18.2	24.6	20.4	20400.0	79.2	
12	4	18.8	24.7	20.7	20700.0	81.1	
13	1	18.5	35.4	22.0	22000.0	90.9	
13	2	18.8	36.4	22.5	22500.0	92.8	
13	3	18.8	36.8	22.9	22900.0	93.2	
13	4	18.5	36.7	22.9	22900.0	92.2	
14	1	18.8	48.4	25.0	25000.0	104.8	
14	2	18.8	49.8	25.2	25200.0	106.2	
14	3	18.8	49.5	25.2	25200.0	105.9	
14	4	18.8	49.6	22.2	2200.02	106.0	← Outlier
15	1	19.1	60.1	26.6	26600.0	117.4	
15	2	18.8	61.3	26.0	26000.0	117.7	

Log Mr = 0.804(log bulk stress) + 2.788
$R^2 = 0.935$

FIGURE 7.19 Example of raw data and analysis of resilient modulus testing of soil.

7.11.3 Dynamic Cone Penetrometer (ASTM D6951)

The dynamic cone penetrometer (DCP) is a device that is used for the determination of resistance against penetration (and thereafter obtain CBR or unconfined strength or modulus, through correlation) of granular materials. It consists of a steel rod with a pointed tip (cone) that is driven into the granular material with a weight that slides along the rod. In an existing pavement, a hole is dug and the DCP is lowered onto the granular layer for carrying out the test. The penetration of the rod per blow of the weight is recorded from a scale attached to the rod, and is then utilized to obtain an indication of the strength/stiffness of the layer (Figure 7.20). A drop of the weight is used to seat the device on the layer. Usually, three people are needed to conduct the test—one person for holding the device perpendicular, the second person to lift and release the weight, and the third person for recording the reading. An 8 kg weight is standard, but a lighter, 4.6 kg weight can be used for soft ground conditions. Readings at every drop of the weight are taken, until the desired depth or refusal depth is reached. The number of blows and the penetration are plotted along the x and y axes, respectively. Best-fit lines are drawn and slope of the DCP data (mm/blows) is determined. Examples of standard table and correlations to determine strength/stiffness are shown in Figure 7.20.

7.12 SUBGRADE SOIL TESTS FOR RIGID PAVEMENTS

When considering the magnitude of stresses induced in a slab under loading, the influence of the subgrade is defined by its modulus of subgrade reaction (k). The modulus of subgrade reaction is defined as the ratio of the pressure applied to the subgrade using a loaded area divided by the displacement experienced by that loaded area. (Figure 7.21 shows a slab with springs deflecting under load.)

7.12.1 Plate Load Test

The plate load test (Figure 7.22) applies a load to a steel plate bearing on the subgrade surface. The plate load is usually applied using a hydraulic jack mounted against a vehicle used for the reaction frame. The resulting surface deflection is read from dial micrometers near the plate edge but away from the loading point. The value of k is found by dividing the pressure exerted on the plate by the resulting vertical deflection, and is expressed in units of N/mm^3, MN/m^3, kg/cm^3, or $lb/in.^3$ (pci). k is established by plate-bearing tests with a load plate diameter of 750 mm (30 in.). A modification is needed if a different plate diameter is used: for a 300 mm diameter plate, k is obtained by dividing the result by 2–3, and for a 160-mm diameter plate, it is divided by 3.8. The modulus of subgrade reaction is determined by the following equation:

$$k = \frac{P}{\Delta}$$

where
   k is the modulus of subgrade reaction
   P is the applied pressure (load divided by the area of the 762-mm [30-in.] diameter plate)
      (or, stress)
   Δ is the measured deflection of the 762-mm (30-in.) diameter plate

Number of blows, A (Number of hammer blows between test readings)	Cumulative penetration, mm, B, Cumulative penetration after each set of hammer blows	Penetration between readings, mm, C, difference in cumulative penetration (B) between readings	Penetration per blow, mm, D, C/A	Hammer Factor, E, 1 for 8 kg weight (hammer), 2 for 4.6 kg weight	DCP Index, mm/blow, F, D*E	CBR, %
0	0	—	—	—	—	—
5	25	25	5	1	5	50
5	55	30	6	1	6	40
15	125	70	5	1	5	50
10	175	50	5	1	5	50
5	205	30	6	1	6	40
5	230	25	5	1	5	50

Type of soil: GW/CL (Adapted from example from ASTM 6951).

Type of soil/CBR	CBR-DCP correlation (DCP in mm/blow)
All soils except for CL soils below CBR 10 and CH soils.	$CBR = 292 / DCP^{1.12}$
For CL soils below CBR 10 and CH soils	$CBR = 1 / (0.017019 * DCP^2)$
CL soils with CBR < 10	$CBR = 1 / (0.002871 * DCP)$

Source: Webster, S.L., Grau, R.H., and Williams, T.P., Description and application of dual mass dynamic cone penetrometer, Report GL-92-3, Department of the Army, Washington, DC, p. 19, May 1992.

FIGURE 7.20 Schematic of dynamic cone penetrometer (DCP), example data table and correlations.

FIGURE 7.21 Slab on springs and spring constant.

FIGURE 7.22 Schematic of plate load test.

Example 7.3

A plate load test was conducted on two types of soils, a subgrade and a 10-in. base over the subgrade. Results from the test are shown in Figure 7.23. Determine (k) for the subgrade and (k) for the subgrade-base combination.

Solution

The modulus of subgrade reaction is determined from the slope of the stress versus deformation curve:

$$(k) = \frac{stress}{deformation}$$

As can be seen from Figure 7.23, the subgrade curve has two slopes: a steep initial slope of 350 pci and a more shallow slope of 133 pci. At this stage, the designer must decide on the level of deformation the subgrade will be subjected to. Otherwise, a more conservative value of 133 should be selected. For the base + subgrade behavior, a relatively constant relationship is observed and the slope is approximately 277 pci.

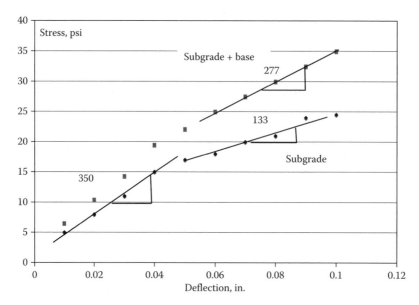

FIGURE 7.23   Example of results from a plate load test.

7.13   SUBBASE AND UNSTABILIZED BASE

Subbase and unstabilized base materials are generally characterized in the same ways. The characterization tests are the same as those mentioned for subgrades, except that organic content is not generally determined.

7.14   SOIL STABILIZATION CONCEPTS AND METHODS: CHEMICAL AND MECHANICAL

Soil stabilization refers to improvement of soil so that it performs a required function. Soil stabilization techniques can be classified into two groups, mechanical and admixture. Mechanical stabilization can be through densification or compaction, addition of granular material and compaction, and the use of reinforcement, such as geotextile. Admixture stabilization (which also includes compaction as part of the process) can be done with the use of additives, such as lime, Portland cement, and asphalt.

The stabilization process or mechanism results in the modification of the properties of the soil. Compaction results in increase in strength or stiffness and decrease in permeability and compressibility. The use of additives (admixtures or modifiers) results in change in basic properties of the soil. For example, the addition of lime and cement and granular materials reduce plasticity. Cementing agents such as asphalt help develop bonds between soil particles, and hence increases cohesive strength, and decreases the absorption of water. Water-retaining agents such as calcium chloride ($CaCl_2$) reduce evaporation rates and help in reducing dust in construction sites as well as unpaved roads.

7.14.1   MECHANICAL STABILIZATION BY DENSIFICATION OR COMPACTION

Compaction is the artificial process of increasing the density or unit weight and decreasing the void ratio of a soil mass. The energy consumed during compaction is called compactive effort.

The mechanism of compaction consists of reorientation of soil particles, fracture of soil particles and breaking of bonds between them, and bending or distortion of soil particles. For coarse-grained soil, the primary mechanism is reorientation with some fracture, and the compactive effort must be high enough so as to overcome the friction between particles to make them reoriented. Water helps in the compaction by lubrication of the contacts. Effective compaction requires reduction in frictional forces between the particles. Vibration is effective in moving particles relative to each other and reducing the normal force and hence the friction.

In fine-grained soil, the primary mechanism of compaction is reorientation and distortion. The compactive effort must be high enough to overcome the cohesive/interparticle forces. Increased water content can decrease cohesion. Effective compaction requires high pressure and shearing or kneading action, to overcome cohesion. Sheepfoot roller (with small contact area), tamping foot roller, and rubber tire roller are effective.

7.14.1.1 Effect of Compaction on Soil Properties

For cohesionless soil (granular soils), the effect of compaction are in terms of increase in strength and stiffness and decrease in compressibility and permeability. The properties are not affected by moisture content. The density that could be achieved with a specific compactive effort influences the properties. For cohesive soils, there can be difference in properties, even at the same dry density (due to difference in the soil structure, which can be either flocculated or dispersed). Generally higher the density, higher is the strength. Note that soils compacted dry of OMC will have higher permeability. The swelling potential is higher when density is high and degree of saturation is low. The potential for shrinkage is higher for low density and high degree of saturation.

7.14.1.2 Field Compaction

It is important to know the desired soil properties and select a laboratory method for controlling density, and hence develop specifications on percent compaction or percent maximum density. The water content should range about the OMC.

The equipment and procedure for compaction should be selected such as to allow the level of compaction at the desired moisture content to be achieved economically.

The factors that influence field compaction are as follows:

1. Type of roller (vibratory rollers may cause fluffing, so the last few passes are made with vibration turned off).
2. Size of rollers—weight and contact area/pressure.
3. Lift (Layer) thickness—thicker the lift less effective is the compaction.
4. Number of passes.
5. Strength and stiffness of the underlying material. (Whether the compaction effort densifies the soil or actually shears it depends on this factor. One can start with lower compactive effort initially, and then gradually go to higher compactive effort.)

7.14.1.3 Field Control

Compaction in the field needs to be controlled to ensure adequate density. This is done with the help of specifications and measuring devices. Specifications are generally written to control density and moisture content. There is a target density and a target water content range about the OMC. The average may be close to the target but there may be some variability. Standard deviation values, from historical data, are used to set tolerances. Generally, a one-sided acceptance limit is set for density, and the specification might read, 100% pay if relative compaction ≥93%.

7.14.1.4 Measuring Devices

For density one can use sand cone or balloon density measurement. It involves direct measurement of mass and volume to calculate density. The disadvantage is that it is time consuming. A better method, which can be used for both density and moisture content, is using a nuclear gage, for which the key is calibration. These measurements are done instantaneously.

The nuclear gage can be used either in direct or backscatter mode. In direct method, a hole is drilled and the probe is inserted in it, and the signal coming out of the probe tip (hydrogen atom) is picked up, measuring average condition in between the depth of the probe and the surface. In backscatter mode, the device is kept on the surface and a definite depth of measurement is associated with this method.

7.14.1.5 Intelligent Soil Compaction System

In recent years, a number of roller companies have developed intelligent soil compaction systems, which continuously measure the stiffness/modulus of the compacted soil in real time, provide automatic feedback to control of vibration amplitude and frequency and have integrated global positioning system to provide a complete geographic information system-based record of the earthwork site. A NCHRP study (NCHRP 21-09, final report available at http://onlinepubs.trb.org/onlinepubs/nchrp/nchrp_rpt_676.pdf) provides details of testing with smooth drum and pad foot rollers (consisting of intelligent soil compaction and *roller-integrated continuous compaction control systems*) of different manufacturers—Ammann (http://www.ammann-group.com/en/home/), Ammann/Case, Bomag (http://www.bomag.com/usa/index.aspx?&Lang = 478), Caterpillar (http://www.cat.com), Dynapac (http://www.dynapac.com/), and Sakai (http://www.sakaiamerica.com/)—and present recommended specifications, fundamentals of the roller measurement systems, relationship between roller-measured stiffness and in situ stress–strain–modulus behavior, evaluation of automatic feedback control-based intelligent compaction, correlation of roller measurement values to spot-test measurements, and case studies of implementation of recommended specifications. Some of the key points that have come out from this study are that the vibration-based roller measurement values, MVs (different MVs are shown in Table 7.8), are correlated to each other but are significantly affected by operational parameters, such as amplitude, and the effect is dependent on the soil and layer conditions. The MVs are representative of a much deeper area than that of common compaction lifts and are a composite indication of the stiffness of a structure, such as subgrade, base, and subbase, and the relative contribution of each layer is dependent on layer thickness, relative stiffness of the layers, vibration amplitude, and drum/soil interaction issues (contact area, dynamics). With the current first-generation technology, recommendations are made for different types of specifications, which may or may not involve the use of MVs for quality assurance, through initial calibration.

7.14.2 Use of Geosynthetics

Geosynthetics (planar products manufactured from polymeric materials) are also used for stabilization of subgrades. Of the different types of geosynthetic, geotextiles and geogrids are primarily

TABLE 7.8
Roller Measurement Values Used

Manufacturer	MV	Drum Vibration Parameters Used for Determining Roller MV
Dynapac, Caterpillar, Hamm, Volvo	Compaction meter value (CMV)	In the frequency domain, ratio of vertical drum acceleration amplitudes at fundamental (operating) vibration frequency and its first harmonic
Compaction control value (CCV)	Sakai	In the frequency domain, algebraic relationship of multiple vertical drum vibration amplitudes, including fundamental frequency, and multiple harmonics and subharmonics
Stiffness, k_s	Ammann, Case	Vertical drum displacement, drum–soil contact force
Vibration modulus, E_{vib}	Bomag	Vertical drum displacement, drum–soil contact force

Source: Adapted from Mooney, M.A. et al., Intelligent soil compaction systems, NCHRP Report 676, Transportation Research Board, Washington, DC, 2010.

used, either for separation of the subgrade from the aggregate base (to avoid contamination of the base with the subgrade materials) and/or for strengthening of the subgrade (particularly for very poor and/or organic soils and unpaved roads). For strengthening of the subgrade, sufficient stiffness and good interlock with the surrounding materials is necessary. Sometimes nonwoven/grids are more effective than woven geotextiles. Hence, the selection of the geosynthetic should be done after careful consideration of the existing soil conditions and objective of stabilization. The construction process consists of rolling out the geotextile on the cleaned and leveled subgrade and pulling out to avoid wrinkles. Adjacent rolls are overlapped (by say 30 in.) and the fabrics are kept in place by pins or staples or base materials. A woven or nonwoven geotexile could be used for strengthening the subgrade, or a geogrid could be used in conjunction with a separator geotextile. The relevant test properties include tensile strength (ASTM D4632 and D4595), tear strength (ASTM D4533), puncture strength (ASTM D 4833), permittivity (ASTM D4491), apparent opening size (ASTM D4751), and resistance to ultraviolet light (ASTM D4355). Geocell is a type of a "three-dimensional" structure formed by welding strips of high-density polyethylene into a honeycombed structure. Different types of thickness of the material, as well as height and aspect ratio, are used for geocells, and they can be backfilled with aggregates. Geocells are utilized as effective means of improving soil confinement and hence improvement of the bearing capacity, for example in poor quality subgrade. Good information is available from the International Geosynthetic Society (http://geosyntheticssociety.org/) as well as from the Geosynthetic Institute (http://www.geosynthetic-institute.org/).

7.14.3 Lime Treatment of Soils

Lime is often used for soil stabilization. (Detailed information can be obtained from the National Lime Association, www.lime.org.) Limestone, $CaCO_3$ or Dolomite, $CaCO_3 + MgCO_3$, is heated in a rotary kiln to produce quicklime, $CaO + CO_2$ or $CaO + MgO + CO_2$. Water is added to it to make it react, $+ H_2O \rightarrow$ producing hydrated lime, $Ca(OH)_2 + Mg(OH)_2$. In a unit amount of quicklime, there are more Ca++ ions than in hydrated lime, and hence it is more effective. There is not much difference between dolomite or limestone, as far as stabilization is concerned.

The mechanisms of soil–lime reaction consist of the following:

1. The modification process happens through cation exchange. Ca++ is substituted for absorbed cations in the clay soil–water system. This process is most effective when the absorbed cations are sodium, Na+. They are replaced by Ca++. Li+ are most easily replaced although they are not that common. It is replaced by Na+, which is the most common, and which is replaced by K+, which can be replaced by Mg++ and Ca++. Two valence materials have stronger replacing power.
   Replacement power:

   $$Li+ < Na+ < K+ \ll Mg++ \ll Ca++ \text{ (Holtz and Kovacs 1981)}$$

2. Flocculation and agglomeration process that allows the end to side contact, and soil particles behave as larger particles.
3. Cementing of soil particles due to carbonation of lime. In the presence of CO_2, $CaCO_3$ is produced from CaO. Also, there is pozzolanic reaction, in which lime reacts with alumina and silica from the clay minerals to form cementitious compounds, similar to the process during hydration of Portland cement. This reaction occurs only when there is excess lime after the completion of the cation exchange process.

The modification process (cation exchange) is a faster reaction, compared to the other mechanisms. The lime content required generally ranges from 3% to 5% (of hydrated lime by weight of the soil). As a result of this mechanism, the PI is decreased, as the liquid limit decreases and the plastic limit increases. The effective particle size increases, and as the moisture content remains the same, it appears that the soil is drier, and more workable. Volume change and swell potential also decreases.

The cementing mechanism can occur only after Ca++ requirements for modification have been satisfied. The requirement generally ranges from 5% to 10% hydrated lime. Adding more than 10% lime is generally not economically viable. This mechanism results in increase of strength, as can be measured with unconfined compressive strength. It is also common to add pozzolona (fly ash) to the soil to provide silica for the cementing action. This process is called lime-fly ash stabilization.

Lime treatment is used in subgrade soils, and subbase and base materials. For subgrade soils (clay), lime can be used to reduce the potential of change in volume, whereas for marginal base and subbase materials it is used for improving their quality. The main benefit from lime is the reduction in plasticity, and hence the reduction in the effect of water. This leads to improvement in both constructability and performance of the pavement. Soils already containing Ca++ are not affected by lime treatment, and the secondary effect of strength gain is, in some cases (such as sand-gravel mix), less than what could be achieved with the addition of Portland cement.

The amount of lime needed depends on whether it is to be used as a modifier alone or also for increasing the strength. The modification process results in reduction of soil plasticity, which is a measurement of the effect of water on the soil, and the applicable tests are the Atterberg limit tests, LL, PL, and PI. Occasionally, swell tests are run to note either pressure or volume change. The cation exchange occurs rapidly and hence these tests can be performed quickly. The lime content is expressed as a percentage of hydrated lime by dry weight of the soil. For the same lime, generally 33% more hydrate lime (by weight) is required compared to quick lime. There may or may not be consideration of the strength and stiffness increase in design, and how permanent the strength increase is, is unknown.

The cementing action results in an increase in strength, and this can be evaluated by running unconfined compressive strength or indirect tensile strength tests. The addition of lime may result in increase in strength in a layer, such as in subbase, which could lead to a reduction of the designed layer. Generally, for subgrade materials, the lime is added purely as a modifier.

The construction process of lime stabilization involves the following steps:

1. Scarification and pulverization, done after cutting down the material to the required slope and grade. The scarification process loosens up the soil and makes mixing easy.
2. Spreading lime, dry or as a slurry, depending on whether the area is windy, as well as costs. Considerations include cost, environmental, production rate, and natural water content of the soil.
3. Preliminary mixing and watering, which can be eliminated for easily workable subgrades and subbase materials. Heavy clays require two mixings, and rotary mixers are the most efficient. Water is required for compaction, to get lime distributed uniformly through the soil, and is needed for the pozzolanic reaction.
4. Preliminary curing, to give time to react throughout the soil. (The earlier two steps are mostly for high-plasticity subgrade, to make them friable and workable.)
5. Second lime application.
6. Final watering and mixing.
7. Compaction.
8. Finishing; subgrade trimmers are used for proper grading.

Curing, which may involve spraying asphalt binder, is required to prevent moisture loss, erosion and damage from traffic, as well as construction traffic. Construction can also be done with a central pugmill mixing, for subbase and base course stabilization. In this process, the mixing of soil, water lime, and fly ash, if needed, are added in proportion and mixed in the pugmill. Spreading is done by trucks and the rest of the processes are similar to in-place application of lime.

Climatic consideration in the stabilization process is important. If temperature is too low, hydration or modification reaction with lime will not occur. The rule of thumb is that lime stabilization should not be started unless the temperature is 40°F–45°F and rising.

Safety consideration is also important. Quicklime is dangerous for the skin and dusting in urban areas could be a problem.

Lime slurry pressure injection (LSPI) is also used for stabilization. In this process, lime slurry is injected into the soil under pressure, ranging from 20 to 200 psi, at spacing of 5 ft, for example, and to a depth of 7–12 ft, depending on the application, but generally deep enough to go below the depth of seasonal moisture fluctuations.

The volume of slurry is generally 10 gal/ft of depth, and 2.5–3 lb of lime per gallon of water. Because of the low permeability of clays (soils for which lime treatment is necessary), fractures, seams, and fissures in the soil are necessary for effective application of this process. This process is most efficient in the dry season when cracks and fissures are the widest.

This process can be used as pretreatment or a posttreatment, for stabilization of building foundations and subgrades for railroads, streets, and airport pavements (e.g., to prevent swelling), and also in levees and slopes to reduce shrinkage potential.

Examples of modern lime spreading and stabilization equipment are shown in Figure 7.24.

7.14.4 CEMENT TREATMENT OF SOIL

In cement treatment, sufficient Portland cement is added to the soil to produce a hardened mixture, resulting from hydration reaction. (Detailed information can be obtained from the Portland Cement Association, www.cement.org, as well as the American Concrete Pavement Association, www.pavement.com.) Portland cement, which is produced from a mixture of limestone, clay, shales, and iron ore, contain Ca, Si, and Al, which reacts with water, to produce hydrated calcium silicate or aluminate. Unlike lime, cement–soil reaction does not require external source of Al, and Si, and hence is more effective in this reaction than lime.

Cement-stabilized soils can be categorized into a few groups:

1. Soil cement, which consists of a mix of natural soil of low or marginal quality (such as natural clay or clayey sand) and cement. The objective of this stabilization is to make such soils usable as subgrade soils. The feasibility of using cement becomes an issue in such cases where the demand for the cement can exceed 20%.
2. Cement treated base (CTB) is a mix of granular materials of reasonably high quality (often they meet base course specifications) and cement, used as high-quality stiff base course in pavements with high loads and traffic volume. The mix can contain wither crushed or uncrushed base course aggregate, or with a blend of both materials. The cement content is generally kept lower than 4% to prevent excessive reflective cracking in pavements.
3. Econocrete, which is a mix of low-quality aggregates (natural or reclaimed aggregates, which do not meet the standard specifications) and cement, used principally as a subbase (such as in airport pavements).

The use of CTB under a relatively thin (say 3 in.) hot mix asphalt (HMA) surface layer would result in reflection cracks. While breaking the bond between the CTB and the HMA with a crushed stone layer is one option, it results in a structure with a weaker layer above a stiffer layer. Since the option of using a very thick HMA is not economical, it is often advised not to use cement-stabilized materials in flexible pavements (with HMA surface layer), or follow guideline regarding the minimum thickness of HMA with CTB. Since the spacing of the cracks is a function of the strength of the CTB, a lower percentage of cement would result in a larger spacing of the cracks. Therefore, the maximum allowable cement content could be determined from the minimum thickness of HMA layer that is needed for controlling the cracks.

Wirtgen SW 16TA lime spreader
Tank capacity: 16.5 m³, Spreading width: 2500 mm
Spreading quantity: 2–50 L/m², Manual operation
spreading auger

Wirtgen SW 16MC lime and cement spreader
Tank capacity: 16.5 m³, Spreading width: 2460 mm
Spreading quantity: 2–50 L/m², Cellular wheel
automatic operation

Lime stabilization with Wirtgen WR 2500
Working width: 8 ft, Working depth: 0–20 in.
Engine output: 690 HP, Weight: 63,000 lb

Lime stabilization with Wirtgen WR 2000
Working width: 6 ft 7 in., Working depth: 0–20 in.
Engine output: 400 HP, Weight: 49,060 lb

Micro-processor controlled
pump for addition of water

Cement pre-spread

Train of equipment

Roller   Grader   Stabilizer   Cement
                    with water   spreader
                      truck

Cutter and mixer of stabilizer

Weigh auger for
accurate cement dosage

2900 gal
water tank

Slurry to stabilizer
via a flexible hose

25 cu yd
cement tank

Pumps for
cement slurry

260 gal/min
slurry mixer

Wirtgen WM 1000 cement slurry mixer
used for cement stablization

FIGURE 7.24 Examples of equipment used in soil stabilization. (Courtesy of Mike Marshall, Wirtgen GmbH, Windhagen, Germany.)

The modification with cement results in three things—reduction in plasticity, decrease in volume change capacity, and increase in strength and stiffness. Because cementation may not be permanent, the reduction in plasticity is perhaps the most important process. The increase in strength, due to cementation, may be lost due to weathering, such as through repeated freeze–thaw cycles. The hydration reaction of Portland cement and water that results in cementitious compounds that cement soil particles together is a time-dependent process. In the initial cementation process, there is rapid strength increase; then there is a slow long-term strength gain. There is release of Ca++ ions as a result of Pozzolanic reaction, and the slow process is the carbonation reaction where Ca reverts back to $CaCO_3$ compound.

7.14.4.1 Mixture Design Process

The mix design process for cement-modified soils (meant for reduction of plasticity and permanent or temporary strength gain) is simple, whereas for cement-stabilized soils (meant for increasing the strength of the soil; change in strength is permanent; durability is important), it is more complex.

7.14.4.1.1 Mix Design for Cement-Modified Soils

Atterberg Limit tests are run on raw and then on modified soils usually at 1, 24, and 48 h after mixing to establish plasticity modification. Note that most of the modification happens within an hour. Moisture density tests are conducted to establish compaction requirements. Strength increase is not considered but compressive early strength test is run to check the requirements of minimum strength.

7.14.4.1.2 Mix Design for Cement-Stabilized Soils

The purposes of the mix design are to determine the amount of Portland cement to adequately harden soil (adequacy is determined by the development and maintenance of strength during weathering), determine the amount of water needed for compaction (OMC is usually sufficient for hydration), and determine how density affects strength, stiffness, and durability of the mix.

Step 1: Selection of preliminary cement content from design guidelines—poorer the soil, the higher is the amount of cement required; the amount needed goes up significantly if there is organic material.

Step 2: Conducting moisture-density tests—different for that of raw soil—this is given in ASTM D558 and AASHTO T134; this step helps in establishing OMC and maximum dry density for compaction durability, strength, and construction compaction control.

Step 3: Conducting durability tests—wet dry tests—ASTM D559, AASHTO T135; freeze–thaw tests—ASTM D560, AASHTO T136. The purpose is to select cement content so that soil cement or CTB will stay hard and maintain strength when subjected to alternate wetting/drying or freezing/thawing cycles. Strength may be considered as secondary requirement but it has been determined that cement content established by durability insure adequate long-term strength. Test procedure calls for three cement contents, and one freeze–thaw and one wet dry specimen at each cement content. Specimens are 4 in. in diameter and $1/30$ ft^3 in volume. The specimens are compacted to OMC and maximum dry density (from step 2) moisture density tests performed on material with median cement content. The specimens are moist cured for 7 days and then subjected to wet dry or freeze/thaw tests, as follows. For wet dry tests, samples are soaked for 5 h at room temperature (77°F), and then oven dried for 42 h at 160°F. Wire brush is used to apply 2 strokes per area. This is repeated for 12 cycles. After 12th cycle, it is dried at 230°F and weighed. The weight loss is calculated as% of the original dry weight of the sample. Note to find the dry weight one needs the moisture content of the sample.

The moisture density curve to get max dry density is plotted for the median cement content. The same compactive effort is needed when preparing wet/dry, or freeze/thaw samples.

For freeze–thaw test, the following steps are used. Freeze at −10°F for 24 h; thaw in moist room at 70°F and 100% relative humidity for 23 h. Brush after each thaw cycle; repeat for 12 cycles; compute weight loss.

Selection of final cement content is based on the weight loss as follows (PCA method). Weight loss for different soil types (AASHTO group):

$$A-1, A-2-4, A-2-5, A-3 \leq 14\%$$

$$A-2-6, A-2-7, A-4, A-5 \leq 10\%$$

$$A-6, A-7 \leq 7\%$$

For compressive strength requirement, supplementary tests to durability tests are conducted at cement content selected on the basis of durability tests.

For this test 4–2 in. diameter specimens are used. The samples are moist cured for 2, 7, and 28 days and tested at various times. Strength criterion is based on minimum strength; additional requirement could be that strength increases with age.

The construction process for soil cement is very similar to that with lime.

The objective include pulverization of the soil, mixing with the proper amount of cement, compacting to maximum optimum water content to get maximum dry density or a percentage thereof, and maintaining moisture to allow hydration of cement. The steps are as follows:

1. Shape to proper line and grade (should be done prior to adding cement since once it starts to hydrate it is difficult to line and grade).
2. Scarify, pulverize, and prewet—important in dry windy condition, since if the interface is dry and windy a significant amount of cement is lost.
3. Spread Portland cement—done with tanker, with spreader bar attached; small jobs may use bags of cement, each bag weighs 94 lb (approximately 1 ft^3).
4. Add water and mix—twin shaft roller may be used just as is used in lime.
5. Compact—initial, usually done with sheep foot or vibratory roller, and finish, done with rubber tired or steel wheel roller.
6. Final shaping and finishing—to reestablish the line and grade—must be accomplished quickly before hydration. Typical finishing operations include shaping with a motor grader, rolling with a steel wheel roller and then with a rubber tired roller to seal the surface. Curing is generally done for 7 days, either with moist cloth or burlap of by spraying with an emulsion (say 0.15–0.3 gal/yd^2). The surface is then sanded to protect surface, if traffic is expected.

The mixing process can also be conducted in a plant. In this process, the subgrade is compacted and shaped to line and grade soil, and the cement and water are mixed in a central plant. A screw conveyor or compressed air is used to get cement from silo to pugmill in the plant. The mix is then transported to the job site in dump trucks. It must be spread by mechanical spreaders for uniform thickness. It is advisable to break up the construction area into smaller elements and do mixing and compaction. Such proper sequencing of construction operation is required to prevent damage to unfinished sections in the case of rain.

Temperature is an important consideration—soil cement work is not started if it is not 40°F and rising (April to October is suitable for most part of the United States). There should not be freezing for 7 days; otherwise preliminary strength gain will be destroyed. The surface must be protected from heavy traffic and erosion from rain. Application of emulsion provides abrasion resistance to protect from constructing equipment. Rain is actually good as it prevents moisture loss; however, a prime coat is needed to resist erosion.

For quality control, checks are required for compaction (dry density), water content, and cement content. Cement content can be checked by putting a 1 yd^2 burlap on the pavement and weighing it after the cement has been spread. Figure 7.23 shows a modern cement stabilization equipment.

7.14.5 Asphalt (Bituminous) Treated Soil

Asphalt viscosity has to be reduced to be mixed with soil. In the hot process, heated asphalt cement is mixed with heated aggregate mixture to produce high-quality HMA. In the cold process, a mix of asphalt emulsion and soil is produced giving a low-quality material. Clean sand and natural gravels are used. (Note that foamed asphalt is also increasing being used for asphalt stabilization—for details, see Section 18.5.2.2.3.)

7.14.5.1 Stabilization Mechanism with Asphalt Treatment

It improves the properties of the soil by the following mechanisms.

1. Waterproofing: The soil particles are coated with asphalt, and water does not get in contact with the soil particles and hence makes the soil less sensitive to water and lowers the absorption of water.
2. Cementation: The asphalt increases the cohesion of the mix.

Asphalt emulsion is used with fine-grained soils primarily to make them waterproof. It may be used in conjunction with lime treatment. For fine-grained soils with maximum LL-40, PI-18, 4%–8% asphalt is used.

Clean sand is also used, to produce sand asphalt, for pavement layers as well as impervious liners or slope protection. Generally 4%–10% asphalt is used, when the primary requirement is cementation. General rules include limiting the PI at 12%, and a also limit on maximum passing the #200 sieve, such as 12% or 25%.

Gravel and sand gravel mixtures are used to improve substandard materials having unacceptably high fine content (–#200 material). The primary function of such mixes is waterproofing and less strength change and volume change potential when exposed to water. General rules include a limit of 12% for the PI, and maximum percent passing the #200 sieve is 15%. Generally 2%–6% asphalt is used.

7.14.5.2 Mix Design Procedure

While the Marshall mix design procedure can be used for HMA (discussed in Chapter 14), the asphalt soil mixes could be designed using the Hubbard–Field mix design procedure (ASTM D915, ASTM D1138—note that both standards have been withdrawn from ASTM). The basic procedure for the Hubbard–Field mix design consists of compaction using different asphalt contents and checking the strength with extrusion test. The extrusion requirements are as follows:

Fine-grained soils
Room temperature—extrusion value before soaking 1000 min
Extrusion value after soaking 7 days—400 min
Expansion, 5% max, absorbed water—7% maximum
Sand and gravel
Sands: Extrusion value minimum, 1200 (tested while submerged at 140°F—there is an air curing period prior to testing)

The properties considered in Marshall mix design are stability, flow, air voids, and voids in mineral aggregate.

Construction considerations for asphalt-treated soils: Asphalt-treated soils are mixed in-place or central mix operation, depending on the location of the material to be stabilized.

The steps in the in-place mixing operation consist of establishing line and grade, pulverizing the material, adding emulsion, mixing, and compaction. The application of emulsion and the mixing can be done with a distributor truck with a spray bar and a mixer, or both steps can be accomplished with one single pass of a specialized equipment. Adequate moisture in the soil is necessary for proper mixing, and the timing of compaction is critical to avoid rutting and shoving.

A travel plant or a traveling pug mill can also be used for mixing and is used mostly for cold mix. Plant mix (central mix) operation is seldom used for fine-grained soils. Emulsion and unheated aggregates are metered and blended in a pugmill. Generally a continuous plant is used, where aggregates and emulsion are fed and mixed continuously. Hauling, spreading, and compaction follow, and timing of the aforementioned operations is critical. The timing should be made relative to the time of the break of the emulsion. The key is the consideration of the type of emulsion used—rapid, medium, or slow setting (determines how fast the breaking occurs).

Rapid setting emulsion is used in surface treatments or chip seals; it involves spraying a layer of asphalt by a distributor truck and putting aggregates with another truck behind it. In many cases, this happens in a trafficked road, so rapid setting emulsion is desirable. Medium setting emulsion is used for short haul jobs, and slow setting emulsions are used for long haul jobs. By varying the type and amount of the emulsifying agent, one controls the setting time (RS, MS, or SS) of the emulsion. For construction control, checking the density and asphalt content (e.g., with nuclear gage) are required.

Examples of modern equipment used for soil stabilization are shown in Figure 7.23.

7.15   DUST CONTROL

Dust control on unpaved roads is necessary because of the following:

1. Dust decreases safety in roads by reducing visibility.
2. Dust in roads causes environmental hazards.
3. Loss of road materials causes increases in cost of maintenance of roads, as well as maintenance of vehicles.

Results of several dust control studies in the United States (The Forest Service, U.S. Department of Agriculture [USDA], see http://www.fs.fed.us/eng/pubs/html/99771207/99771207.html) as well as other countries such as South Africa (Jones, 1999) and Australia (Foley et al., 1996) are available for guidance. The amount of dust that is generated depends on a number of factors such as silt content of the soil, speed, weight, and number of vehicles as well as rainfall. Dust palliatives can be grouped into different types, such as water, water-absorbing chemicals (salts), organic petroleum products (e.g., asphalt and resins), organic nonpetroleum product (such as lignosulfate), electrochemical products (e.g., enzymes), synthetic polymer products (such as polyvinyl acetate), and clay (e.g., bentonite).

Water is readily available and is easy to apply but it evaporates quickly and does not last long. Asphalt emulsions are also relatively easy to apply, but are relatively expensive and could run off to side ditches. Hygroscopic materials, such as salts (NaCl, $CaCl_2$, and $MgCl_2$), draw water from the air and keep the ground moist and hence reduce the potential of formation of dust. NaCl is not very effective since it requires relative humidity >75% to work. $CaCl_2$ and $MgCl_2$ are effective in relative humidity of as low as 30%–40%. They are capable of drawing moisture from atmosphere in low relative humidity and keep the soil more moist than it would actually be without the salts. Generally $CaCl_2$ or $MgCl_2$ are in put down in solution by distributor truck and mixed in with the soil with motor grader. The advantages of $CaCl_2$, $MgCl_2$ are that they are relatively inexpensive and easy to apply. The disadvantages are that they are corrosive and cause rusting of equipment/vehicles, and salts in ground water and surface water are not desirable, particularly in agricultural areas. Also, salts may dissolve in rain and get washed away, and hence not provide a permanent solution to the dust problem. Sometimes the salts could make a road surface more slippery.

Lignosulfates are processed pulp liquor from paper mills, typically consisting of 50% solids. When sprayed down it acts as a binder and holds the particles together. Advantages are that it is relatively inexpensive, effective, noncorrosive, environmentally safe, and gives some hardening of the surface and makes it more trafficable. The disadvantages are that it may leach away, and the treated material may not be easily degradable. Emulsified petroleum resins are also used as a dust palliative. Containing about 60% resin, this product is generally diluted with water and applied at the rate of ½ to ¾ gal/yd². These commercial products are relatively expensive, but provide the advantages of the ability to control the depth of penetration by varying the amount of concentrate, and the ability to store products for long term (such as through winter). The buildup of resin residue in the soil allows decreasing the percentage of concentrate over succeeding applications. (For example, see http://www.goldenbearoil.com/PDF%20Files/Coherex.pdf.)

New generations of dust control additives are being developed on the basis of sophisticated technology. One such example is TerraLOC®, based on nanotechnology, that has been developed by MonoSol. The material is environmentally safe, biodegradable, cures quickly, and can be applied without modification of conventional spraying equipment. The application of water can extend the life of TerraLOC that is already in the soil, in between applications. This product has been successfully used in military applications such as landing areas for helicopters, and civil airports, and can be used for unpaved roads and industrial areas for dust control (for details, see http://www.terraloc.com/).

QUESTIONS

7.1   What are the different soil-forming minerals?

7.2   Classify the following inorganic soil according to the AASHTO procedure:

Sieve	% Passing	Atterberg Limits
No. 40	96	LL = 35
No. 200	55	PL = 18

7.3   Compute void ratio, degree of saturation, and dry density for the following soil:
   Volume = 0.37 ft³
   Mass = 50 lb
   Mass after drying = 46 lb
   Specific gravity = 2.700

7.4   For a soil with a specific gravity of 2.650, with the following results from a laboratory compaction test, determine the optimum moisture content.

Moisture Content (%)	Total Density (lb/ft³)
10	95
15	120
20	130
25	125
30	108

7.5   What are the different types of soil compaction rollers?

7.6   Describe some soil stabilization techniques.

8 Aggregates for Asphalt and Concrete Mixes

8.1 DEFINITION, PARENT ROCK, AND TYPES

Aggregate is defined as processed soil. Natural aggregates are obtained from quarries or river-beds. The process of obtaining aggregates consists of blasting or dredging. Large-size particles are crushed to obtain usable sizes that range from 50 mm to less than 0.075 mm for most pavement mixes. Since aggregates are derived from parent rocks, their characteristics depend on the properties of the parent rocks. The parent rocks can be classified in terms of their geologic origins, which dictate their chemical composition and hence many of the other key properties that affect their behavior and performance. Artificial aggregates are obtained as by-products of other industrial processes (such as slag from the production of steel).

Elements combine to form minerals, and minerals form different types of rocks. There are five groups of minerals that make up most rocks: silicates (containing primarily silicon and oxygen), carbonates (with carbon, oxygen, and other elements), oxides (oxygen and various metals), sulfates (sulfur, oxygen, and metals), and sulfides (sulfur and metals). Examples of minerals are quartz, SiO_2; mica, $K(Mg, Fe)_3 (AlSi_3O_{10})(OH)_2(Mg, Fe)_3(OH)_6$ (Chlorite); and feldspar, $KAlSi_3O_8$.

Geologically, parent rocks can be divided into three types: igneous, sedimentary, and metamorphic. Extrusive igneous rocks are formed from magma—a viscous liquid composed of silicates, which is erupted onto the earth's surface as ash, lava flows, or solid chunks. These rocks are classified according to their texture and mineralogy and are generally finer grained than the intrusive igneous rocks (which are not visible on the earth's surface, but remain below it). Extrusive rocks include andesite, basalt, and rhyolite. In the case of intrusive igneous rocks, magma is forced into other rocks as crosscutting or parallel (to layers of other rocks) and may consist of large areas consisting of thousands of square miles. Such large masses tend to contain a greater amount of silica (50%–60%), compared to smaller dark-colored intrusions, which are sometimes referred to as *trap rock*. Intrusive rocks include granite, diorite, gabbro, and peridotite. Generally, good-quality aggregates can be produced from different types of igneous rocks, which may contain a wide variety of minerals.

Sedimentary rocks are those that are formed as a result of the consolidation of sedimentary materials formed by the reduction (clastic processes) of rock particle size through weathering and/or abrasion or through the consolidation of chemical precipitates such as marine plant and animal deposits (carbonate rocks). In the clastic process, the weathering and abrasion action can be caused by wind, water, ice, or gravity, whereas the cementation of the materials is caused by deposition of silica or carbonate materials carried by groundwater and compressed by the weight of the overlying materials. Examples of sedimentary materials formed by the clastic process are shale (from clay), siltstone (from silt), sandstone (from sand), and conglomerate (from gravel).

Examples of carbonate rocks are limestone and dolomites, formed by the consolidation and cementation of shells of marine plants and animals and also from fine carbonate mud from marine water. Both silica-rich materials, such as sand and silt, and clastic sedimentary rocks can be found within or interlayered with carbonate rocks.

Sedimentary rocks may also form under specific chemical conditions, forming materials such as iron and gypsum.

Metamorphic rocks are formed by the alteration (recrystallization) of igneous or sedimentary rocks through high pressure and temperature. Parallel orientation and platy appearance for such rocks are common, as a result of shearing action. Sometimes, metamorphic rocks are formed by the action of heat resulting from their vicinity to large intrusive magma or other sources of heat. An example is the formation of marble from carbonate rock. A metamorphic equivalent of igneous rocks such as granite is gneiss, whereas metamorphic equivalents of sedimentary rock such as shale are slate and schist.

Natural sand and gravel are unconsolidated sedimentary materials, which are formed from the breakdown of rocks through the action of ice, wind, or water, and generally consist of smooth and rounded particles. The nature of the parent rock dictates the quality of sand and gravel that result from its weathering. Hence, sands produced from igneous and metamorphic rocks or those produced from the ice- or water-weathering action of stronger rocks at higher elevations tend to be stronger than those produced from shales or siltstones or those derived from weaker rocks in low-lying areas. Climate, under which weathering takes place, also has an important effect on the quality of sands and gravels. For example, those resulting from prolonged and deep chemical weathering under humid conditions are of better quality than those formed by the weathering of rocks by a continental ice sheet.

Weathering can be physical, chemical, or combined. Physical weathering takes place as a result of changes in temperature, humidity, and cycles of freezing and thawing or wetting and drying. Chemical weathering always results in the release of salts and colloids, which may be deposited and influence the quality of some rocks.

8.2   SUITABILITY FOR APPLICATION

For evaluation of suitability of aggregates from different parent rocks for different applications, it is important to understand that within the similar type of rocks (such as igneous), there could be wide-ranging variations, such as in composition and texture. For example, igneous rocks can be porous or dense, and the hardness and abrasion resistance depend on the size of the individual crystals that make up these rocks (they are greatest when crystals are less than 2 mm in size). Compositional factors are important; for example, certain high-silica glassy materials are unsuitable for use in Portland cement applications because of their tendency to react with the alkali in the cement.

Rocks of the same group with similar materials but different texture produce different-quality aggregates. For example, although all of them belong to the igneous group, quartz have tightly interlocked mineral grains (giving good-quality aggregates), whereas micas are without interlocking and are present as weak layer deposits, and olivines consist of structures with rounded grains. Variations during the crystallization process of low-viscosity and low-silica igneous material can also result in segregation of minerals (magmatic differentiations), resulting in variation in rocks. Variations in metamorphic rocks are also common. For example, slates, with platy structure, are formed as a result of shearing action at low temperature, whereas schists formed at higher temperature may have more sound structure. For sedimentary rocks, variations can be in the form of amount of clay (shale has high clay content) as well as in cementation, often within formations, the extent of cementation being in many cases dictated by the age of the sedimentary rock formation.

In most cases, pavement engineers have to use locally available aggregates. There are certain characteristics of different types of aggregates that must be considered during the mix and structural design of pavements. Petrographic examination (ASTM C-295) of aggregates can be conducted to identify different parent rock materials and minerals in an aggregate sample. Such examination can provide valuable information regarding the suitability of aggregates for asphalt and concrete mixtures, particularly with respect to their durability, such as resistance against breakdown under freeze–thaw or moisture actions. Petrographic examination may include tests ranging from detailed visual examination to investigation using a polarizing microscope, differential thermal analysis, and x-ray diffraction. Figure 8.1 shows an example of results of petrographic examination of an aggregate sample.

8.3 PRODUCTION

Natural aggregates are either obtained from rock quarries or dredged from river beds. The production of aggregates is a complex process consisting of many steps. The process starts with the identification of a viable deposit by a geologist. Many factors are considered before setting up a quarry. These factors include development of the facility, hauling, processing, regulatory expenses and administrative costs, as well as expected sales and profits. A thorough step-by-step exploration is generally performed to identify the location of the quarry.

Standard topographical quadrangle maps and aerial photos are used to identify sand and gravel deposits, and such deposits with coarser aggregate sizes are preferred over those with finer sizes. Maps of geological bedrocks are studied for the identification of different strata before deciding on a quarry location for crushed aggregates (stones). Information from aerial photos and geophysical methods, and information from existing quarries in the area, may also be used. Once a potential site is located, it is important to study and understand outcrops as well as to investigate the uniformity of the different strata (outcropping and not outcropping), and geologic mapping may be conducted, especially if it is of a complex nature. To determine the reserves in a potential quarry, sampling can be taken from outcrops or pits, or by taking a core after drilling.

One important step is the determination of the total amount of overburden, which is the material above the mineable aggregate resource that should be removed. Such determination can be made with the help of exploration with drilling and/or with geophysical methods such as seismic or radar technology.

The amount of aggregate reserve (volume) in a proposed quarry can be calculated by simply multiplying the thickness of the layers by the area or, as in the case of layers of variable thickness, calculated from the sum of reserve for separate cross sections, drawn at regular intervals through the quarry. The tonnage of crushed aggregate is calculated by multiplying the volume by weight of usable rock per unit volume. Typical specific gravities of common raw rock are shown in Table 8.1.

$$\text{Yield}(\text{ft}^3/\text{ton}) = \frac{2000}{62.4 * \text{Specific gravity}}$$

$$\text{Reserve tonnage} = \frac{\text{Volume of reserve}(\text{ft}^3)}{\text{Yield}(\text{ft}^3/\text{ton})}$$

The usable tonnage is the in-place reserve less the unrecoverable material from buffer zones, pit slopes, waste, haul roads, and ramps.

The production of aggregates from rocks in a quarry consists of two primary steps: removal of rock from the quarry by blasting and sizing and separating different-sized particles with the help of a series of crushers and screens. The blasted particles, for example, 8 in. in diameter, are fed into a jaw crusher, after screening the minus-1-inch material that results from the breaking of the weathered rocks around the edges of the sound rock. The crushed material from the jaw crusher is screened and then separated into different size stockpiles, whereas the larger fraction (e.g., plus-1-inch material) is fed into secondary cone crushers for further crushing. Although the type of crushing equipment used is dictated by the rock and the properties of the end product that is desired, all crushers primarily use four types of mechanisms for reducing the size of particles—impact, attrition, shearing, and compression.

Lithology	#4	#8	#16	#30	#50	#100	#200
Igneous							
Granitic	17.6	23.8	21.9	22.3	21.3	20.2	23.0
Felsite	2.6	0.7	0.6	1.3	0.0	0.0	0.0
Gabbro	0.0	0.0	1.2	2.0	0.0	0.0	0.0
Basalt	0.7	0.0	1.2	0.0	0.0	0.0	0.0
Metamorphic							
Meta-granite	26.0	27.2	15.0	10.5	6.3	2.6	0.0
Quartzite	8.5	7.9	9.2	8.5	5.7	3.9	0.0
Meta-sediment	5.2	4.6	4.0	3.3	2.5	0.0	0.0
Phyllite	3.3	4.0	2.3	1.3	1.9	0.0	1.3
Phyllite, friable	0.7	0.0	0.0	0.0	0.0	0.0	0.0
Schist, hard	5.9	1.3	0.6	3.3	1.3	0.7	0.0
Schist, friable	1.3	0.0	0.0	0.0	0.0	0.0	0.0
Mylonite	0.7	1.3	0.6	0.0	0.0	0.0	0.0
Gneiss	0.7	0.7	0.0	0.0	0.0	0.0	0.0
Sedimentary							
Sillstone	3.9	4.6	13.2	12.4	12.6	13.1	12.5
Sillstone, friable	0.0	1.3	0.0	0.0	0.0	0.0	0.0
Sillstone, iron oxide cemented	0.0	1.3	0.6	0.0	0.0	0.0	0.0
Sandstone	13.1	11.3	9.2	5.2	3.8	6.5	2.6
Sandstone, iron oxide cemented	0.0	1.3	0.6	2.0	1.3	0.0	0.0
Greywacke	6.5	4.0	2.3	7.8	3.1	0.0	0.0
Shale	2.6	2.6	6.4	3.3	0.6	1.3	3.9
Chert, hard	0.7	0.7	1.2	0.6	5.0	0.7	0.0
Iron oxide	0.0	0.7	0.6	0.6	0.0	0.0	0.0
Mineral							
Feldspar	0.0	0.7	5.2	7.2	16.4	19.6	17.8
Quartz, undulatory extinction	0.0	0.0	3.5	7.2	13.9	14.4	17.8
Quartz, unit extinction	0.0	0.0	0.0	0.0	1.9	3.9	9.9
Mica	0.0	0.0	0.6	0.6	3.1	12.4	7.2
Heavy mineral	0.0	0.0	0.0	0.6	0.0	0.7	1.3
Amphilbole	0.0	0.0	0.0	0.0	0.0	0.0	1.3
Pyroxene	0.0	0.0	0.0	0.0	0.0	0.0	0.7
Magnetite	0.0	0.0	0.0	0.0	0.0	0.0	0.7
Total	100.0	100.0	100.0	100.0	100.0	100.0	100.0

(a)

(b)

FIGURE 8.1 Example of results of petrographic examination of a sand. (a) Percent of rock type retained in each sieve size. The table (a) shows classification by parent rock and minerals, and the figure (b) shows a thin section of material less than 0.075 mm in diameter under plane polarized light—majority particles with quartz, feldspar, and mica.

TABLE 8.1
Specific Gravities of Different Rock Type

Rock Type	Specific Gravity
Andesite	2.4–2.8
Basalt	2.7–3.2
Traprock	2.8–3.1
Dolomite	2.7–2.8
Gabbro	2.9–3.1
Granite	2.6–2.7
Limestone	2.7–2.8
Marble	2.6–2.9
Sandstone	2.0–3.2

Source: Barksdale, R.D., Ed., *The Aggregate Handbook*, National Stone Association, Washington, DC, 1991.

Generally, once production is started from a quarry, the aggregates are tested for their different properties (called *source properties*), and depending on the results of the tests, the aggregates are approved for use in specific applications, such as by the state department of transportation.

Excellent information on all types of aggregates used in the construction industry can be obtained from the National Stone, Sand & Gravel Association (n.d.; http://www.nssga.org).

8.4   OVERVIEW OF DESIRABLE PROPERTIES

Specific properties that are critical for good performance in different applications have been identified, and tests have been developed to measure them. Such tests have also been standardized by agencies, such as the ASTM. Although there are common desirable features, specific applications require specific properties of aggregates. For pavements, such applications include use as structural layers (in roads as well as railroads), drainage layers, asphalt mixtures, and cement concrete.

8.4.1   PROPERTIES CRITICAL FOR STRUCTURAL LAYERS

Aggregates could be used as base layers to decrease the stress coming from traffic (above) to the subbase and subgrade layers (below). For proper functioning, aggregates must be of sufficient strength, and if used as unsurfaced roads, they must possess good wearing resistance as well. When used as a ballast layer, directly below the ties in a railroad, to provide support to the ties and reduce the stress to the layers underneath, aggregates must be resistant to degradation action of traffic and maintenance equipment, cycles of wet–dry and freeze–thaw, accumulation of fines or binding of the aggregate layers, and the potential of swelling or frost heaves.

8.4.2   PROPERTIES CRITICAL FOR DRAINAGE LAYERS

When used as drainage layers, the proportion of different-sized particles in the aggregate (gradation) must be such as to allow sufficient drainage of water and work as a filter to prevent erosion and clogging due to finer materials.

8.4.3 PROPERTIES CRITICAL FOR ASPHALT MIX LAYERS

For use in asphalt mixes, the general requirements include cleanliness, proper shape, and resistance against the effects of traffic- and environment-related abrasion and weathering.

8.4.4 PROPERTIES CRITICAL FOR CEMENT CONCRETE LAYERS

In the blends used for cement concrete mixes, limitations are generally applied on the amount of fine aggregates to prevent the presence of clay particles. The aggregate must be resistant to the effects of abrasion and weathering due to the traffic as well as the environment and must be such as to make the mix sufficiently workable and provide adequate strength and durability.

The different properties that are relevant to the different applications and tests are presented next.

8.5   GRADATION FOR ASPHALT PAVEMENTS

Gradation refers to the relative proportion of different-sized particles in an aggregate blend. Gradation of a blend is determined from sieve analysis. Conversely, a blend can be separated into different sizes by sieving. Generally, in a quarry, the gradation of each stockpile aggregate material is determined first, and then the gradations are analyzed to determine in what proportion the aggregates from the different stockpiles can be combined to produce a *specific* gradation. This gradation is generally specified (in specifications) by pavement project owners (such as state departments of transportation). The specifications are developed on the basis of experience and results of laboratory and field studies. Figure 8.2 shows a variety of gradations for dense graded asphalt mixes with the same nominal maximum aggregate size (NMAS), which is defined as one sieve size larger than the first sieve to retain more than 10%. The maximum aggregate size is defined as one sieve size larger than the NMAS. The variation in gradation for the different NMASs is due to the fact that gradations are specified to remain close to the maximum density gradation.

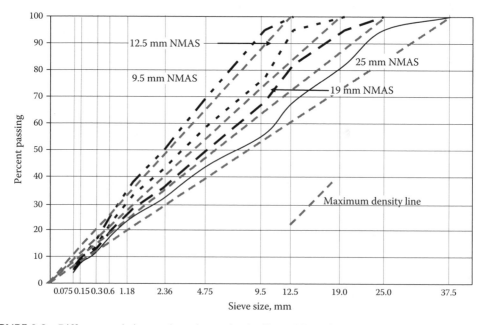

FIGURE 8.2   Different gradations and maximum density lines. (Note: sieve size raised to the power 0.45.)

The maximum density line can be obtained from the FHWA 0.45 power gradation chart by joining the origin to the maximum aggregate size by a straight line. Figure 8.2 shows the maximum density lines for the different NMASs in the FHWA 0.45 power plot. The maximum density lines are based on the concept proposed by Fuller and Thompson (1907, also known as *Fuller's curve equation*), which is as follows:

$$P = 100 * \left(\frac{d}{D} \right)^n$$

where
    P is the percentage passing
    d is the diameter of sieve
    D is the maximum size of the aggregate
    n is the coefficient, first proposed as 0.5 and then modified to 0.45

Note that gradations are specified so that they remain approximately parallel to this maximum density line but remain a few percentage points above or below—to remain close to the maximum gradation line so as to provide a stable structure but also retain enough space for accommodating adequate asphalt binder for durability. More sophisticated methods of developing gradations for providing aggregate interlock and adequate space for asphalt cement binder (to provide durability) have been developed. These methods, such as the voids in coarse aggregate (VCA) method as well as the Bailey method, are based on volumetric calculations with proper consideration of unit weight of individual aggregates.

Gradations for different types of mixes are shown in Figure 8.3. Note that the gradations are different for different types of asphalt mixes, which are prepared for different applications. For common paving applications, dense graded mixes are used. To improve drainage of water and

FIGURE 8.3 Gradations for different types of asphalt mixes.

friction, open graded friction course (OGFC) is used, whereas stone matrix asphalt (SMA) is used as a high-rut-resistant mix. The gap gradation (certain fractions missing) helps in achieving a high permeability in the OGFC mixes. In the case of SMA, the percentage passing the 4.75 mm sieve is important; it is kept at a relatively low number to allow the interlock of coarse aggregate particles and hence develop strong resistance against rutting. Note, however, that the percentage passing the 0.075 mm sieve is relatively high, to provide a sufficient amount of filler material to stiffen the relatively large amount of asphalt binder used to provide a stable fine matrix, and prevent draindown of asphalt binder.

8.5.1 Aggregate Tests

In the ASTM C136: Standard Method for Sieve Analysis of Fine and Coarse Aggregates test, a weighed sample of dry aggregates is separated by running them through a set of sieves. The sieves are stacked in such a way so that the openings decrease from top to bottom. After separating the particles of the different sizes, the percentage of particle passing each sieve is calculated to determine the gradation, or particle size distribution. The weight of the test sample depends on the size of the aggregates used. Generally mechanical sieve shakers are used, which could impart vertical and lateral motion to the sieves, during the sieving process. Guidelines for determining the adequacy of the sieving process are provided in the standard. Generally, 8 in. diameter round sieves are used. Dry sieve analysis provides an estimate of particle size distribution, whereas washed sieve analysis provides an accurate indication. A stack of sieves in a sieve shaker is shown in Figure 8.4.

The steps required to combine aggregates from different stockpiles to produce a specific gradation are presented in Figure 8.5, with an example.

Note that, in an asphalt mix, the surface of the aggregates is covered with asphalt binder. The total surface area of the aggregates dictates the amount of asphalt that is needed for covering the

FIGURE 8.4 Stack of sieves in sieve shaker.

Blending of aggregates in a mix design consists of the following steps.
Step 1: Consider all the aggregates make up 100% of a batch
Step 2: Determine the gradation of the individual stockpile aggregaes, for
example,

Sieve	Percent Passing for Different Stockpiles				
Sizes	3/8	1/2	Sand	Stone Sand	Dust
19.00	100.0	100.0	100.0	100.0	100.0
12.50	100.0	81.0	100.0	100.0	100.0
9.50	93.0	15.0	100.0	100.0	100.0
4.75	24.0	1.7	98.7	95.3	100.0
2.36	7.5	1.5	79.0	77.4	84.5
1.18	5.8	1.4	57.4	50.0	57.4
0.60	5.1	1.3	39.7	31.0	40.5
0.30	4.5	1.2	16.9	15.7	29.2
0.150	3.5	1.0	6.9	5.1	21.1
0.075	1.6	0.7	2.8	1.6	14.2

Step 3: Based on approved designs and/or experience, select percentages of a
batch from the different stockpiles, for example:

Batch Size (lb):		100.00
	Percentage of Batch	Cumulative
3/8	15	15.00
1/2	20	35.00
Sand	30	65.00
Stone sand	25	90.00
Dust	10	100.00

Step 4: Compute the contribution of each stockpile to each aggregate size by
multiplying the percentage passing each aggregate of a stockpile with the
percentage of the stockpile in the total aggregate batch, for example:

	3/8 in. Stockpile		
Sieve Size	Percent Passing A	Percent in Batch B	Percent of Total Batch (=A*B%)
19.00	100.0	15	15.0
12.50	100.0	15	15.0
9.50	93.0	15	14.0
4.75	24.0	15	3.6
2.36	7.5	15	1.1
1.18	5.8	15	0.9
0.60	5.1	15	0.8
0.30	4.5	15	0.7
0.150	3.5	15	0.5
0.075	1.6	15	0.2

FIGURE 8.5 Steps to combine different stockpile aggregates to produce a specific blend.

(*continued*)

1/2 in. Stockpile			
Sieve Size	Percent Passing A	Percent in Batch B	Percent of Total Batch (=A*B%)
19.00	100.0	20	20.0
12.50	81.0	20	16.2
9.50	15.0	20	3.0
4.75	1.7	20	0.3
2.36	1.5	20	0.3
1.18	1.4	20	0.3
0.60	1.3	20	0.3
0.30	1.2	20	0.2
0.150	1.0	20	0.2
0.075	0.7	20	0.1

Sand Stockpile			
Sieve Size	Percent Passing A	Percent in Batch B	Percent of Total Batch (=A*B%)
19.00	100.0	30	30.0
12.50	100.0	30	30.0
9.50	100.0	30	30.0
4.75	98.7	30	29.6
2.36	79.0	30	23.7
1.18	57.4	30	17.2
0.60	39.7	30	11.9
0.30	16.9	30	5.1
0.150	6.9	30	2.1
0.075	2.8	30	0.8

Stone Sand Stock Pile			
Sieve Size	Percent Passing A	Percent in Batch B	Percent of Total Batch (=A*B%)
19.00	100.0	25	25.0
12.50	100.0	25	25.0
9.50	100.0	25	25.0
4.75	95.3	25	23.8
2.36	77.4	25	19.4
1.18	50.0	25	12.5
0.60	31.0	25	7.8
0.30	15.7	25	3.9
0.150	5.1	25	1.3
0.075	1.6	25	0.4

Dust Stockpile			
Sieve Size	Percent Passing A	Percent in Batch B	Percent of Total Batch (=A*B%)
19.00	100.0	10	10.0
12.50	100.0	10	10.0
9.50	100.0	10	10.0
4.75	100.0	10	10.0
2.36	84.5	10	8.5
1.18	57.4	10	5.7
0.60	40.5	10	4.1
0.30	29.2	10	2.9
0.150	21.1	10	2.1
0.075	14.2	10	1.4

FIGURE 8.5 (continued)   Steps to combine different stockpile aggregates to produce a specific blend.

Step 5: Compute the final blend by adding up the contribution to each aggregate size from each of the different stockpiles, in a horizontal layout table, and compare with the limits in the specification, for example:

Stockpile	Percent Batched	Combined Gradation, % Passing Sieves, mm									
		19.0	12.5	9.5	4.75	2.36	1.18	0.60	0.30	0.15	0.075
3/8	15	15.0	15.0	14.0	3.6	1.1	0.9	0.8	0.7	0.5	0.2
½	20	20	16.2	3	0.34	0.3	0.28	0.26	0.24	0.2	0.14
Sand	30	30	30	30	29.61	23.7	17.22	11.91	5.07	2.07	0.84
Stone sand	25	25.0	25.0	25.0	23.8	19.4	12.5	7.8	3.9	1.3	0.4
Dust	10	10.0	10.0	10.0	10.0	8.5	5.7	4.1	2.9	2.1	1.4
Total	100.0	100	96	82	67	53	37	25	13	6	3.0
Upper limit		100.0	100.0	90.0		58.0					10
Lower limit		100.0	90.0			28.0					2

Step 6: If any of the percentages are outside the acceptable limits then the blend percentages can be readjusted to produce a more acceptable gradation. For example, in the data shown here, the blend can be modified as follows:

Batch size (lb):		100.00
	Percentage of Batch	Cumulative
3/8	15	15.00
1/2	20	35.00
Sand	35 (from 30)	65.00
Stone sand	20 (from 25)	90.00
Dust	10	100.00

Stockpile	Percent Batched	Combined Gradation, % Passing Sieves, mm									
		19.0	12.5	9.5	4.75	2.36	1.18	0.60	0.30	0.15	0.075
3/8	15	15.0	15.0	14.0	3.6	1.1	0.9	0.8	0.7	0.5	0.2
½	20	20	16.2	3	0.34	0.3	0.28	0.26	0.24	0.2	0.14
Sand	35	35.0	35.0	35.0	34.5	27.7	20.1	13.9	5.9	2.4	1.0
Stone sand	20	20.0	20.0	20.0	19.1	15.5	10.0	6.2	3.1	1.0	0.3
Dust	10	10.0	10.0	10.0	10.0	8.5	5.7	4.1	2.9	2.1	1.4
Total	100.0	100	96	82	68	53	37	25	13	6	3.1
Upper limit		100	100	90		58					10
Lower limit		100	90			28					2

FIGURE 8.5 (continued) Steps to combine different stockpile aggregates to produce a specific blend.

aggregates adequately, that is, as stated commonly, with an adequate film thickness. Although the concept is purely theoretical, as there is no uniform or one film thickness for the entire mix, it is a logical one, and can be used to ensure an optimum "coating" of aggregates. Fine aggregates have larger surface area, and hence blends with a higher amount of fine aggregates will have a higher surface area compared to those with a low amount of fines. Relatively thin films of asphalt can negatively affect the durability of the mix, whereas thick films can adversely affect its resistance against rutting. An optimum film thickness in the range of 8–12 μm has been mentioned in the literature.

The surface area of aggregates could be determined from surface area factors, as presented in Table 8.2. The factors are multiplied by the percentages and then added up. An example of calculation is also shown.

Aggregates have an important role in proportioning concrete. Aggregates affect the workability of fresh concrete tremendously. The aggregate particle size and gradation, shape, and surface texture will influence the amount of concrete that is produced with a given amount of paste (cement plus water). The selection of the maximum size aggregate is governed by the thickness of the slab and by the closeness of the reinforcing steel. The maximum aggregate size should not be obstructed to flow easily during placement and consolidation.

TABLE 8.2

Surface Area Factors and Calculations

Sieve Size	Surface Area Factors (ft²/lb)
% passing maximum sieve	2
% passing No. 4	2
% passing No. 8	4
% passing No. 16	8
% passing No. 30	14
% passing No. 50	30
% passing No. 100	60
% passing No. 200	160

Example Calculation

Sieve Size	% Passing	Calculation	Surface Area (ft²/lb)
1/2	100	1 * 2	2
3/8	90		
No. 4	75	0.75 * 2	1.5
8	62	0.62 * 4	2.5
16	50	0.5 * 8	4.0
30	40	0.4 * 14	5.6
50	30	0.3 * 30	9
100	10	0.1 * 60	6
200	4	0.04 * 160	6.4
		Σ = 37.0 ft²/lb	

Source: Reprinted from Roberts, F.L. et al., *Hot Mix Asphalt Materials, Mixture Design, and Construction*, National Asphalt Pavement Association (NAPA) Education Foundation, Lanham, MD, 1996.

The maximum size of coarse aggregate should not exceed one-fifth of the narrowest dimension between sides of forms or three-fourths of the clear space between individual reinforcing bars or the clear spacing between reinforcing bars and form walls. For unreinforced slabs on grade, the maximum aggregate size should not exceed one-third of the slab thickness.

8.6 SPECIFIC GRAVITIES AND ABSORPTION

Specific gravity of a material is defined as the ratio of the weight of a unit volume of the material to the weight of an equal volume of water at 23°C. This parameter is needed to calculate volumetric properties in a mix, and any calculation that involves determination of mass from volume or vice versa. There are three different specific gravities that could be used for determination of volumetric properties in asphalt mixes. The different specific gravities are because of the effect of aggregate pores and absorption. The bulk and apparent specific gravities are determined from the same test, whereas the effective specific gravity is calculated. The applicable ASTM standards are C-127 and C-182. Bulk and apparent specific gravities are explained in Figure 8.6.

Bulk-specific gravity is defined as the ratio of the oven dry weight in air of a unit volume of a permeable material (including both permeable and impermeable voids normal for the material)

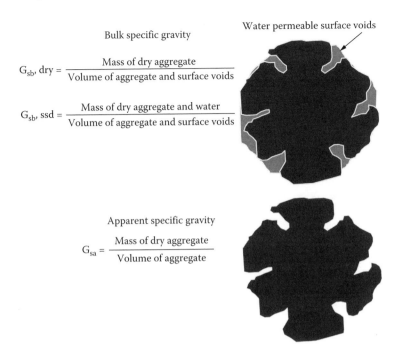

Bulk specific gravity

$$G_{sb}, dry = \frac{\text{Mass of dry aggregate}}{\text{Volume of aggregate and surface voids}}$$

$$G_{sb}, ssd = \frac{\text{Mass of dry aggregate and water}}{\text{Volume of aggregate and surface voids}}$$

Water permeable surface voids

Apparent specific gravity

$$G_{sa} = \frac{\text{Mass of dry aggregate}}{\text{Volume of aggregate}}$$

FIGURE 8.6 Bulk and apparent specific gravities.

at a stated temperature to the weight of an equal volume of gas-free distilled water at the stated temperature:

$$\text{Bulk specific gravity, } G_{sb} = \frac{\text{Weight of dry aggregate}}{\text{Volume of dry aggregate} + \text{External voids}}$$

$$= \frac{\text{Weight in air (dry)}}{\text{Weight in SSD condition} - \text{Weight in submerged condition}}$$

Apparent specific gravity is the ratio of the oven dry weight in air of a unit volume of an imperme-able material at a stated temperature to the weight of an equal volume of gas-free distilled water at the stated temperature:

$$\text{Apparent specific gravity, } G_{sa} = \frac{\text{Weight of dry aggregate}}{\text{Volume of dry aggregate}}$$

$$= \frac{\text{Weight in air (dry)}}{\text{Weight in air} - \text{Weight in submerged condition}}$$

Effective specific gravity is the ratio of the oven dry weight in air of a unit volume of a permeable material (excluding voids permeable to asphalt) at a stated temperature to the weight of an equal volume of gas-free distilled water at the stated temperature:

Effective specific gravity, G_{se}

$$= \frac{\text{Weight of dry aggregate}}{\text{Volume of dry aggregate} + \text{Volume of voids between asphalt and aggregate}}$$

Effective specific gravity is calculated from the theoretical maximum specific gravity of asphalt mix.

The *absorption* of an aggregate is defined as the water content of the aggregate in its saturated surface dry condition:

$$\text{Absorption, \%} = \frac{\text{Saturated surface dry weight of aggregate, g} - \text{oven dried weight of aggregate, g}}{\text{Oven dried weight of aggregate, g}}$$

This parameter is needed to determine how much of the total asphalt added is absorbed in the aggregates, and hence how much remains on the surface of the aggregate as effective asphalt. Generally, the water absorption of aggregate is determined and used as an indicator of the asphalt absorption.

The AASHTO T-84 (ASTM C-128) Specific Gravity and Absorption of Fine Aggregates test procedure consists of the following steps. About 1000 g of oven-dried fine aggregate is immersed in water in a flask (pycnometer) with known weight. After 24 h of immersion, the aggregate is dried on a flat surface, using warm air, until saturated surface dry (SSD) condition is reached. The SSD condition is detected at that moisture content when a lightly compacted mass of fine aggregate in a cone first slumps when the cone is removed. About 500 g of the SSD aggregate is then placed in the flask and weighed. The flask is then filled with water, and its weight is recorded. Finally, the fine aggregate is removed from the flask, oven dried, and weighed.

In the AASHTO T-85 (ASTM C-127) Specific Gravity and Absorption of Coarse Aggregates test procedure, about 5 kg of washed coarse aggregate (retained on a 4.75 mm sieve) is oven dried and weighed and then immersed in water for 24 h. The aggregates are then removed from water, drained and surface dried (saturated surface dry) until there is no visible film of water on the surface, and weighed. The SSD aggregates are next submerged underwater in a wire-mesh basket and weighed.

The absorption of the aggregate blend (combined coarse and fine aggregates) is determined on the basis of a percentage of the two fractions. Maximum limits on absorption are specified, which may range from 0.5% to 5%.

8.7 CLEANLINESS AND DELETERIOUS MATERIALS

Deleterious or undesirable materials can be of different types, such as organic materials, vegetation (roots, sticks), clay lumps, clay coating on aggregates, and metal oxides. Such impurities are undesirable since they negatively affect the performance of asphalt mixtures. For example, organic materials can absorb water and lead to deterioration of the mix, clay lumps can form during production and subsequently disintegrate in the asphalt mix, clay dust can coat aggregates and hence negatively affect the bond between asphalt and aggregate, and metal oxides can react with water, causing swelling, popouts, and staining in pavements. Petrographic examination can identify impurities in many cases, and washing the aggregates reduces the impurities to a great extent. However, tests must still be conducted to ensure that the aggregates are clean to a minimum acceptable level, and/or the amount of impurities is below a maximum acceptable level.

In the AASHTO T-176 (ASTM D-2419) Plastic Fines in Graded Aggregates and Soils by Use of the Sand Equivalent test, aggregates passing the 4.75 mm sieve are placed and stirred in a graduated, transparent cylinder filled with water and a flocculating agent (a mixture of calcium chloride, glycerin, and formaldehyde). After settling, the sand in the fine aggregate separates from the clay, if any; the heights of the sand and clay columns in the flask are read off; and the ratio of the former to the latter is determined as the sand equivalent. Depending on the layer, minimum sand equivalents are specified, which can range from 25 to 60.

The AASHTO T-90 (ASTM D-4318) Determining the Plastic Limit and Plasticity Index of Soils test involves the determination of liquid limit and plastic limit, and the plasticity index (PI), which is the difference between liquid limit and plastic limit, for materials passing the 0.42 mm sieve. Maximum limits on PI (such as 4) are specified on either the fraction passing the 0.42 mm sieve or that passing the 0.075 mm sieve.

In the AASHTO T-112 (ASTM C-142) Clay Lumps and Friable Particles in Aggregates test, washed aggregates are dried and sieved to remove the fraction passing the 0.075 mm sieve. Samples are then soaked in water for 24 h, after which particles are tested for removal by wet sieving and breaking with fingers. If a particle can be removed by wet sieving or breaking, it is classified as a clay lump or friable particle, and the total percentage of such particles on the basis of the weight of the sample is determined. Maximum limits, such as 1%, are generally specified.

The AASHTO T-113 Light Weight Pieces in Aggregate test involves the determination of the percentage of lightweight pieces in coarse and fine aggregates by separating (floating) them in a liquid of suitable specific gravity. Liquid such as zinc chloride solution (specific gravity = 2.0) can be used to separate particles such as those of coal.

The steps in the AASHTO T-11 Material Finer than 75 μm (No. 200 Sieve) in Mineral Aggregate by Washing test are washing and wet sieving aggregates to determine the percentage of material passing the 75 μm sieve. Allowable ranges of material finer than 75 μm are generally specified, and depending on the layer and type of mix, they can range from 2% to 10%.

In the AASHTO T-P57 Methylene Blue Value test, a sample of material passing the 75 μm sieve is dispersed in distilled water using a magnetic stirrer, and the methylene blue solution is titrated into the dispersed mix until the sample cannot absorb any more reagent. This end point is indicated by a blue ring when a drop of the mix is placed on a filter paper. The methylene blue value (amount, MBV) is proportional to the amount of organic material or clay in the aggregate, and a maximum limit is generally specified.

8.8 TOUGHNESS OR RESISTANCE AGAINST ABRASION LOSS

For both asphalt and concrete mixes, the resistance against degradation (breaking down of larger particles into smaller sizes) by abrasion, or toughness, can be measured by the Los Angeles (LA) abrasion (also sometimes referred to as the LA degradation) test.

In the AASHTO T96 (ASTM C131) Resistance to Degradation of Small Size Coarse Aggregate by Abrasion and Impact in the Los Angeles Machine test, a 5000 g blend sample of coarse aggregate is placed in a steel drum, containing 6–12 steel balls (each weighing 420 g). The drum is rotated for 500 revolutions, at a speed of 30–33 rpm, and during this time, the steel balls and aggregates are lifted and dropped about 69 cm from a shelf within the drum. The tumbling action leads to the shattering of brittle particles by impact, and degradation of smaller particles by surface wear and abrasion. At the end of tumbling, the aggregates are sieved dry over a 1.77 mm sieve (through which none of the aggregates passed before the sieve), and the percentage of the material passing this sieve is reported as the Los Angeles degradation value. A high value would indicate the potential of generation of dust, and breakdown during construction and in the field, although it may fail to identify such potential in lightweight aggregates such as slag. Maximum limits are specified, and may range from 30% to 50%. Values may range from 10 (for aggregates from hard igneous rock) to 60 (for aggregates from soft limestone).

The ASTM D-6928, AASHTO T-P58 MicroDeval Abrasion Loss test consists of placing aggregates in ajar with water and 3/8 in. diameter steel balls and rotating the jar (Figure 8.7) at 100 rpm for 2 h. The breakdown of the aggregate is then evaluated. The aggregate sample is soaked in water for 24 h before tumbling in the jar. At the end of the tumbling, the sample is washed and dried, and the amount of material passing the No. 16 sieve is determined. This value is then used as a

FIGURE 8.7   MicroDeval equipment.

percentage of the original sample weight to determine the percentage of abrasion loss. A maximum limit, such as 18%, is specified.

8.9   PARTICLE SHAPE AND SURFACE TEXTURE

Angular particles tend to provide greater interlock and internal friction (and hence shear strength), and particles with rough texture (as found mostly in crushed aggregates) are desirable because of increased strength and voids for accommodating more asphalt (and hence more durability). Also, the occurrence of flat and/or elongated particles is undesirable, since they affect compaction and tend to increase voids, break down during construction and under traffic, and produce particles with uncoated (with asphalt) surfaces and negatively alter the volumetric properties.

For coarse aggregates, the ASTM D-3398 Index of Aggregate Particles Shape and Texture can be used. The volume of voids between packed, uniform-size particles is an indicator of the combined effect of shape, angularity, and surface texture of the aggregates. Clean, washed, and oven-dried aggregates are separated into different sizes, and aggregates of each size are compacted in a separate mold with a tamping rod, using first 10 blows per layer of three layers and then 50 blows per layer of three layers. The voids are determined from the weight of the mold with the aggregate and the bulk-specific gravity of the aggregates. The particle index is calculated for each fraction, and then a combined index is calculated on the basis of the relative proportion (percentage of the blend by weight) of the different sizes:

$$I_a = 1.25V_{10} - 0.25V_{50} - 32.0$$

where
   I_a is the particle index value
   V_{10} is the percentage voids in the aggregate compacted with 10 blows per layer
   V_{50} is the percentage voids in the aggregate compacted with 50 blows per layer

Particle index values of blends may range from 6 (for rounded aggregates with a smooth surface) to 20 (for angular crushed aggregates with a rough texture). For the same air voids, mixes with aggregates with a higher particle index value tend to show higher strength.

The ASTM D-4791 Flat or Elongated Particles in Coarse Aggregates test is also used. Flat or elongated particles are defined as those aggregates by particles that have a ratio of width to thickness or length to width greater than a specified value, such as 3:1, 4:1, or 5:1 (most common). Tests are conducted on particles of each size, using a proportional caliper, on a representative sample, and then the total value is calculated as a percentage, in terms of number or mass. Maximum limits on the percentages of flat and elongated particles are specified, which may range from 5 to 20, for example, for a ratio of 5:1.

FIGURE 8.8   Testing for uncompacted voids for fine aggregates.

The ASTM D-5821 Determining the Percentage of Fractured Particles in Coarse Aggregates test involves the determination of percentage of coarse aggregates (that are retained on the 4.75 mm sieve) with one or more fractured face(s), which is defined as a face that exposes the interior surface of a particle. A particle is said to have a fractured face if, for example, at least 25% of the area (ASTM standard) of the face is fractured. During the test, from a specific sample, the aggregate particles identified as those with fractured faces are separated from those that are not, and the percentage of particles with one (or two) fractured faces is determined. Minimum limits on the percentage of fractured particles in an aggregate blend are specified, and may range from 30 to 90, for both one and two or more fractured faces, dependent on the layer, with surface layers having requirements for higher percentages.

In the AASHTO T-P56 Uncompacted Voids in Coarse Aggregates test, coarse aggregate is dropped through a hopper into a cylinder with known weight and volume. The excess aggregate on top of the cylinder is struck off, and the weight of the filled cylinder, along with the known bulk-specific gravity of the aggregates, is used for determination of the uncompacted void content.

For fine aggregates, the ASTM C-1252, AASHTO TP-33, Uncompacted Void Content test is used (Figure 8.8). The test consists of pouring the fine aggregate blend of specified gradation through a funnel into a cylinder with a volume of 100 cm³. The cylinder is weighed after striking off the excess particles. Using this weight, the bulk-specific gravity of the aggregate and the known volume and weight of the cylinder, the uncompacted void content is determined. A high uncompacted void content of, say, 45% indicates a fine aggregate with angular shape and rough texture, and tends to produce asphalt mixes with relatively high resistance against the potential of rutting.

8.10   DURABILITY/SOUNDNESS

Aggregates could be subjected to repeated cycles of weathering action due to wetting and drying or freezing and thawing of trapped water (due to expansion upon freezing). Tests are conducted to determine the potential of breakdown when subjected to such stresses.

In the AASHTO T-104 (ASTM C-88) Soundness of Aggregate by Use of Sodium Sulfate or Magnesium Sulfate test, the sample is immersed in a solution of sodium or magnesium sulfate of specified strength for a period of 16–18 h at a temperature of 21°C ± 1°C. The sample is removed from the solution and drained for 15 ± 5 min, and then dried at 110°C ± 5°C to a constant weight. During immersion, the sulfate salt solution penetrates into the permeable pores of the aggregate, and drying dehydrates the sulfate salts in the pores. Upon further immersion, the sulfate salts rehydrate, and the resulting expansion and stress simulate the freezing of water and associated stress. The same sample is then subjected to four more cycles (for a total of five cycles) of immersion and drying. The sample is then sieved, and the weight loss for each fraction and then the weighted average loss of the sample are determined. A higher loss indicates a potential of loss of durability through weathering action of the environment. Maximum limits (such as 12% loss after five cycles) are specified.

There are three different versions of the AASHTO T-103 Soundness of Aggregate by Freezing and Thawing test. In procedure A, samples are immersed in water for 24 h and then subjected to freezing and thawing, while in procedure B the samples are vacuum (25.4 mm of mercury) saturated, using a 0.5% (by mass) solution of ethyl alcohol and water, before subjecting them to freezing and thawing (in the same solution). Procedure C is similar to B, except that water is used instead of the alcohol–water solution. Samples are dried and sieved, at the end of the final cycle (cycles may range from 16 to 50 cycles, depending on the procedure), and the weight loss for each size is determined.

In the AASHTO T-210 (ASTM D-3744) Aggregate Durability Index test, a washed and dried sample of coarse aggregate is agitated in water for a period of 10 min. The washed water and fraction passing the 75 μm sieve are mixed with calcium chloride solution and placed in a plastic sand equivalent cylinder for 20 min. Next, the level of the sediment column is read, which is used for calculating the durability index.

8.11 EXPANSIVE CHARACTERISTICS

Expansion of aggregates (such as steel slag) as a result of hydration can lead to swelling in asphalt mixes and resulting disintegration of asphalt pavements. Tests can be conducted to identify the potential of expansion.

The ASTM D-4792 Potential Expansion of Aggregates from Hydration Reaction test method involves compaction of the aggregate to its maximum density in a CBR mold, and submerging it in water at 71°C ± 3°C for a period of at least 7 days. A perforated plate is kept on top of the compacted aggregate. Percentage expansion of the sample is determined from dial readings of the height of the sample submerged in water, and the initial height of the sample. Maximum permissible expansion, such as 0.5%, is specified.

8.12 POLISHING AND FRICTIONAL CHARACTERISTICS

Polishing is defined as the loss of microtexture and the gradual smoothing and rounding of exposed aggregates. High potential of polishing in aggregates by traffic is undesirable as it leads to lowering of friction and hence decrease in skid or frictional resistance, which is a measure of how quickly a vehicle can be stopped, say, under adverse/wet conditions.

In the AASHTO T-278 (ASTM E-303) Measuring Surface Frictional Properties Using the British Pendulum Tester test, aggregate particles are placed in a small mold along with a bonding agent (such as epoxy). A thin film of water is applied over the sample, and then a pendulum with a rubber shoe is released such that it touches and goes over the sample. The pendulum goes past the sample to a certain height, depending on the friction offered by the aggregate sample. This height, or the length of the arc, is read off. This zero-polishing reading is then used, along with readings taken after polishing the sample surface (for, say, 9 h) with a British polish wheel (AASHTO T-279, ASTM D-3319) or other polishing device, to determine the polishing resistance.

In the ASTM D-3042 Insoluble Residue in Carbonate Aggregate test, the percentage of noncarbonated (acid-insoluble) materials in carbonate aggregates is determined. Five hundred grams of a sample (retained on a 4.75 mm sieve) is placed in a glass beaker with 1 L of hydrochloric acid solution, and the mix is agitated until effervescence stops. This is repeated with an additional 300 mL of acid. Then the beaker is heated to 110°C, with new acid added in increments until effervescence stops. Next, the aggregate residue is washed over a 75 μm sieve, the plus-75-μm material is weighed, and the weight is expressed as insoluble residue (as a percentage of the original sample weight). The higher the amount of insoluble residue, the higher the amount of siliceous material, and hence the higher the resistance against polishing.

8.13 AGGREGATE TESTS SPECIFICALLY FOR CONCRETE

8.13.1 FINENESS MODULUS (FM; ASTM C125)

This property assesses the coarseness of the fine aggregate, and it is used in combination with the maximum nominal size coarse aggregate to estimate the proportions of both fine and coarse aggregates used in a concrete mix, to satisfy workability requirements. The fineness modulus is calculated by summing the cumulative percentage retained in the following sieves and dividing the sum by 100: 9.5, 4.75, 2.36, 1.18, 0.6, 0.3, and 0.15 mm. It varies between 2.5 and 3.1 usually. The higher the FM, the harsher and rougher the concrete mix is. However, it is more economical to have a mix with a higher FM, because the mix will not require as much water, and therefore, a relatively lower amount of cement could be used for a specific water–cement ratio.

8.13.2 GRADATION

The gradation requirements for aggregates used in concrete are provided in ASTM C-33 and AASHTO M-80 for coarse aggregates, and ASTM C-33 and AASHTO M-6 for fine aggregates. These gradations assist the designer in providing a workable concrete in which the coarse aggregates are coated and suspended in the mortar (sand, cement, and water). The gradations differ on the basis of nominal maximum size of the coarse aggregate. The maximum aggregate size is defined as the smallest sieve through which 100% of the aggregate passes. The nominal maximum aggregate seize is defined as the smallest sieve that can retain up to 15% of the aggregates. The larger the maximum size, the greater the percentage of coarse aggregates in a mix. The limits are imposed by the thickness of the slab and the gap between the reinforcements, if any. Figure 8.9 shows examples of gradations for coarse and fine aggregates, along with ASTM limits.

FIGURE 8.9 Examples of gradations for coarse and fine aggregates for PCC pavement.

8.13.3 Bulk Density and Voids in Aggregates Test

The Bulk Density and Voids in Aggregates (AASHTO T-19; also known as the *dry-rodded unit weight*) test is conducted for the volumetric mix design of PCC. The steps in this are filling a container with a known volume (0.5 ft³) with dry coarse aggregate using a specified procedure, and then measuring the weight of the filled container and determining the bulk density by dividing the weight of the aggregates needed to fill the container by the volume of the container.

8.14 AUTOMATED AGGREGATE ANALYSIS (AASHTO TP81 AND PP64)

Modern imaging systems and image analysis software are being used to develop automated aggregate image analysis for determining aggregate properties. Figure 8.10 shows one such equipment. Both coarse and fine aggregate could be analyzed, and texture for coarse aggregate could also be determined. The system consists of a hardware that captures the 3D image of aggregate particles and software that analyzes the image and estimates the angularity, form, and texture.

8.15 ARTIFICIAL AGGREGATES

Artificial aggregates are those that are not naturally occurring, but rather are formed from waste products or as by-products of industrial processes. The feasibility of producing lightweight aggregates (synthetic lightweight aggregates, or SLA, weighing less than 70 lb/ft³) suitable for use in pavements, from waste fly ash and mixed waste plastics, has been proven. Such products are generally formed through intense heating of several materials, such as heating shale/clay/slate in excess of 1000°C to produce ceramic SLA. Slag, a by-product of the iron and steel industry, is used in pavement mixtures and is noted for its good skid resistance (and relatively high absorption).

FIGURE 8.10 Automated aggregate imaging system. (Courtesy Pine Instruments, Grove City, PA.)

Blast furnace (or iron) slags, most widely used in the pavement industry, are formed as a by-product in blast furnaces, whereas steel slags are formed in basic oxygen and electric arc furnaces. While blast furnace slag is lighter than natural aggregates, steel slags are heavier. The light weight of the blast furnace slag provides the advantage of fewer truckloads for the same volume of materials. Steel slags, on the other hand, have a high angle of internal friction, and are very suitable for rut-resistant pavement layers. The National Slag Association (n.d.; http://www.nationalslag.org) provides excellent information on this topic.

QUESTIONS

8.1   What are the primary types of rocks? How are they formed?

8.2   For the given information from sieve analysis, prepare a 0.45 power plot.

Sieve Size (in.)	% Passing
2	100
1 1/2	100
1	100
3/4	100
1/2	97
3/8	75
No. 4	20
No. 8	15
No. 16	10
No. 30	8
No. 50	7
No. 100	5
No. 200	3

8.3   Four different aggregates are to be combined to produce a blend. The percentages of the different aggregates in the blend are 20, 50, 10, and 20, and the corresponding bulk-specific gravities are 2.650, 2.720, 2.600, and 2.690. Compute the bulk-specific gravity of the combined aggregates in the blend.

8.4   Determine the bulk-specific gravity of an aggregate whose apparent specific gravity is 2.690 and absorption is 1.5%.

8.5   Combine the following aggregates in the right proportion to produce a blend with the given target gradation:

Sieve Size (mm)	CA7 % Passing	Washed Sand % Passing	Dry Sand % Passing	% Passing Target
19.00	100	100	100	100.0
12.50	98.8	100	100	100.0
9.50	62.1	100	100	93.0
4.75	2.1	98.1	100	62.0
2.36	0	81	99	45.0
1.18	0	69.9	80.7	35.0
0.60	0	50.6	59.9	25.0
0.30	0	20.7	40	15.0
0.150	0	5.1	24.2	9.0
0.075	0	1.9	17.2	4.0

8.6   What is the importance of the fineness modulus for PCC?

8.7   Determine the fineness modulus for the following aggregate, and comment on the effect on fresh concrete.

Sieve Size (mm)	Cumulative % Passing
9.50	100.00
4.75	98.00
2.36	85.00
1.18	65.00
0.60	45.00
0.30	21.00
0.15	3.00

9 Asphalt and Emulsions

9.1 ASPHALT BINDER

In the United States, *asphalt* refers to the binder, also known as *bitumen*. Asphalt can refer to a mixture of bitumen and aggregates (as in Europe), which is usually called *asphalt mix* in the United States. In this book, the U.S. definition will be followed—asphalt will mean the binder.

Asphalt is a material that is made up of predominantly hydrocarbons; small amounts of sulfur, nitrogen, and oxygen; and trace amounts of metals such as vanadium, nickel, iron, magnesium, and calcium. Analysis of a typical asphalt would indicate the following: carbon, 82%–88%; hydrogen, 8%–11%; sulfur, 0%–6%; oxygen, 0%–1.5%; and nitrogen, 0%–1%.

Traditionally "penetration," "viscosity," and "shear modulus" tests have been used to characterize asphalts. While discussing different types of asphalt, results from one of these tests are provided to make a relative comparison of the "softness/hardness" or "stiffness" of the material. In layman's language, the higher the penetration or the lower the viscosity/modulus, the "softer" the asphalt binder is, and vice versa.

9.2 NATURALLY OCCURRING ASPHALTS

Asphalt can be obtained in different forms, although refined asphalt from fractional distillation of crude oils is the most commonly used. The naturally occurring asphalts are lake asphalts, rock asphalt, and gilsonite.

9.2.1 LAKE ASPHALT

The primary source of lake asphalt is a deposit in a lake in the southern part of Trinidad. This asphalt has been deposited as a result of surface seepage in the bed of the lake. Asphalt is excavated from the lake bed and transported to various places. This asphalt is a mixture of different materials, including pure asphalt binder. This asphalt is also known as Trinidad lake asphalt (or TLA). The refinement of the excavated material is done by heating it to 160°C and vaporizing the water. The material is then sieved to remove large-size/other particles and organic matter. This refined product has about 54% binder, 36% mineral matter, and 10% organic material. TLA has a penetration of about 2 mm, with a softening point of 95°C. This material was once used alone or in combination with softer asphalt for regular use in asphalt pavement construction. These days, its use is confined to mostly high-stiffness asphalt mixes for pavements and roofing applications.

9.2.2 ROCK ASPHALT

Rock asphalts are obtained from rocks that are impregnated with asphalts. They are formed by the entrapment of asphalt in impervious rock formations. These asphalts are obtained from mines or quarries. Their deposits can be found in Switzerland, France, and Italy, as well as in Utah and Kentucky in the United States. The concentration of asphalt in such deposits can range from 7% to 13%. Although used in some of the earliest roads in Europe and the United States, it is used very little at present, mostly in waterproofing and high-stiffness asphalt mix applications.

9.2.3 GILSONITE

Gilsonite is found in Utah in the United States. It has a penetration of zero with a softening point of 115°C–190°C. It is used in combination with regular asphalt to alter its characteristics to reduce the penetration and increase its softening point. Its use is mostly limited to bridge and waterproofing applications and high-stiffness asphalt mixes.

9.3 REFINED ASPHALT FROM CRUDE OIL

Refined, regularly used asphalt for paving applications is produced from crude oil. Crude oil is formed as a result of the action of heat from within the earth's crust and weight of many layers of deposits of remains of marine organisms and vegetation over millions of years. Natural gas and oil reservoirs are detected by surveys and recovered by drilling through impermeable rocks. The oil is produced in mainly four regions of the world—the Middle East, Russia, the United States, and Central America. The characteristics of the crude oil differ significantly from source to source. Of the 1500 different sources of crude oils in the world, only a few are actually found to be suitable for obtaining asphalt.

Asphalt is manufactured from crude oil starting with a process called *fractional distillation*. This process separates the different hydrocarbons with different boiling points that make up crude oil. Fractional distillation is carried out in tall steel towers. The crude oil is first heated in a furnace to a temperature around 350°C, and the mix of liquid and vapor is then directed to the lower part of the tower, which has a pressure greater than the atmospheric pressure. The vapors rise up in the tower through holes in the horizontal layers that divide the tower vertically into segments. The vapors lose heat as they rise through the holes in the trays. When the temperature of the vapor falls below its boiling point, it condenses and the liquid that accumulates in one of the trays is then drawn off. The lightest fractions of the crude oil that remain as vapor are taken off from the top of the tower and include propane and butane. Further down the tower, kerosene and then gas oil are recovered. The heaviest fraction (long residue) that comes from the bottom of the tower is a complex mixture of hydrocarbons and is distilled again at a reduced pressure (10–15 mmHg and 35°C–45°C) in a vacuum distillation tower. One of the products of this step is the short residue, which is used as the feedstock for producing different grades of asphalt. As the viscosity of the short residue is dependent on the source of the crude and the temperature and pressure during the process, the conditions are adjusted on the basis of the source of the crude oil to produce an asphalt with a penetration in the range of 35–300 dmm (1 dmm is 0.1 mm).

Air blowing or oxidation is used to modify the proportion of the short residue at a temperature of 240°C–320°C. The result of such oxidation is a decrease in penetration, increase in softening point, and decrease in temperature susceptibility. Such modification, which can be in continuous, semi, or full (oxidation) mode, results in the formation of asphaltenes and an increase in size of the existing asphaltenes.

The three processes that take place are oxidation, dehydrogenation, and polymerization. In the continuous blowing process, the penetration and softening point of the asphalt depend on the viscosity of feedstock, temperature in the blowing column, residence time in the blowing column, origin of the crude oil, and air-to-feed ratio. Fully blown or oxidized asphalts are produced by vigorous air blowing or by blending with relatively soft flux.

The properties in this case are dependent on the viscosity and chemical nature of the feed. The specific grade of asphalt is produced by selecting a suitable feedstock, controlling viscosity of the feed, and the conditions in the tower.

9.4 SAFE DELIVERY, STORAGE, AND HANDLING OF ASPHALTS

Considerations must be made regarding delivery, storage, and handling of asphalts to make sure that adequate safety is maintained, and that asphalt properties are not adversely affected. The safety factors are discussed first, followed by factors that could affect asphalt properties adversely.

9.4.1 CAUSES OF HAZARDS AND PRECAUTIONS

The main hazard from asphalt results from the fact that it is usually applied at high temperatures, which facilitate the emission of fumes of hydrocarbons and H_2S (the amount doubles for every $10°C–12°C$ increase in temperature). Exposure to even small amounts of H_2S is fatal. Although relatively very high temperatures are required for spontaneous burning of asphalt ($400°C$), small amounts of H_2S from asphalts can react with rust on tank walls to form iron oxide. This material reacts readily with O_2 and can self-ignite and also ignite carbonaceous deposits on the roof and walls of the tank. Also, if water comes in contact with hot asphalt, it is converted into steam, and foam is produced in the asphalt with a volume increase of approximately 1400 times. In order to prevent these three hazardous conditions, the following steps must be taken:

1. Test areas where H_2S may be present, and check if they are gas-free before entering.
2. Openings in asphalt tanks such as areas to roofs should be restricted, and the manhole should be kept closed to prevent entry of O_2.
3. Prevent water from entering asphalt tanks.

9.4.2 HEALTH HAZARDS

Some benzene compounds are found in polycyclic aromatic compounds (PACs), which are present in crude oils—however, the concentration in asphalt that is refined from crude oil is very low. Intimate and prolonged skin contact of asphalt as well as asphalt emulsion should be avoided to prevent risk of irritation of skin and eyes.

Specific instructions on how to treat skin for first-, second-, and third-degree burns; eye burns; and inhalation should be clearly displayed and available in areas where asphalt is handled. Under normal working conditions, exposure to particulate asphalt, hydrocarbons, and H_2S that are emitted during heating of asphalt is well below allowable maximum limits. However, inhalation can cause usually temporary irritation to the eyes, nose, and respiratory tract; headache; and nausea. Such exposure should be minimized, and if it occurs, the affected person should be removed to fresh air and, if necessary, given medical attention as soon as possible.

9.4.3 PRECAUTIONS AND GOOD PRACTICES

Proper equipment should be used for protecting against burns from hot asphalt. Standard gear includes helmet and neck aprons, heat-resistant gloves, safety boots, and overalls. Soiled garments should be replaced/dry cleaned as soon as possible. Persons handling asphalts can use barrier creams to help subsequent cleaning, for which only approved skin cleaners and warm water should be used.

Personnel should be properly equipped and trained to handle fires should one break out. Direct water jets should not be used since they may cause foaming and spread the hot asphalt. Dry chemical powder, foam, vaporizing liquid, or inert gases can be used to extinguish small fires. Portable fire extinguishers must be placed at strategic, permanent, and conspicuous areas. The local fire station should be consulted to formulate a plan if initial efforts fail to extinguish a fire.

Spills and splashes of hot asphalt should be avoided during sampling. Sampling areas should be well lit and have safe in and out areas, and adequate protective clothing must be worn. Sampling by dipping a weighted can should be done for only small samples and be avoided for cutback tanks because of the presence of flammable atmosphere in tank vapor spaces. The best way is to use properly designed sample valves from pipelines and tanks. The proper sample valve remains hot by the product so that there is no blockage in closed position, and also such that the plunger of the valve extends into fresh asphalt and allows obtaining fresh liquid asphalt when opened.

Regarding leaching of components from asphalt, it has been shown that prolonged contact with water leads to leaching of PACs, but the levels reach equilibrium rapidly and the numbers are one order of magnitude below the limits of potable water.

Users of asphalt (laboratories, plants, and job sites) should review materials safety data sheets and follow all Occupational Safety and Health Administration (OSHA) standards such as the OSHA Hazard Communication Standards (e.g., 29 CFR 1910.1200; see http://www.ilpi.com/msds/osha/1910_1200.html; OSHA, n.d.).

9.5 ASPHALT BINDER PROPERTIES

Since asphalt binders are by-products of the petroleum distillation process, by nature, if heated to a very high temperature, they can give off vapors that can get ignited and pose safety hazards in plants. They can also become stiffer at low temperatures and relatively soft at high temperatures. Furthermore, asphalts lose components through volatilization as well as oxidation, when exposed to air/oxygen and/or high temperature, and as such, they may undergo a drastic change in properties. In fact, asphalts tend to become "harder" with time—what is commonly known as *age hardening*. Therefore, tests have been developed to characterize asphalts in terms of their relevant properties, such as those just described—safety, temperature susceptibility, and age hardening. These tests are presented in the following.

9.5.1 SPECIFIC GRAVITY: ASTM D-70

Specific gravity of a material is defined as the ratio of the mass of the material at a given temperature to the mass of an equal volume of water at the same temperature. For asphalt binder, the ASTM D-70 procedure, Density of Semi-Solid Bituminous Materials (Pycnometer Method), is used for the determination of specific gravity and density. In this procedure, a calibrated pycnometer with a sample of known weight is first weighed, and then filled with water. The setup is then brought to the test temperature and weighed. The density (Specific Gravity * Density of Water) is then calculated from the mass of the sample and the water displaced by the sample in the pycnometer, and the specific gravity is calculated. Note that the density of water varies between 999.1 and 997 kg/m^3 between a temperature of 15°C and 25°C. If the density of water is approximately taken as 1 gm/cm^3, then the density is numerically equal to specific gravity. The specific gravity of the asphalt is expressed along with the temperature at which both asphalt and water were used in the test. Specific gravity is used in converting weight to volume and vice versa, in paying asphalt suppliers/contractors, as well as in determining the effective specific gravity of aggregate.

9.5.2 CLEVELAND OPEN CUP METHOD (FLASH POINT): ASTM D-92

This test involves heating the asphalt in a brass cup, and passing a small flame over the cup periodically, until a quick flash occurs. The temperature at which the flash occurs is called the *flashpoint*. It is conducted to determine the temperature range that can be used during the production of HMA without causing safety hazards.

9.5.3 SOLUBILITY TEST: ASTM D-2042

In this test, a sample of asphalt is mixed with a solvent and then poured through a glass fiber pad filter. The retained impurities are washed, dried, and weighed. This test is run to minimize the presence of such impurities as dust and organic materials in the asphalt.

9.5.4 SPOT TEST: AASHTO T-102

In this test, visual evaluation of a drop of asphalt and solvent mix on a filter paper is made. Results are reported as acceptable or not acceptable, based on the absence or presence of a dark area in the spot on the filter paper. This test was developed to identify asphalts that have been damaged due to overheating (cracking—molecules thermally broken apart) during the asphalt production process.

9.5.5 PENETRATION: ASTM D-5

In this test, a container of asphalt maintained at 77°F is placed under a needle. The needle, under a weight of 100 g, is allowed to penetrate the asphalt in the container for 5 s. The depth of penetration, expressed in dmm, is reported as the penetration. The purpose of this test is to classify asphalt, check consistency, and evaluate the overheating of asphalt. Generally five different grades are available: 40–50, 60–70, 85–100, 120–150, and 200–300. A 60–70 grade means an asphalt binder that has a penetration value ranging from 60 to 70 dmm and is harder than the 85–100 grade and softer than the 40–50 grade. A binder with a lower penetration grade such as 40–50 would be used in a project in a warmer location, whereas those with higher grades (such as 120–150) would be used in a project in a colder region.

9.5.6 VISCOSITY TESTS

Viscosity is defined as the ratio of applied shear stress to applied shear strain. It can be measured by direct tests (such as a sliding plate viscometer), but is generally measured by indirect procedures. This test helps in measuring the resistance of the asphalt to flow. Two types of viscosity tests are generally run: absolute viscosity (expressed in poise) (flow in vacuum) (ASTM D-2171, AASHTO T-202) at 140°F, and kinematic viscosity (expressed in centistokes) (by gravity flow) (ASTM D-2170, AASHTO T-201) at 275°F.

In the absolute viscosity test, a sample of asphalt is introduced through one side of a capillary tube, maintained in a bath at 140°F. After reaching 140°F, the asphalt is made to flow by applying a vacuum on the other side of the tube, and the time to flow through two marked points in the tube is measured. The time is then used, along with the calibration constant of the tube, to determine the viscosity.

The kinematic viscosity test is similar to the absolute viscosity test, but runs at a higher temperature, at which the asphalt is sufficiently fluid to flow. A different type of tube is used for this test, and after reaching the test temperature, only a small vacuum is applied on the other side of the tube to initiate the flow of asphalt due to gravity. The kinematic viscosity is then determined by multiplying the time taken by the asphalt to flow between two timing marks and the calibration factor of the tube.

Viscosity tests are used for classification of asphalts (absolute viscosity), as well as on conditioned asphalts to evaluate overheating/overaging of asphalt. ASTM D-3381 presents the specifications for the different types of viscosity-graded asphalt. Viscosity grades are generally expressed as an AC, followed by a number expressing its viscosity divided by 100. For example, an AC 20 asphalt binder means a binder with absolute viscosity of 2000 P. Generally higher viscosity asphalts are used in warmer climatic regions, whereas lower viscosity materials are used in colder regions.

Kinematic viscosity is used to determine suitable mixing and compaction temperature for HMA. An equiviscous temperature (temperature at which viscosities are similar) is used for mixing and compacting different types of mixes, such that the effect of the asphalt can be made uniform. Generally, the viscosity for mixing and compaction is specified as 170 ± 20 and 280 ± 30 cSt (for mixing and compaction, respectively; Asphalt Institute 1984). Generally, the absolute viscosity is determined at 60°C, and the kinematic viscosity is measured at 135°C, and then a plot of temperature (in Rankine) versus kinematic viscosity (log–log) is plotted with two data points, one with the measured value at 135°C and another one calculated from the absolute viscosity at 60°C as follows:

$$\text{Kinematic viscosity at } 60°C = \frac{\text{Absolute viscosity at } 60°C * 100}{0.98 * G_b}$$

where
    Kinematic viscosity is in centistokes
    Absolute viscosity is in poise
    G_b is the specific gravity of the asphalt binder at 15°C

9.5.7 Softening Point (Ring and Ball) Test: ASTM D-36

This test is conducted to determine the temperature at which asphalt starts flowing. This test involves the placement of a steel ball over a ring of asphalt in a bath, and heating the assembly at 5°C/min. The temperature at which the asphalt cannot support the ball anymore and sinks is reported as the softening point. This is not used for paving asphalts that much, but is used for roofing asphalts.

9.5.8 Fraass Breaking Point Test: BS EN 12593, BS 2000-8

This test is conducted to determine the very low temperature stiffness of asphalt binder. In this test, a steel plate coated with a thin layer of asphalt binder is slowly bent and released repeatedly, while the temperature is reduced constantly at the rate of 1°C/min, until the asphalt binder cracks.

9.5.9 Ductility: ASTM D113

In this test, a sample of asphalt is stretched until it breaks or until the testing equipment limit is reached. The asphalt sample is stretched at a rate of 5 cm/min at 77°F, and the ductility is measured in cm. If run at a lower temperature, the rate of stretching is reduced. This test is run to ensure the use of nonbrittle asphalts.

9.5.10 Thin Film Oven Test (TFOT): ASTM D1754

This is a conditioning procedure, in which a 50 g asphalt sample is placed in a flat-bottom pan on a shelf in a ventilated oven maintained at 325°F. The shelf is rotated at 5–6 revolutions/min for 5 h. The loss in penetration or increase in viscosity as well as change in weight of the sample are usually measured after the heating. The use of this test is to evaluate the effect of heat and air on asphalts, and identify those that would harden excessively due to volatilization and/or oxidation.

9.5.11 Rolling Thin Film Oven Test (RTFOT): ASTM D-2872, BS EN 12591, AASHTOT-240

In this test, eight bottles with asphalt are placed in a rack inside an oven maintained at 325°F (Figure 9.1), and the rack is rotated around a heated air-blowing orifice for 75 min. As the rack rotates, asphalt binder inside each bottle flows inside, exposing the fresh surface to the air. The mass loss, as well as viscosity, after this process is checked. This test is run to classify asphalts (aged residue, or AR, process) and to ensure minimization of mass loss and change in properties.

9.6 ASPHALT BINDER PROPERTIES AND PAVEMENT DISTRESS AND PERFORMANCE

For asphalt pavements, rutting at high temperature and cracking at low temperature are the two primary topics of interest regarding asphalt. During the production of HMA and during its life on the pavement, the asphalt could become too brittle. It is important to select an asphalt that would not become too brittle because of the asphalt itself, because of the mix components, or because of construction. Hence, there are several factors that must be checked to ensure this, such as the properties of asphalt, using the TFOT, using the correct amount of miner filler (aggregates less than 0.075 mm in diameter) in the mix, and compacting the HMA to a minimum density, such that the voids are below a maximum allowable level.

FIGURE 9.1 Rolling thin film oven test (RTFOT) equipment.

9.6.1 Aging of Asphalt Binder

The composition of asphalt binder is affected by air (oxygen), ultraviolet radiation, and fluctuations in temperature. The primary effect is the hardening or stiffening of the asphalt binder, both during the production process as well as during the service life of the pavement. This hardening needs to be "predicted" and taken into consideration during the selection of the asphalt binder, since excessive hardening of the binder increases the potential of cracking of asphalt mixes. Hardening of the binder takes place on the surface of the pavement as well as in the mix inside the pavement.

Different mechanisms, caused by the different environmental factors, are responsible for the hardening of asphalt binder. Although 15 different factors have been identified (as listed in the following), the first four are considered to be the more important ones:

1. Oxidation in the dark
2. Volatilization
3. Steric or physical factors

4. Exudation of oils
5. Photo-oxidation by direct light
6. Photo-oxidation in reflected light
7. Photochemical reaction by direct light
8. Photochemical reaction by reflected light
9. Polymerization
10. Changes by nuclear energy
11. Action of water
12. Absorption by solid
13. Absorption of components at a solid surface
14. Chemical reactions
15. Microbiological deterioration The four important factors are explained as follows:
    a. *Oxidation*: Asphalt binder components oxidize and form heavier and more complex molecules, increasing its stiffness and decreasing its flexibility. The rate of oxidation is affected by the temperature and the asphalt binder film thickness, with higher temperature and thinner films facilitating oxidation. For temperatures above 100°C, the rate of oxidation doubles for every 10°C increase in temperature.
    b. *Volatilization*: Depending on the temperature and the surface area exposed, the asphalt binder loses lighter components by volatilization.
    c. *Steric or physical hardening*: Reorientation of molecules and slow crystallization of waxes are responsible for causing steric hardening of asphalt binder at ambient temperatures.
    d. *Exudation of oils*: Depending on its chemical nature as well as the porosity of the aggregate, oils from the asphalt binder are exuded into the aggregates in an asphalt mix to a different extent.

The hardening of asphalt occurs at different extents and because of different factors during storing, during mixing (during production), and in the pavement. When stored inside a tank in bulk, and not exposed to oxygen, very little hardening occurs, even at high temperatures. Proper layout of storage tanks should be used to ensure avoiding exposure of a larger surface area of asphalt binder to oxygen during circulation of the binder in the tank. During the mixing process, asphalt binder is subjected to relatively high temperature, and is spread over aggregates in films with thickness generally ranging from 5 to 15 μm. The time of mixing can be different for different types of mixing plants (shorter for drum plants, and hence shorter exposure to oxygen and less hardening effect on the asphalt binder), and for specific asphalt binder selected, the temperature and the amount of asphalt should be controlled properly to make sure that the asphalt film is of sufficient thickness to minimize hardening. Hardening increases with a decrease in film thickness. If the asphalt mix is stored in a silo, some hardening will take place as a result of the introduction of air along with the mix. The silo should, therefore, be airtight, and be as full as possible, to minimize the presence of air. Air within the silo may be removed by the introduction of inert gases, and any oxygen inside the silo could also react and form carbon dioxide and prevent further oxidation. If no fresh air is allowed inside the silo during storage, very little or no oxidation could occur between this time and transportation and laydown.

It is important to note that any gain in lowering the viscosity of the asphalt binder by overheating the binder during mixing will be lost by the subsequent increased rate of oxidation and resulting high viscosity, and the increased viscosity will be detrimental to the performance of the asphalt mix.

The hardening of the asphalt binder during storing, mixing, transportation, and laydown is grouped under *short-term aging*, whereas the hardening of the asphalt binder in the pavement, during its service life, is called *long-term aging*. This hardening is primarily affected by the amount of air voids in the compacted asphalt mix in the pavement. Since a higher amount of air voids would introduce a higher amount of air, oxidation and hence hardening can be minimized by providing adequate compaction and lowering voids, preferably below 5%. In practice, it may not be possible to compact to such an extent to lower the voids below 5% for all types of asphalt binder, and in general the average air voids of dense graded mixes are kept below 8%. Mixes with similar air voids may not have the same rate of hardening if the distribution of air voids in the mixes is different—mixes with interconnected voids would have higher air permeability and a faster rate of hardening.

Note that the hardening of the asphalt binder in the asphalt mix on the surface of the pavement occurs at a faster rate than that of the asphalt binder in the mix inside the pavement, because of the relatively higher amount of oxygen available, higher temperatures, and photo-oxidation by ultraviolet radiation. This top oxidized layer of asphalt could be eroded by water and, while present, could prevent further oxidation. However, adequate thickness of the asphalt film should be ensured through the provision of an adequate amount of asphalt binder. Generally, thicker films (i.e., higher asphalt content) are provided for compacted asphalt mixes with relatively high amounts of air voids. Optimum film thickness can range from 9 μm to above 12 μm for dense and open graded asphalt mixes, respectively.

Note that while excessive hardening of the surface mix is always detrimental to the performance of the pavement, the hardening of binder, resulting in the increase in stiffness of the asphalt mix *base* course ("curing"), is also considered to be a beneficial effect with respect to the structural design of the pavement.

9.6.1.1 Hardening

There are parameters, based on penetration and viscosity tests, that could be used to evaluate the "hardening" of asphalt due to volatilization as well as oxidation.

$$\text{Retained penetration, } \% = \frac{\text{Penetration after TFOT}}{\text{Penetration before TFOT}}$$

$$\text{Viscosity ratio} = \frac{\text{Viscosity after TFOT}}{\text{Viscosity before TFOT}}$$

In order to ensure good performance of an asphalt pavement, the temperature susceptibility of asphalt must be controlled, since it is always desirable to have an asphalt that would possess sufficient stiffness at high temperature to prevent rutting and possess sufficient ductility at low temperature to prevent cracking.

There are three parameters that could be used to measure temperature susceptibility of asphalts: the penetration index (PI), the pen-vis number (PVN), and viscosity–temperature susceptibility (VTS), as explained next:

$$\text{Penetration index (PI)} = \frac{20 - 500\,A}{1 + 50\,A}$$

$$A = \frac{\text{Log}\left(\text{penetration at } T_1\right) - \text{Log}\left(\text{penetration at } T_2\right)}{T_1 - T_2}$$

The PI of asphalt generally ranges between −1 and +1, and a higher PI indicates lower temperature susceptibility.

$$\text{Pen-vis number (PVN)} = \frac{L - X}{L - m}(-1.5)$$

where
   L is the log of viscosity at 275°F for PVN = 0.0
   m is the log of viscosity at 275°F for PVN = −1.5
   X is the log of viscosity at 275°F

Typical values of the PVN range between −2 and 0.5, and a higher PVN indicates a lower temperature susceptibility.

$$\text{Viscosity-temperature susceptibility (VTS)} = \frac{\text{Log}\big(\log \text{ viscosity at } T_2\big) - \text{Log}\big(\log \text{ viscosity at } T_1\big)}{\text{Log } T_1 - \text{Log } T_2}$$

Typical values of VTS range from 3.36 to 3.98. The lower the VTS, the lower the temperature susceptibility.

9.7 STIFFNESS

A significant amount of research has been conducted on defining, measuring, and evaluating stiffness of asphalt. Some of these are indirect, while the rest are direct measurement techniques, although the approach varies significantly. One method is to determine the stiffness of the asphalt from the stiffness of the mix from the nomograph shown in Figure 9.2.

Another of the earliest developed methods is the use of the softening point (ring and ball test) and the PI, as shown in Figure 9.3.

FIGURE 9.2   Determination of stiffness of asphalt from stiffness of mix. *Note:* VMA refers to voids in mineral aggregate, which is the amount of total voids in a compacted mix with respect to the total volume of the mix. (Reprinted from Van der Poel, C., *J. Appl. Chem.,* 4, 221, May 1954; Roberts et al., *Hot Mix Asphalt Materials, Mixture Design, and Construction,* Lanham, MD, 1996, by permission of National Asphalt Pavement Association Research and Education Foundation.)

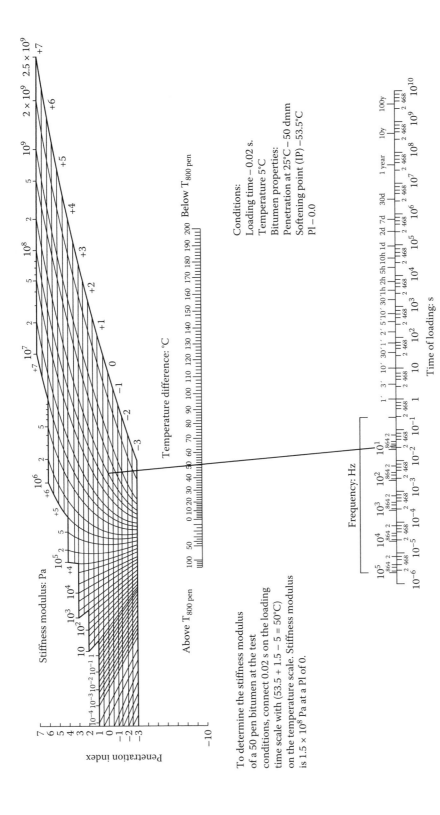

FIGURE 9.3 Determination of stiffness from softening point and penetration index. (Reproduced from Shell, *Shell Pavement Design Manual: Asphalt Pavements and Overlays for Road Traffic*, London, U.K., Shell International Petroleum, 1978. With the kind permission of Shell International Petroleum Company, part of the Royal Dutch Shell Group.)

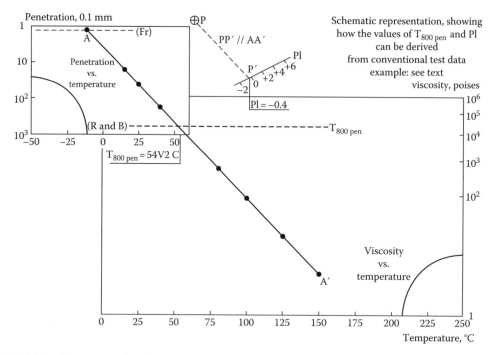

FIGURE 9.4 Bitumen test data chart. (Reprinted from Heukelom, W., An improved method of characterizing asphaltic Bitumens with the aid of their mechanical properties, in *Proceedings, Association of Asphalt Paving Technologists*, Vol. 42, 1973. With kind permission from the Association of Asphalt Paving Technologists.)

Van der Poel, who developed this procedure, showed that for the same time of loading, two asphalt binders with the same PI have the same stiffness at temperatures that are equidistant from their respective softening points. A modification of this method is to use the corrected PI and the corrected softening point, which is done as follows. The penetration is measured at three temperatures, and the corrected PI (pen/pen) is determined from the plot of temperature versus penetration on the bitumen test data chart (BTDC). The BTDC is a chart that can be used to present penetration, softening point, Fraass breaking point, and viscosity data on the same chart. The chart (Figure 9.4), with two horizontal scales for the temperature and Fraass breaking point, and two vertical scales for penetration and viscosity, is prepared in such a way that data from asphalt binders with "normal" temperature susceptibilities plot as straight lines on the chart.

The corrected softening point is determined from the point at which the extended penetration straight line and the 12,000 poise ordinate on the BTDC intersect. A further modification of the process is the use of the PVN instead of the PI, as shown in Figures 9.5 and 9.6.

9.7.1 Viscosity for Stiffness

For asphalts, the viscosity parameter is often used as a measure of stiffness, using the definition of absolute viscosity—the ratio of shear stress to the rate of shear. A large number of viscometers have been developed over the years, which may be of rotational or capillary tube or sliding plate type. While high-temperature (>140°F) viscosity is suitably measured with a capillary tube–type viscometer, as mentioned earlier, for low temperatures cone and plate viscometers and sliding plate viscometers are most widely used.

In the sliding plate viscometer, a sample of asphalt is sandwiched between two plates, one of which is clamped to a frame. A load applicator needle is placed onto the asphalt sample, using

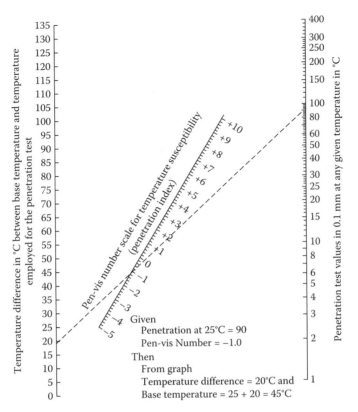

FIGURE 9.5 Relationship between temperature, penetration, and pen-vis number. (Reprinted from McLeod, N.W., *Am. Soc. Test. Mater. J. Test. Evaluat.*, 4, July 4, 1976; Roberts et al., *Hot Mix Asphalt Materials, Mixture Design, and Construction*, Lanham, MD, 1996, by permission of National Asphalt Pavement Association Research and Education Foundation.)

an adapter. At the start of the test, the support beneath the asphalt sample and the unclamped plate is removed and the asphalt sample is subjected to shearing load. The shearing stress and the shear displacement, as measured by a transducer, are used to determine the viscosity of the asphalt.

9.8 VISCOELASTIC NATURE OF ASPHALT AND DIRECT MEASUREMENT OF STIFFNESS

The viscoelastic nature of an asphalt mix comes primarily from the viscoelastic behavior of the asphalt binder. Depending on the temperature and rate of loading, an asphalt binder could behave as an elastic material or as a viscous material, or in a way that reflects both its elastic and viscous nature. At low temperatures and short periods of loading the response is elastic, whereas at high temperature and long periods of loading the response is viscous. At intermediate temperatures and periods of loading, asphalt binders show viscoelastic behavior. The lag between the shear stress and shear strain (Figure 9.7) is called the *phase angle*, δ, and can range from 0, for a perfectly elastic material, to 90° (one-quarter of a cycle) for a perfectly viscous material. Several models that have been developed by researchers capture the dependence of asphalt binder stiffness on temperature and periods of loading.

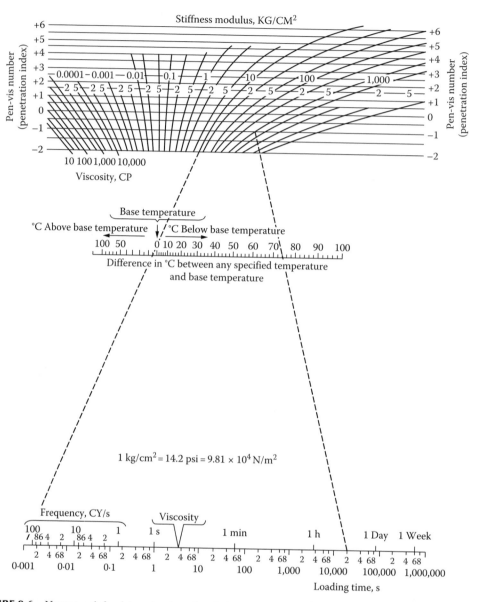

FIGURE 9.6 Nomograph for determination of asphalt binder stiffness. (Reprinted from McLeod, N.W., *Am. Soc. Test. Mater. J. Test. Evaluat.*, 4, July 4, 1976; Roberts et al., *Hot Mix Asphalt Materials, Mixture Design, and Construction*, Lanham, MD, 1996, by permission of National Asphalt Pavement Association Research and Education Foundation.)

Conceptually, this can be expressed as follows:

$$\text{Stiffness, } S_{t,T} = \frac{\sigma}{\varepsilon_{t,T}}$$

where

$S_{t,T}$ is the stiffness at loading time t at temperature T

σ is the stress

$\varepsilon_{t,T}$ is the strain at loading time t at temperature T

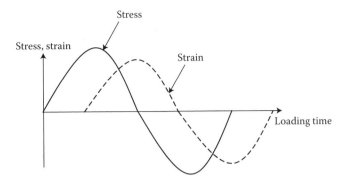

FIGURE 9.7   Concept of lag between stress and strain.

The stiffness relevant for asphalt binder is related to shear failure. To determine stiffness or modulus related to shear failure, one needs to measure shear stress and shear strain. The shear stress during a test can be applied in different ways, to be consistent with the specific mode of failure in the pavement. For example, it can be applied as a sinusoidal wave (dynamic loading to simulate the effect of traffic), with varying stress, using a constant amplitude and frequency. Or, the load can be applied as a constant load for a long period of time (as in a creep test to simulate the effect of the environment). The shear modulus can be determined as the ratio of shear stress to strain at a specific frequency. Note that the frequency is related to the period or time of loading. Under dynamic conditions of loading, the stiffness (or modulus) is equal to three times the shear modulus.

The dynamic tests could be conducted for shorter loading periods, and creep tests for longer loading periods, and the combined test data could be used for obtaining stiffness versus loading-time information, over the range of loading times expected in a pavement (Figure 9.8). Note that stiffness versus loading time plots for a specific asphalt binder obtained at different temperatures would coincide if shifted by different factors on the same plot. This property of asphalt binder is often referred to when asphalt binders are said to be "thermorheologically simple."

9.9   TENSILE BEHAVIOR

The behavior of asphalt binder in tension is usually expressed in terms of two parameters—breaking stress and strain at failure. At higher stiffness, the strain at failure is a function of the elastic modulus. High stress due to low temperatures occurs when the asphalt binder has relatively high stiffness, and the strain at failure is very low. The strain at failure at a specific temperature can be predicted from the PI, the softening, and a specific time of loading.

Toughness and tenacity: in this test (Figure 9.9), which is similar to the tensile strength test, the stress to cause the elongation as well as the area under the stress/strain curve are determined. Generally this test is useful for polymer-modified asphalt binders. The test results show two distinct areas under the stress/strain curve—the relatively small one under the beginning to the yield point of the asphalt, and the second larger area under a milder peak and prolonged elongation. The area of total curve (first and second combined) is called *toughness*, whereas the area under the second (large) part, which comes mainly from the contribution of the polymer modification, is known as the *tenacity*.

9.10   SUPERPAVE (SUPERIOR PERFORMING ASPHALT PAVEMENTS)

Superpave is the HMA mix design system that resulted from the $150 million, 5-year-long Strategic Highway Research Program (SHRP) in the United States between 1987 and 1993. The major objectives behind the development of Superpave were to identify and define properties of asphalt

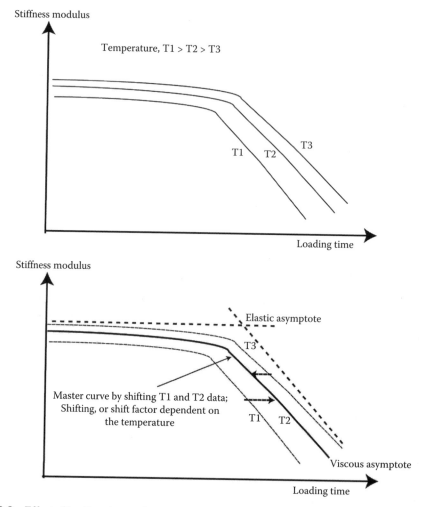

FIGURE 9.8 Effect of loading time and temperature on stiffness.

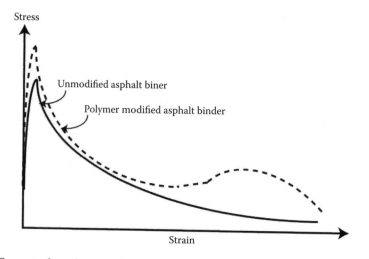

FIGURE 9.9 Concept of toughness and tenacity test. (Redrawn from Read, J. and Whiteoak, D., Shell bitumen. *The Shell Bitumen Handbook*, 5th edn. London, U.K.: Thomas Telford, 2003.)

binders, aggregate, and hot mix asphalt that influence pavement performance, and develop test methods for performance-based specifications. Even though research continues to improve the Superpave system, at present, the Superpave system is being used by most of the state departments of transportation in the United States.

The major elements of Superpave are (1) Superpave performance-graded (PG) asphalt binder specifications, (2) aggregate properties, and (3) the Superpave mix design and analysis system. The asphalt binder specifications and tests are described in the following paragraphs.

The two major outcomes in the area of asphalt binder specifications are the introduction of performance-related tests and the shift from the traditional approach of testing asphalt at the same temperature for different specified values of test results to testing asphalt at different temperatures for the same specified values of test results. The new tests relate to three asphalt binder characteristics—temperature susceptibility, viscoelastic behavior, and age hardening. The condition of the asphalt binders during the tests simulates the condition of the asphalt binder at supply, during production of HMA, and in service in the pavement. The modes of failure for which the asphalt binder characteristics are specified include rutting and cracking (temperature and fatigue related). Several relatively new asphalt tests are used routinely now, or are being implemented, as a result of the introduction of the Superpave system.

9.10.1 HIGH-TEMPERATURE VISCOSITY

High-temperature viscosity measurements, such as those required for determination of suitable mixing and compaction temperature and to check for suitability of pumping during mixing, can be conducted with a rotational viscometer. The test consists of rotating a spindle with a sample of asphalt binder maintained at the test temperature, measuring the torque required to maintain a constant rotation speed, and calculating the viscosity from this measurement. For plotting the kinematic viscosity versus temperature data, this test could be used to determine the viscosities at 135°C and 165°C (and avoid the need to convert absolute viscosity to kinematic viscosity, as discussed earlier). Kinematic viscosity is measured in centipoises and reported in the Pa-s (SI) unit, and the viscosities used for determining the equiviscous temperatures are 0.17 ± 0.02 Pa-s for mixing and 0.28 ± 0.03 Pa-s for compaction. The different parameters used for the calculation of kinematic viscosities are as follows.

Viscosity at 60°C, absolute viscosity, expressed in poises (P, cgs unit) or centipoises (CP), where 100 cP = 1 P. The SI unit is Pa-s:

$$1\,\text{Pa-s} = \frac{1\,\text{kg}}{\text{ms}}\,(1\,\text{P} = 0.1\,\text{Pa-s})$$

$$\text{Kinematic viscosity} = \frac{\text{Absolute viscosity}}{\text{Density}} = \frac{\text{Pa-s}}{\text{kg}/\text{m}^3} = \frac{\text{kg}/\text{ms}}{\text{kg}/\text{m}^3} = \frac{\text{m}^2}{\text{s}} = \text{m}^2\text{s}^{-1}$$

In rotational viscometer, the absolute viscosity is measured as the ratio of shear stress to rate of shear, and the kinematic viscosity is expressed in the same unit as absolute viscosity (Pa-s), as follows:

$$\frac{\text{Absolute viscosity}}{\text{Density}} = \frac{\text{Pa-s}}{\text{CF} * G_b * \left(1000\,\text{kg}/\text{m}^3\right)} = \frac{0.001\,\text{Pa-s}}{\text{CF} * G_b * \left(1\,\text{kg}/\text{m}^3\right)} = \frac{0.001\,\text{Pa-s}}{\approx 1 * \left(1\,\text{kg}/\text{m}^3\right)}$$

where
  CF is a correction factor for the test temperature, $\text{CF} = 1.0135 - 0.0006 * (\text{T})_{\text{test}}$.
  G_b is the specific gravity of asphalt binder at 15°C (approximately 1.03)
  The product of CF and G_b for a temperature range of 135°C–165°C is approximately equal to 1

Note that the field compaction temperature should be based on the aforementioned guidelines but adjusted according to a number of factors such as air and base temperature, wind speed, haul distance, roller type, and lift thickness. One excellent tool for simulating different conditions and determining appropriate compaction temperature is the PaveCool software developed by the Minnesota Department of Transportation (2000–2006; http://www.dot.state.mn.us/app/pavecool/).

9.10.2 Complex Modulus and Phase Angle

The dynamic shear rheometer (DSR; test method: AASHTO T-315; see Figure 9.10) is used to characterize the viscoelastic behavior of asphalt binder, and evaluate its rutting and cracking potential. The basic principle used for DSR testing is that asphalt behaves like elastic solids at low temperatures and like a viscous fluid at high temperatures. These behaviors can be captured by measuring the complex modulus (G^*) and phase angle (δ) of an asphalt binder under a specific temperature and frequency of loading. The G^* and δ parameters (Figure 9.10) are measured by the DSR by applying a torque on the asphalt binder between a fixed and an oscillating plate, and measuring the resultant strain. G^* is defined as the ratio of shear stress to shear strain, and δ is the time lag between the stress and the strain in the asphalt binder. These properties are determined at high and intermediate pavement service temperatures. The high temperature is determined from the average of maximum HMA pavement temperature over a 7 day period through summer, which is obtained from data from weather stations. The intermediate testing temperature is determined from an average of 7 day average maximum and minimum design temperature. Figure 9.11 shows results of DSR tests conducted on different asphalts at different temperatures. Dividing G^* by the angular velocity (10 rad/s used in DSR testing), one can obtain complex dynamic shear viscosity from DSR tests.

9.10.3 Aging Tests

Short-term aging is done by subjecting the asphalt binder to the rolling thin film oven (RTFO) test, explained earlier. Long-term aging is simulated with the use of the pressure aging vessel (PAV; ASTM-D6521, AASHTO R-28; Figure 9.12). The PAV test consists of subjecting the RTFO residue to high pressure and temperature inside a closed vessel. The PAV is placed inside an oven for maintaining the specified high temperature, and a cylinder of clean, dry compressed air is used to supply and regulate air pressure.

9.10.4 DSR Tests Conducted on Aged Asphalt

The asphalt binder in its unaged condition, short-term aged condition (RTFO conditioned), and long-term aged condition (PAV conditioned) is tested with the DSR for evaluation of its rutting and fatigue characteristics. The rutting potential is evaluated by the rutting factor, defined as $G^*/\mathrm{Sin}\,\delta$, of which G^* and δ can be obtained from DSR testing. Since higher G^* means greater resistance to rutting, and lower δ means a more elastic asphalt binder, higher G^* and lower δ values are desirable to make the HMA less susceptible to rutting. Considering rutting as a stress-controlled, cyclic loading phenomenon, it can be shown that rutting potential is inversely proportional to $G^*/\mathrm{Sin}\,\delta$. Hence, a minimum limit is set on the value of $G^*/\mathrm{Sin}\,\delta$. Testing asphalt for the rutting factor prior to long-term aging is critical, since with aging the stiffness of the asphalt binder increases and resistance to rutting increases (Bahia and Anderson, 1995). Hence, the DSR test for rutting potential is conducted on RTFO residue. It is also conducted on unaged asphalt binder, since in some cases RTFO may not indicate the actual aging of the binder during the production and placement of HMA.

The asphalt binder is also tested with the DSR to determine the fatigue cracking potential of HMA. The fatigue potential is measured by the fatigue factor, defined as $G^* \mathrm{Sin}\,\delta$. This is done because

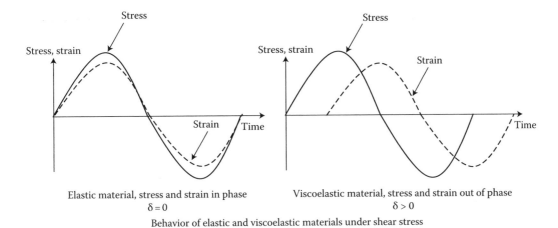

Elastic material, stress and strain in phase
$\delta = 0$

Viscoelastic material, stress and strain out of phase
$\delta > 0$

Behavior of elastic and viscoelastic materials under shear stress

Twisting plate with asphalt binder

Fixed plate

Schematic of DSR test

DSR equipment

Asphalt binder on testing plate

Complex modulus, G*

Viscous modulus, G″

δ

Storage modulus, G′

Complex modulus, viscous modulus, storage modulus and δ

FIGURE 9.10 Dynamic shear rheometer (DSR) test for asphalts.

it can be shown that under a strain-controlled process, the work done for fatigue is proportional to G* Sin δ, and hence a lower value of G* Sin δ means a lower potential of fatigue cracking. This can be explained by the fact that as G* decreases, the asphalt binder becomes less stiff and is able to deform without building up large stresses, and as δ value decreases, the asphalt binder becomes more elastic and hence can regain its original condition without dissipating "energy." An upper limit is set on the G* Sin δ value in the Superpave specification, in order to minimize the potential of fatigue cracking (Bahia and Anderson, 1995). Since the asphalt binder becomes stiffer due to aging during its service life and becomes more susceptible to cracking, the DSR test for fatigue factor determination is conducted on PAV-aged samples.

FIGURE 9.11 DSR text results from different asphalts. (a) Complex modulus (G*) versus temperature and (b) phase angle (δ) versus temperature.

FIGURE 9.12 Pressure aging vessel (PAV) equipment with air supply cylinder. (a) PAV equipment and (b) racks for asphalt that are placed inside the PAV.

FIGURE 9.13 Bending beam rheometer.

9.10.5 Low-Temperature Stiffness (ASTM D-6648, AASHTO T-313)

The bending beam rheometer (BBR) is used to evaluate the stiffness of asphalt binder, and its cracking potential at low temperature. The BBR (Figure 9.13) consists of a loading mechanism in which an asphalt binder beam is subjected to a transient creep load at a constant low temperature. The creep stiffness and the rate at which the stiffness changes with loading time give an indication of its cracking potential at the test temperature. The higher the stiffness (s), the lower the rate (m), and the greater the potential for cracking. A lower slope or rate of change in stiffness means a reduced ability of the asphalt binder to relieve thermal stress by flowing. Hence, in the Superpave specification, a maximum limit is set on the value of stiffness, and a minimum limit is set on the value of slope. Since the asphalt binder becomes stiffer with aging and more susceptible to thermal cracking, the BBR test is conducted on PAV-aged samples.

9.10.6 Direct Tension Test (ASTM D-6723, AASHTO T-314)

In some cases, such as modified binders, a measure of stiffness is not adequate to characterize the cracking potential. This is because these binders may have a higher stiffness, but they may not be susceptible to cracking, as they are more ductile and can stretch without breaking. The direct tension test (DTT; Figure 9.14) is used to characterize such binders. It is used for testing any binder that has a stiffness exceeding the specified stiffness at low temperatures. The DTT consists of a tensile-testing machine, which measures the strain at failure of a dog bone–shaped asphalt binder specimen, which is subjected to tensile load at a constant rate. A minimum limit is set on the strain at failure.

9.10.7 Superpave Requirements

In the Superpave system of specifications, the requirements for results from tests on asphalt binders are the same. However, the temperature at which the requirements are met is used to classify the

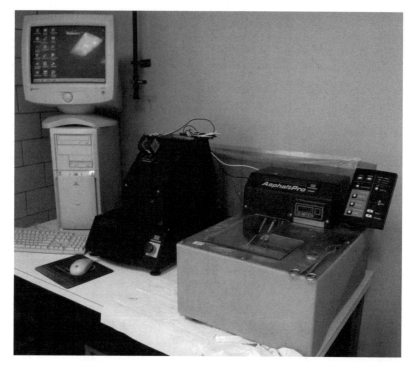

FIGURE 9.14 Direct tension testing equipment. (Courtesy of Denis Boisvert and Alan Rawson, New Hampshire Department of Transportation, Concord, NH.)

asphalt binder. For example, a binder classified as PG 64-22 means that the binder meets high-temperature physical property requirements up to a temperature of 64°C, and low-temperature physical property requirements down to −22°C. There is also a provision of using higher grade asphalt (e.g., PG 76-22) for mixes to be used in slow transient or standing load areas.

A specific asphalt is classified (according to the PG classification system) through the following steps:

1. Check flashpoint (minimum = 230°C).
2. Check viscosity (ASTM 4402) at 135°C; maximum: 3 Pa-s.
3. Check at which highest temperature, for the neat asphalt, $G^*/\text{Sin } \delta \geq 1.00$ kPa, when tested with the DSR at 10 rad/s; note temperature (high pavement temperature: 46/52/58/64).
4. Run RTFOT (AASHTO T240) or TFO (AASHTO T-179).
5. Check mass loss; maximum: 1.00%.
6. Check at which highest temperature, for the RTFOT/TFO aged asphalt, $G^*/\text{Sin } \delta \geq 2.2$ kPa, when tested with the DSR at 10 rad/s; note temperature (high pavement temperature: 46/52/58/64).
7. Run pressure aging vessel test with the RTFOT/TFO-aged binder at 90°C for PG 46/52 (as identified earlier), 100°C (58/64/70/76/82), and 110°C (desert climate, 70/76/82).
8. Check at which temperature, for the RTFOT/TFO-PAV-aged asphalt, $G^* \text{ Sin } \delta \leq 5000$ kPa, when tested with the DSR at 10 rad/s; note temperature (intermediate pavement service temperature: 10/7/4 for PG 46, 25/22/19/16/13/10 for PG 52, 25/22/19/16/13 for PG 58, 31/28/25/22/19/16 for PG 64, 34/31/2825/22/19/16/13 for PG 70, 37/34/31/2825/22 for PG 76, and 40/37/34/31/28 for PG 82).

FIGURE 9.15 Plots of complex modulus and phase angle versus temperature for different grades. (a) Complex modulus (G*) versus temperature (°C) and (b) phase angle (δ) versus temperature.

9. Check lowest temperature at which the S value is ≤300 MPa and m value is ≥0.300 in the test BBR run for 60 s (low pavement temperatures: −36 to 0, in 6° decrements).
10. If stiffness exceeds 300 MPa in step 9, run DTT and check minimum temperature at which failure strain is ≥1%, when run at 1.0 mm/min (low pavement temperatures: −36 to 0, in 6° decrements).

To illustrate the concept, Figure 9.15 is presented. It shows the complex modulus and phase angles at different temperatures for different asphalt grades. Note that the lower the high-temperature grade, the lower the complex modulus, and the higher the phase angle at a specific temperature. Also note that the PG 82-22 binder is a polymer-modified binder, and hence shows significantly different properties than the rest. Generally, asphalt binders with high range (*range* means the difference between the high and low temperatures) are polymer modified.

The asphalt binder to be used in a specific project is selected on the basis of the geographical location of the project, the pavement temperature and air temperature, and the specific reliability that the user wants to have in selection of the binder. The process is illustrated next.

Example 9.1

Consider a project located in Worcester, MA, and another one in Auburn, AL. The selection of the binder could be done with the help of freely available software, LTPPBIND, which is based on the original SHRPBIND software. The LTPPBIND (version 3.1 beta, accessed November 20, 2007) software can be obtained from http://www.fhwa.dot.gov/pavement/ltpp/ltppbind.cfm (FHWA, n.d.).

Project No. 1: Surface layer; needs reliability of 98% for both high and low temperatures.

- Location: Worcester, MA; latitude: 42.27, longitude: 71.87.
- Weather station: Worcester Municipal Airport.
- High air temperature: mean: 29.9°C, standard deviation: 1.4, minimum: 27.4, maximum: 33 (from 34 years of data).
- Low air temperature: mean: −21.7°C, standard deviation: 2.4, minimum: −25, maximum: −16 (from 35 years of data).
- Low air temperature drop: mean: 24.2°C, standard deviation: 2.4, minimum: 20.5, maximum: 31 (from 35 years of data).
- Degree days over 10°C: mean: 2119, standard deviation: 105, minimum: 1893, maximum: 2345 (from 34 years of data).
- Pavement temperature: for high temperature, reliability level at 48.6°C is 50%, whereas reliability level at 52°C is 98%. Choose 52°C. For low temperature, reliability at −16°C is 56%, whereas reliability at −22°C is 98%. Choose −22°C. Appropriate binder will be PG 52-22, or any binder that has a high-temperature level above 52°C and low-temperature level below −28°C (e.g., a PG 64-28).

Project No. 2: Surface layer: needs reliability of 98% for high temperature and low temperature.

- Location: Auburn, AL; select nearest weather station at Opelika, AL; latitude: 32.63; longitude: 85.38.
- High air temperature: mean: 35.1°C, standard deviation: 1.4, minimum: 32.7, maximum: 38.5 (from 31 years of data).
- Low air temperature: mean: −12.1°C, standard deviation: 3.5, minimum: −21.5, maximum: −6 (from 33 years of data).
- Low air temperature drop: mean: 28.6°C, standard deviation: 2.6, minimum: 25, maximum: 36 (from 33 years of data).
- Degree days over 10°C: mean: 3699, standard deviation: 141, minimum: 3414, maximum: 3957 (from 31 years of data).
- Pavement temperature: for high temperature, reliability level at 61.9°C is 50%, whereas reliability level at 64°C is 98%. Choose 64°C. For low temperature, reliability at −10°C is 90%, whereas reliability at −16°C is 98%. Choose −16°C. Appropriate binder will be PG 64-16, or any binder that has a high-temperature level above 64°C and low-temperature level below −16°C (e.g., a PG 70-22).

9.10.7.1 Explanation

The steps in the software are as follows. A high-temperature algorithm is used for the determination of the high-temperature PG. An equation, developed from a database of pavement temperatures and rutting damage, is used to determine a PG damage (PGd, or a "base PG" with 50% reliability at the surface layer), then an estimate of variability of pavement temperature (CVPG) is obtained using an equation, and the base PG is adjusted for any other desired level of reliability. Finally, the PG is adjusted for any other depth of the pavement, using another equation. Note that the determined PG grade is valid for fast-moving traffic with a design volume of three million equivalent single-axle loads (ESAL), and that adjustments (known as *grade bumping*) could also be made to accommodate other traffic volumes and speed. The adjustment factor (a calculated average for different PG, traffic volume, and speeds), shown in Table 9.1, needs to be added to the base PG to determine the design PG:

$$PG_d = 48.2 + 14\ DD - 0.96\ DD^2 - 2\ RD$$

$$CVPG = 0.000034\ (Lat\text{-}20)^2\ RD^2$$

$$PG_{rel} = PG_d + Z\ PG_d \frac{CVPG}{100}$$

TABLE 9.1

Average Grade Adjustments "Grade Bumps"

Speed	Base Grade	Traffic Loading Equivalent Single-Axle Loads, Millions			
		<3	3–10	10–30	>30
Fast	52	0	10.3	16.8	19.3
	58	0	8.7	14.5	16.8
	64	0	7.4	12.7	14.9
	70	0	6.1	10.8	12.9
Slow	52	3.1	13	19.2	21.6
	58	2.9	11.2	16.8	19
	64	2.7	9.8	14.9	17
	70	2.5	8.4	12.9	14.9

where

PG_{rel} is the PG at a reliability, °C

PG_d is the damage-based performance grade, °C

CVPG is the yearly PG coefficient of variation, %

DD is the average yearly degree-days air temperature over 10°C, ×1000

RD is the target rut depth, mm

Lat is the latitude of site, degrees

Z is the from standard probability table, 2.055 for 98% reliability

For adjusting PG with depth:

$$T_{pav} = 54.32 + 0.78T_{air} - 0.0025Lat^2 - 15.14 \log_{10}(H + 25) + z(9 + 0.61S_{air}^2)^{0.5}$$

where

T_{pav} is the high AC pavement temperature below surface, °C

T_{air} is the high air temperature, °C

Lat is the latitude of the section, degrees

H is the depth to surface, mm

S_{air} is the standard deviation of the high 7 day mean air temperature, °C

z is the standard normal distribution table, z=2.055 for 98% reliability

The low pavement temperature for different reliabilities is determined using the following equation:

$$T_{pav} = -1.56 + 0.72T_{air} - 0.004Lat^2 + 6.26 \log_{10}(H + 25) - z(4.4 + 0.52S_{air}^2)^{0.5}$$

where

T_{pav} is the low AC pavement temperature below surface, °C

T_{air} is the low air temperature, °C

Lat is the latitude of the section, degrees

H is the depth to surface, mm

S_{air} is the standard deviation of the mean low air temperature, °C

z is the standard normal distribution table, z=2.055 for 98% reliability

TABLE 9.2

Typical Grades Available in the United States

High-Temperature Grade	Low-Temperature Grade						
	−10	−16	−22	−28	−34	−40	−46
46							
52							
58							
64							
70							
76							
82							

Table 9.2 shows the different grades of asphalts generally available in the United States. Note that higher range asphalt binders are mostly modified binders (such as with polymer, discussed in Chapter 23), and grades wider than the typical ranges presented in the table can be produced for specialty applications. Polyphosphoric acid (PPA), is often used (generally up to a maximum of 1.5%) to stiffen asphalt binders. The stiffening effect is dependent on the specific asphalt binder and its crude source. PPA can increase the moisture sensitivity of certain binders; however, it has been reported that such increased sensitivity can be counteracted with the use of lime treatment of aggregates. The use of PPA can enhance the cross-linking and performance of styrene butadiene styrene (SBS) polymer in asphalt modification.

The steps in the selection of an asphalt binder for a specific project can thus be summarized as follows:

1. Determine the type of traffic loading that is expected on the constructed pavement (standing/slow, 50 km/h or less/fast, 100 km/h or more).
2. Determine the high pavement temperature.
3. Determine the low pavement temperature.
4. Determine the design volume of traffic.
5. Select the asphalt grade—slow or standing traffic, and higher traffic volumes require a higher grade.

9.10.8 MULTIPLE STRESS CREEP RECOVERY TEST (AASHTO TP70, SPECIFICATION, AASHTO MP19)

The Multiple Stress Creep Recovery Test has been developed as a high temperature binder characterization test. It can be utilized for modified binders that need an elasticity evaluation test, as well as for unmodified binders. The test is conducted with DSR equipment on the RTFO-aged material at the appropriate climate grade—the test temperatures are shown in Table 9.3. The steps in testing and evaluation of binder are as follows:

1. Use 10 creep/recovery cycles for stress levels of 0.1 and 3.2 kPa—each cycle consisting of 1 s loading and 9 s recovery. Determine two parameters—the nonrecoverable creep compliance, J_{nr}, and (percentage of) recovery—average of results from 10 cycles for each stress level.
2. From the test data, determine J_{nr} values at 0.1 and 3.2 kPa shear stress and the corresponding MSCR Recovery values at the same stress levels.

TABLE 9.3

Recommended MSCR Testing Temperatures

Grade[b]	States[c]	MSCR Test Temperature[a] (°C) 46	52	58	64	67	70
PG 46-28	1	X					
PG 52-28	3						
PG 52-34	4		X				
PG 58-22	9			X			
PG 58-28	25			X			
PG 58-34	12		X[d]	X[d]			
PG 64-10	1				X		
PG 64-16	4				X		
PG 64-22	38				X		
PG 64-28	31			X[e]	X[e]		
PG 64-34	7		X[d]	X[d]			
PG 67-22	5					X	
PG 70-10	2						X
PG 70-16	3						X
PG 70-22	22				X		
PG 70-28	22			X[e]	X[e]		
PG 70-34	4		X[d]	X[d]			
PG 76-16	1						X
PG 76-22	30				X		
PG 76-28	12			X[e]	X[e]		
PG 76-34	2		X[d]	X[d]			
PG 82-16	1				X[f]		X[f]
PG 82-22	6				X		
PG 82-28	2			X[e]	X[e]		

Source: Guidance on the Use of the MSCR Test with the AASHTO M320 Specification, Asphalt Institute Technical Advisory Committee 2 December, 2010, downloaded from: http://amap.ctcandassociates.com/wp/wp-content/uploads/Guidance-on-Using-MSCR-with-AASHTO-M320_Final.pdf

[a] All MSCR testing is performed on the asphalt binder after RTFQ-aging.

[b] AASHTO M320 Table 1. "Premium" grades (defined as those grades where the temperature differential is 92° or greater) are shown in gray.

[c] Number of states listing the grade in the Asphalt Institute binder specification database (www.asphaltinstitute.org).

[d] Test at either 52°C or 58°C depending on the climate of the project. Users can test at both temperatures if desired.

[e] Test at either 58°C or 64°C depending on the climate of the project. Users can test at both temperatures if desired.

[f] Test at either 64°C or 70°C depending on the climate of the project. Users can test at both temperatures if desired.

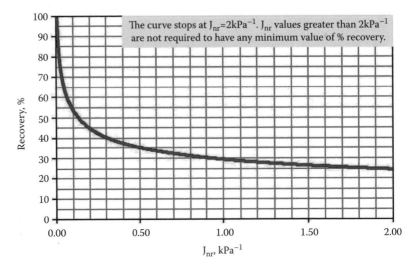

FIGURE 9.16 Comparison of MSCR Jnr and recovery to assess delayed-elastic response. (From Guidance on the Use of the MSCR Test with the AASHTO M320 Specification, Asphalt Institute Technical Advisory Committee, December 2, 2010, downloaded from: http://amap.ctcandassociates.com/wp/wp-content/uploads/Guidance-on-Using-MSCR-with-AASHTO-M320_Final.pdf)

3. With the data from the 3.2 kPa shear stress portion of the test, plot the MSCR Recovery as a function of J_{nr}, and compare with Figure 9.16. If the data points lie above the curve, then the inference is that there is sufficient delayed elastic response for an elastomeric-modified asphalt binder.
4. Calculate stress sensitivity parameter, $J_{nr,diff}$, from

$$J_{nr,diff} = \frac{\left(J_{nr,3.2\,kPa} - J_{nr,0.1\,kPa}\right)}{J_{nr,0.1\,kPa}}$$

5. If $J_{nr,diff} > 0.75$, then the asphalt binder is considered to be stress sensitive; a maximum ratio of 0.75 is permitted by AASHTP MP19; it is to be reported if the MSCR is being used with AASHTO M320.

9.11 RECOVERY OF ASPHALT BINDER FROM ASPHALT MIX

The asphalt binder from an asphalt mix could be recovered for testing, with the help of the solvent extraction method, using a rotary evaporator (Figure 9.17). The asphalt mix is mixed with a solvent (such as methylene chloride) to remove the asphalt binder from the aggregates to the solution. In the next step, the mineral filler and very fine aggregates are separated from the solution by filtration and centrifuging. In the final step, the asphalt binder–solvent mix is heated in a flask maintained at a specific temperature in an oil bath, such that the solvent is evaporated and condensed elsewhere, while the pure asphalt binder remains in the flask. If the asphalt binder is not required for testing, then the amount of asphalt binder present in a mixture of asphalt and aggregate can be determined by the use of the ignition test (AASHTO T-308, Determining the Asphalt Binder Content of Hot Mix Asphalt [HMA] by the Ignition Method), in which the mix is heated to a very high temperature (500°C–600°C) to burn the asphalt off, and the difference between the initial and final weight of the mix is determined (Figure 9.18).

FIGURE 9.17 Rotary evaporator used for extraction of asphalt binder.

FIGURE 9.18 Ignition test equipment.

9.12 ADHESION PROPERTIES

The primary function of the asphalt binder is to bind the aggregate particles together, or work as an adhesive. The aggregate-asphalt binder adhesion is affected by a number of factors, and if not considered carefully, water can destroy it and adversely affect pavement performance. These factors are related to the properties of the aggregate, the asphalt binder, and the asphalt mix. External factors, such as rainfall and temperature, are also important.

The surface energy concept has been used and researched to explain the adhesive bond between aggregate and asphalt binder, and its failure. The use of liquid with a polarity opposite to that of an aggregate with unbalanced surface charge, and energy, leads to the formation of an adhesive bond. Water, if present in addition to asphalt binder, could meet the surface energy requirement of an aggregate, and hence displace asphalt binder (stripping) as the preferred coating.

The nature of the aggregate is of primary importance. Generally, siliceous quartz and granite, which are also classified as "acidic type," have less affinity for asphalt binder (and hence make it more difficult to provide a bond between the aggregate and the binder) than basic rocks such as basalts and limestone. A significant increase in the adsorption of asphalt can be caused by the presence of a fine microstructure of pores, voids, and microcracks on the aggregate surface. The surface texture of the aggregate, viscosity of asphalt during mixing, and polarity of asphalt affect the adhesion significantly.

9.13 ASPHALT EMULSIONS

Asphalt emulsions are used in many applications as they allow the introduction of asphalt at ambient temperature, without the need for high temperature, because of the low viscosity (0.5–10 P at 60°C, compared to 100–4000 P at 60°C for asphalt). Furthermore, emulsions can be diluted with water for various applications, such as dust control and priming and they are compatible with additives such as cement, lime or latex.

Emulsions are produced by dispersing asphalt binder as droplets of small diameter with the help of an emulsifying agent. The emulsifying agent(s) and other additives, if needed, provide the electrical charges to the droplets of asphalt binder so that they repel each other and remain dispersed in water. Common anionic emulsifiers, which impart negative charges, are made from the saponification of fatty acid–type materials, such as rosins and lignins (wood by-products) with sodium or potassium hydroxide. Cationic emulsifiers are generally made from fatty amines (such as diamines) and hydrochloric acids, as well as from fatty quaternary ammonium salts.

Emulsifiers are surfactants with nonpolar lipophilic and polar hydrophilic portions in the same molecule, along with a counterion. When applied, the emulsifier concentrates on the interface between water and asphalt, and as the counterion diffuses into the water, the asphalt droplets are left with a net charge, which also depends on the pH. Cationic emulsifiers have ammonium compounds with positively charged N atom, whereas anionic emulsifiers contain negatively charged O atoms. The presence of "excess" or free emulsifier in the water phase prevents coalescence after emulsification, during storage and transport, and affects the setting process. Manufacturers select a proper emulsion making "recipe" (emulsifier/asphalt/water/stabilizer) to meet specific demands of aggregates (reactivity) and temperature during application.

The successful use of a specific emulsion depends on the type of the aggregate with which it is used. Depending on the charge, there are primarily two types of emulsions that are used in pavement mixes—anionic, containing negatively charged droplets of asphalt binder, and cationic, containing positively charged droplets. The emulsifier keeps the emulsion in equilibrium (stabilizes the emulsion) and dictates the breaking (coalescing of droplets) rate and adhesion of the asphalt binder to the aggregate. Nonionic emulsions (without any charge) can also be produced.

Emulsions are produced in a colloid mill, containing a high-speed (1000–6000 revolutions/min) rotor, through which hot asphalt binder and emulsifier solutions are passed at specific temperatures.

The shearing action of the rotor forms asphalt binder droplets, which get coated with emulsifier and as a result get charged, which prevents them from coalescing. High-pressure mills are used to facilitate the production of emulsion from higher viscosity grade asphalt binders, and static mixers with baffle walls, through which the materials can be pumped at high speed, can also be used instead of a colloid mill. Generally an emulsion may contain about 28% droplets smaller than 1 μm in size, 57% droplets 1–5 μm in size, and the rest 5–10 μm in size. In general, emulsions contain 50%–75% of asphalt.

9.13.1 PROPERTIES

The desirable properties of emulsion include adequate stability, proper viscosity and breaking characteristics, and strong adhesion. The emulsion must be stable during storage, transportation, and applications, and must be sufficiently fluid (but not too fluid, to be optimum) to facilitate application. The emulsion should break in a timely manner, in contact with the aggregates, and form strong bonds.

Stability of an emulsion refers to the ability to keep the asphalt droplets well dispersed throughout the emulsion. Because the density of asphalt is slightly higher than the aqueous phase of the emulsion at ambient temperature, the droplets would tend to settle down over time, form clumps by agglomeration, and ultimately form larger globules by coalescing. Allowable ranges of storage temperatures vary from 20°C–60°C (minimum) to 50°C–85°C. Slow- and medium-setting emulsions as well as high floats have a higher range of storage temperature. For storage, small diameter vertical storage tanks with adequate protection against freezing should be used.

The viscosity of emulsion can be increased by increasing the amount of asphalt binder; increasing the viscosity of the continuous phase, such as by increasing the emulsifier content; and decreasing the droplet sizes, such as by increasing the flow rate through the colloid mill or decreasing the viscosity of the asphalt binder used for emulsion. Emulsion viscosity is measured with an orifice viscometer, in which viscosity is determined from the time required for a specific amount of emulsion to flow through a standard orifice.

The breaking characteristics of an emulsion can be affected by the asphalt content, composition of the aqueous phase, droplet size distribution, environmental conditions, type of aggregates, and use of breaking agents. The rate of breaking can be increased with an increase in asphalt and emulsifier content, and by reducing the size of asphalt droplets. Note that the breaking process is accelerated by the evaporation of water from the emulsion, as well as by the absorption of the emulsifier into the aggregates. High temperatures and low humidity are favorable to breaking, while the presence of dust coating on the aggregate surface could accelerate the breaking process without causing any adhesion of the asphalt binder to the aggregate surface. The presence of calcium and magnesium ions on the aggregate surface would also accelerate the breaking process. If the emulsifier is absorbed from the solution into the aggregates at the desired rate (such as through a rough-textured porous surface), the lowering of the charge on the droplets causes rapid coalescence and good adhesion of the asphalt binder to the aggregate surface. The breaking can also be accelerated with the use of breaking agents, sprayed along with the emulsion or just after the application of the emulsion, as well as with the help of mechanical compaction (using a roller).

The adhesive nature of emulsion is made possible by good "wetting" of the aggregate surface (high surface tension on the aggregate surface) and the creation of a thermodynamically stable system of minimum surface energy. There are many factors that affect this process, including the aggregate type (such as acidic or basic), type and amount of emulsifier (such as cationic or anionic), grade of the asphalt, and size of the droplets.

It is important to note that although the viscosity of the asphalt binder has no effect on the viscosity of an emulsion formed with it, there are several properties of the asphalt binder that do affect the emulsion properties significantly. These include its electrolyte content, which can cause an increase as well as a decrease in viscosity depending on its amount, density (a higher density causes lower stability), and acid content (depending on the type of the emulsion, it may improve or deteriorate stability).

Emulsions are also made with polymer-modified binders (polymer is generally added to reduce temperature susceptibility), generally requiring higher temperature for manufacturing. However, one beneficial effect of polymer is the ability to use a higher asphalt content in the emulsion without affecting its viscosity (and hence ease of use) adversely.

9.13.2 TESTS FOR ASPHALT EMULSIONS

Particle charge test: In this test, a current is sent through an anode and cathode immersed in an emulsion. If the cathode shows an appreciable layer of asphalt binder at the end of the test, then the emulsion is characterized as cationic.

Viscosity test: The Saybolt Furol viscosity test is conducted to measure viscosity, either at 25°C or 50°C.

Demulsibility test: This test is conducted to check the rate of breaking of rapid-setting emulsions when applied on aggregates or soils. The test consists of mixing an emulsion with a solution of water and either calcium chloride or dioctyl sodium sulfosuccinate (for anionic and cationic emulsions, respectively), and sieving the mix. The concentration of the solution and the minimum amount of asphalt retained on the sieve (as a result of coalescence) are generally specified. A similar test is conducted for the SS grades, using finely ground Portland cement, and washing over a 1.40 mm sieve. Generally, limits are specified on the amount of material that is retained on the sieve, to ensure the ability of the emulsion to mix with the soil/aggregate with high surface area before breaking.

Identification test for CRS (ASTM D244): In this test, sand washed with hydrochloric acid and alcohol is mixed with emulsion for 2 min, and then the uncoated area is compared to the coated area.

Identification test for CSS (ASTM D244): In this test, washed and dried silica sand is mixed with the emulsion until they are completely coated, and the mix is cured for 24 h and then placed for 10 min in a beaker with boiling water. Then the sample is spread on a level surface and the part of the mix coated, as percentage of the mix is determined.

Settlement and storage stability tests: In this test, a specified volume of emulsion is made to stand in a graduated cylinder for either 24 h (for the storage test) or 5 days (for the settlement test), and then samples are taken from the top and bottom of the cylinder, weighed, and heated to evaporate the water. The difference in weight of the residue from the top and the bottom parts, if any, is determined.

Sieve test: In this test, an emulsion is poured over a 850 μm sieve, and the sieve is rinsed with mild sodium oleate solution and distilled water for anionic emulsions and distilled water only for cationic emulsions. Then the sieve and retained asphalt binder are dried and weighed, and the amount of asphalt binder that is retained on the sieve is determined.

Coating ability and water resistance test: This test is conducted for MS emulsions to check their ability to coat aggregates, remain as a film during mixing, and resist washing action of water after mixing. The test consists of two steps. In the first step, the aggregate is coated with calcium carbonate dust and then mixed with emulsion. Half of the mix is placed on an absorbent paper for visual inspection of coating. The rest is sprayed with water and rinsed until the water runs clear, and then placed on the absorbent paper for inspection of coating. In the second step, the first step is repeated but the emulsion is added to the aggregate only after coating the aggregates with water.

Field coating test: In this test, measured amounts of aggregate and emulsion are mixed with the hands, and the coating is observed during the 5 min mixing process. Then the coated aggregates are placed in a container, which is filled with water, and the water is drained off five times (after refilling). Next the aggregates are checked for coating. A good coating indicates a full coating of aggregate, a fair coating indicates an excess of coated over uncoated area, and a poor coating indicates an excess of uncoated over coated area.

Unit weight test: The weight of an asphalt emulsion in a standard measure of a known volume is found out to determine its unit weight.

Residue and oil distillate by distillation: The asphalt, oil, and water can be separated in an emulsion for subsequent testing by distillation, which consists of heating the emulsion in an aluminum alloy still at a temperature of 260°C for 15 min.

Residue by evaporation: Residue from the emulsion can also be obtained by evaporating it in an oven at 163°C. This test should not be run if the residue needs to be tested for the float test. The residue can be tested for specific gravity (ASTM D-70, AASHTO T-228).

Solubility in trichroloethylene (ASTM D-2042, AASHTO T-44): In this test, the residue is dissolved in a solvent, and the insoluble (contaminants) part is separated by filtration.

Penetration test (ASTM D-5, AASHTO T-49), ductility test (ASTM D-113, AASHTO T-51), and ring and ball softening point test (ASTM D-36, AASHTO T-53): Similar to that described for asphalt binders.

Float test (ASTM D-139, AASHTO T-50): This test, conducted to determine resistance to flow at high temperatures, consists of forming a test plug of residue asphalt (from emulsion) in a brass collar and screwing the collar at the bottom of an aluminum float placed in a 60°C water bath. The time required by the water to break through the plug is determined, and a higher time indicates a higher resistance. Asphalt emulsions showing high float test numbers (called *high floats*) possess a gel-like structure and are resistant to flow from high temperatures during the summer.

Breaking index: In this test, which is run for rapid-setting (RS) emulsion, silica sand is added and mixed with the emulsion under controlled conditions of rate and temperature. *Breaking index* is defined as the weight of the sand needed in grams to break the emulsion.

Vialit test: This test simulates the formation of chip seal—aggregate embedded by rolling onto an emulsion-coated surface—in a small metal plate. After curing, the aggregates are treated with water and dried. Finally, a 500 g steel ball is dropped from a specified height three times on the reverse side of the plate, and the weight loss of the aggregate is measured.

Zeta potential: In this test, the speed of movement of asphalt binder droplets in an aqueous medium is measured to determine the intensity of charge.

Elastic recovery after ductility: This test follows the ductility test, and involves determining the percentage recovery of the thread that has been used in the ductility test, when cut into half.

Force ductility: The test is similar to the ductility test, but the force required to cause the elongation of the asphalt thread is measured.

Tensile strength test: In this test, the force (stress) required to pull an asphalt thread in the vertical direction to a specified elongation (strain) is determined.

Torsional recovery: In this test, a shaft and arm assembly is immersed in a heated sample of the residue asphalt, and is twisted 180° inside the sample at a lower temperature. The percentage recovery after a specified time is then determined.

9.13.3 CLASSIFICATION OF EMULSIONS AND SELECTION

Emulsions are classified according to their charge (cationic [C] and anionic [no symbol]), as mentioned earlier. They are also classified according to their rate of breaking or setting—rapid (rapid setting, or RS), quick (QS), medium (MS), or slow (SS). A number is used for denoting the relative viscosity of the emulsion. Hence, *CRS-1* means a cationic RS emulsion with less viscosity than an MS-2 emulsion, which is anionic and medium setting. A h or a s following the emulsion name would indicate the use of hard and soft asphalt binder, respectively, whereas HF indicates "high float" properties as measured in the float test.

The separation of water from the emulsion is called *breaking*, the coalescing of asphalt binder droplets is called *setting*, whereas the formation of a continuous adhesive layer of asphalt binder (through breaking) on the surface of the aggregates is called *curing*. Breaking is generally accompanied by the formation of black and sticky material from the more fluid and brownish emulsion. Too early coalescence of asphalt droplets can lead to the formation of "skin" that can prevent the evaporation of water that is also needed for the curing phase. Such a situation can arise in the case of an emulsion with smaller asphalt droplets with narrow size distribution.

Emulsions are graded according to their rate of breaking (SS, MS, QS, and RS), the nature of their charge (cationic [C] and anionic [no symbol]), the presence of hard and soft asphalt (H or S), and the results of the float test (HF if high results). The following are the ASTM and AASHTO specified grades of emulsion.

ASTM D-977, AASHTO M-140: RS-1, RS-2, HFRS-2, MS-1, MS-2, MS-2H, HFMS-1, HFMS-2, HFMS-2H, HFMS-2S, SS-1, SS-1H
ASTM D-2397, AASHTO M-208: CRS-1, CRS-2, CMS-2, CMS-2h, CSS-1, CSS-1h

Generally, rapid setting emulsions are used for aggregates with low surface area (low reactivity) such as those for chip seals, whereas as slow setting emulsions are used for gradations with sufficient surface area, such as for cold dense mixes. In the case of tack coat, a slow setting emulsion or medium setting emulsion is used, after diluting with water. For stability and to prevent bleeding, rapid setting emulsions are not used in this case; however, because of low application rate (0.1 g/square yard) and hence a resulting thin layer, it breaks quickly.

Emulsions can be selected for specific applications according to the following guidelines.
Cape seal: RS-2, CRS-2
Dust palliative, mulch treatment, and crack filler: SS-1, SS-1h, CSS-1, CSS-1h
Fog seal: MS-1, HFMS-1, SS-1, SS-1h, CSS-1, CSS-1h
Immediate-use maintenance mix: HFMS-2s, CRS-2, CMS-2
Microsurfacing: CSS-1h
Mixed in-place open-graded aggregate mix: MS-2, HFMS-2, MS-2h, HFMS-2h, CMS-2, CMS-2h
Mixed in-place well-graded aggregate mix, sand mix, and sandy soil mix: HFMS-2s, SS-1, SS-1h, CSS-1, CSS-1h
Plant mix hot or warm mix asphalt: MS-2h, HFMS-2h
Plant mix cold open-graded aggregate mix: MS-2, HFMS-2, MS-2h, HFMS-2h, CMS-2, CMS-2h
Plant mix cold dense-graded aggregate mix and sand mix: HFMS-2s, SS-1, SS-1h, CSS-1, CSS-1h
Prime coat: MS-2, HFMS-2, SS-1, SS-1h, CSS-1, CSS-1h
Sand seal: RS-1, RS-2, HFRS-2, MS-1, HFMS-1, RS-1, CRS-2
Sandwich seal: RS-2, HFRS-2, CRS-2
Single- and multiple-surface treatments: RS-1, RS-2, HFRS-2, RS-1, CRS-2
Slurry seal: HFMS-2s, SS-1, SS-1h, CSS-1, CSS-1h
Stockpile maintenance mix: HFMS-2s
Tack coat: MS-1, HFMS-1, SS-1, SS-1h, CSS-1, CSS-1h

For a good discussion on asphalt emulsion technology, the reader is requested to review the Transportation Research Board Circular, E-C102, available at http://onlinepubs.trb.org/onlinepubs/circulars/ec102.pdf

QUESTIONS

9.1 What are the different sources of asphalt binder?

9.2 Of the three different asphalts for which DSR results are shown next, pick the one that is suitable for a project in a climate where the maximum pavement temperature is 76°C, on the basis of results from unaged asphalt only:

Asphalt	Temperature (°C)	Complex Modulus, kPa	Phase Angle, °
A	76	1.20	87
B	76	0.90	86
C	76	1.08	88

9.3 For a BBR test, the stiffness is calculated as follows:

$$S = \frac{PL^3}{4bh^3\delta(t)}$$

where

S is the creep stiffness, MPa
P is the applied load, N
L is the distance between supports of the beam, mm
b is the width of the beam, mm
h is the thickness of the beam, mm
$\delta(t)$ is the deflection of the beam, mm

If P=0.981 N, L=102 mm, b=12.5 mm, h=6.25 mm, and δ=0.3 mm, determine S.

9.4 Download the LTPPBIND software, and determine the most suitable asphalt binder grade for your town/city.

9.5 What are the different steps in curing of asphalt emulsion?

9.6 What are the different tests for asphalt emulsions?

10 Concrete Fundamentals for Rigid Pavements

10.1 CONCRETE

Portland cement concrete (PCC) is the most commonly used construction material in the world. Approximately half of all PCC used in the United States is consumed in the construction of PCC pavements.

The constituents of PCC include coarse aggregates (natural stone or crushed rock), fine aggregate (natural and manufactured sand), Portland cement, water, and mineral or chemical admixtures. The paste phase is made up of the Portland cement, water, and air (entrapped and entrained). The paste phase is the glue or bonding agent that holds the aggregates together. The mortar phase is the combination of paste and fine aggregate. Aggregates occupy approximately 60%–75% of the total volume of the concrete. The aggregates should be well graded to efficiently utilize the paste. It is desirable that each aggregate particle is coated with paste. The quality of the concrete is therefore highly dependent on the quality of the paste, the aggregates, and the interface or bond between the two.

The quality of the hardened concrete is strongly influenced by the quality of the paste, which in turn is highly dependent on the amount of water used in mixing the concrete. Therefore, the water-to-cement ratio by mass (W/C) is an important parameter that relates to many concrete qualities. Excessive amounts of mix water (a higher W/C ratio) will increase the porosity, lower the strength (compressive, tensile, and flexural), increase the permeability and lower water tightness and durability to aggressive chemicals, decrease bond strength between concrete and reinforcement, and increase volume changes (more cracking) due to moisture changes (wetting and drying) and temperature changes. Therefore, it is important to use the least amount of water for mixing the concrete, provided that it can be placed and consolidated properly. Stiffer mixes with less water are more difficult to work with; however, mechanical vibration is available for proper consolidation.

To produce a concrete that is strong, resists abrasion, and is watertight and durable, the following sequence should be followed: proper proportioning, batching, mixing, placing, consolidating, finishing, and curing.

The ease of placing, consolidating, and finishing concrete without segregation of the coarse aggregates from the mortar (paste and sand) phase is called *workability*. The degree of workability is highly dependent on the concrete type and method of placement and consolidation. Mechanical placement and consolidation can handle much stiffer mixes. Other properties related to workability include consistency, segregation, pumpability, bleeding, and finishability. The slump test is used to measure consistency, which is closely related to workability. A low-slump concrete has a stiff consistency and may be difficult to place and consolidate properly. However, if the mix is too wet, segregation or separation of large-size aggregates and honeycombing could occur. The consistency should be the driest that is most practical for placement using the available equipment (Powers, 1932).

Concrete mixtures with the larger aggregate size and a continuous gradation will have less space to fill with paste and less aggregate surface area to coat with paste; therefore, less cement and water are needed. Stiffer mixtures usually result in improved quality and economy, granted the concrete is properly consolidated.

10.2 AGGREGATES

Aggregates should be of sufficient quality with adequate strength, soundness, and environmental durability so as not to negatively impact the performance of the hardened concrete. The larger portions of crushed rock, stone, and sand are called *coarse aggregate* with sizes larger than the No. 4 sieve (4.75 mm). *Fine aggregate* is natural sand or crushed rock that has a particle size smaller than the No. 4 sieve.

For concrete pavements subject to cycles of freezing and thawing, aggregate and cement paste durability is critical to the pavement performance. For aggregates, the frost resistance is related to the porosity, absorption, permeability, and pore structure. If an aggregate particle absorbs enough water to reach critical saturation, then the pore structure is not able to accommodate the hydraulic pressures associated with the freezing of water. If enough inferior aggregates are present in the concrete, concrete distress and disintegration may occur. If the problem aggregate particles are present at the surface, popouts can occur. Popouts appear as conical breaks that are removed from the concrete surface.

Cracking of concrete pavements caused by freeze–thaw deterioration of the aggregate within the concrete is called D-cracking (see Figure 11.2). The visual distress of D-cracking resembles the letter D, where closely spaced cracks emerge parallel to the transverse and longitudinal joints and progress outward toward the center of the slab. Assessing aggregates for freeze–thaw durability can be done by past performance or by laboratory testing of concrete. The Freeze–Thaw Testing of concrete (ASTM C666, AASHTO T161) involves conditioning concrete beams through cycles of freeze–thaw (freeze in air and thaw in water, or freeze and thaw in water), and testing for dynamic modulus of the beam before and after conditioning. If the conditioned modulus falls below 60% of the unconditioned modulus, then the mix is not accepted. Other tests include resistance to weathering using immersion cycles in sodium or magnesium sulfate solutions. Salt crystal growth in the aggregate pores creates a pressure similar to the hydraulic pressure from freezing water. A rapid pressure-release test developed by Janssen and Snyder (1994) has also been correlated to D-cracking potential. In this test, aggregates are placed in a high-pressure chamber. The pressure is released rapidly, and unsound aggregates with ineffective pore structure fracture.

10.3 CEMENT

Portland cements are hydraulic cements that chemically react with water to produce calcium silicate hydrates, which are considered to be the glue that gives PCC its strength. Portland cement is produced by combining raw material (or feed) containing calcium, silica, alumina, and iron compounds. Appropriate amounts of this raw feed are combined in pulverized or slurry form and sintered or burned in an inclined rotary kiln. (Figure 10.1 shows a schematic of the cement production process.)

The raw feed is introduced into the kiln, and as the kiln rotates, the materials will slide down the kiln at a slow rate. The kiln temperature ranges from approximately 200°C at the entry to 1450°C at the exit and burner location. Table 10.1 shows the clinkering (solid solution) reactions in the kiln. At the end of the clinkering process, clinker nodules measuring 1–25 mm are produced. To produce Portland cement, the clinker and approximately 5% gypsum are combined and ground to a very fine powder averaging 15 µm in diameter (and a surface area of about 300–600 m^2/kg).

Since the chemical equations for cement chemistry are long and complex, a shorthand (abbreviated) notation was developed and is in use today by cement and concrete technologists, as shown in Table 10.2.

A clinker chemical analysis is typically given in oxide form, as shown in Table 10.3. From the chemical analysis, the quantity of each of the four main minerals (C_3A, C_4AF, C_3S, and C_2S) may be calculated using the "Bogue" equations (Taylor, 1997).

FIGURE 10.1 Schematic of cement production process.

TABLE 10.1

Clinkering Process for Cement Production

Up to 700°C	Raw materials are free flowing; water is lost, and clay compounds recrystallize
700°C–900°C	Calcination continues; CO_2 is liberated; reactive silica combines with CaO to start formation of C_2S (belite)
1150°C–1200°C	Solid solution; particles interact; small belite crystals form
1200°C–1350°C	Solid solution; interaction of belite and CaO to form C_2S (alite)
1350°C–1450°C	Agglomeration and layering continue; belite crystals decrease in amount and increase in size; alite increases in size and amount
Cooling	Upon cooling, C_3A and C_4AF crystallize in liquid phase; belite develops lamellar structure

TABLE 10.2

Shorthand (Abbreviation) for Cement Chemistry

Symbol	Compound	Symbol	Compound
C	CaO_3	M	MgO
S	SiO_2	T	TiO_2
A	Al_2O_3	H	H_2O
F	Fe_2O_3	$\bar{S}$	SO_3
K	K_2O	$\bar{C}$	CO_2
N	Na_2O		

TABLE 10.3

Example of a Typical Clinker Analysis (Oxide Weight%)

SiO_2	21.6	K_2O	0.6
Al_2O_3	5.3	Na_2O	0.2
Fe_2O_3	2.9	SO_3	1.0
CaO	66.6	LOI	1.4
MgO	1.0	IR	0.5
Total	98.9		

TABLE 10.4

Compound Name, Approximate Chemical Formula, Abbreviated Name, and Range for the Compound Composition for Typical ASTM 150 Cements

Compound Name	Approximate Chemical Formula	Abbreviation	Potential Compound Composition Range (%)
Alite	$3CaO \cdot SiO_2$	C_3S	45–55
Belite	$2CaO \cdot SiO_2$	C_2S	15–30
Tricalcium aluminate	$3CaO \cdot Al_2O_3$	C_3A	4–10
Tetracalcium aluminoferrite	$4CaO \cdot Al_2O_3 \cdot Fe_2O_3 \cdot Fe_2O_3$	C_4AF	8–15

Note: Composition balance is made up of alkali sulfates and minor impurities. Free lime = 1.0% CaO; the balance is due to small amounts of oxides of titanium, manganese, phosphorus, and chromium.

Clinker contains four main compounds that constitute over 90% of Portland cement. Table 10.4 shows the compound name, the approximate chemical formula (Note: cement chemistry stoichiometry is not exact), the shorthand or abbreviated name, and a range for the compound composition for typical ASTM C-150 cements.

10.3.1 Types of Portland Cement

There exist many different types of Portland cements in the world to meet the needs for specific purposes. Manufactured Portland cements usually meet the specifications of ASTM C-150 (Standard Specifications for Portland Cement), AASHTO M85 (Specification for Portland Cement), or ASTM 1157 (Performance Specification for Hydraulic Cements). ASTM C-150 provides standards for eight different Portland cements, as shown in Table 10.5.

Type I. General-purpose cement suitable for all uses where special properties or other cement types are not required.

Type II. Used where precaution against moderate sulfate attack is warranted. Type II cements should be specified where water-soluble sulfate (SO_4) in soil is between 0.1% and 0.2% or the concentration of SO_4 in water is between 150 and 1500 ppm. The moderate sulfate resistance of Type II cement is due to its limited C_3A content (<8%). Sulfate degradation of hydrated Portland cement is due to the formation of additional ettringite when external sulfate is introduced. Additional formation of ettringite and other calcium aluminate products in the paste may cause cracking and degradation of the concrete.

TABLE 10.5

Types of Cements

Type I	Normal
Type IA	Normal, air entraining
Type II	Moderate sulfate resistance
Type IIA	Moderate sulfate resistance—air entraining
Type III	High early strength
Type IIIA	High early strength—air entraining
Type IV	Low heat of hydration
Type V	High sulfate resistance

Type III. Provides higher strength at an early age (1 week or less) compared to Types I and II. The strength is comparable to that of Type I in the long term (months and years). Type III cement is chemically similar to Type I; however, it is ground much finer. The increase in surface area allows for more reaction contact sites between the cement and water and, therefore, results in a faster hydration product formation.

Type IV. Used where the rate and amount of heat generated from hydration must be minimized. Strength rate development is much lower than in other types. This type of cement is specially ordered and is not readily available.

Type V. Used where precaution against severe sulfate exposure is warranted. Type V cements should be specified where the water-soluble sulfate (SO_4) in soil is greater than 0.2% or the concentration of (SO_4) in water is greater than 1500 ppm. The high sulfate resistance of Type V cement is due to its limited C_3A content (<5%). The sulfate degradation mechanism is similar to that for Type II cement. ASTM C-150 and AASHTO M85 allow both chemical and physical standards to assure the sulfate resistance of Type V cement. However, only one approach may be specified but not both. ASTM C-452 is used to assess the potential expansion of mortar specimens exposed to sulfate.

10.4 WATER

Potable water without any pronounced taste or odor is suitable for making PCC. If the water is not potable, it can also be used if the 7 day compressive strength of mortar cubes with this water is at least 90% of the 7 day compressive strength of mortar cubes made with distilled water (ASTM C-109 and AASHTO T106). The suitability of the water can also be checked using tests for setting time of the cement according to ASTM C-191 and AASHTO T131.

10.5 HYDRATION

Cement reacts with water to form a hydration product that is the basis for the "glue" that gives hydrated Portland cement its bonding characteristics. This is a chemical reaction that involves four main cement compounds: tricalcium silicate, dicalcium silicate, tricalcium aluminate, and tetracalcium aluminate ferrite.

When water is added to cement and mixed together, the reaction is exothermic and heat is generated. Monitoring the rate of heat generation produced is a good indicator of the rate of reaction of Portland cement compounds. An illustrative example of the heat evolution curve produced by cement using a calorimeter (ASTM C-186) is shown in Figure 10.2. It shows the stages of heat evolution during Portland cement hydration as determined by a calorimeter.

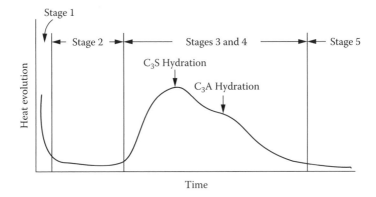

FIGURE 10.2 Schematic of heat evolution as a function of time for Portland cement.

FIGURE 10.3 Scanning electron micrograph showing cement hydration products at various ages; phases include CSH, Ca(OH)$_2$, and Ettringite.

As water is added to cement, some of the clinker sulfates and gypsum dissolve, producing an alkaline and sulfate-rich solution. The C$_3$A phase reacts vigorously with water to form an aluminate-rich gel (stage 1 in Figure 10.2). The gel reacts with the sulfates in solution to form small rodlike crystals of ettringite (see Figure 10.3). The hydration of C$_3$A is a strongly exothermic reaction but only lasts a few minutes (approximately 7 min) and is followed by a period of a few hours of relatively low heat evolution. This is called the dormant, or induction, period (stage 2). The first part of the dormant period corresponds to when concrete can be placed, and it lasts a couple of hours. As the dormant period progresses, the paste becomes stiff and unworkable. This period is related to initial set in concrete.

At the end of the dormant period, the alite (C$_3$S) and belite (C$_2$S) compounds start to hydrate, with the formation of calcium silicate hydrate (CSH) and calcium hydroxide (CH). This corresponds to the main period of cement hydration (stage 3), rate of strength increase, and final set for concrete. The cement grains react from the surface inward as water diffuses through layers of formed hydration product, and the anhydrous particles decrease in size. In stage 4, C$_3$A hydration continues as well, as water diffuses through cement particles and newly exposed crystals become accessible to water. The rate of early strength gain is influenced in this period. In addition, conversion of ettringite crystals to monosulfate aluminate crystals is also occurring. This peak occurs between 12 and 90 h. In stage 5, the slow steady formation of hydration products establishes the rate of later strength. The ferrite compounds continue their slow hydration at this phase, although their contribution to strength gain is limited. The period of maximum heat evolution for Portland cement mixes occurs approximately between 12 and 20 h after mixing, and then dissipates. For blended cements and the use of pozzolans such as fly ash, silica fume, and slag, the heat evolution periods may be different. Table 10.6 shows the hydration reactions.

Figure 10.3 shows the cement microstructure during hydration.

TABLE 10.6

Portland Cement Compound Hydration Reactions (Oxide Form)

$2C_3S + 11H = C_3S_2H_8 + 3CH$

Tricalcium silicate + water = Calcium silicate hydrate (CSH) + calcium hydroxide

$2C_2S + 9H = C_3S_2H_8 + CH$

Dicalcium silicate + water = Calcium silicate hydrate (CSH) + calcium hydroxide

$C_3A + 3C\overline{S}_3H_2 + 26H = C_6A\overline{S}_3H_{32}$

Tricalcium aluminate + gypsum + water = Ettringite

$2C_3A + C_6A\overline{S}_3H_{32} + 4H = 3C_4A\overline{S}_3H_{12}$

Tricalcium aluminate + Ettringite + water = Calcium monosulfoaluminate

$C_3A + CH + 12H = C_4AH_{13}$

Tricalcium aluminate + calcium hydroxide + water = Tetracalcium aluminate hydrate

$C_4AF + 10H + 2CH = C_6AFH_{12}$

10.6 STEEL IN CONCRETE

Steel-reinforcing bars (commonly known as *rebar*) are produced by pouring molten steel into casters and then running it through a series of stands in the mill. Most bars are deformed, which means that a raised pattern is rolled along the perimeter. The deformed ribs improve the bond between the steel and the concrete, and help transfer the loads between the two materials. Plain bars are also fabricated. Some are used for dowels, since at least one-half of a dowel bar should be smooth and lubricated to allow for free expansion and contraction. The specifications for concrete rebar are given in the following ASTM standards:

- A615/A615M-05a: Standard Specification for Deformed and Plain Billet-Steel Bars for Concrete Reinforcement (covers grades 40 and 60/soft metric grades 420 and 520)
- A616: Standard Specification for Rail-Steel Deformed Bars for Concrete Reinforcement (covers grades 50 and 60)
- A617: Standard Specification for Axle-Steel Deformed Bars for Concrete Reinforcement (covers grades 40 and 60)
- A706/A706M-96b: Standard Specification for Low-Alloy-Steel Deformed and Plain Bars for Concrete Reinforcement (grade 60 only)

The commonly used steel grades are given in Table 10.7. The allowable stress is usually given as two-thirds of the yield strength.

The size designations (Table 10.8) are the number of eighths of an inch in the diameter of a plain round bar having the same weight per foot as the deformed bar. So, for example, a No. 6 bar would have the same mass per foot as a plain bar 6/8 in. in diameter. The metric size is the same dimension expressed to the nearest millimeter.

Table 10.9 shows steel type, yield strength (40–75 ksi, 300–520 MPa), and ASTM standard for the different types of steels used for reinforcement.

Welded wire mesh is steel wire welded together to form a flat sheet with a square grid pattern. A common grid size is 150 mm × 150 mm (6 in. × 6 in.), and a common steel wire thickness is 4 mm (1/8 in.). Wire fabric or bar mats are used in PCC slabs to control temperature cracking. The stresses developed in the concrete slab during these volume changes are mainly due to friction that develops. However, these reinforcements do not increase the structural capacity of the slab. The wire fabric aids in holding the cracked concrete pavement together and allows for load transfer through aggregate interlock. Additionally, the reinforcement allows for greater spacing between joints.

TABLE 10.7
Commonly Used Grades of Steel

English Grade	International System of Units (SI) Grade	Minimum Yield Strength	
		psi	MPa
Grade 40	Grade 280	40,000	280
Grade 50	Grade 350	50,000	350
Grade 60	Grade 420	60,000	420
Grade 75	Grade 520	75,000	520

TABLE 10.8
Sizes and Dimensions

Bar Designation Number	Nominal Diameter (in.; Does Not Include the Deformations)	Metric Designation Number	Weight (lb/ft)
3	0.375	10	0.376
4	0.500	13	0.668
5	0.625	16	1.043
6	0.750	19	1.502
7	0.875	22	2.044
8	1.000	25	2.670
9	1.128	29	3.400
10	1.270	32	4.303
11	1.410	36	5.313
14	1.693	43	7.650
18	2.257	57	13.60

TABLE 10.9
Steel Type, Strength, and Applicable Standard

Mark	Meaning	Applicable ASTM Standard by Grade					
		40 and 50	60	75	300 and 350	420	520
S	Billet	A615	A615	A615	A615M	A615M	A615M
I	Rail	A616	A616	—	A996M	A996M	—
IR	Rail meeting supplementary requirements S1	A616	A616	—	—	—	—
A	Axle	A617	A617	—	A996M	A996M	—
W	Low alloy	—	A706	—	—	A706M	—

QUESTIONS

10.1 What is the importance of cement chemistry with respect to strength, durability, and setting time of concrete?

10.2 Why is gypsum added to cement clinker?

10.3 What are the different applications for the different types of cement?

10.4 Describe the hydration process.

10.5 Why is steel used in PCC pavements?

11 Distress and Performance

A brand-new pavement at the start of its design life is expected to be one without any "distress" or undesirable features. Such features, which include rutting, cracking, patching, or roughness, are "undesirable" from the point of view of performance of the pavement—the more distress, the shorter the pavement's life—and at some point, the distresses are so great in intensity (e.g., deep ruts) as well as extent (e.g., 75% of the wheelpath area in a project area has cracks) that the pavement is considered to be "failed" or at the end of its design life.

The following sections explain the possible distresses and their relationship to the performance of a pavement.

11.1 DISTRESSES IN ASPHALT PAVEMENTS

There are different forms of distresses in asphalt pavements, each tied to a specific reason (such as poor mix design, construction, or environmental conditions) or a combination thereof, and most happening as a result of traffic. Figure 11.1 lists the common distresses in asphalt pavements. The different distresses are described in the following sections, in alphabetical order.

11.1.1 BLEEDING

Bleeding is the appearance of asphalt binder on the surface of the pavement. This is a surface defect caused by excessive asphalt binder in the surface asphalt mix layer. It is measured in square meters. The different conditions of bleeding include discoloration, covering of aggregate with a thin reflective surface, and loss of texture.

11.1.2 BLOCK CRACKING

Block cracking refers to a pattern of cracks that divide the surface into approximately rectangular pieces (>0.1 m²). Such cracks occur due to shrinkage of asphalt mix because of volume changes in the base or subgrade. It is measured in square meters, and its severity can be described as low (crack width ≤6 mm or sealed cracks whose width cannot be measured, with sealant in good condition), moderate (crack width >6 mm but ≤19 mm or any crack with mean width ≤19 mm and adjacent low-severity random cracking), and high (crack width >19 mm or any crack with mean width ≤19 mm and adjacent moderate- to high-severity random cracking).

11.1.3 CORRUGATIONS

Corrugations are ripples formed laterally across an asphalt pavement surface. These occur as a result of lack of stability of the hot mix asphalt (HMA) at a location where traffic starts and stops or on hills where vehicles brake downgrade. The causes of the lack of stability are too much or too soft asphalt, a high sand content, and an excessive presence of smooth and rounded aggregate in the mix.

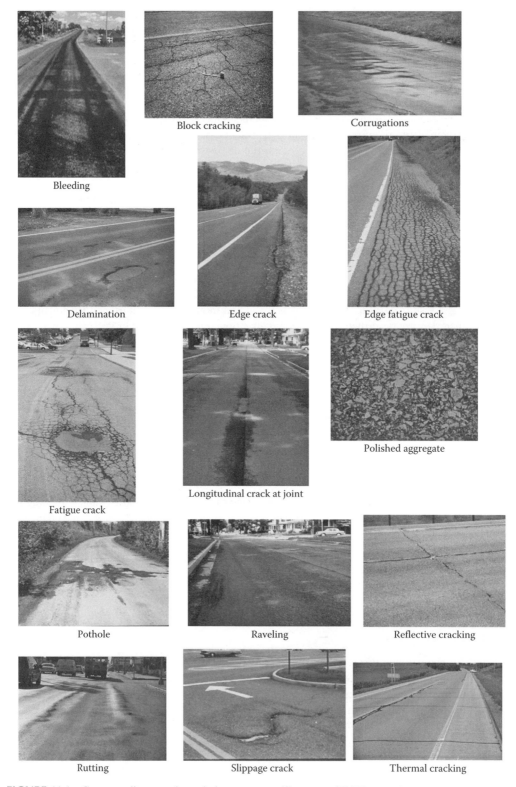

FIGURE 11.1 Common distresses in asphalt pavements. (Courtesy of Ed Kearney.)

11.1.4 DELAMINATION

Delamination is the separation of the top wearing layer from the layer underneath. It is caused by poor bond or by failure of the bond between the two layers. The poor bond can be due to improper surface preparation or tack coat before the application of the wearing layer and/or a relatively thin wearing layer. The loss of bond can be caused by environmental factors such as ingress of water and repeated freeze–thaw cycles.

11.1.5 EDGE CRACKS

Edge cracks are found in pavements with unpaved shoulders. They are crescent shaped or fairly continuous cracks that intersect the pavement edge. The cracks are located within 0.6 m of the pavement edge adjacent to the shoulder. Longitudinal cracks outside of the wheelpath and within 0.6 m of the pavement edge are also classified as edge cracks.

The cracking is due to the lack of lateral (shoulder) support, base or subgrade weakness caused by frost action, and inadequate drainage. It is measured in meters, and severity is reported as low (cracks without breakup or loss of material), moderate (loss of material and breakup in up to 10% of the cracked area), or high (loss of material or breakup in more than 10% of the cracked area).

11.1.6 FATIGUE CRACKS AND EDGE FATIGUE CRACKS

Fatigue cracks generally begin at the wheelpath. Edge fatigue cracks are formed due to poor underlying support at the edge of the pavements with paved shoulders. This distress, found on wheelpaths, begins as a series of interconnected cracks and develops into a chicken wire/alligator pattern. Interconnected cracks give rise to many-sided, sharp-angle pieces, usually with a maximum length of 0.3 m. Repeated tensile stress/strain at the bottom of the asphalt mix layer, caused by traffic, leads to fatigue cracking. It starts from the bottom and moves upward. The cracked area must be quantifiable in order to be counted as a fatigue crack. The cracked area should be measured in square meters, for each of the three severity levels: low, medium, and severe. If, within an area, there are different severity levels and they cannot be distinguished, the entire area should be rated as for the highest severity rating. A *low*-severity area is one in which there are no or few connecting cracks, the cracks are not spalled or sealed, and there is no evidence of pumping. *Moderate*-severity areas are those in which the cracks are interconnected, forming a complete pattern, and cracks may be slightly spalled or sealed, with no evidence of pumping. In a *high*-severity area, the cracks are interconnected with moderate or severe spalls, forming a complete pattern; pieces may move under traffic; cracks may be sealed; and pumping may be evident.

11.1.7 LONGITUDINAL JOINT CRACKS

Longitudinal joint cracks develop at construction joints because of poorly paved joints or improper construction techniques. They are measured in meters, and their severity is expressed in the same manner as in block cracking.

11.1.8 POLISHED AGGREGATE

Polished aggregate is the exposure of coarse aggregate due to wearing away of the asphalt binder and fine aggregates from the surface asphalt mix. It is measured in square meters.

11.1.9 POTHOLES

Bowl-shaped holes, with a minimum plan dimension of 150 mm in the pavement surface, are called *potholes*. There are four main causes of potholes: (1) insufficient thickness of the pavement to support traffic through winter–spring freeze–thaw cycles; (2) poor drainage, leading to accumulation of excess water; (3) failures at utility trenches and castings; and (4) paving defects and unsealed cracks. For a distress survey, the number of potholes and the square meters of affected area are recorded. Severity levels can be reported as low (<25 mm deep), moderate (25 mm < depth < 50 mm), or high (>50 mm deep).

11.1.10 RAVELING

Raveling refers to wearing away of the pavement surface by loss of asphalt binder and displacement of aggregates. The process starts as loss of fines and can continue to a situation with loss of coarse aggregate and a very rough and pitted surface. Raveling is caused by the action of water that finds its way through the surface of the pavement because of poor compaction and hence low density and relatively high voids, and it can initiate the potholing process. It is measured in square meters.

11.1.11 REFLECTIVE CRACKING

Reflective cracks are those in asphalt overlays caused by discontinuities in the pavement structure underneath. This can be due to old cracked asphalt pavement or joints in concrete pavement underneath. The severity levels are recorded in the same manner as in block cracking.

11.1.12 RUTTING

A *rut* is defined as a longitudinal depression in the wheelpath, with or without transverse displacement. It can be measured with a straight edge or a profiler at regular intervals. A rut is a physical distortion of the surface, and it also prevents the cross drainage of water during rains, leading to accumulation of water in the ruts and increasing the potential of hydroplaning-related accidents. Generally a rut depth of 0.5 in. is considered a rutting failure. Rutting is the result of repeated loading, which causes accumulation, and increase of permanent deformations. Rutting can be (1) low- to moderate-severity rutting—one-dimensional densification or vertical compression near the center of the wheelpath, caused by densification of mixes with excessive air voids in the in-place mix under traffic; (2) moderate- to high-severity rutting—a depression in the wheelpath along with humps on either side of the depression, caused by lateral flow due to plastic deformation, resulting from shear failure of the mix under traffic, and generally associated with very low air voids in the mix; and (3) rutting accompanied by cracks on the surface of the pavement, caused by rutting in underlying layers, such as the subgrade or subbase.

11.1.13 SLIPPAGE CRACK

Slippage cracks are typically crescent- or half-moon-shaped cracks produced when vehicles brake or turn, which causes the pavement surface to slide or push. This is caused by a low-strength HMA or a lack of bond between the surface and lower courses.

11.1.14 THERMAL CRACKS

Thermal cracking occurs in the form of transverse cracking, which is defined as cracks that are predominantly perpendicular to the pavement centerline. These cracks occur at regular intervals. Thermal cracks can be caused by the fracture of asphalt mix due to a severe drop in temperature or by thermal fatigue caused by repeated low- and high-temperature cycles. The severity of thermal cracking is measured by the width of the crack. A low-severity transverse crack is one with a mean width of ≤6 mm, or a sealed crack with sealant material in good condition and a width that cannot be determined. A medium-severity transverse crack is one with a mean width >6 and ≤19 mm, or any crack with a mean width ≤19 mm and adjacent low-severity random cracking, while a high-severity crack has a mean width >19 mm, or ≤19 mm but with adjacent moderate- to high-severity random cracking. Only cracks that are >0.3 m in length are counted. The number and length of transverse cracks at each severity level are recorded. The entire transverse crack is to be rated at the highest severity level that is present for at least 10% of the total length of the crack. The length, in m, is the total length of the crack. If the cracks are sealed, then the length of cracks with sealant in good condition (for at least 90% of the crack) should be measured at each severity level. Part of the thermal transverse crack extending into a load-induced fatigue crack area is not counted.

11.2 DISTRESSES IN CONCRETE PAVEMENTS

The different types of distresses found in PCC pavements are described next. Figure 11.2 summarizes the different distresses.

11.2.1 CORNER BREAKS

A corner portion of the slab is separated by a crack. The crack intersects the adjacent transverse and longitudinal joints, and is approximately at a 45° angle with the direction of traffic. The length of the sides is from 0.3 m to one-half the width of the slab on each side of the corner. Corner breaks are measured by counting the number that is encountered per unit length. For a low-severity level, the crack is not spalled more than 10% of the length of the crack, and there is no faulting. For a moderate-severity level, the crack is spalled >10% of its total length, or the faulting of the crack or joint is <13 mm and the corner piece is not broken into two or more pieces. For a high-severity level, the crack is spalled at moderate to high severity >10% of its total length, the faulting of the crack or joint is ≥13 mm, or the corner piece is broken into two or more pieces or contains patch material.

11.2.2 DURABILITY CRACKING (OR "D" CRACKING)

In this type of distress, the cracking is due to freezing and thawing and usually initiates at construction joints. These are usually closely spaced crescent-shaped hairline cracks that appear like the letter D. D-cracks are measured by recording the number of slabs with "D" cracking and square meters of area affected at each severity level. The slab and affected area severity rating is based on the highest severity level present for at least 10% of the area affected. For low-severity distress, D-cracks are close together, with no loose or missing concrete pieces, and no patches in the affected area. For moderate severity, D-cracks are well defined, and some small concrete pieces are loose or have been displaced. For high severity, D cracking has a well-developed pattern, with a significant amount of loose or missing material, and may have been patched.

11.2.3 LONGITUDINAL CRACKING

These are stress cracks that are predominantly parallel to the pavement centerline and are measured in meters. For low-severity distress, crack widths are <3 mm in width, with no spalling and no measurable faulting, or are well sealed and with a width that cannot be determined. For moderate-severity distress, crack widths range between 3 and 13 mm, with spalling <75 mm, or faulting up to 13 mm. For high-severity distress, crack widths are <13 mm, with spalling <75 mm, or faulting <13 mm.

Blowups   Corner break   D cracking   Faulting

Joint seal damage   Longitudinal cracking   Map cracking   Patch

Polished aggragate   Popouts   Pumping   Punchout

Separation   Shoulder drop-off   Spalling of joints   Transverse cracking

FIGURE 11.2   Distress in PCC pavements. (From Distress Identification Manual for the Long term Pavement Performance Program, FHWA-RD-03-031, June 2003.)

11.2.4   TRANSVERSE CRACKING

These are stress cracks that are predominantly perpendicular to the pavement centerline and are measured in meters. For low-severity distress, crack widths are <3 mm, with no spalling and no measurable faulting, or are well sealed similar to longitudinal cracking. For moderate-severity distress, crack widths are >3 and <6 mm, with spalling <75 mm, or faulting up to 6 mm. For high-severity distress, crack widths are >6 mm, with spalling >75 mm, or faulting >6 mm.

11.2.5   SPALLING OF TRANSVERSE JOINTS

This distress consists of cracking, breaking, chipping, or fraying of slab edges within 0.3 m from the face of the transverse joint, and is measured in frequency of occurrence and meters. For low-severity distress, spalls are <75 mm wide, measured to the face of the joint, with loss of material, or spalls with no loss of material and no patching. For medium-severity distress, spalls are 75–150 mm wide, measured to the face of the joint, with loss of material. For high-severity distress, spalls are >150 mm wide, measured to the face of the joint, with loss of material, or broken into two or more pieces or contain patch material.

11.2.6   MAP CRACKING AND SCALING

Map cracking: A series of interconnected cracks that extend into the upper surface of the slab. Usually, larger cracks are oriented in the longitudinal direction of the slab and are interconnected by finer transverse or random cracks. These are measured in frequency of occurrence and in square meters.

Scaling: Scaling is the deterioration and flaking of the upper concrete slab surface, normally in areas of 3–13 mm, and may occur anywhere over the pavement. This is measured by the number of occurrences and the square meters of affected area.

11.2.7   POLISHED AGGREGATE

Exposed and polished coarse aggregate occur due to surface mortar and paste loss, which results in a significant reduction of surface friction. Distress is measured in square meters of affected surface area. The level of severity is not applicable for this distress. Diamond grinding also removes surface texturing, but this condition should not be recorded as polished aggregate.

11.2.8   POPOUTS

Popouts are small pieces of pavement broken loose from the surface, ranging in diameter from 25 to 100 mm, and ranging in depth from 13 to 50 mm.

11.2.9   BLOWUPS

Blowups are slab length changes resulting in localized upward movement of the pavement surface at transverse joints or face cracks, and are usually accompanied by shattering of the concrete in that area. Distress is measured by recording the number of blowups. Severity levels are not applicable. However, severity levels can be defined by the relative effect of a blowup on ride quality and safety.

11.2.10   FAULTING OF TRANSVERSE JOINTS AND CRACKS

This occurs as a result of elevation difference across a joint or crack. Distress is measured in millimeters, to the nearest millimeter at a location 0.3 and 0.75 m from the outside slab edge (this is approximately the location of the outer wheelpath). For a widened lane, the wheelpath location will be 0.75 m from the outside lane edge stripe. Faulting is recorded as positive (+) if the "approach" slab is higher than the "departure" slab; if the approach slab is lower, faulting is recorded as negative (−). Faulting on PCC pavements can be measured using a Federal Highway Administration (FHWA)-modified Georgia faultmeter.

11.2.11   LANE-TO-SHOULDER DROPOFF

This is a difference in elevation between the edge of the slab and the outside shoulder, and usually occurs when the outside shoulder settles. Distress is measured at the longitudinal construction joint between the lane edge and shoulder. Distress is recorded to the nearest millimeter at 15–25-m intervals. The recorded value is negative (−) if the traveled surface is lower than the shoulder.

11.2.12   LANE-TO-SHOULDER SEPARATION

This is a widening of the joint between the edge of the slab and the shoulder. Distress is measured to the nearest millimeter at intervals of 15–25 m along the lane-to-shoulder joint. It should be documented if the joint is well sealed or not at each shoulder location. Severity levels are not applicable.

11.2.13 Patch/Patch Deterioration

A patch is a portion greater than 0.1 m², all of the original concrete slab that has been removed and replaced, or additional material applied to the pavement after original construction. For a low-severity-level distress, the patch displays low-severity distress of any type, there is no measurable faulting or settlement, and pumping is not evident. For a moderate-severity level, the patch has moderate-severity distress of any type or faulting or settlement up to 6 mm, and pumping is not evident. For a high-severity level, the patch has a high-severity distress of any type, or faulting or settlement >6 mm, and pumping may be evident.

11.2.14 Water Bleeding and Pumping

This is the ejection or seepage of water from beneath the pavement through cracks. The residue of very fine materials deposited on the surface that were eroded (or pumped) from the support layers, causing staining, aids in recognizing pumping phenomenon. This is measured by recording the number of occurrences of water bleeding and pumping and the length in meters of affected pavement with a minimum length of 1 m. The combined length of water bleeding and pumping cannot exceed the length of the test section.

11.2.15 Punchouts

Punchouts are broken areas enclosed by two closely spaced (commonly <0.6 m) transverse cracks, a short longitudinal crack, and the edge of the pavement or a longitudinal joint. They also include "Y" cracks that exhibit spalling, breakup, or faulting. The number of punchouts should be recorded at each severity level. It should be noted that the cracks that outline the punchout are also recorded under "longitudinal cracking" and "transverse cracking." Punchouts that have been completely repaired by removing all broken pieces and replacing them with patching material (rigid or flexible) are rated as patches. For low-severity distress, transverse cracks are tight and may have spalling <75 mm or faulting <6 mm, with no loss of material and no patching. For moderate distress, spalling >75 mm and <150 mm or faulting >6 mm and <13 mm exists. For a high-severity rating, spalling is >150 mm, or concrete within the punchout is punched down by >13 mm, is loose and moves under traffic, is broken into two or more pieces, or contains patch material.

11.2.16 Joint Seal Damage

Joint seal damage is any condition that enables incompressible materials or a significant amount of water to infiltrate the joint from the surface. Common types of joint seal damage are extrusion, hardening, adhesive failure (bonding), cohesive failure (splitting), complete loss of sealant, and the intrusion of foreign material in the joint such as weed growth in the joint. This distress is measured by recording the number of longitudinal joints that are sealed and the length of sealed longitudinal joints with joint seal. The severity level is not applicable.

11.3 CONSIDERATION OF PERFORMANCE

A pavement is constructed for the safe and smooth passage of traffic. There is a need to quantify the extent to which the pavement is serving its purpose, or its "performance." Such quantification can be done with respect to distress conditions exhibited by the pavement at any time after its construction.

The AASHO Road Test introduced the concept of serviceability to measure performance. It was necessary to characterize the pavement sections in terms of their condition, in order to develop relationships between the performance and the factors affecting the performance. First, the concept of present serviceability rating (PSR) was introduced. PSR (Figure 11.3) is defined as the judgment of an observer as to the current ability of a pavement to serve the traffic it is meant to serve.

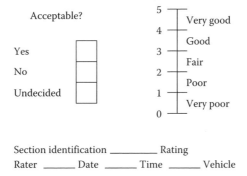

FIGURE 11.3 Concept of present serviceability rating.

The PSR ranges from 1 to 5, starting from a perfect 5 and decreasing with the passage of traffic. To characterize pavements in terms of the serviceability rating, the pavement serviceability index (PSI) was then introduced on the basis of observed distress conditions:

$$PSI = 5.03 - 1.91\log\left(1 + \overline{SV}\right) - 1.38RD^2 - 0.01\sqrt{C+P} \text{ (flexible pavement)}$$

$$PSI = 5.41 - 1.80\log\left(1 + \overline{SV}\right) - 0.09\sqrt{C+P} \text{ (rigid pavement)}$$

where
  $\overline{SV}$ is the mean of the slope variance in the two wheelpaths (measured with the CHLOE profilometer or Bureau of Public Roads [BPR] roughometer)
  C, P are the measures of cracking and patching in the pavement surface
  C is the total linear feet of Class 3 and Class 4 cracks per 1000 ft^2 of pavement area

A Class 3 crack is defined as opened or spalled (at the surface) to a width of 0.25 in. or more over a distance equal to at least one-half the crack length. A Class 4 crack is defined as any crack that has been sealed. RD is the average rut depth (in.). P is expressed in terms of ft^2 per 1000 ft^2 of pavement surfacing. The basic idea was that just like PSR (numerically the same value as PSI), PSI would drop over time (i.e., with the passage of traffic), starting from an initial p_0 to a terminal p_t. The curve can be expressed with an equation as follows (Figure 11.4):

$$p_o - p = (p_0 - p_t)\left(\frac{W}{\rho}\right)^\beta$$

where
  β and ρ depend on pavement structure (thickness and stiffness) and loading
  β determines the shape of the graph
  β is the number of loads at which $p = 1.5$
  W is the cumulative load

Note that performance at any time can be expressed as the area under the serviceability curve from the beginning to the point under consideration. The serviceability of the pavement at any time can be "raised" by rehabilitation, such as an asphalt mix overlay.

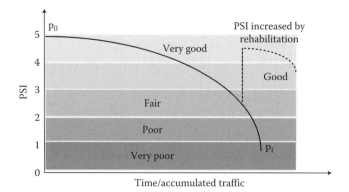

FIGURE 11.4   Plot of PSI versus time.

11.4 DAMAGE

In the AASHTO design process, the loss of serviceability (ΔPSI) is referred to as *damage*. This damage is considered to be caused by traffic, environmental conditions, and age. In mechanistic–empirical pavement design processes, the performance of the pavement is expressed in terms of distresses. The distresses are predicted on the basis of models (commonly referred to as *transfer functions*) relating mechanistic response (stress/strain) to observed distress. Note that such models need to be calibrated for specific regions, climate, materials, and construction conditions. In more sophisticated methods, a damage index is calculated from a mechanistic response, and the damage index is used to compute accumulation of distress with time. Damage is defined as the deterioration of the pavement due to the action of traffic over different environmental conditions. Such deterioration happens as a result of change in the engineering properties of the pavement layer materials.

This damage does not happen altogether at the same time, but rather progressively—or, more precisely, in increments—with the passage of every vehicle or, if expressed in time, every hour during its service. Note that the damage at every hour is not the same—it can be higher, for example, due to a heavier vehicle moving over it at that hour, or due to a lowering of the modulus of the asphalt mix layer due to a high temperature that hour. Therefore, the most rational approach is to consider and compute the damage in each and every increment by considering the relevant traffic (class of vehicle) and the pavement material properties (with respect to environmental conditions such as temperature) for that increment period (say, 1 h). The damage at any increment can be expressed as follows:

$$D = \frac{n}{N}$$

where
  n is the calculated load applications
  N is the allowable number of load applications

Note that the allowable number of loads is dependent on the condition of the pavement layer at any increment period—hence, N is different for different increment periods.

The total damage at any point is computed by summing up all of the damages over time, up to that time, as follows (commonly referred to as *Miner's hypothesis*; Miner, 1945):

$$\text{Total damage} = \sum_{i=1}^{n}\sum_{j=1}^{m}\sum_{k=1}^{o}\frac{n_{ijk}}{N_{ijk}}$$

where i, j, and k are the different categories over which the summation of the damage is made. Such categories can include different time increments, traffic wander, and truck class. The basic idea is that for each and every time increment (i.e., for a unique set of material properties tied to a unique set of environmental conditions), the damage is determined for each and every type and movement of the vehicle. The more categories used in the calculation of increments, the more precise is the damage calculation.

This total damage over time can then be related to distress, and then finally the distress can be expressed as a function of time. For any specific type of distress, when total damage (sometimes referred to as the *cumulative damage factor* or CDF) reaches 1, the pavement is said to have failed (say, by cracking).

In a simpler approach, instead of considering the damage in each and every increment individually, effective or weighted values of each variable could be used to determine the total damage. For example, the concept of effective soil (roadbed/subgrade) resilient modulus is recommended in the AASHTO design procedure. The method consists of the laboratory or in-place determination (or prediction through the correlation of other properties) of resilient modulus of the soil for different moisture conditions (wet/dry) that are expected in the different moisture seasons. Then an effective resilient modulus value is calculated that has the same (equivalent) effect as the combined effect of the different moduli under different moisture conditions. This effective modulus is calculated on the basis of relative damage due to variation of the soil modulus in different seasons. The procedure is illustrated with an example in Figure 11.5.

11.5 FORENSIC INVESTIGATION FOR DETERMINATION OF TYPE AND CAUSE OF DISTRESS

Forensic investigation to determine the cause of a specific distress is very important in order to determine an effective rehabilitation plan. Because no two pavement sections are alike and the same distress can be due to a number of (or a combination thereof) problems, an appropriate forensic investigation plan must be made and implemented in each case. Prior to the development of the investigation plan, a copy of the report of distress, compilation of existing data, and as assessment of missing data need to be made available. A decision depending on the benefit of the forensic investigation, approximate cost and the assurance of the highway agency to cooperate with the forensic investigation has to be made by the forensic team, and preferably the deliberations be noted down for future reference.

11.5.1 FORENSIC INVESTIGATION PLAN

Based on a preliminary site inspection, a forensic investigation plan needs to be developed—the plan should aim at two things: (1) to determine the mechanism(s) and the cause(s) of the distress and (2) to obtain the desired data, as well as missing and questionable data. The different activities that could be utilized for the investigation of different distress are summarized in Table 11.1. The plan for obtaining the missing data should include specific tests that need to be conducted and samples that need to be retrieved to accomplish the testing. As far as it is possible, testing and sampling plan should be based on existing standard protocols. A project manual with specific data collection sheets should be developed. The investigation should also include sampling and testing of materials from good performing areas of the same pavement to determine and confirm the absence of factors that contributed to distress. The different steps in a forensic investigation as outlined by different researchers/organizations are shown in Table 11.2.

11.5.2 NONDESTRUCTIVE AND DESTRUCTIVE TESTS

For site evaluation, important considerations are topography, geology, drainage pattern, traffic volume and composition, history of construction and maintenance, and effectiveness of maintenance

Month	Roadbed soil modulus, M_R (psi)	Relative damage, U_f
Jan.	20,000	0.01
Feb.	20,000	0.01
Mar.	2,500	1.51
Apr.	4,000	0.51
May.	4,000	0.51
June	7,000	0.13
July	7,000	0.13
Aug.	7,000	0.13
Sept.	7,000	0.13
Oct.	7,000	0.13
Nov.	4,000	0.51
Dec.	20,000	0.01
Summation: Σu_f =		3.72

Average: $\bar{u}_f = \dfrac{\Sigma u_f}{n} = \dfrac{3.72}{12} = 0.31$

Effective roadbed soil resilient modulus, M_R (psi) = 5,000 (corresponds to $\bar{u}_f$)

FIGURE 11.5　Calculation of effective soil modulus based on relative damage. (From American Association of State Highway and Transportation Officials (AASHTO), *AASHTO Guide for Design of Pavement Structures*, AASHTO, Washington, DC, © 1993. Used with permission.)

work. Deflection testing, such as those with the falling weight deflectometer (FWD), is commonly used as a nondestructive test, although a ground penetrating radar (GPR), spectral analysis of surface waves (SASW) equipment, as well as a profilometer could be used. The location of the water table should be ascertained, and drainage conditions need to be evaluated.

Coring and trenching are the two commonly used destructive sampling/evaluation methods. While coring allows visual evaluation and subsequent laboratory testing, trenching allows the investigators to determine exact location of failures (such as rutting), actual thickness of layers, presence of moisture in layers, if any, and bond between different layers.

TABLE 11.1
Tests for Determination of Distress Mechanism

Structure Type	Distress	Actions	Comments
AC, AC/AC	Fatigue cracking	Core	Within and outside area
	Longitudinal WP	Core	
	Longitudinal NWP	Core	
	Rutting	Trench	Layer profile
	Transverse cracking	Core	
AC/JCP, AC/CRCP	Longitudinal WP	Core	
	Longitudinal NWP	Core	
	Rutting	Core	Within and outside to note differences
	Transverse	Mill off AC	Observe joint/crack condition
JCP	Transverse	Core	
	Corner break	Core	
	Longitudinal cracking	Core	
	Pumping	Test pit	Void detection using NDT
	Faulting		
CRCP	Punchout	Core	
	Pumping	Test pit	Void detection using NDT

Source: Federal Highway Administration (FHWA), Framework for LTPP Forensic Investigations—Final, April 2004.

Note: Site-specific conditions not listed here should be investigated at the discretion of the forensic team. NDT, nondestructive testing; NWP, nonwheelpath; WP, wheelpath.

TABLE 11.2
Different Steps in Pavement Forensic Investigation

Number	Step	Details
1	Planning	Preliminary meeting, prepare for investigation, identify need and purpose of rehabilitation
2	Review of data	Review of project and other records, background information gathered, historical and other data reviewed, desktop evaluation, desktop investigation
3	Interview	Interviews with people familiar with project or road, interview related employees
4	Nondestructive evaluation	Visual investigation, detailed condition survey, initial site visit and nondestructive testing, examine functional, moisture and structural condition, site evaluation visual inspection and deflection testing
5	Destructive evaluation	Materials sampling and testing, subsurface investigation destructive testing materials testing, examination of pavement, materials pavement materials investigations
6	Determination of cause(s)	Analyze data to find likely failure cause, analyze collected data in a systematic way
7	Determination of rehabilitation method	Included in final report, select alternative options for analysis, decide on treatment
8	Report	Final report to detail entire forensic investigation, forensic report prepared

Source: Adapted from Smith, R., *Forensic Investigation of Pavement Failures*, University of Southern Queensland, Darling Heights, Queensland, Australia, 2004.

For soil and aggregate layers, the dynamic cone penetrometer (DCP) test could be conducted, while in the laboratory, California bearing ratio (CBR), the repeated load triaxial, or the resilient modulus test could be conducted. In addition, determination of moisture content, aggregate gradation, dry density, and Atterberg limit tests could be carried out.

One suggested method of determination of cause(s) of failure is to match the matrix of results of field and laboratory evaluation against the different failure hypotheses, and determine which one is the most satisfied. Based on the cause of failure, several alternatives for rehabilitation should be examined, and the best option should be selected with consideration of economics, design, and construction.

QUESTIONS

11.1 List the distresses for asphalt and concrete pavements, along with their causes.

11.2 From a review of information from your local newspapers, county, and/or municipality, find out the most common pavement distress in your locality. Can you suggest some corrective actions?

11.3 Determine the effective resilient modulus of a soil if the moduli at different months are as follows:

Month	Modulus (psi)
January	18,000
February	18,000
March	3,000
April	5,000
May	5,000
June	8,000
July	8,000
August	8,000
September	8,000
October	8,000
November	8,000
December	10,000

12 Consideration of Major Distress Mechanisms and Material Characterization for Asphalt Pavements

Different types of distresses have been presented in Chapter 11. The major distresses are considered in the mix and structural design of asphalt mixes. This consideration consists of three basic parts: explaining and formulating the problem, conducting appropriate tests to determine the relevant material properties (characterization), and using models to relate the amount/extent of distress to the relevant material properties, and hence predict the distress. All of these three parts are presented in the following sections.

12.1 FATIGUE CRACKING

The mechanism of fatigue cracking can be divided into two parts: (1) the occurrence of tensile stress/tensile strain in the asphalt mix layer, and (2) the repetitive occurrence of such tensile stress/strain under traffic repetitions. The occurrence of tensile stress/strain results from the elastic behavior of the pavement structure under traffic loading, whereas the initiation and formation of cracks result from "fatigue" behavior of the layer. The high tensile stresses/strains responsible for causing fatigue cracks can occur because of a relatively thin or weak asphalt mix layer, very high load or tire pressure, or a relatively weak subgrade and/or base and/or subbase, due to moisture or low density due to inadequate compaction.

If a beam is subjected to loading from above, then directly under the load, the beam would tend to assume a convex shape, with compressive stresses/strains on the top part and tensile stress/strain in the bottom part. Due to the elastic nature of the pavement, in such a situation, it springs back to its original position, without the stresses/strains, when the load is removed. However, under the next load, it undergoes the same cycle. Under the heaviest possible load, if the pavement structure is such that the tensile stress/strain (which is responsible for causing cracks) is very small, then the pavement can sustain many such cycles/repetitions without any damage. If the tensile stress/strain is moderate, then it can sustain a certain number of repetitions without damage, and if the tensile stress/strain is very high, then it can sustain a relatively small number of repetitions before permanent damage.

There are two analysis steps that are necessary to predict how many repetitions the pavement can sustain before it gets damaged. Assuming a standard (high load), one can determine the tensile stress/strain at the bottom of the asphalt mix layer of an assumed pavement structure (if there is tensile stress/strain in the layer, then it is the maximum at the bottom), using layered elastic analysis, as explained in Chapter 2. Then, conducting an experiment to determine how many repetitions the pavement can sustain before there is damage, one can determine how many repetitions it can sustain for this specific tensile stress/strain. Now let us get into the details of these steps.

If one assumes a pavement structure, the elastic moduli, thickness, and Poisson's ratio of the different layers are needed to determine the tensile stress/strain under a specific load. These properties can be used from experience. The load can be considered as the highest allowable load—say, 9,000 lb—by considering a maximum allowable load of 18,000 lb per axle. The Poisson's ratio of the generally used pavement materials does not differ greatly, and a value of 0.35–0.4 can be assumed.

Now if the pavement structure is such that the asphalt mix layer on the top is very thin, one may determine that there is no tensile stress/strain in the layer, and all through its depth there is only compressive stress/strain. In that case, there is no potential of forming fatigue cracks, as explained here. There can be situations between two extremes: (1) the asphalt mix layer is such that it shows a very high level of tensile stress/strain at the bottom, and (2) the asphalt mix layer and the pavement structure are such that there is a very small amount of tensile stress/strain at the bottom. Note that whether or not there will be a high tensile stress/strain is dictated by the thickness and stiffness of the asphalt mix layer as well as the stiffness of the layers underneath (and supporting) the asphalt mix layer.

Next comes the experimentation part to determine how many repetitions of the tensile stress/strain the pavement can sustain before there is damage. This needs the development of a model, based on statistical analysis of experimental data, of the following form:

$$N_f = f(\sigma \text{ or } \varepsilon)$$

where
  N_f is the number of repetitions to failure
  $f(\sigma \text{ or } \varepsilon)$ indicates "function" of initial stress or strain

From this relationship, one can determine an N_f for a specific σ or ε as determined with the layered elastic analysis. Before proceeding any further, note that N_f is a function of the tensile (σ or ε) for a specific material or, more precisely, for a material with a specific modulus (or stiffness). A simple analogy to illustrate this concept is that a stiffer paper clip can be bent fewer times before it snaps, compared to a more flexible paper clip, provided they are both "bent" (or strained) to the same extent. Indeed, a more complete way to express N_f would be to say,

$$N_f = f(\sigma \text{ or } \varepsilon \text{ and } E)$$

where E is the dynamic modulus.

Note that the same asphalt mix has different E at different temperatures. So a test run with the same mix, but at different temperatures, would yield the required data for constructing the N_f model. How does one arrive at this model?

Suppose one conducts an experiment with an asphalt mix beam over four supports and loaded at two points in the middle, commonly referred to as the *third-point loading test*. This loading configuration results in a uniform bending moment over the middle portion of the beam. The loading is applied repetitively, to simulate the repeated loading of traffic. Note that the intent of this testing is to cause cracking failure in the beam through initiation and propagation of the cracks by loading. The test can be conducted in two ways. Either a load is applied such that the stress in the beam remains constant from the beginning, or the load is varied such that the strain in the beam remains constant. If a constant stress is used, as the beam gets damaged, the strain will keep on increasing, while in the constant strain model, as the beam gets damaged, the load will need to be decreased to keep the strain at the same level.

Whether to use constant stress or constant strain depends on the thickness of the asphalt mix layer that is being considered for design. For thicker layers, strains would quickly increase with a decrease in the stiffness of the asphalt mix layer, and the constant stress mode is applicable; whereas for thinner layers, where strains would remain constant and are primarily dictated by the stiffness of the layers underneath the asphalt layer, constant strain is applicable. In a constant stress test in the laboratory, the strain will increase and the beam would fail relatively quickly after the initiation of the crack. In a controlled strain test, the time for crack propagation is long, and failure would take a relatively long time. To end the test, sometimes "failure" is defined as the point where the modulus of the mix has decreased to one-half of its original modulus. For asphalt pavement layers less than 6 in. in thickness, typically constant strain is generally used.

Center-point loading is used for tests with a rubber pad underneath the beam to simulate an elastic foundation, while third-point loading (most commonly, AASHTO T321, Method for Determining the Fatigue Life of Compacted Hot Mix Asphalt [HMA], Subjected to Repeated Flexural Bending) is generally used in tests without an elastic foundation. Note that other tests include the indirect tension test, where a compressive load is used on the cylindrical specimen to develop tensile stresses along the diametral axis; tests with either rectangular or trapezoidal section cantilever beams; and the direct tension test of rectangular beams.

Typically, 12–15 in. long and 2–3 in. wide and deep asphalt mix beam samples are used in this test, where the load is applied at the third points in the form of a haversine wave, with 0.1 s duration and then 0.4 s rest (Figure 12.1). Tests are run to span a range of stress such that the beams fail within a range of a small (1,000) to large (say, 1,000,000) number of repetitions. Each generates one data point, consisting of a strain and the number of repetitions to failure.

The common form of equation used to express N_f as a function of the initial stress or strain and modulus is

$$N_f = K_1(\varepsilon_t)K_2(E)K_3$$

where K_1, K_2, and K_3 are regression constants.

Beam

FIGURE 12.1   Fatigue tests of asphalt mix beam.

Maximum tensile stress and strain are calculated as follows:

$$\sigma_t = \frac{(0.357\ P)}{(bh^2)}$$

$$\varepsilon_t = \frac{(12\delta h)}{(3L^2 - 4a^2)}$$

where
σ_t is the maximum tensile stress, Pa
P is the load applied, N
b is the width of beam, m
h is the thickness of beam, m
ε_t is the maximum tensile strain, m/m
δ is the maximum deflection at center of beam, m
a is the space between inside clamps in the beam (0.119 m)
L is the length of beam between outside clamps (0.357 m)

12.1.1 MATERIAL CHARACTERIZATION TESTS

Before moving on to the various fatigue cracking models, it is appropriate to discuss the three important tests that are run to characterize HMA.

12.1.1.1 Indirect Tensile Strength: Test Method

In the indirect tensile test (ASTM D-4123, SHRP Protocol P07), a load is applied diametrically to a cylindrical sample of HMA until it fails. The load is applied at a deformation rate of 2 in./min (Figure 12.2).

$$\text{Indirect tensile strength (or ITS)} = \frac{2P}{\pi dt}$$

where
P is the load, lb-f
d is the diameter, in.
t is the thickness, in.

For the example shown in Figure 12.2, the peak load is 1880 lb for a 4 in. diameter sample with a thickness of 2 in.

$$\text{ITS} = \frac{2 * 1880}{\pi * 4 * 2} = 149.6\ \text{psi}$$

12.1.1.2 Resilient Modulus

In resilient modulus testing (ASTM D-4123, SHRP Protocol P07), a haversine loading is applied to a sample to determine the resulting deformation, and hence calculate resilient modulus, which is defined as the ratio of stress and resilient strain (as opposed to viscous strain) in an asphalt mix sample. Generally, for asphalt mix samples, the test is carried out in indirect tensile mode, as shown in Figure 12.3. The haversine load utilized in the protocol has a period of 0.1 s, followed by an appropriate rest period, generally 0.9 s. The tensile strength of each replicated set is determined prior to testing by performing an indirect tensile test on a companion specimen. The magnitude of the applied load causes tensile stress levels within the specimen equivalent to

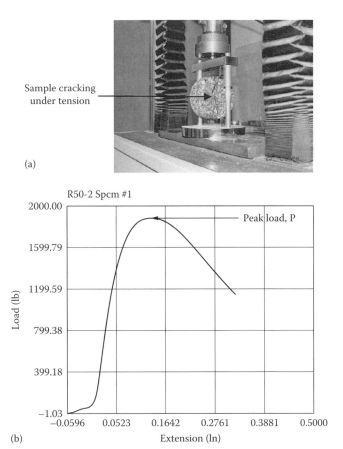

(a)

(b)

FIGURE 12.2   Indirect tensile strength test and results. (a) Indirect tensible strength test sample in equipment and (b) plot of results from test.

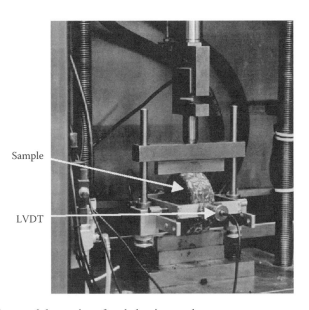

FIGURE 12.3   Resilient modulus testing of asphalt mix sample.

30%, 15%, and 5% of the tensile strength at 25°C, at 5°C, 25°C, and 40°C, respectively; and the seating load is 3%, 1.5%, and 0.5% (10% of the applied load) of the specimen tensile strength measured at 25°C, at each of the three test temperatures, respectively. The sample is placed under a diametrical repeated loading equipment with linear variable displacement transducers (LVDTs) placed on both sides of the horizontal diameter to measure horizontal deformations. The loading is applied on top through a horizontal metal bar, and the sample is rested at the bottom on a similar bar. The horizontal bar provides the distributed load along the edge of the sample. An example is shown in Figure 12.4.

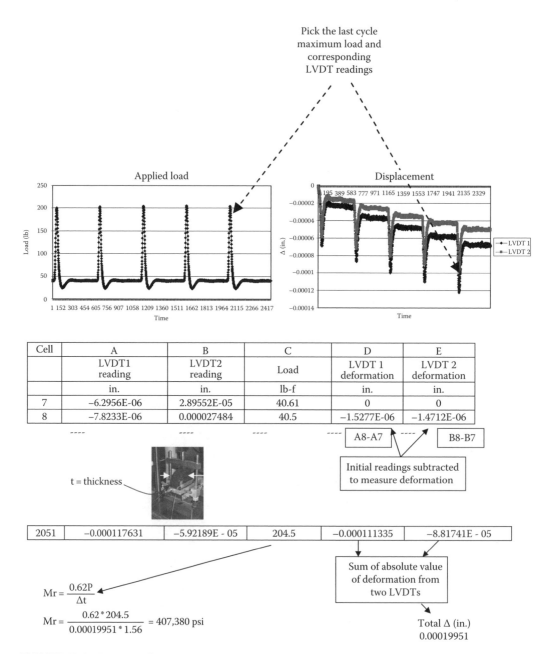

FIGURE 12.4 Example of resilient modulus calculation.

12.1.1.3 Dynamic Modulus (AASHTO TP62-03)

Dynamic modulus (E*) is defined as the absolute value of the complex modulus calculated by dividing the maximum (peak-to-peak) stress by the recoverable (peak-to-peak) axial strain for a material subjected to sinusoidal (repeated vertical) loading (Figure 12.5).

The dynamic modulus is the absolute value of the complex modulus

$$|E^*| = \sqrt{\left(\frac{\sigma_0 \cos \phi}{\varepsilon_0}\right)^2 + \left(\frac{\sigma_0 \sin \phi}{\varepsilon_0}\right)^2} = \frac{\sigma_0}{\varepsilon_0}$$

where σ_0 and ε_0 are the stress and strain amplitudes, respectively. An example is shown in Figure 12.6.

The dynamic modulus of asphalt mixes can also be determined from an empirical equation that relates it to other properties of the mix as well as temperature and loading times. The following is the widely used equation, which has been modified a number of times, and was originally developed by Fonseca and Witczak (1996).

FIGURE 12.5 Dynamic modulus testing setup and results. (a) Sample in equipment and (b) screenshot of deformation and load plots.

Project: 158913
Specimen diameter:
4 in.
height   6 in.

Area for stress = 12.56 in.^{2}

Remarks:
Your remarks here

A	B	C	D	E	F	G	H	I	J	K
LVDT1 reading, inch	LVDT2 reading, inch	Actuator inch	Load lb-f	Count	LVDT1 deformation, inch AX-A1	LVDT2 deformation, inch BX-B1	Average deformation, inch (F+G)/2	Stress psi (D/12.56)	Strain inch/inch (H/6)	E* psi (I/J)
7.09E−02	6.17E−02	−2.98E+00	0	1	0.00E+00	0.00E+00	0.00E+00	0	0.00E+00	#DIV/0!
7.13E−02	6.25E−02	−2.98E+00	1.82E+01	14028	3.86E−04	7.78E−04	5.82E−04	1.45E+00	9.70E−05	14948.5

Note: Determine the average stress and average strain for a number of readings and then calculate the dynamic modulus

FIGURE 12.6 Example calculation of dynamic modulus.

$$\log E^* = 3.750063 + 0.02932 * (p_{200}) - 0.001767 * (p_{200})^2 - 0.002841 * (p_4) - 0.058097 * (V_a)$$

$$-0.802208 \left(\frac{V_{beff}}{V_{beff} + V_a} \right)$$

$$+ \frac{3.871977 - 0.0021 * (p_4) + 0.003958 * (p_{38}) - 0.000017 * (p_{38})^2 + 0.005470 * (p_{34})}{1 + e^{\left(-0.603313 - 0.313351 \log(f) - 0.393532 \log^{(n)}\right)}}$$

where
  E^* is the dynamic modulus
  p_{200} is the percentage passing the No. 200 sieve
  p_4 is the percentage retained on the No. 4 sieve
  V_a is the percentage of air voids
  V_{beff} is the effective binder content by volume
  p_{38} is the percentage retained on 3/8 in. sieve
  p_{34} is the percentage retained on 3/4 in. sieve
  f is the loading frequency
  η is the binder viscosity

Generally a (log) T (in Rankine) versus log–log viscosity plot is generated from asphalt test data obtained at a few temperatures, and then A (intercept) and VTS (slope) of the plot are determined to develop a regression equation relating viscosity to temperature. This equation is then used to predict the viscosity at temperatures spanning the relevant range. An example is shown in Table 12.1.

Note that it is important to consider both temperature and loading time because of the fact that asphalt mixes are sensitive to those factors, as well as the fact that the temperature does vary over the months of a year (with changing seasons) at any specific depth and, over the depths, at any specific time. The loading time (which can be represented as the "width" of the loading pulse, in a time scale) varies with the depth of the pavement for any selected design speed of the vehicle. An example of variation of temperature with depth and months (for a pavement location in the state of Maine) is shown in Table 12.2, and the corresponding dynamic modulus values are shown in Table 12.3 (for a frequency of 10 Hz). Note that for practical purposes, all of the surface-layer

TABLE 12.1
Change in Viscosity with a Change in Temperature

T_c	T_r	log log η	η (Poise)	η (10^{-6} Poise)
14.21	517.25	0.97687	30285832.77	30.28583277
15.65	519.85	0.96938	20852148.05	20.85214805
19.86	527.42	0.94776	7357533.81	7.35753381
24.93	536.56	0.92211	2281143.336	2.281143336
30.42	546.43	0.89486	707609.29	0.70760929
34.18	553.19	0.87649	334804.3767	0.334804377
36.45	557.28	0.8655	217146.7636	0.217146764
35.74	556.01	0.86889	247867.73	0.24786773
31.82	548.96	0.88797	532454.8361	0.532454836
27.08	540.42	0.9114	1427172.935	1.427172935
22.67	532.47	0.93353	3808745.994	3.808745994
17.18	522.59	0.9615	14179717.73	14.17971773
		A	VTS	
		10.312	−3.44	

TABLE 12.2
Temperature in Different Months

	Temperature (°C)					
	Surface HMA Layer (h=75 mm)			HMA Base Layer (h=125 mm)		
Month	Top	1/3 h	2/3 h	Top	1/3 h	2/3 h
January	11.07	22.90	20.99	19.46	17.52	15.92
February	13.01	24.73	22.74	21.15	19.11	17.44
March	18.68	30.08	27.84	26.05	23.76	21.87
April	25.51	36.54	33.99	31.95	29.36	27.22
May	32.90	43.52	40.64	38.34	35.41	33.00
June	37.96	48.29	45.19	42.72	39.56	36.96
July	41.01	51.18	47.94	45.36	42.06	39.35
August	40.07	50.28	47.09	44.54	41.29	38.61
September	34.79	45.30	42.34	39.98	36.96	34.48
October	28.40	39.27	36.59	34.45	31.72	29.48
November	22.46	33.65	31.24	29.31	26.85	24.83
December	15.07	26.67	24.59	22.92	20.80	19.05

TABLE 12.3
Dynamic Modulus (in psi) for Different Months

	Surface HMA Layer (h=75 mm)			HMA Base Layer (h=125 mm)		
Month	Top	1/3 h	2/3 h	Top	1/3 h	2/3 h
January	9,166,279	4,452,691	5,029,908	5,537,231	6,249,742	6,892,441
February	8,200,535	3,958,450	4,501,614	4,982,925	5,664,571	6,284,646
March	5,821,436	2,794,771	3,236,174	3,636,507	4,217,280	4,758,755
April	3,763,643	1,837,941	2,167,159	2,473,542	2,930,351	3,368,561
May	2,323,931	1,182,623	1,414,822	1,636,158	1,974,936	2,309,144
June	1,678,406	887,065	1,068,462	1,243,758	1,516,244	1,789,631
July	1,384,429	750,503	906,627.6	1,058,590	1,296,773	1,537,953
August	1,465,851	788,345	951,551.9	1,110,067	1,357,902	1,608,160
September	2,056,430	1,060,497	1,272,352	1,475,392	1,788,054	2,098,531
October	3,117,721	1,543,560	1,831,574	2,102,279	2,510,245	2,906,070
November	4,582,654	2,214,560	2,591,529	2,938,459	3,449,581	3,933,799
December	7,253,065	3,486,796	3,992,318	4,444,117	5,089,618	5,682,087

(and all of the base-layer) HMA could be considered as one single layer, and dynamic modulus values are determined at a range of temperature that spans the highest and lowest temperatures that can be expected in that layer.

Temperatures in pavements can be determined from available data, thermocouples, and/or equations, such as those shown in Table 12.4.

12.1.2 MODELS

Fatigue cracking models relating the number of load repetitions to failure have been developed by different agencies, as shown in Table 12.5. Note that the models use different parameters, and each one is applicable for the specific definition of failure, based on which model has been developed.

TABLE 12.4

Temperature Equations for Asphalt Pavements

Temperature at the surface	$T_{surf(max)} = T_{air(max)} - 0.00618 \, lat^2 + 0.2289 \, lat + 24.4$, where T is expressed in °C and the latitude is in degrees	Solaimanian and Kennedy (1993), Huber (1994), Solaimanian and Bolzan (1993)
Temperature at different depths	$T(d) = T(surf) \, (1 - 0.063 \, d + 0.007 \, d^2 - 0.0004 \, d^3)$, where T(d) and T(surf) are in °F and the depth, d, is in inches	

TABLE 12.5

Fatigue Cracking Models

The Asphalt Institute (AI; 1991) fatigue cracking model is as follows:

$N_f = 0.00432 \, C \, (\varepsilon_t)^{-3.291} \, (E)^{-0.854}$

$C = 10^M$

$$M = 4.84 \left(\frac{V_b}{V_a + V_b} - 0.69 \right)$$

where

N_f is the number of load repetitions to failure, which is defined as fatigue cracking over 20% of the entire pavement area

C is the correction factor

ε_t is the tensile strain at the bottom of the asphalt mix layer

V_a is the air void (%) in the asphalt mix

V_b is the asphalt content (%) by volume in the asphalt mix

E is the dynamic modulus of the asphalt mix, psi

The shift factor, needed to transform the laboratory fatigue data to field data, for the AI equation is 18.4

For a standard asphalt mix with 5% air voids and 11% asphalt (by volume), the AI equation reduces to the following:

$N_f = 0.0796(\varepsilon_t)^{-3.291}(E)^{-0.854}$

The Shell equation (Shell, 1978) is as follows:

$N_f = 0.0685(\varepsilon_t)^{-5.671}(E)^{-2.363}$

where

N_f is the number of load repetitions at constant strain to failure

ε_t is the tensile strain at the bottom of the asphalt mix layer

E is the dynamic modulus of the asphalt mix, psi

Shell equations for allowable tensile strains are as follows:

For constant stress

$\varepsilon_t = (4.102 \, PI - 0.205 \, PI \, V_b + 1.094 V_b - 2.7807) * E_{mix}^{-0.36} * N^{-0.2}$

For constant strain $\varepsilon_t = (0.300 \, PI - 0.015 \, PI \, V_b + 0.080 \, V_b - 0.198) * E_{mix}^{-0.28} * N^{-0.2}$

where

N is the number of equivalent single-axle loads

ε_t is the allowable permissible tensile strain, mm/mm

PI is the penetration index

V_b is the bitumen content,% by volume of the mix

E_{mix} is the stiffness modulus of the asphalt mix, N/m^2

Modified Shell equation

$$N_f = A_f \, F'' K_{1\sigma} \left(\frac{1}{\varepsilon_t} \right)^5 E^{-1.4}$$

TABLE 12.5 (continued)
Fatigue Cracking Models

$$F'' = 1 + \frac{F}{1 + \exp^{(1.354h_{ac} - 5.408)}}$$

where

N_f is the number of load repetitions to cause fatigue cracks

ε_t is the tensile strain at critical location

E is the dynamic modulus of asphalt mix, psi

$K_{1\sigma}$ is the laboratory calibration parameter

A_f is the laboratory to field adjustment factor

h_{ac} is the thickness of the HMA layer, inches

Modified Shell equation considering total strain

$$Nji = FK_\alpha{}^5 \left(\frac{1}{\varepsilon_t}\right)^5 \left(\frac{1}{E}\right)^{1.4} Nj$$

where

$$F = \frac{1 + (13908E^{-0.4}) - 1}{1 + e^{1.354h_{ac} - 5.408}}$$

$K_\alpha = [0.0252\,PI + 0.00126\,PI\,(V_{be}) - 0.0167]$

ε_t is the tensile strain

E is the dynamic modulus of asphalt mix, psi

h_{ac} is the thickness of asphalt mix layer, in.

PI is the penetration index of binder

V_{be} is the effective binder content, % by volume

For the surface sublayer (top and bottom)

$\varepsilon_t = \varepsilon_{tl} + \varepsilon_{tt}$

where

ε_t is the total tensile strain

ε_{tl} is the tensile strain due to the load

ε_{tt} is the tensile strain due to temperature drop

The FHWA cost allocation (Rauhut et al., 1984) study model is as follows (note that this model uses the resilient modulus instead of the dynamic modulus):

$N_f = K_1(\varepsilon_t)^{K2}$

where

$$K_1 = K_{1R} \left[\frac{E_r}{E_{Rr}}\right]^{-4}$$

$K_2 = 1.75 - 0.252\,[\text{Log}\,(K_1)]$

K_{1R} are the coefficients determined from fatigue tests at a reference temperature of 70°F, $7.87 * 10^{-7}$

E_r is the resilient modulus (total) from indirect tensile test, at a specific test temperature, psi

E_{Rr} is the reference resilient modulus (total) for a test temperature of 70°F, 500,000 psi

The Transportation and Road Research Laboratory (Powell et al., 1984) model

$N_f = 1.66 * 10^{-10}(\varepsilon_t)^{-4.32}$

Maupin and Freeman (1976)

$N_f = K_1\,(\varepsilon_t)^{-n}$

where

$n = 0.0374(\sigma_t) - 0.744$

$\text{Log}\,(K_1) = 7.92 - 1.122\,(\sigma_f)$

σ_f is the indirect tensile strength at 70°F, psi

(continued)

TABLE 12.5 (continued)
Fatigue Cracking Models

Illinois procedure (Thompson, 1987)

$N_f = 5.0 * 10^{-6}(\varepsilon_t)^{-3.0}$

where N_f is the number of strain repetitions to failure (initiation of fatigue cracks) in terms of 18-kip ESALs

$N_{18} = 5.6 * 10^{11} (\delta_{18})^{-4.6}$

where

  N_{18} is the number of 80 kN (18-kip) single-axle loads to fatigue failure

  δ_{18} is the surface deflection for a moving 80 kN (18-kip) single-axle load (mils)

Belgian Road Research Center (Verstraeten et al., 1977)

$N_f = 4.92 * 10^{-14} (\varepsilon_t)^{-4.76}$

Probabilistic distress models for asphalt pavements (PDMAP, 1986; Monismith et al., 1972; Finn et al., 1973) are as follows:

$$\text{Log} N_f(\text{Laboratory}) = 14.82 - 3.291 \text{Log}\left(\frac{\varepsilon_t}{10^{-6}}\right) - 0.854 \text{Log}\left(\frac{E}{10^3}\right)$$

$$\text{Log} N_f(\text{fatigue cracks over 10\% of wp}) = 15.947 - 3.291 \text{Log}\left(\frac{\varepsilon_t}{10^{-6}}\right) - 0.854 \text{Log}\left(\frac{E}{10^3}\right)$$

$$\text{Log} N_f(\text{fatigue cracks over 45\% of wp}) = 16.29 - 3.291 \text{Log}\left(\frac{\varepsilon_t}{10^{-6}}\right) - 0.854 \text{Log}\left(\frac{E}{10^3}\right)$$

where

  N_f is the load applications of constant stress to cause fatigue failure

  ε is the initial horizontal tensile strain on the bottom of the HMA layer, in./in.

  E is the HMA modulus, psi

MICH-PAVE program (Baladi, 1989):

$$\text{Log(ESAL)} = -2.544 + 0.154 h_{AC} + 0.069 h_{EQ} - 2.799 \text{Log}\delta_0 - 0.261 V_a + 0.917 \text{Log} E_{base}$$

$$= 0.0000269 M_R - 1.0694 \text{Log}\varepsilon_t + 1.173 \text{Log}\varepsilon_v - 0.001 \eta_K + 0.064 \text{ANG}$$

$$h_{EQ} = h_{base} + h_{subbsase}\left(\frac{E_{subbase}}{E_{base}}\right)$$

where

  ESAL is the number of 80 kN (18 kip) equivalent single-axle loads to failure

  h_{AC} is the thickness of the HMA layer, in.

  h_{EQ} is the equivalent thickness of base materials, in.

  h_{base} is the actual thickness of base materials, in.

  $h_{subbase}$ is the actual thickness of subbase materials, in.

  E_{base} is the resilient modulus of the base materials, psi

  $E_{subbase}$ is the resilient modulus of the subbase, psi

  δ_0 is the surface deflection, in.

  V_a is the air voids (%) in the mix

  M_R is the resilient modulus of subgrade, psi

  ε_t is the tensile strain at the bottom of the HMA layer, in./in.

  ε_v is the compressive strain at the bottom of the HMA layer, in./in.

  η_K is the kinematic viscosity at 135°C (275°F), Cst

  ANG is the aggregate angularity (4 for 100% crushed material, 2 for 100% rounded river-deposited material, and 3 for a 50% mix of crushed and rounded aggregate)

Dense graded (Monismith et al., 1967) base courses

Controlled stress: $N_f = 2.0 * 10^{-7}(\varepsilon_t)^{-3.38}$

Controlled strain: $N_f = 7.5 * 10^{-6}(\varepsilon_t)^{-2.79}$

Asphalt-treated base (Kallas and Puzinauskas, 1972; test temp = 70°F)

$N_f = 2.520 * 10^{-9}(\varepsilon_t)^{-3.58}$

Relationships between log failure strain and log resilient modulus have been developed, and a longer fatigue life is indicated by a relatively larger tensile strain at failure for a specific modulus value.

The models shown in Table 12.5 are based on empirical test methods and parameters. The second approach for characterizing fatigue behavior of asphalt mix is with the use of the concept of dissipated energy under dynamic loading, using either a flexural center-point or a third-point fatigue test. The following equation can be used (Tangella et al., 1990):

$$N_f = \left(\frac{W}{A}\right)^{-z}$$

where

W is the total dissipated energy $= W_{total} = \sum_{i=1}^{n} W_i$

A, Z are the mixture characteristic constants

Dissipated energy, ith load cycle

$$w_i = \pi * \sigma_0 * \varepsilon_0 * \sin \phi_0$$

where

σ_0 is the amplitude of initial stress

ε_0 is the amplitude of initial strain

ϕ_0 is the phase angle, indicating lag between stress and strain in viscoelastic materials

Another method is the use of concepts of *fracture mechanics*, in which stress and fracture toughness are utilized. The steps are based on the original work by Paris et al. (1961), and are as follows:

1. Rate of growth of crack is proportional to stress intensity factor:

$$\frac{dc}{dN} = A(K)^n$$

   where

   c is the length of crack

   N is the number of repetitions of load

   A, n are the constants related to fracture behavior of the material

   K is the stress intensity factor $= \sigma\sqrt{\pi a}$, where a is the stress and a is the radius of the cylindrical specimen

2. Lytton et al.'s (1993) modified model for crack propagation (note that the crack initiation can be determined from laboratory beam fatigue tests, as described earlier):

$$N_{fp} = \left(\frac{1}{A}\right)\int_{c_0}^{h_{ac}} \frac{dc}{K^n}$$

   where

   N_{fp} is the number of load applications to propagate a crack of length 8 mm (0.3 in.) to the surface

   h_{ac} is the asphalt mix layer thickness

   c_0 is the initial crack length

   K is the stress intensity factor

   A is the asphalt mix properties

Majidzadeh et al.'s (1970) equation for determination of A

$$A*10^9 = 0.23213 + 2613\left(\frac{\sigma_t}{E*}\right) - 3.2334(K_{IC})$$

where
  σ_t is the indirect tensile strength of the mix
  $E*$ is the complex modulus of the mix
  K_{IC} is the fracture toughness

Molenaar's (1983) equation for determination of A:

$$|LogA| = 0.977 + 1.628(n)$$

$$LogA = 4.389 - 2.52(LogE)*(\sigma_t)(n)$$

where
  σ_t is the indirect tensile strength of the asphalt mix at a given loading condition, MPa
  E is the modulus of mix at a given loading condition, MPa

Schapery's (1986) equation for determination of n:

$$n = 2\left(\frac{1}{m} + 1\right)$$

$$n = \frac{2}{m}$$

$$n = 0.8\left(\frac{1}{m} + 1\right)$$

where m is the slope of the log creep compliance curve.

12.1.3 DEFINITION OF FAILURE

Note that the number of repetitions at which the pavement will fail in the field is significantly different from the number of repetitions that is indicated in the laboratory, because of discrepancies between field and laboratory loading conditions. These discrepancies include lateral wandering of the traffic in the field versus loading over the same spot in the laboratory and rest periods in the field being much longer than those in the laboratory. Also, while crack initiation and a resulting decrease in modulus, for example, are taken as failures in the laboratory, it is the visual manifestation of cracks that is considered as a sign of failure in the field. To be visible on the surface, the cracks after initiation at the bottom need to propagate to the surface, and the process takes time. A fundamental discrepancy could also be that while according to elastic theory the tensile stresses/strains are always maximum at the bottom of the asphalt mix layer, due to rapid cooling of the surface and formation of a rigid layer over a relatively soft layer, the cracks could initiate somewhere in the middle of the layer in the field.

Whatever the discrepancies may be, actually observed N_f could be 10–100 times that of laboratory-obtained N_f. Hence, the laboratory-obtained N_f values are multiplied by an appropriate "shift factor" derived from field experience. One question at this point is as follows: to determine N_f in the field (which is required for predicting the shift factor for a specific location and class of asphalt mixes used by an agency), how would one determine whether a pavement has failed or not—in other words, what is the criterion for failure?

Fatigue failure has been defined earlier. Structurally, a pavement is said to have failed by fatigue when one or more of the following conditions are noted within the scope of the project. Note that the more conservative criteria are used for interstate and freeway pavements, whereas the less conservative ones are used for primary or secondary routes.

1. Over 45% of the wheelpath shows fatigue cracking and crack deterioration around the edges.
2. Over 20% of the wheelpath area has severely deteriorated cracks, with pieces moving under traffic.
3. Around 20% of the entire pavement area has fatigue cracks.
4. Around 37% of the wheelpath area has fatigue cracks.

12.1.4 USE OF MODELS

The next step is to utilize the fatigue equations in design processes. Note that with changing seasons, critical properties such as moduli of layers change, causing a change in the tensile stresses/strains in the asphalt mix layer, and hence causing a change in the way damage occurs. Furthermore, if the traffic is different in different seasons, then the relative proportion of damage throughout a year differs also. To consider all these factors, the concept of cumulative damage (Miner, 1945) is used. It is assumed that accumulation of damage causes ultimate "failure" of a pavement, damage by fatigue failure is cumulative, and a cumulative damage factor or index is defined such that when it reaches a value of 1, the pavement is assumed to "fail."

The damage factor for the ith loading is defined as the number of actual repetitions (n_i) of load divided by the "allowable" repetitions (N_i) that would cause failure. The total damage index, or the cumulative damage factor (CDF), for the parameter is given by summing the damage factors over all of the different loadings in the different seasons.

Mathematically, this can be expressed as follows:

$$\text{Total damage index (or DI)} = \sum_{t=1}^{m} \Delta DI_{ij}$$

$$\Delta DI_{ij} = \left(\frac{n_i}{N_f} \right)_{ij}$$

where
n_i is the number of actual load repetitions in a specific season (specific season being defined in terms of temperature and moisture conditions), i, for load of j, for a specific year
$N_{f(ij)}$ is the allowable number of load repetitions for a specific season (i) and load (j) determined from fatigue equations
ΔDI_{ij} is the Damage Index for a specific year

The concept of endurance limit has been proposed, which assumes that under a very small repetitive strain, an asphalt mix layer can sustain infinite or a very large number of loads (such as >10 million). The concept of long life or perpetual pavement is based on this theory. Suggested values of such low strains include 100 and 65 microstrains. Details of a study conducted by NCHRP can be found at: http://onlinepubs.trb.org/onlinepubs/nchrp/nchrp_rpt_646.pdf.

12.1.5 Relationship between Mix Design and Fatigue Performance

The fatigue characteristics of an asphalt mix are dictated by its materials—types and relative proportions, as well as mix volumetric properties—and are reflected in its mechanical properties. The different materials and properties are discussed next:

Aggregate: The effect of percentage of material passing the No. 200 sieve is significant—this material increases the fatigue resistance up to an optimum content and then reduces it beyond that content. The more angular the fine aggregate is, the more resistant it is to fatigue failure.

Asphalt: Both asphalt properties and amount (asphalt content in the mix) are significant. As discussed in Chapter 9 ("Asphalt and Emulsions"), the stiffness and the phase angle are controlled [(G*) * (Sinδ)] to maintain sufficient fatigue resistance, through limiting the stiffness (G*) and the phase angle (δ). A higher asphalt content provides greater fatigue resistance.

Air voids/density: A high density (low air voids) is better for fatigue cracking resistance.

Note that some of the properties also affect the aging of the binder and hence affect its fatigue properties indirectly, since a higher rate of aging increases stiffness and reduces the resistance against fatigue failure. For example, a higher asphalt content and/or a higher density reduces the potential of aging of the binder and hence reduces the potential of deterioration of fatigue resistance.

12.1.6 Relationship between Pavement Structure and Fatigue Performance

The fatigue resistance of the asphalt mix layer is also affected by the structure of the asphalt pavement—the thickness of the asphalt mix and base course layer. Depending on the magnitude of the load and tire pressures, a very thin asphalt mix layer (e.g., less than 2 in.) will not generate tensile stresses/strains and will not be susceptible to bottom-up fatigue failure; a moderately thin asphalt layer will be susceptible to fatigue failure; whereas a thick asphalt pavement (e.g., >8 in.) will definitely be very resistant to fatigue failure, because of the low tensile stress/strain. Both thickness and modulus of the base course are important, since they contribute to the overall stiffness of the pavement structure—a high modulus and a thicker layer improve fatigue resistance.

Finally, the wheel load is an important factor—a higher load naturally increases the potential of fatigue cracking.

12.1.6.1 Steps for Avoiding Premature Fatigue Cracking

To consider all of the important factors, the following steps can be taken, in general, to provide adequate resistance to fatigue cracking:

1. Adopt a standard maximum load.
2. Select appropriate materials for different layers so that they have desirable fatigue resistance properties.
3. Design the mixes (for the base as well as asphalt mix layers, for example) such that the volumetric properties are at the desirable level for adequate fatigue resistance.
4. Considering the mechanical properties of the designed mixes, design the pavement structure such that the overall stiffness of the structure is adequate.
5. Construct the pavement such that the controllable properties are brought up/down to desirable levels, as determined in Step 3.

12.2 THERMAL CRACKING

Thermal cracking occurs in two ways: (1) when the thermal stresses due to a drop in temperature exceed the fracture strength of the material (low-temperature cracking), and (2) when, due to repeated thermal cycles, the strain in the asphalt layers causes thermal fatigue cracking. The potential

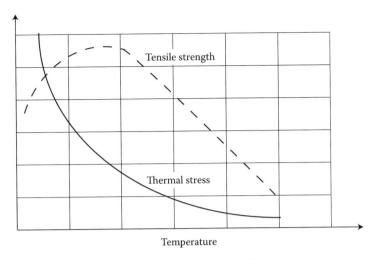

FIGURE 12.7   Concept of change in tensile strength and stress with temperature.

of thermal cracking of a mix is evaluated by evaluating the mix stiffness and fracture strength characteristics with respect to temperature and time of loading.

The tensile strengths of asphalt mixes change with a change in temperature. As the temperature in the pavement drops (from the surface, which is in contact with the cold air, that causes the drop in temperature), the tensile strength keeps increasing, but beyond a point, due to micro-cracks at the aggregate–asphalt interface, it does not increase anymore and actually begins to decrease (Figure 12.7). On the other hand, as the temperature drops the asphalt mix layer tends to shrink, but is restrained by friction with the layers underneath. If the stress builds up relatively slowly, then it can be dissipated through stress relaxation in the viscoelastic asphalt mix at temperatures above 70°F. However, at lower temperatures, depending on the rate at which temperature drops and the properties of the asphalt mix components and the mix, the built-up stress exceeds the tensile strength (at that specific temperature), causing a crack on the surface.

Similar to load-related fatigue cracking, repetitive alternate cycles of high stress (due to a drop in temperature at night) and low stress (due to warmer temperatures during the day) can cause cracking (thermal fatigue cracking). The number of stress cycles required to cause failure depends on the magnitude of the repeated high stress and the thermal fatigue–related properties of the mix, which include the effects of viscoelastic properties as well as aging.

12.2.1 MATERIAL CHARACTERIZATION

For low-temperature cracking, the drop in temperature and the rate at which it drops are both important. The effect can be quantified with the creep modulus parameter—a higher value indicates a lower relaxation of stress, and hence a higher potential of buildup of thermal stress and hence cracking. The creep modulus is affected primarily by the low-temperature properties of the asphalt binder, and appropriate tests have been discussed in Chapter 9. The tests include the bending beam rheometer test (to evaluate stiffness and rate of relaxation) and direct tension test (to evaluate strain at failure).

> *Asphalt mix properties*: The low-temperature indirect tensile creep test is used to determine the two properties that are used, the indirect tensile strength and the creep modulus. The test is run by applying a load at the rate of 1.25 mm/min, at a temperature ranging from 0°C to −20°C, for 100–3600 s.

FIGURE 12.8   Measurement of deformation with extensometers on the asphalt mix sample.

Indirect tensile strength/creep compliance test: This test (AASHTO T-315, T-322) is conducted to determine the tensile strength (as discussed earlier) as well as the indirect tensile creep compliance, which is defined as the ratio of time-dependent strain to the applied stress. The sample setup is shown in Figure 12.8. In the test for creep compliance, a stress is applied quickly and then maintained constantly throughout a period of time. The applied load and the resulting vertical and horizontal deformations are measured (as a function of time), and the creep compliance D(t) is calculated as follows:

$$D(t) = \frac{\Delta X_t * d * b}{P * GL} * C_{cmpl}$$

where
  ΔX_t is the mean horizontal deflection, mm
  d is the diameter of the sample, mm
  b is the thickness of the sample, mm
  P is the creep load, kN
  GL is the gage length over which deformation is measured, mm
  C_{cmpl} is the compliance factor

$$C_{cmpl} = 0.6354 \left(\frac{X}{Y} \right)^{-1} - 0.332$$

where X/Y is the ratio of horizontal to vertical deflection.
  The limits of C_{cmpl} are as follows:

$$\left[0.704 - 0.213 \left(\frac{b}{d} \right) \right] \leq C_{cmpl} \leq \left[1.566 - 0.195 \left(\frac{b}{d} \right) \right]$$

$$0.20 \leq \frac{b}{d} \leq 0.65$$

The compliance, D(t), is used to calculate the relaxation modulus, E, which is used for predicting thermal stress in the asphalt mix. Paris' law is used then to determine the development of cracks.

The coefficient of *thermal contraction* of asphalt mixes can be estimated from the following equation (Jones et al., 1968):

$$B_{mix} = \frac{VMA * B_{AC} + V_{AGG} * B_{AGG}}{3 * V_{TOTAL}}$$

where
B_{mix} is the linear coefficient of thermal contraction of the asphalt mix (1/°C)
B_{AC} is the volumetric coefficient of thermal contraction of the asphalt binder in the solid state (1/°C)
B_{AGG} is the volumetric coefficient of thermal contraction of the aggregate (1/°C)
VMA are the voids in mineral aggregate (% by volume)
V_{AGG} is the volume of aggregate in the mix (%)
$V_{TOTAL} = 100\%$

Typical values of coefficient of thermal contraction of asphalt mixes have been reported as $1.17–2.05 * 10^{-5}/°F$ ($2.11–3.69 * 10^{-5}/°C$).

12.2.2 MODELS

The thermal cracking prediction models are based on the approach of predicting a critical temperature below which an asphalt mix layer is expected to crack, and most methodologies combine the effect of low temperature and thermal fatigue cracking. The use of these methods requires the determination of environmental conditions (for the project site) and asphalt mix properties, some of which are determined, while the others are commonly taken from available values in the literature.

12.2.2.1 Environmental Conditions

The pavement temperature, in response to air temperature, could be determined with the help of the enhanced integrated climatic model (EICM; developed by Larson and Dempsey, 1997), based on the integrated climatic model (ICM; developed by Lytton et al., 1990). The model is based on the balance of heat transfer between the pavement and the environment, considering the three different modes of heat transfer—conduction, convention, and radiation. The equation relating these factors is as follows:

$$T_{(i,t+\Delta t)} = T_{(i,t)} + \left(\frac{K\Delta t}{\gamma_d C \Delta z^2} \right) \left[T_{(i+1,t)} + T_{(i-1,t)} - 2T_{(i,t)} \right]$$

where
T is the temperature, °F
z is the axis along the depth of the pavement
K is the thermal conductivity, BTU/(h-ft²-°F)
C is the mass specific heat in BTU/(h-°F)
γ_d is the dry density

Using the predicted conditions and the asphalt mix properties, either regression equations or fracture mechanics–based equations could be used to predict the extent of low-temperature-related cracking in an asphalt mix layer.

12.2.2.2 Regression Equation Approach (Hajek and Haas, 1972)

$$I = 30.3974 + (6.7966 - 0.8741h + 1.338a)\log(0.1S_{bit}) - 2.15165d - 1.2496m + 0.06026S_{bit}\log d$$

where

I is the Cracking Index (≥ 0); number of full cracks plus one-half of the half-transverse cracks per 500 ft section of two-lane roads (does not consider cracks shorter than half-lane widths)

S_{bit} is the stiffness modulus, kg/cm^2, of the original asphalt; determined from Van der Poel (1954) nomograph with a loading time of 20,000 s and the winter design temperature. Penetration index (PI) and ring and ball softening point determined from penetration at 25°C and kinematic viscosity at 275°F (135°C), according to McLeod's method (1970)

h is the total thickness of asphalt mix layer, in.

a is the age of pavement, years

m is the winter design temperature, without considering the "negative" sign, °C

d is the code for subgrade type: sand, 5; loam, 3; and clay, 2

Haas et al.'s (1987) model (based on cracking in airfields):

$$\text{TCRACK} = 218 + 1.28 * \text{ACTH } 2.54 \text{ MTEMP} + 30 * \text{PVN} - 60 * \text{COFX}$$

where

TCRACK is the transverse crack average spacing, m

MTEMP is the minimum temperature recorded on site, °C

PVN is the dimensionless pen-vis number

COFX is the coefficient of thermal contraction, mm/100 mm/°C

ACTH is the thickness of asphalt mix layer, cm

12.2.2.3 Fracture Mechanics Approach: SHRP Thermal Cracking Model

There are three components of the thermal cracking distress model, which work on the basis of each other. The stress intensity model determines the stress at the tip of a local vertical crack, which is used to predict the depth of crack propagation, and based on that, the crack amount model predicts the number of thermal cracks per unit length of the pavement.

The time- and temperature-dependent relaxation modulus of asphalt mix, which reflects its viscoelastic property, dictates the thermal stress developed in an asphalt mix layer during temperature drop. Mathematically, this is expressed as follows:

$$\sigma(\xi) = \int_0^\xi E(\xi - \xi') \frac{d\varepsilon}{d\xi} d\xi'$$

where

$\sigma(\xi)$ is the stress at reduced time ξ

$E(\xi - \xi')$ is the relaxation modulus at reduced time $(\xi - \xi')$

ε is the strain at reduced time $(\xi - \xi') = \alpha(T(\xi') - T_0))$
α is the linear coefficient of thermal contraction
$T(\xi')$ is the pavement temperature at reduced time ξ'
T_0 is the pavement temperature when $\sigma = 0$
ξ' is the variable of integration

The relaxation modulus of asphalt mix can be represented by the Maxwell model for viscoelastic materials:

$$E(\xi) = \sum_{i-1}^{N+1} E_i e^{-\xi/\lambda_i}$$

where
    ξ is the reduced time $= t/a$
    $E(\xi)$ is the relaxation modulus at reduced time
    T is the real time
    a_T is the temperature shift factor
    E_i, λ_i are the Prony series parameters

The creep test is performed at multiple temperatures to determine the creep compliance D(t), from which the relaxation modulus function is obtained.

    The data are shifted from different temperatures to one continuous "master" creep compliance curve using a shift factor.

$$D(\xi) = D(0) = \sum_{i-1}^{n} D_i \left(1 - e^{-\xi/\tau_i}\right) + \frac{\xi}{\eta_v}$$

where
    $D(\xi)$ is the creep compliance at reduced time ξ
    ξ is the reduced time $= t/a_T$
    a_T is the temperature shift factor
    $D_{(0)}$, D_i, τ_i, η_v are the Prony series parameters

A series of steps are needed to develop the master curve for creep compliance, using regression.

1. Creep tests are conducted for different times of loading and temperature, and the results are plotted as shown, with log D(t) and log(time), for the different temperatures (T).
2. A reference temperature is selected, and all of the curves at the other temperatures are shifted onto this curve to result in one smooth master curve. The determination of the shift factors (as a function of the temperature) as well as of Prony series parameters is done with regression, using a least-square criterion. Four Prony series coefficients are determined for data obtained at three temperatures. To complete the master curve, shift factors for temperatures at which tests were run are determined on the basis of the regression relationship between the shift factor and temperature.
3. An equation for the compliance curve is then developed to determine m, the slope of the linear portion of the log time versus log compliance curve:

$$D(\xi) = D_0 + jD_1\xi^m$$

where

  D_0, D_1, and m are the coefficients of the equation
  m is the primary parameter that characterizes the low-temperature cracking potential of an asphalt mix

The next step is the determination of the relaxation modulus, which enables one to determine the stress for causing thermal cracks. While the relaxation modulus can be approximated as just the inverse of the creep compliance, the more precise relationship is the following:

$$sL[D(t)] * sL[E(t)] = 1$$

where

  L[D(t)] is the Laplace transform of creep compliance, D(t)
  L[E(t)] is the Laplace transform of relaxation modulus, D(t)
  s is the Laplace parameter
  t is the time or reduced time, ξ

Using a computer program, the master relaxation modulus can be determined from the master creep compliance equation.

  Boltzmann's superposition principle (1874) for linear viscoelastic materials is used to determine thermal stresses (as mentioned earlier):

$$\sigma(\xi) = \int_0^{\xi} E(\xi - \xi') \frac{de}{d\xi} d\xi'$$

where

  $\sigma(\xi)$ is the stress at reduced time ξ
  $E(\xi - \xi')$ is the relaxation modulus at reduced time, $\xi - \xi'$
  ε is the strain at reduced time ξ $(= \alpha(T(\xi') - T_0))$
  $T(\xi')$ is the pavement temperature at reduced time ξ'
  T_0 is the pavement temperature when $\sigma = 0$
  ξ' is the variable of integration

The assumptions are that the HMA material is a thermorheologically simple material, allowing the use of the time-temperature superposition principle, and that the layer behaves as a uniaxial rod.

  The form of the equation with real time, t, is as follows:

$$\sigma(t) = \int_0^{\xi} E(\xi(t) - \xi'(t)) \frac{de}{dt'} dt'$$

The use of Prony series representation of $E(\xi)$ results in the following finite difference solution:

$$\sigma(t) = \sum_{i=1}^{N+1} \sigma_i(t)$$

where

$$\sigma_i(t) = e^{-\Delta/\lambda_i}\sigma_i(t-\Delta t) + \Delta E_i(\varepsilon)\frac{\lambda_i}{\Delta\xi}\left(1-e^{-\Delta\xi/\lambda_i}\right)$$

where

$\Delta\varepsilon$ is the change in strain over time $t - \Delta t$ to t

$\Delta\xi$ is the change in reduced time $t - \Delta t$ to t

12.2.2.4 Models for Cracking

Stress intensity model 2 d FE model (Chang et al., 1976):

$$K = \sigma\left(0.45 + 1.99C_0^{0.56}\right)$$

where

K is the stress intensity factor

σ is the far-field stress from the pavement response model at depth of the crack tip

C_0 is the current crack length

The change in depth of a local crack subjected to a cooling cycle is determined by the Paris law:

$$\Delta C = A(\Delta K)^n$$

where

ΔC is the change in crack depth due to a cooling cycle

ΔK is the change in stress intensity factor due to a cooling cycle

A, n are the fracture parameters for the specific asphalt mix

The stress intensity factor can be used, in Paris' equation, to determine the change in crack length due to a drop in temperature:

$$\Delta C = A(\Delta K)^n$$

A and n can be determined from fracture tests and the following equations, where m is the slope of the linear portion of the log compliance–log time curve and is determined from the creep test:

$$n = 0.8\left[1 + \frac{1}{m}\right]$$

Then, using n and σ_m (which is determined from the indirect tensile strength test), A is determined using the following formula:

$$A = 10^{(\beta*(4.389 - 2.52*\log(E*\sigma_m*n))}$$

where

E is the modulus of the asphalt mix

σ_m is the tensile strength of the asphalt mix, psi

β is the parameter from calibration

The total crack depth is estimated by computing and summing up the change in crack depth (ΔC) with time. The HMA layer is subdivided into four layers. A relationship between the probability distribution of the log of crack depth to layer thickness ratio and the percentage of cracking is used to predict the degree of cracking.

$$C_f = \beta_1 * N\left(\frac{\log C/h_{ac}}{\sigma} \right)$$

where

C_f is the observed amount of thermal cracking

β_1 is the coefficient of regression equation for field calibration

$N(z)$ is the standard normal distribution evaluated at (z)

σ is the standard deviation of the log of the depth of cracks in the pavement = 0.769

C is the crack depth

h_{ac} is the thickness of asphalt mix layer

12.2.3 CRACKING AND PROPERTIES OF ASPHALTS AND AGGREGATES

The resistance of an asphalt mix layer to thermal cracking can be controlled by a variety of parameters, which affect tensile strength and relaxation properties. Apart from the asphalt grade, which has the primary influence, parameters such as asphalt content, air voids, and aggregate type and gradation have significant influence.

The effect of asphalt type has been discussed in Chapter 9. A high asphalt content results in a low creep modulus and hence lower potential of thermal cracking. Note that the effective asphalt content can be lowered by an absorptive aggregate (if not considered properly during mix design) and hence increase the potential of cracking directly (or indirectly, by allowing the asphalt to age faster). The aging can also be accelerated by the presence of relatively high air voids, which can also decrease the tensile strength, and hence increase the potential of cracking. The increase in tensile strength can be achieved by using appropriate gradation as well as aggregate with crushed faces.

12.3 RUTTING OR PERMANENT DEFORMATION

Rutting is the result of repeated loading, which causes accumulation, and increase of permanent deformations. The one-dimensional densification–consolidation rutting, resulting from a decrease in air voids, occurs with volume change and is vertical deformation only (primary rutting), whereas the two-dimensional rutting is caused by shear failure and is accompanied by both vertical and lateral movement of the material (secondary and tertiary rutting).

The densification–consolidation rutting may be caused by the action of high stress load at high temperatures near the surface of the pavement, leading to a decrease in air voids of the asphalt mix, especially if they are too high compared to desirable air voids; and/or due to the densification of the underlying layers if they are at densities lower than the maximum dry density or desirable densities; or due to the consolidation of fine-grained soils in the subgrade, for example, with high levels of moisture. The one-dimensional rutting mechanism can be relatively easily simulated in the laboratory.

In the structural design of pavement, the rutting in a pavement is assumed to be caused by either or both of two causes—excessive strain in the subgrade or permanent deformation in any of the layers in the pavement.

12.3.1 MATERIAL CHARACTERIZATION

Any element in any layer in a pavement subjected to a moving load experiences a variety of stresses, which change as the load moves. Figure 12.9 shows the different stresses acting on an element—vertical, horizontal, and shear stress—as well as the phenomenon of stress reversal as the load changes position.

For example, when the load is "near" any point near the bottom of a stiff surface layer, the point experiences compressive stress in the horizontal direction—this stress changes to a tensile stress when the load is directly above it. Because of the variety of stresses and stress reversals, it is extremely difficult, if not impossible, to completely simulate such a stress state in the laboratory with any test procedure. As such, any one or more of different types of tests, varying in complexity and ability of simulation of in-place conditions, are used, depending on the level of sophistication required. Such tests may be "fundamental" or "simulative." Fundamental tests create in-place loading conditions with the help of complex loading equipment and allow the determination of fundamental parameters such as modulus, whereas simulative tests simulate in-place loading conditions, not necessarily with consideration of scale effects, but specifically to "compare" the performance between different asphalt mixes.

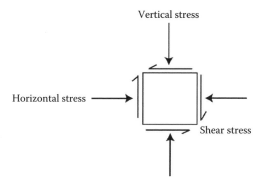

Different stresses acting on a pavement element under load

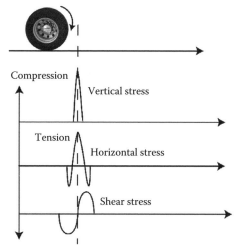

Stress reversals in a pavement element due to moving load

FIGURE 12.9   Different stresses due to a moving wheel load.

The relevant parameter that resists the assumed shear failure due to shear stress is the shear modulus:

$$G = \frac{\tau}{\gamma} = \frac{E}{2(1+\mu)}$$

where
  G is the shear modulus of an elastic material
  τ is the shear stress
  γ is the shear strain—angle, in radians, of deformation of the material due to the shear stress
  E is the Young's modulus
  μ is the Poisson's ratio

Since G is affected by both time of loading (longer time of loading lower is G) and temperature (higher temperature lower is G), rutting is also affected by these parameters.

12.3.1.1 Creep Testing

The deformation of a pavement material in any of the layers begins with elastic deformation, then changes over to partly elastic and partly plastic (elastoplastic) and then completely plastic deformation. In the case of asphalt mixes, which consist of asphalt, which is a viscoelastic material, the elastic and plastic deformations also consist of viscoelastic and viscoplastic deformations. The deformations resulting from the temperature and time of loading-dependent viscoelastic and viscoplastic behavior cause rutting, and can be simulated in the laboratory using either static or repeated load creep tests. In the creep test, the vertical strain in an asphalt mix sample subjected to a constant load (or repeated load), at a specific temperature, is measured during and after the application of the load. The static creep tests can be done either in uniaxial mode or with triaxial mode, which applies confining stress.

The total strain, ε_T, consists of different components, as shown in Figure 12.10:

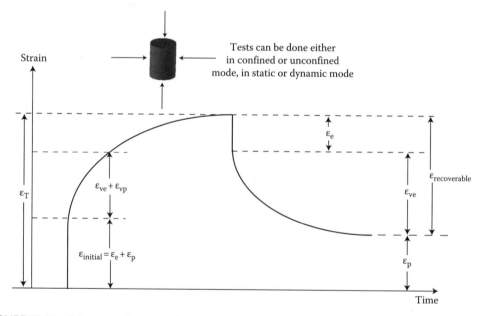

FIGURE 12.10 Schematic of creep testing.

where

ε_T is the total strain

ε_e is the elastic strain, recoverable and time independent

ε_p is the plastic strain, irrecoverable and time independent

ε_{ve} is the viscoelastic strain, recoverable and time dependent

ε_{vp} is the viscoplastic strain, irrecoverable and time dependent

Note that in creep tests, it is very important to have the loading faces of the sample smooth, parallel, and perpendicular to the loading head. The creep test can be run with one load–unload cycle or with incremental load–unload cycles, and the results provide the different deformations mentioned earlier. From the creep test, the creep modulus is calculated to determine the increase in elastic and plastic deformations at specific loading times. The creep test is used specifically for asphalt mixes (and not for other layer materials such as soils) because only the asphalt mixes generally have viscous components with time-dependent stiffness and deformation properties:

$$E_c(t) = \frac{\sigma_c}{\varepsilon_T(t)}$$

where

$E_c(t)$ is the compressive creep modulus at time t

σ_c is the compressive stress

$\varepsilon_T(t)$ is the compressive strain at time t

Also measured is the percentage of recovery of the deformation in the elastic–plastic region of the deformation, for which a higher number indicates lower permanent deformation or lower potential of rutting:

$$X_R = \frac{\varepsilon_t(t)}{\varepsilon_r}$$

where ε_r is the recoverable strain.

For viscoelastic material, the term *compliance* D(t) is used rather than *modulus*, since D(t) allows the separation of the different deformation components. D(t) is approximately equal to the inverse of the modulus. Just like the deformation, the D(t) can be divided, with respect to time, into a primary, secondary, and tertiary component (Figure 12.11).

The secondary part of the curve can be modeled as follows, using a power regression curve:

$$D' = D(t) - D_0 = a(t)^m$$

where

D' is the viscoelastic compliance

D(t) is the total compliance at time t

D_0 is the instantaneous compliance

t is the loading time

a, m are the coefficients for specific asphalt mix

The coefficients a and m are used to predict the mix's rutting potential. A low a value is desirable since a large a value means a large D value and lower modulus value, and also, for a specific a, the higher the m value, the higher is the potential of permanent deformation.

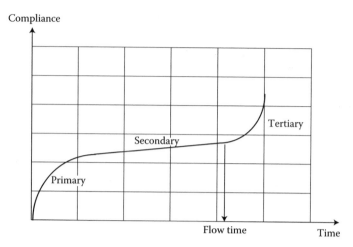

FIGURE 12.11   Different parts of the compliance curve.

12.3.1.2 Triaxial Test

From the triaxial repeated load tests, resilient modulus, E_R; total plastic strain, ε_p; plastic strain per load cycle, $\varepsilon_{p(N)}$; and strain ratio, $\varepsilon_p/\varepsilon_r$, can be obtained. The resilient strain, ε_r, is the recoverable strain during the rest period of the load cycle. The resilient shear modulus is the ratio of compressive stress to resilient axial strain. The strain ratio is the ratio of plastic strain to resilient strain.

In the triaxial shear strength test, a modification of AASHTO T-167 can be used with confining stress to develop a Mohr–Coulomb failure envelope, which is expressed as follows:

$$\tau = c + \sigma(\tan\varphi)$$

where
   τ is the shear stress
   σ is the normal stress
   c is the cohesion
   φ is the angle of internal friction, or the slope of the failure envelope

This simulation is based on the assumption that the shear strength of HMA develops from cohesion (c) from the asphalt binder and interlock of the aggregates matrix, which is characterized by the angle of internal friction (φ). The relative contributions of c and φ toward the shear strength of the mix are dependent on the time of loading, temperature, and volumetric properties of the mix.

The results of triaxial tests can also be used for characterizing the behavior of the asphalt mix according to the Drucker–Prager failure envelope. In this model, the combined effect of failure stresses is represented in terms of the first invariant stress tensor and the second invariant deviatoric stress tensor:

$$J_2 = \gamma^{1/2}I_1 + k$$

$$J_2^{1/2} = \left(\frac{1}{3}\right)(\sigma_1 - \sigma_3)$$

$$I_1 = \sigma_1 + 2\sigma_3$$

where

  I_1 is the first invariant stress tensor
  $J_2^{1/2}$ is the second invariant deviatoric stress tensor
  k is the intercept
  $\gamma^{1/2}$ is the slope

12.3.2 MODELS

The use of the *subgrade strain model* ensures that there is sufficient cover over the subgrade to resist plastic deformation of the subgrade and, hence, resulting "ride-on" deformation in all of the layers above it. This is based on the assumption that the layers above the subgrade are designed and constructed properly so that there is no rutting in any of those layers. The form of the model is as shown next:

$$N_f = b_1\beta(\varepsilon_v)^{-b_2}(M_R)^{b_3}$$

where

  N_f is the number of load repetitions causing failure that is causing deformation exceeding a specific surface deformation
  M_R is the design resilient modulus of the subgrade soil (resilient modulus has been discussed in Chapter 7)
  ε_v is the vertical compressive strain at the top of the subgrade
  b_1, b_2, b_3 are the coefficients derived from triaxial tests
  β is the field calibration factor

The predominant way of using this model is to limit the vertical subgrade strain to a specific value such that the N_f matches or exceeds the anticipated load (traffic) repetition. Different agencies have developed their own models based on this model—some of which do contain M_R, and the rest contain ε_v only. It is important to note that the models are developed for a specific range of site, materials, and environmental conditions, and that failure criteria (or, more specially, the specific depth of rutting which is called a rutting failure) are different for the different models, and one must use the same criteria for failure as those in the specific model that is being used. Examples of such models are shown in Table 12.6.

TABLE 12.6
Examples of Rutting Models Based on Subgrade Strain

Asphalt Institute (1984)
N_f for 13 mm (0.5 in. rutting) = $1.365 * 10^{-9}(\varepsilon_v)^{-4.477}$
Shell (50% reliability; Shell, 1978)
N_f for 13 mm (0.5 in. rutting) = $6.15 * 10^{-7}(\varepsilon_v)^{-4.0}$
Shell (95% reliability)
N_f for 13 mm (0.5 in. rutting) = $1.05 * 10^{-7}(\varepsilon_v)^{-4.0}$
Transport Road Research Laboratory (TRRL, UK; 85% reliability; Powell, 1984)
N_f for 10 mm (0.4 in. rutting) = $6.18 * 10^{-8}(\varepsilon_v)^{-3.95}$
Belgian Road Research Center (Verstraeten et al., 1977)
N_f for 10 mm (0.4 in. rutting) = $3.05 * 10^{-9}(\varepsilon_v)^{-4.35}$
U.S. Army Corps of Engineers (modified by Rauhut et al., 1984; Von Quintas et al., 1991)
N_f for 13 mm (0.5 in. rutting) = $1.259 * 10^{-11}(\varepsilon_v)^{-4.082}(M_R)^{0.955}$

12.3.2.1 Consideration of Rutting in Asphalt Mix Only

Rutting in the asphalt mix layer is caused specifically by high temperature and standing traffic. At high temperatures, such as those above 60°C, the effect of repeated loading on the mix can be considered to be primarily dictated by the properties of the binder, such as viscosity. The influence of binder and the measures taken to select an appropriate binder with adequate resistance against rutting have been discussed.

The penetration index (PI) has been shown to relate (inversely) well with rutting. Rutting results from wheel-tracking tests have been correlated well with creep stiffness measured from a uniaxial unconfined creep test. The deformation is inversely proportional to the stiffness of the mix, which is proportional to the stiffness of the binder, that can be derived from the PI of the binder.

$$\text{Deformation} = k_1 \frac{1}{S_{mix}}$$

$$S_{mix} = k_2 (S_{asphalt\,binder})^{k_3}$$

where k_3 has been suggested as 0.25. Increasing the PI results in a significant increase in viscosity and decrease in rutting potential.

For strains that cause rutting, the binder is assumed to remain in zero shear viscosity regime, with Newtonian fluid characteristics, in which stress and strain are linearly proportional to each other and the stiffness modulus is independent of the stress or strain level and the viscosity is independent of the shear rate. Such viscosity can be measured by a dynamic shear rheometer. A practical upper limit of zero shear viscosity has been proposed as 10^5 Pa s. The value of the zero shear viscosity or the shear complex modulus and elastic component of the binder, as represented by G*/ sin delta, for example, can be used to evaluate the rutting potential of an asphalt binder.

The effect of asphalt mix properties can be modeled for rutting prediction by different ways, such as through the use of statistical (regression) equations, or through the use of material characterization parameters, as shown next.

12.3.2.2 Statistical Predictive Models on the Basis of Different Properties (Baladi, 1989)

$$\text{Log RD} = -1.6 + 0.067V_a - 1.4\log h_{AC} + 0.07T_{avg} - 0.000434\eta_k + 0.15\log\text{ESAL} - 0.4\log M_R$$

$$- 0.50\log E_{base} + 0.1\log\delta_0 + 0.01\log\varepsilon_v - 0.7\log h + 0.09\log(50 - h_{AC} + h_{EQ})$$

where
  RD is the rut depth, in.
  ESAL is the number of 80 kN (18 kip) ESAL corresponding to the rut depth
  T_{avg} is the average annual temperature, °F
  h_{AC} is the equivalent thickness of base material, in.
  E_{base} is the effective resilient modulus of base materials, psi (*effective* means it is influenced by freezing index and seasonal variations)
  δ_0 is the surface deflection, in.
  V_a are the air voids (% in mix)
  M_R is the effective resilient modulus of subgrade, psi
  ε_v is the compressive strain at the bottom of the asphalt mix layer, in./in.
  η_k is the kinematic viscosity of the asphalt binder at 135°C (275°F), cSt

The Shell International Procedure (Shell, 1978) uses a compressive creep test:

$$\Delta h = C_m h_{AC} \left(\frac{\sigma_c}{E_{mix}} \right)$$

where

Δh is the reduction of thickness in asphalt mix layer, mm

C_m is the correction factor for dynamic effect, depending on the type of mix

h_{AC} is the thickness of asphalt mix layer, mm

σ_c is the average vertical compressive stress in the asphalt mix layer, kPa

E_{mix} is the modulus of the asphalt mix, kPa

12.3.2.3   Layered Vertical Permanent Strain Approach

In the layered vertical permanent strain approach, the permanent strain in each layer is determined as a function of load repetitions, the deformation is calculated by multiplying the strain times the layer thickness, and then the deformation in different layers is added up to determine total rutting which is visible on the surface of the pavement. One such model recommended by Barenberg and Thompson (1992) is of the form:

$$Log(\varepsilon_p) = a + b[log(N)]$$

where

ε_p is the permanent strain

a, b are the empirical coefficients

N is the number of load repetitions

As suggested by Huang (1993), the steps consist of (1) dividing the pavement structure into a number of layers, (2) estimating vertical and radial stresses at the mid-depth of each layer, (3) conducting repeated load tests in the lab using the estimated vertical stress and radial stress plus overburden stress as confining stress, (4) determining permanent strain, (5) computing vertical deformation by multiplying permanent strain by thickness, and (6) adding the permanent strains in all of the layers to obtain rutting at the surface.

Examples of such models are shown in Table 12.7.

12.3.2.4   Permanent Strain Rate Method

In this method, the strain per load application is expressed as a power model:

$$\frac{\partial \varepsilon_p}{\partial N} = \varepsilon_{pn} = \frac{\partial(aN^b)}{\partial N}$$

$$\varepsilon_{pn} = abN^{(b-1)}$$

$$\frac{\varepsilon_{pn}}{\varepsilon_r} = \left(\frac{ab}{\varepsilon_r}\right)N^{b-1}$$

$$\text{If } \mu = \frac{ab}{\varepsilon_r}$$

and

$$\alpha = 1 - b$$

then,

$$\frac{\varepsilon_{pn}}{\varepsilon_r} = \mu N^{-\alpha}$$

TABLE 12.7

Examples of Layered Vertical Permanent Strain Approach Models

Asphalt Institute (May and Witczak, 1992), for pavements with >3% air voids, deviator stress ≤90 psi; strain from other layers is negligible.

$$\text{Log}\varepsilon_p = -14.97 + 0.408\text{Log}(N) + 6.865\text{Log}(T) + 1.107\text{Log}(S_d) - 0.117\text{Log}(V) + 1.908\text{Log}P_{eff} + 0.971(V_v)$$

where

ε_p is the permanent axial strain

N is the number of load repetitions to failure

T is the temperature,°F

S_d is the deviator stress, psi

V is the viscosity at 21°C (70°F), Ps * 106

P_{eff} is the percentage by volume of effective asphalt

V_v is the air voids (%)

Allen and Deen (1986), for any layer $\text{Log}\varepsilon_p = C_0 + C_1[\text{Log}(N)] + C_2[\text{Log}(N)]^2 + C_3[\text{Log}(N)]^3$

where

ε_p is the permanent axial strain

N is the number of load applications

C_0, C_1, C_2, C_3 are the coefficients depending on the type of material

For asphalt mixes

$C_0 = -0.000663T^2 + 0.1521T - 13.304 + (1.46 - 0.00572T) * \log\Phi_1$

$C_1 = 0.63974, C_2 = -0.10392, C_3 = 0.00938$

For dense-graded aggregate base

$C_0 = -4.41 + (0.173 + 0.003w) * \Phi_1 - (0.00075 + 0.0029w) * \Phi_3$

$C_1 = 0.72, C_2 = -0.142 + 0.092 (\log w), C_3 = 0.006 - 0.004 (\log w)$

For subgrade

$C_0 = -6.5 + 0.38w - 1.1 (\log\Phi_3) + 1.86 (\log\Phi_1)$

$C_1 = 10^{(-1.1+0.1w)}, C_2 = 0.018w, C_3 = 0.007 - 0.001w$

where

T is the temperature,°F

Φ_1 is the deviator stress, psi

Φ_3 is the confining stress, psi

w is the moisture content (%)

FHWA VESYS model (Moavenzadeh et al., 1974; Brademeyer, 1988):

Axial repeated load creep test is conducted to determine the resilient strain, resilient modulus, and permanent deformation:

$\varepsilon_p = IN^s$

where

I is the linear intercept on the permanent strain axis

N is the number of load applications

S is the slope of the linear portion of the log log curve

The two derived parameters used to characterize the rutting potential of mixes are gnu and alpha:

$$\text{Gnu (or } \mu) = \frac{IS}{E_r}$$

alpha (or α) = I − S.

12.3.2.5 Plastic–Elastic Vertical Strain Ratio Method

In this method, the plastic strain is expressed as a function of the elastic or resilient strain:

$$\frac{\varepsilon_p}{\varepsilon_r} = \beta_r a N^b$$

where

ε_p is the accumulated plastic strain after N repetitions of load
ε_r is the resilient strain depending on mix properties
N is the number of load applications
a, b are the nonlinear regression coefficients
β_r is the field adjustment factor

The different models are shown in Table 12.8.

TABLE 12.8
Different Plastic–Elastic Vertical Strain Ratio Models

Leahy (1989)

$$\log\left(\frac{\varepsilon_p}{\varepsilon_r}\right) = -6.631 + 0.435\log N + 2.767\log T + 0.110\log S + 0.118\log \eta$$
$$+ 0.930\log V_{beff} + 0.50111\log V_a$$

where

ε_p is the accumulated permanent strain

ε_r is the resilient strain

N is the number of load repetitions

T is the mix temperature (°F)

S is the deviatoric stress (psi)

η is the viscosity at 70°F (10^6 poise)

V_{beff} is the effective asphalt content (% by volume)

V_a is the air void content (%)

Ayres (1997)

$$\log\left(\frac{\varepsilon_p}{\varepsilon_r}\right) = -4.80661 + 2.58155\log T + 0.429561\log N$$

Kaloush and Witczak (2000)

$$\frac{\varepsilon_p}{\varepsilon_r} = 10^{-3.15552\beta_{r1}} T^{1.734 * \beta_{r2}} N^{0.39937 * \beta_{r3}}$$

where

β_{r1}, β_{r2}, β_{r3} are the factors obtained from calibration of rut model,
and where the resilient strain can be determined as follows:

$$\varepsilon_{rz} = \frac{1}{|E^*|}\left(\sigma_z - \mu\sigma_x - \mu\sigma_y\right)$$

where

E* is the dynamic modulus, which is determined in the laboratory,
as a function of time of loading and temperature using a master
curve. E* can also be determined from a regression equation
relating asphalt, aggregate and asphalt mix volumetric
properties, and time of loading

12.3.2.6 Rutting Rate Method (Majidzadeh, 1981)

$$\frac{\varepsilon_p}{N} = A(N)^m$$

where

ε_p is the permanent strain
N is the allowable number of load applications
A, m are the constants derived from experiments

12.3.2.7 Alternate Model Relating Tertiary Flow Characteristics to Mix Properties

Since many of the aforementioned equations use regression coefficients derived from the secondary rutting part of tests, the rutting resulting from tertiary rutting is ignored, resulting in a lower than actual rutting prediction. An alternative method is as follows.

In the tertiary zone, the material does not undergo any volume change, but there is a large increase in compliance, D(t). The starting point of tertiary deformation, known as *flow time* (F_T; see Figure 12.11; see also Witczak et al., 2000), has been shown to be a significant indicator of the mix's rutting potential. Flow time is defined as the time at which the shear deformation without any volume change starts.

Another parameter that has been used to relate to the mix's permanent deformation potential is the flow number, F_N (Figure 12.12), which is defined as the number of cycles at which tertiary permanent strain begins, when the permanent strain is plotted against the number of cycles. This type of plot is generated from data obtained in an unconfined or confined repeated load test, in which a haversine load is applied for, say, 3 h of 10,000 cycles, with a loading time of 0.1 s and rest period of 0.9 s. The results of these tests are plotted on a log (N) versus log ε_p scale, and the linear portion is modeled as a power curve with intercepts as a and slope as b:

$$\varepsilon_p = aN^b$$

The number of cycles of loading at which tertiary flow would occur has been correlated to mix properties (Kaloush and Witczak, 2000):

$$F_N = (1.00788E5)T^{-1.6801}S^{-0.1502}\eta^{0.2179}V_{beff}^{-3.6444}V_a^{-0.9421}$$

where S is the stress level.

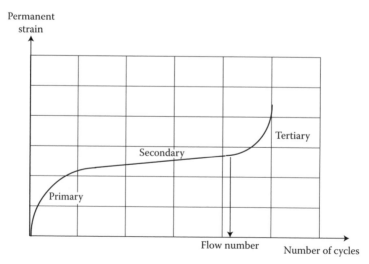

FIGURE 12.12 Concept of flow number.

TABLE 12.9
Rutting Models for Unbound Materials

Granular material

$$\log\left(\frac{\varepsilon_0}{\varepsilon_r}\right) = 0.80978 - 0.06626W_c - 0.003077\sigma_\theta + 0.000003E_r$$

$$\log\beta = -0.9190 + 0.03105W_c + 0.001806\sigma_\theta - 0.0000015E_r$$

$$\log\rho = -1.78667 + 1.45062W_c + 0.0003784\sigma_\theta^2 - 0.002074W_c^2 - 0.0000105E_r$$

Subgrade material

$$\log\left(\frac{\varepsilon_0}{\varepsilon_r}\right) = -1.69867 + 0.09121W_c - 0.11921\sigma_d + 0.91219\log E_r$$

$$\log\beta = -0.9730 + 0.0000278W_c^2\sigma_d + 0.017165\sigma_d - 0.0000338W_c^2\sigma_\theta$$

$$\log\rho = 11.009 + 0.000681W_c^2\sigma_d - 0.40260\sigma_d + 0.0000545W_c^2\sigma_\theta$$

where

W_c is the water content, %

σ_d is the deviator stress, psi

σ_e is the bulk stress, psi

E_r is the resilient modulus of the layer/sublayer, psi

12.3.2.8 Models for Unbound Materials

For unbound materials, the rutting models developed are shown in Table 12.9. The basic form is as follows (Tseng and Lytton, 1986):

$$\delta_a(N) = \left(\frac{\varepsilon_0}{\varepsilon_r}\right)e^{-(\rho/N)^\beta}\varepsilon_v h$$

where

δ_a is the permanent deformation for the layer/sublayer

N is the number of traffic repetitions

ε_0, β, and ρ are the material properties

ε_r is the resilient strain imposed in laboratory tests to obtain material properties ε_0, β, and ρ

ε_v is the average vertical resilient strain in the layer/sublayer as obtained from the primary response model

h is the thickness of the layer/sublayer

β_1 is the calibration factor for the unbound granular and subgrade materials

$\varepsilon_0/\varepsilon_r$ depend on the type of material, granular, or subgrade soil.

12.3.2.9 Ayres Combined Model for Subgrade and Granular Materials (NCHRP, 2004)

$$\log\left(\frac{\varepsilon_0}{\varepsilon_r}\right) = 0.74168 + 0.08109W_c - 0.00001215M_r$$

where

$(\varepsilon_0/\varepsilon_r)$ is multiplied by ADJ_Strain_Ratio

ADJ_Strain_Ratio $= 1.2 - 1.39 * e^{-0.058*(M_r/1000)}$

If $M_r/1000 < 2.6$, use 2.6
If ADJ_Strain_Ratio $<$ le-7, use ADJ_Strain_Ratio $=$ le-7
$\log \beta = -0.61119 - 0.017638W_c$
β is multiplied by 0.7 as a correction factor
$\log \rho = 0.622685 + 0.541524W_c$
$W_c = 51.712 * CBR^{-0.3586*GWT^{0.1192}}$

$$CBR = \left(\frac{M_r}{2555}\right)^{(1/0.64)}$$

where
  W_c is the water content (%)
  CBR is the CBR ratio of the unbound layer
  GWT is the groundwater table (ft)
  M_r is the resilient modulus of the layer/sublayer (psi)

Since the subgrade is often modeled as an infinite layer, it is not practical to divide it into sublayers, and it is better to adopt an alternative approach as suggested by Ayres:

$$\varepsilon_p(z) = (\varepsilon_{p,z=0})e^{-kz}$$

where
  $\varepsilon_p(z)$ is the plastic vertical strain at depth z, measured from the top of the subgrade, for an infinite subgrade
  $(\varepsilon_{p,z=0})$ is the plastic vertical strain at the top of the subgrade ($z=0$)
  k is the regression constant

Total permanent deformation in the subgrade, $\delta = \displaystyle\int_0^{h_{bedrock}} \varepsilon_p(z)dz$

Or

$$\delta = \varepsilon_{p,z=0} * \int_0^{h_{bedrock}} e^{-kz}dz = \left(\frac{1 - e^{-kh_{bedrock}}}{k}\right) * \varepsilon_{p,z=0}$$

where
  δ is the total plastic deformation of the subgrade
  $h_{bedrock}$ is the depth to bedrock

Cumulative damage concept—using subgrade vertical compressive strain, using Miner's law:

$$D_k = \sum_{j=1}^{k} \sum_{i=1}^{m} \frac{n_{ij}}{N_{fij}}$$

where
  D_k is the subgrade distortion damage through season k; when it reaches 1, the pavement is assumed to have reached a state where it needs rehabilitation
  n_{ij} is the actual number of load repetitions for load class I ($I = 1 \ldots m$) during season j
  N_{fij} is the allowable number of load repetitions for load class I and season j to reach failure

The equivalent annual resilient modulus for unbound materials can be used to reduce the number of response computations.

12.3.2.9.1   For Aggregate Base and Subbase

$$M_{R(Aggregate)} = \frac{\sum[(M_{RA})_i * (UF)_i]}{[\sum UF_i]}$$

$$UF_i = 1.885 * 10^3 [M_{Ri}]^{-0.721}$$

where

$M_{R(Aggregate)}$ is the equivalent annual resilient modulus for unbound aggregate base and subbase materials, psi

UF_i is the damage factor for unbound aggregate base and subbase materials in season i

M_{RAi} is the resilient modulus measured in the laboratory for a moisture content in season i, psi

12.3.2.9.2   For Subgrade Soil

$$M_{R(Soil)} = \frac{\sum[(M_{RS})_i * (UF)_i]}{[\sum US_i]}$$

$$US_i = 4.022 * 10^7 [M_{RSi}]^{-1.962}$$

where

$M_{R(Soil)}$ is the equivalent annual or design resilient modulus of the subgrade soil, psi

M_{RSi} is the resilient modulus of subgrade soil measured in the laboratory for a moisture content or physical condition within season i, psi

US_i is the damage factor for the subgrade soil in season i

The AASHTO 1993 method to determine a representative resilient modulus of soil, considering the damage due to different seasons, was presented in Chapter 11.

12.3.2.10   Equivalent Temperature Concept

Models are used to convert traffic (number of load applications) at a given temperature to an equivalent traffic at a reference temperature:

$$\sum TEF_i * ESAL_i = \text{Eqivalent ESAL}_i$$

where

TEF_i is the temperature equivalency factor for the ith temperature interval

$ESAL_i$ is the design equivalent single-axle loads that accumulate during the temperature interval i

The critical location for rutting is selected close to the surface, such as 5 cm (2 in.) from the surface. The temperature gradient is given as follows:

$$G = \frac{(T_B - T_{2''})}{D}$$

where

T_B is the temperature at the bottom of the asphalt mix layer

$T_{2''}$ is the temperature at a depth of 5 cm (2 in.)

D is the thickness of the asphalt mix layer less than 5 cm (2 in.)

A model relating number of load repetitions to a distortion failure to the temperature of interest is as follows:

$$\ln(N) = A_1 + A_2 * T + A_3 * G + A_4 * T^2 + A_5 * G^2 + A_6 * T * G$$

where
   N is the number of load repetitions to failure
   T is the temperature at critical pavement location
   G is the temperature gradient
   A_x is the coefficients form regression

Equivalency factors

$$TEF_i = \frac{N_s}{N_i}$$

where
   N_s is the number of load repetitions to failure at a standard reference temperature
   N_i is the number of load repetitions to failure at the ith temperature

The TEF_i is then used in the following equation mentioned earlier:

$$\sum TEF_i * ESAL_i = \text{Equivalent ESAl}_i$$

12.3.2.11   El-Basyouny and Witczak Mode! (NCHRP, 2004)
Form

$$\frac{\varepsilon_p(N)}{\varepsilon_r} = \left(\frac{\varepsilon_0}{\varepsilon_r} \right) e^{-(\rho/N)^\beta}$$

For traffic level, $N = 1$, $\dfrac{\varepsilon_p(1)}{\varepsilon_r} = a_1 E^{b_1}$

For traffic level, $N = 2$, $\dfrac{\varepsilon_p(10^9)}{\varepsilon_r} = a_9 E^{b_9}$

Assuming values for $\varepsilon_p/\varepsilon_r$ at different E to calculate the a and b coefficients,

$$\left(\frac{\varepsilon_0}{\varepsilon_r} \right) = \frac{\left(e^{(\rho)^\beta} * a_1 * E^{b_1} \right) + \left(e^{\left(\rho/10^9\right)^\beta} * a_9 * E^{b_9} \right)}{2}$$

where

$$\text{Log}\beta = -0.61119 - 0.017638 W_c$$

$$\rho = 10^9 \left(\frac{C_0}{\left[1 - (10^9)^\beta \right]} \right)^{\overline{\beta}}$$

$$C_0 = \left(\ln \frac{a_1 * E^{b_1}}{a_9 * E^{b_9}} \right)$$

The equilibrium moisture content of an unbound material can be predicted from the groundwater table depth and in-place modulus, using regression models which were developed on the basis of variables used with the enhanced integrated climatic model (EICM). The EICM uses the soil water characteristic curve (SWCC; variation of water storage capacity within the macro and micro pores of a soil, with respect to suction), generally plotted as variation of water content with soil suction. The SWCC prediction can be made by the use of equations proposed by Fredlund and Xing (1994). To avoid direct measurement of soil suction, the parameters of the Fredlund and Xing equation were fitted to soil index properties—percentage passing the No. 200 sieve (P_{200}), D_{60}, and plasticity index (PI). When the PI of a soil >0, the SWCC parameters are correlated with P200 * PI; if the PI is zero, the parameters correlated with D_{60}. Other important properties are specific gravity of solids (G_s) and saturated hydraulic conductivity (k_{sat}), both of which can be estimated from P_{200} * PI and D_{60} values if no field or laboratory value is available.

The steps in the calculation of equilibrium moisture corresponding to different CBR and GWT depths content are shown in Table 12.10 (NCHRP, 2004).

12.3.3 DEFINITION OF FAILURE

With consideration of serviceability, an asphalt pavement is said to have failed by rutting when the rut depth at the surface exceeds 13 mm or 0.5 in.

The concept of incremental rutting in each layer and summing up rutting in all of the layers to get the total rutting has been used in several models. The steps include determination of the resilient strain, then the plastic strain, and then the rut depth at each depth, and finally summing up the ruts through the layer to get the total rutting in the layers:

1. $\varepsilon_{rz} = \dfrac{1}{|E^*|}(\sigma_z - \mu\sigma_x - \mu\sigma_y)$
2. $\varepsilon_p = f(\varepsilon_{rz})$
3. $\Delta R_{d_i} = \varepsilon_{pi} \cdot \Delta h_i$
4. $R_d = \displaystyle\sum_{i=1}^{n} \Delta R_{d_i}$

12.4 SMOOTHNESS CONSIDERATION

Smoothness is a measure of ride quality as perceived by the user. Smoothness is affected significantly by rutting, rut depth variance, and fatigue cracking. The smoothness of the initial (newly constructed) pavement is an indicator of the future smoothness, since it indicates the overall quality of construction of the pavement as well as the rate of deterioration of smoothness over time. Different researchers have come up with different lists of factors affecting smoothness. They include initial smoothness, ESAL, age, base thickness, freezing index, initial international roughness index (IRI), type of subgrade, thickness of overlay, maximum and minimum temperatures, and annual number of wet days. Researchers have also listed the various distresses that affect smoothness. They include rut depth, potholes, depression and swells, transverse cracks, standard deviation or variance of rut depth, patching, and fatigue cracking.

The different models for smoothness are shown in Table 12.11.

12.5 TOP-DOWN CRACKING

Top-down cracking initiates at the surface, generally in the wheelpath. It can extend up to 1 in. or the depth of the top layer, and leads to ingress of water and raveling and formation of potholes.

The primary cause of top-down cracking is the failure of the surface mix as a result of excessive surface shear and tensile stresses from tires (generally truck tires). The excessive stress can be due

TABLE 12.10

Steps in Determination of Equilibrium Moisture Content

1. $M_r = 2555(CBR)^{0.64}$

2. $P_{200}PI = \dfrac{(75/CBR) - 1}{0.728}$

$D_{60} = \left(\dfrac{CBR}{28.091}\right)^{2.792516}$

3. Specific gravity, $G_s = 0.041(P_{200}\,PI)^{0.29} + 2.65$

4. Degree of saturation at optimum moisture content,

$S_{opt} = 6.752(P_{200}\,PI)^{0.147} + 78$

5. Gravimetric moisture content at optimum condition,

$W_{opt} = 1.3(P_{200}\,PI)^{0.73} + 11$

When PI = 0,

$w_{opt(T99)} = 8.6425(D_{60})^{-0.1038}$

For base courses,

$w_{opt} = w_{opt(T99)} - \Delta w_{opt}$

$\Delta w_{opt} = 0.0156[w_{opt(T99)}]^2 - 0.1465 w_{opt(T99)} + 0.9$

For nonbase course layers,

$w_{opt} = w_{opt(T99)}$

6. Maximum compacted dry unit weight,

$\gamma_{d\,max\,comp} = \dfrac{G_s \gamma_{water}}{1 + (w_{opt}G_s/S_{opt})}$

7. Volumetric water content at optimum condition,

$\theta_{opt} = \dfrac{w_{opt}\,\gamma_{d\,max}}{\gamma_{water}}$

8. Saturated volumetric water content,

$\theta_{sat} = \dfrac{\theta_{opt}}{S_{opt}}$

9. Volumetric water content is calculated from Fredlund and Xing (1994) equation:

$\theta_w = C(h)*\left[1 - \dfrac{\ln(1 + (h/h_r))}{\ln(1 + (1.45*10^5/h_r))}\right]$

10. The SWCC coefficients are calculated as follows:

$a_f = \dfrac{0.00364(P_{200}PI)^{3.35} + 4(P_{200}PI) + 11}{6.895}\,psi$

$\dfrac{b_f}{c_f} = -2.313(P_{200}PI)^{0.14} + 5$

$c_f = 0.0514(P_{200}PI)^{0.465} + 0.5$

$\dfrac{h_r}{a_f} = 32.44e^{0.0186(P_{200}PI)}$

11. The equilibrium gravimetric moisture content,

$w_{equ} = \dfrac{\theta_w \gamma_{water}}{\gamma_{d\,max}}$

Using these calculations, the final regression equation was obtained as follows:

w_{equi}, in % $= 51.712*CBR^{0.3586*GWT^{0.1192}}$

TABLE 12.11

Different Models for Considering Smoothness

AASHO Road Test model,

$PSR = 5.03 - 1.91 Log(1 + SV) - 0.01(C + P)^{0.5} - 1.38 RD^2$

where

   PSR is the present serviceability rating (mean rating of the panel)

   SV is the slope variance

$$SV = \frac{\sum Y^2 - (1/n)(\sum Y^2)}{n - 1}$$

   Y is the difference between two elevations 9 in. apart

   n is the number of elevation readings

   C is the major cracking in ft per 1000 ft^2 area

   P is the asphalt mix patching in ft^2 per 1000 ft^2 area

   RD is the average rut depth of both wheelpaths in in., measured at the center of a 4 ft span in the most deeply
      rutted part of the wheel path

Zero-Maintenance Pavements Study model (Darter and Barenberg, 1976)

$PSR = 4.5 - 0.49 RD - 1.16 RDV^{0.5}(1 - 0.087 RDV^{0.5}) - 0.13 Log(1 + TC) - 0.0344(AC + P)^{0.5}$

where

   RD is the rut depth in both wheelpaths of the pavement, in.

   RDV is the rut depth variance, in.^{2} * 100

   AC is the Class 2 or Class 3 alligator or fatigue cracking, ft^2/1000 ft^2

   TC is the transverse and longitudinal cracking, ft^2/1000 ft^2

   P is the patching, ft^2/1000 ft^2

World Bank HDM III model (World Bank, 1995),

ΔRI is the $134 e^{mt} MSNK^{-0.05} \Delta NE4 + 0.1114 \Delta RDS + 0.0066 \Delta CRX + 0.003 h \Delta PAT + 0.16 \Delta POT + mRI_t \Delta t$

where

   ΔRI is the increase in roughness over time period t, m/km

   MSNK is the factor related to pavement thickness, structural number, and cracking

   $\Delta NE4$ is the incremental number of equivalent standard-axle loads in period t

   ΔRDS is the increase in rut depth, mm

   ΔCRX is the increase in area of cracking (%)

   ΔPAT is the increase in surface patching (%)

   ΔPOT is the increase in total volume of potholes, m^3/lane km

   M is the environmental factor

   RI_t is the roughness at t years

   Δt is the incremental time period for analysis, years

   t is the average age of pavement or overlay, years

   h is the average deviation of patch from original pavement profile, mm

FHWA/Illinois Department of Transportation Study (Lee and Darter, 1995) model:

PSR is the $4.95 - 0.685 D - 0.334 P - 0.051 C - 0.211 RD$

where

   D is the number of high-severity depressions (number per 50 m)

   P is the number of high-severity potholes (number per 50 m)

   C is the number of high-severity cracks (number per 50 m)

   RD is the average rut depth, mm

Models Suggested in the 2002 Design Guide:

The model suggested by work conducted under NCHRP 1-37A (Development of the 2002 Guide for Design of
   New and Rehabilitated Pavements) on the basis of LTPP data was used to modify existing models.

(continued)

TABLE 12.11 (continued)
Different Models for Considering Smoothness

$IRI = IRI_0 + \Delta IRI_d + \Delta IRI_F + \Delta IRI_S$

where

IRI_0 is the initial IRI

ΔIRI_d is the IRI due to distress

ΔIRI_F is the IRI due to frost heave potential of the subgrade

ΔIRI_S is the IRI due to swell potential of the subgrade

The models were then separated for asphalt pavements with different types of bases and subbases, as follows:
Unbound aggregate base and subbase

$$IRI = IRI_0 + 0.0463\left[SF\left(e^{age/20} - 1\right)\right] + 0.00119(TC_L)_T + 0.183\left(COV_{RD}\right)$$

$$+ 0.00384(FC)_T + 0.00736(BC)_T + 0.00155\left(LC_{SNWP}\right)_{MH}$$

where

IRI_0 is the initial IRI

SF is the site factor

$$SF = \left(\frac{(R_{SD})(P_{0.075} + 1)(PI)}{2*10^4}\right) + \left(\frac{\ln(FI + 1)(P_{02} + 1)[\ln(R_m + 1)]}{10}\right)$$

R_{sd} is the standard deviation of the monthly rainfall, mm

$P_{0.075}$ is the percentage passing the 0.075 mm sieve

PI is the percentage plasticity index of the soil

FI is the average annual freezing index

P_{02} is the percentage passing the 0.02 mm sieve

R_m is the average annual rainfall, mm

$e^{age/20}$ = age term

$(FC)_T$ is the fatigue cracking in wheelpath

COV_{RD} is the coefficient of variation of rut depths = SD_{RD}/RD

SD_{RD} is the standard deviation of rut depth along the pavement

RD is the average rut depth

$(TC_L)_T$ is the length of transverse cracks

$(BC)_T$ is the area of block cracking as a% of the total lane area

$(LCS_{NWP})_{MH}$ is the length of sealed longitudinal cracks outside wheelpath, ft/mile (moderate and high severity)

Asphalt-treated base

$$IRI = IRI_0 + 0.0099947(Age) + 0.0005183(FI) + 0.00235(FC)_T + 18.36\left[\frac{1}{(TC_S)_H}\right] + 0.9694(P)_H$$

where

$(TC_S)_H$ is the transverse crack spacing, m (high severity level)

$(P)_H$ is the area of patching as a% of total area (high severity)

Cement or pozzolanic-treated bases

$IRI = IRI_0 + 0.00732(FC)_T + 0.07647(SD_{RD}) + 0.0001449(TC_L)_T + 0.00842(BC)_T + 0.0002115(LC_{NWP})_{MH}$

where

$(LC_{NWP})_{MH}$ is the length of longitudinal cracks outside wheelpath, m/km (moderate and high severity)

Model to predict SD_{RD}

$SD_{RD} = 0.665 + 0.2126(RD)$

where

SD_{RD} is the standard deviation of rut depths, mm

RD is the mean rut depth, mm

HMA overlays placed on flexible pavements

TABLE 12.11 (continued)
Different Models for Considering Smoothness

$$IRI = IRI_0 + 0.011505(Age) + 0.0035986(FC)_T + 3.4300573\left(\frac{1}{(TC_S)_{MH}}\right)$$

$$+0.000723(LC_S)_{MH} + 0.0112407(P)_{MH} + 9.04244(PH)_T$$

where
$(TC_S)_H$ is the average spacing of medium and high severity transverse cracks, m.
(LC_S) MH is the medium and high severity sealed longitudinal cracks in the wheel path, m/km.
$(P)_{MH}$ is the area of medium and high severity patches, percent of total lane area, %.
$(PH)_T$ is the potholes, percent of total lane area, %.
HMA overlays placed on rigid pavements

$$IRI = IRI_0 + 0.0082627(Age) + 0.0221832(RD) + 1.33041\left(\frac{1}{(TC_S)_{MH}}\right)$$

where RD is the average rut depth, mm

to one or many factors, including anomalies in load distribution, axle configuration, axle loadings, tire types, and rigidity as well as tire condition and pressures. The failure could be accelerated because of poor mix qualities that could result from many factors, including segregation, inadequate density, and premature aging of asphalt binder.

The best approach to avoid such problems, where high tensile and shear stresses on the surface are expected, is to use HMA mixes with adequate resistance against such stresses—such as those with polymer-modified binders. Top-down cracks, once found, should be sealed as soon as possible, to avoid ingress of moisture and further rapid deterioration of the layer.

A fracture mechanics–based design method to consider top-down cracking has been developed in NCHRP project 01-42A (final report available at http://onlinepubs.trb.org/onlinepubs/nchrp/nchrp_w162.pdf).

The researchers have proposed two separate mechanisms to explain and predict the initiation of top-down cracking in thin and medium HMA layers, and thicker HMA layers. The mechanism governing the initiation of crack in thin HMA is related to bending-induced surface tension away from the tire, while the other that influences the initiation of crack in thicker HMA is related to shear-induced near-surface tension at the tire edge. For both mechanisms, the exacerbating factor is aging of the mix. The other influential factors are the presence of thermal stresses and damage zones. Two models have been presented by the researchers as part of an unified predictive system: a viscoelastic continuum damage (VECD) model to predict crack initiation by modeling damage zones and their effect on response prior to cracking (i.e., damage zone effects), and an HMA fracture mechanics (HMA-FM) model to predict crack propagation by modeling the presence of macro cracks and their effect on response (i.e., macro crack effects). The models are yet to be developed to a state at which they could be integrated into a mechanistic–empirical pavement design method.

12.5.1 Pavement Surface Characteristics

The important pavement surface characteristics include pavement surface friction, pavement–tire noise, and splash and spray during rains. Macropavement and micropavement surface texture have the most significant effect on these characteristics, whereas megatexture and roughness have more influence on ride quality, rolling resistance, vehicle wear, and in-vehicle noise.

Pavement microtexture is largely influenced by the surface properties of the aggregate particles within the asphalt or concrete paving material, and is characterized by wavelength of 1 μm–0.5 mm and amplitude of less than 0.2 mm. Two portable friction testers are utilized to get a surrogate indication of microtexture in pavements: the British portable tester (BPT), and the dynamic friction tester (DFT). In the BPT test, (AASHTO T278) a pendulum with a rubber shoe is swung while touching the surface of aggregates that are embedded in a holder. The British pendulum number (BPN) gives an indication of the friction of the aggregates (and hence its microtexture) in terms of the resistance offered by the aggregates in slowing the swinging pendulum down across its surface. In the DFT, torque required to rotate three small, spring-loaded rubber pads in a circular path over a wetted pavement surface at different speeds (ASTM E1911) is measured. Correlations between torque measurements and BPN are available.

Pavement macrotexture wavelengths range from 0.5 to 50 mm with vertical amplitudes between 0.1 and 20 mm. Macrotexture in HMA pavements is determined by the shape, size, and gradation of aggregates, whereas that in PCC pavements is influenced by small surface channels, grooves, or indentations that are intentionally formed (plastic concrete) or cut (hardened concrete) to allow water to escape from beneath a vehicle's tires. A high mean texture depth (MTD) (such as >1.5 mm) is necessary for avoiding traffic-related accidents. Macrotexture can be measured by traditional volumetric methods as well as sophisticated high-speed laser profilers. In the "sand patch" method (ASTM E-965), a specified volume of very small glass spheres (ASTM D1155) is spread on the surface, and the MTD is estimated by dividing the known volume of the glass spheres used, by the area of the circle (whose average diameter is calculated by measuring on four axes). For laser profiling, the MTD is estimated from the mean profile depth (MPD), which is determined by the measurement of pavement surface elevations (ASTM E-1845). Generally, a combination of a horizontal distance measuring device, a very high-speed laser triangulation sensor, and a portable computer are utilized. In the outflow meter method (OFM, ASTM E2380), the time required for a known volume of water to pass from a vertical cylinder resting on a rubber annulus through the pavement is determined. The outflow time (OFT) is used as an indicator of macrotexture (smooth or rough). Another method of measuring the macrotexture is the use of the circular texture (track) meter (CTM) (E 2157). In this method, a noncontact laser device is used to measure the profile of a 11.2 in. diameter circle. The MPD is calculated after dividing the measured profile into eight segments of 4.4 in.. The MPD is highly correlated to MTD.

QUESTIONS

12.1 Determine the indirect tensile strength of a sample from the following test data.
Thickness of the sample: 2.9 in., breaking load = 4908 lb, diameter = 6 in.

12.2 Determine the resilient modulus of a sample from the following test data.

Load = 250 lb, total deformation = $1.9567 * 10^{-4}$ in., thickness = 2 in.

12.3 Compute the dynamic modulus values at different frequencies and temperatures for the following mix:
Asphalt test data

T (°C)	η (10⁻⁶ P)
−28	237870351.6
−22	11729449.93
−16	783062.0737
−10	68238.2326
−4	7521.773921
2	1020.675635
8	166.5412517
14	32.01494597

20	7.122800176
26	1.805759608
32	0.514587666
38	0.162871888
44	0.056654775
50	0.021457668
56	0.008776083
62	0.003847795
64	0.002965245

Mix test data:

Air Void Content (%)	Effective Binder Content, % by Volume	Cumulative % Retained on the 3/4 in. Sieve	Cumulative % Retained on the 3/8 in. Sieve	Cumulative % Retained on the No. 4 Sieve	% Passing on the No. 200 Sieve
5.8	12	0	3	37	6

12.4 Using the Asphalt Institute (AI) and Shell equations, estimate the number of load repetitions to fatigue cracking failure for a HMA pavement with a tensile strain of 250 microstrain and a modulus of 3500 MPa. Air voids = 5%, asphalt content = 11% (by volume).

12.5 Using the Transportation Research Laboratory (TRL) and the U.S. Army Corps of Engineers models, determine the number of repetitions to subgrade rutting failure for a pavement with a vertical subgrade strain of 200 microstrain, with a resilient modulus of 7252 psi.

13 Distress Models and Material Characterization for Concrete Pavements

The major distresses of concrete pavement against which the pavement is designed consist of cracking, faulting, pumping, and punchouts. Degradation of smoothness and ride quality over time will severely affect the functionality of the pavement. These distresses are commonly caused by excessive stresses, strains, and deflections. The goal of the mechanistic-empirical (M-E) approach to pavement design is to reduce these stresses, strains, and deflections, and maintain the pavement below a "critical" threshold to minimize deterioration and prevent failure.

To prevent cracking, the stresses are limited by a certain value. Stresses develop as a combination of factors such as expansive stresses due to temperature and moisture changes, or curling and warping, or such as fatigue stresses caused by repeated loading. To counter the cracking problem, some of the major test properties required are modulus of elasticity, the static and dynamic flexural strength, and coefficients of thermal expansion. The tests are discussed in the second part of this chapter.

Faulting is prevented by providing a stiff base with relatively low deformation under load. Minimizing curling and warping also helps in preventing faulting. Pumping is prevented by providing a free draining base, preventing moisture intrusion into the joints, and minimizing curling. Tests on fresh concrete are conducted for quality control and achieving consistency in the mix. The purposes are also to ensure good placement and finishing of the concrete.

Hence, the pavement design features that require the appropriate selection of slab thickness, base type and thickness, joint spacing, dowel design, and drainage, all contribute to reducing induced stresses, strains, and deflections, and minimizing observed distress.

13.1 DISTRESSES AND MODELS

13.1.1 CRACKING

Cracking in concrete pavements can occur due to excessive static stresses or strains due to fatigue stress/strain failure. The high stresses that can cause cracking can be induced by a combination of factors, including restraint forces that are developed due to temperature and/or moisture volume changes that cause warping and curling and traffic loads. Figure 13.1 shows the deflected pavement shape at the onset of cracking with a combination of attributes that contribute to cracking.

13.1.1.1 Fatigue Cracking in JPCP

For JPCP, contraction joints are provided at close intervals to prevent the development of midslab cracks resulting from excessive stresses induced by loading, temperature, and moisture changes. The consequences of these stresses is an accumulation of damage in a critical portion of the slab. Enough accumulated stress will eventually initiate cracking, and further propagation of cracks

FIGURE 13.1 Deflections in concrete pavement.

will manifest at the bottom or surface of the PCC slab over time. Depending on the critical factors involved in crack propagation, fatigue cracking in JPCP can be divided into four major categories:

- Bottom-up transverse cracks
- Top-down transverse cracks
- Longitudinal cracks
- Corner breaks

Bottom-up transverse cracks occur when a critical combination of loading and temperature curling creates summative stresses. For example, when truck axles are applying loads near the longitudinal edge of a slab and midway between the transverse joints, a critical tensile bending stress occurs at the extreme bottom fiber of the slab. This bending stress increases greatly when coupled with a positive thermal gradient that produces downward curling. The process initiates micro-cracks that eventually grow and propagate with repeated loading stresses that manifest into visible transverse cracks that affect the pavement performance. Bottom-up cracking models must therefore account for the accumulation of fatigue damage caused by every truck axle load within a specified time increment, including appropriate thermal stresses and summed over the total design period.

Top-down transverse cracks also occur when a critical combination of loading and temperature curling creates summative stresses. In this case, when the truck front axle (steering axle) approaches a transverse joint within 10–20 ft, a large tensile stress is induced in the top of the slab between axles. This top tensile stress increases greatly when there is a negative thermal gradient through the slab, a built-in negative gradient from the slab construction, or significant drying shrinkage at the top of the slab. *Negative thermal gradient* is defined when the top of the slab is cooler than the bottom and upward curling occurs. Similarly to the aforementioned case, repeated loading stresses will eventually manifest into visible transverse cracks that affect the pavement performance. Models for top-down transverse cracks should account for similar stresses as presented for bottom-up transverse cracks.

Longitudinal cracks and corner breaks also occur due to combined stresses. The mechanism for this type of distress is similar to that of top-down transverse cracking except for the location of the critical stresses that develop. The critical induced stresses occur along both the longitudinal and transverse joints but near the corner.

Assessment of damage accumulation is essential for structural design to properly consider all the different load magnitudes that are applied to highway pavements. The fatigue damage concept as presented by Miner (1945) allows the designer to sum the fatigue damage from different loads of various magnitudes and to compute the resulting combined fatigue damage, as shown next:

$$\text{Fatigue damage} = \sum_{ijk=1}^{p} \frac{n_{ijk}}{N_{ijk}}$$

where

n_{ijk} is the number of actual load applications under conditions represented by i, j, and k

N_{ijk} is the number of allowable load applications under conditions represented by i, j, and k

The subscripts i, j, and k represent factors such as PCC strength and modulus, load, axle type, temperature gradient, location of applied load, and k, among others. PCC pavements almost never fail due to a single application of load. Field and laboratory experience has shown that fatigue failure of PCC slabs and beams occurs due to repeated flexural stresses. Representative modeling of fatigue stresses should include more than loading, and this includes critical fatigue location, critical traffic load stresses, critical curling and warping stresses, seasonal variation in support, load transfer, warping and curling, concrete strength over time, and appropriate use of the concept of accumulated damage.

Based on numerous published data, it has been observed that the concrete will not reach failure by fatigue if the stress ratio (applied stress divided by maximum flexural stress) is less than 50% after approximately 20 million cycles or repetitions (Mindess and Young, 1981, Neville, 1981).

Since concrete continues to gain strength with age, the flexural strength of normal concrete reaches only about 70% of its long-term potential strength at 28 days and approximately 90% at 90 days. Based on Wood (1991), a 20% increase in flexural strength above the 28 day strength can be used for analysis purposes.

Fatigue damage in PCC pavements can be determined using Miner's equation when the allowable number of load repetitions is known. Various models have been developed to determine the allowable number of load repetitions as a function of the stress ratio (σ/S_c). These models are presented next.

13.1.1.2 Zero-Maintenance Design Fatigue Model

The zero-maintenance design model presents the relationship between the number of stress applications to failure and the stress ratio in JPCP. The model was developed by Darter and Barenberg (1976, 1977), and is presented in the following equation:

$$\log N_f = 17.61 - 17.61 \left(\frac{\sigma}{S_c} \right)$$

where

N is the number of stress applications

(σ/S_c) is the stress ratio

σ is the applied cyclic stress

S_c is the maximum static flexural strength

This relationship represents a 50% probability of failure, which is too high a risk for design purposes. Another equation was later developed that reduced the probability of failure to 24%. Developing an expression with a lower probability of failure in fatigue life was not possible due to the inherent variation in strength:

$$\log N_f = 16.61 - 17.61 \left(\frac{\sigma}{S_c} \right)$$

13.1.1.3   Calibrated Mechanistic Design Fatigue Model

This model was developed from data obtained from the U.S. Army Corps of Engineers' full-scale accelerated traffic tests and from the AASHTO Road Test. The model presents the relationship between the number of edge stress repetitions, the stress ratio, and the probability level (P) as presented by Salsilli et al. (1993):

$$\log N = \left[\frac{-(\sigma/S_c)^{-5.367} \log(1-P)}{0.0032} \right]^{0.2276}$$

13.1.1.4   ERES-COE Fatigue Model

This model was developed by Darter (1988) using the Corps of Engineers' test results from full-scale field sections. This model was originally developed for airport pavements but was used for other applications later:

$$\log N = 2.13 \left(\frac{\sigma}{S_c} \right)^{-1.2}$$

13.1.1.5   PCA Fatigue Model

The PCA fatigue model is similar to the zero-maintenance model. However, the PCA model assumes that no fatigue failure will occur below a stress ratio of 45% even at an infinite number of cycles. The PCA equations as presented by Packard and Tayabji (1985) are as follows:

$$\text{For } \frac{\sigma}{S_c} \geq 0.55 : \log N_f = 11.737 - 12.077 \left(\frac{\sigma}{S_c} \right)$$

$$\text{For } 0.45 < \frac{\sigma}{S_c} \geq 0.55 : N_f = \left[\frac{4.2577}{(\sigma/S_c) - 0.4325} \right]^{3.268}$$

$$\text{For } \frac{\sigma}{S_c} \leq 0.45 : N_f = \text{unlimited}$$

13.1.1.6   ARE Fatigue Model

This model was developed based on the AASHTO Road Test data from all sections that developed Class 3 and 4 cracking. The traffic loads were converted to 18 kip ESALs, and the maximum mid-slab stresses were computed using elastic layer theory:

$$N_f = 23,440 \left(\frac{\sigma}{S_c} \right)^{-3.21}$$

13.1.1.7   Vesic Distress Model

The Vesic and Saxena (1969) model was also based on the AASHO Road Test using a Serviceability Index of 2.5 as the failure criterion:

$$N_f = 225,000 \left(\frac{\sigma}{S_c} \right)^{-4}$$

13.1.1.8 RISC Distress Function

The RISC model is also dependent on the AASHTO Road Test, and failure was defined as the number of 18 kip ESALs required to reach a terminal Serviceability Index of 2.0:

$$N_f = 22,209 \left(\frac{\sigma}{S_c} \right)^{-4.29}$$

13.1.1.9 Transverse Cracking

Transverse cracks occur in both JPCP and JRCP. The cracking mechanism is, however, different for each type. Transverse cracks in JPCP are commonly caused by thermal curling or fatigue loading induced by traffic loads. However, transverse cracks in JRCP are commonly caused by thermal curling and shrinkage. As described in the previous section, the incremental damage induced by traffic and environmental conditions is converted to physical pavement distress such as transverse cracks, longitudinal cracks, and corner breaks using a calibrated damage-to-distress correlation model. Some of the models developed for predicting transverse cracking are presented next.

13.1.1.9.1 Transverse Cracking Prediction Models

13.1.1.9.1.1 RIPPER The RIPPER model is based on a fatigue life depletion approach similar to the one used in the zero-maintenance design model. The maximum number of allowable repetitions "N" was computed using fatigue damage. The cumulative fatigue damage (n/N) is then computed and plotted against the corresponding percentage of slabs cracked. The model is as follows:

$$P = \frac{1}{0.01 + 0.03 \left[20^{-\log(n/N)} \right]}$$

where
P is the percentage of slabs cracked
n is the actual number of 18 kip ESALs at slab edge
N is the allowable 18 kip ESALs from the stress ratio from the ERES-COE fatigue model

Calibrated mechanistic design

$$P = \frac{1}{0.01 + 0.0713 \left[2.5949^{-\log(FD)} \right]}$$

where
P is the percentage of slabs cracked
FD is the fatigue damage

13.1.1.9.1.2 RIPPER 2 This model is the follow-up to RIPPER and was developed by Yu et al. (1998) from the largest national-scale PCC pavement performance study. The approach was similar to that of RIPPER. In addition, the effect of wander was given more emphasis. A ratio termed the *pass-to-coverage* (p/c) *ratio* was defined, which gives the number of traffic passes needed to produce the same amount of fatigue damage at the critical location as one pass that causes the critical loading condition (e.g., edge location). For example, if the (p/c) has a value of 100, this means that it takes 100 traffic passes to cause the same amount of damage as one load placed directly at the edge. The (p/c) is represented mathematically as follows:

$$\frac{p}{c} = \frac{FD_{Dii}}{\sum P(COV_{Dj})FD_{Dij}}$$

where

FD_{Dii} is the fatigue damage at a location D_i due to load at D_j

$P(COV_{Dj})$ is the probability that the load will pass through location D_j

FD_{Dij} is the fatigue damage at a location D_i due to load at D_j

The results from RIPPER 2 show that fatigue cracking is sensitive to slab thickness, joint spacing, and shoulder type. A reduction in fatigue cracking in JPCP is achievable by increasing the slab thickness and/or reducing the joint spacing. A tied PCC shoulder and wider slabs can significantly reduce the amount of slab cracking. A stabilized bonded base can also reduce slab cracking. Hotter climates can cause more cracking because of the greater temperature gradients induced in the slab. The flexural strength or modulus of rupture (MOR) of the PCC will affect cracking. A higher MOR will result in less slab cracking. The modulus of subgrade reaction has a minor role, except for extreme conditions.

13.1.2 TRANSVERSE JOINT FAULTING IN JOINTED PLAIN CONCRETE PAVEMENTS

Transverse joint faulting is the difference in elevation between adjacent slabs across a joint. Excessive faulting reduces the performance of JPCP by increasing roughness and user discomfort and resulting in ultimate cracking and corner breaks. Faulting is caused by the erosion beneath the leave slab and the buildup on the approach slab of base fines by the action of pumping (Figure 13.2). For faulting to occur, the right combination of factors must exist, and they include heavy axle loading, poor load transfer across joints, base materials that are erodible, and free water in the base (water is the medium to transport fines under the slab). Faulting is commonly measured using a faultmeter (such as the Georgia Digital Faultmeter).

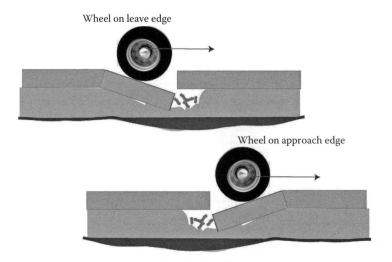

FIGURE 13.2 Faulting in jointed plain concrete pavements without dowels.

Although it is commonly agreed upon that pumping is the basic mechanism that leads to faulting, other factors can still influence faulting. Faulting increases in the cold months since joints open up due to PCC contraction and aggregate interlock is reduced. Joints with adequate dowel support increase joint load transfer and decrease faulting. However, if doweled joints are deficient or become so, faulting can result and has been observed in doweled joints. In addition, the base can usually have a higher moisture content that contributes to pumping. Conversely, in the warmer months the concrete expands and the joints are tighter, increasing aggregate interlock. Upward curling can contribute to slab corner deflection when subjected to heavy axle loads.

13.1.2.1 Models to Predict Faulting

M-E faulting models tend to incorporate the following parameters that contribute to faulting: slab corner deflections, base or subbase erodibility, and free water within the pavement structure. For example, a typical model is as follows:

$$Fault = C_l DE^\alpha EROD^\beta WTR^\gamma$$

where
  Fault is the average joint faulting, in. (or total joint faulting per mile)
  DE is the factor representing slab corner deflection
  EROD is the factor representing base/subbase erodibility
  WTR is the factor representing the amount and frequency of free water within the pavement
    system
  $C_l \alpha \beta \gamma$ are the regression constants

The models developed for faulting use mechanistic principles to predict slab corner deflections; however, base or subbase erodibility and moisture conditions are estimated using empirical methods.

13.1.2.2 Slab Corner Deflections

13.1.2.2.1 NAPCOM Approach

The Purdue method for analysis of rigid pavements (PMARP) uses energy methods to determine the volume of materials pumped from beneath a loaded slab given a deformation energy induced by wheel loads:

$$TE = \sum_{i-1}^{n} k_i A_i \Delta_i^2$$

where
  TE is the deformation energy imposed by an axle load, psi/in.
  K_i is the modulus of subgrade reaction, psi/in.
  A_i is the area associated with node i, in.^{2}
  Δ_i is the deflection at node i, in.

The deformed area under the loaded PCC slab with the potential for pumping is shown in Figure 13.3.

To model the energy that contributes to water movement under a slab, the differential deformation energy was defined as the difference in energy (pore water pressure) experienced at the interface of adjacent slabs beneath the joint:

FIGURE 13.3 Deflection under load.

$$DE = E_L - E_{UL}$$

where

DE is the differential elastic deformation energy per unit area between the loaded and the unloaded slab, kN/mm

E_L is the elastic deformation energy per unit area imposed on the loaded slab, KN/mm

E_{UL} is the elastic deformation energy imposed on the unloaded slab, kN/mm

The expression for DE can be rewritten as follows:

$$DE = \frac{1}{2} k(W_L + W_{UL})(W_L - W_{UL})$$

where the term $(W_L + W_{UL})$ is the differential deflection between the loaded and unloaded slabs. The higher the difference, the higher the probability of movement of material from beneath the leave slab to beneath the approach slab.

DE is used as the mechanistic parameter for computing the allowable number of axle load applications, N:

$$\log N_{ijk} = \alpha_1 - \alpha_2 \log(DE + 1)$$

where

N_{ijk} is the allowable axle load repetitions of axle "type i" and load j for foundation condition k

α_1, α_2 are the regression constants

DE is the differential slab deflection due to axle "type i" and load j for foundation condition k

The damage due to repeated differential slab deflections and pumping can be calculated using Miner's damage equation:

$$DAMAGE_{ijk} = \sum_{ijk} \frac{n_{ijk}}{N_{ijk}}$$

where

$DAMAGE_{ijk}$ is the damage for axle type i and load j for foundation condition k

n_{ijk} is the expected number of axle load repetitions for each axle group i and load j for foundation condition k

N_{ijk} is the allowable number of repetitions for each axle group i and load j for foundation condition k to a maximum faulting

13.1.2.2.2 PCA Model

In the PCA model, erosion damage is related to power. Power is defined as the rate of work each axle pass over the PCC slab joint exerts on the underlying pavement base/subbase materials. The PCA defines power exerted by each axle pass at the slab corner as the product of corner deflection (w) and pressure (p) at the slab base-subbase interface divided by the length of the deflection basin, which is a function of the radius of relative stiffness. The equation is as follows:

$$P = 268.7 \frac{p^2}{hk^{0.73}}$$

where
P is the power (rate of work)
p is the pressure at slab-foundation interface, psi
h is the slab thickness, in.
k is the modulus of subgrade reaction, psi/in.

The allowable axle load repetitions for a given axle type and foundation condition can then be determined using the following:

$$N_{ijk} = 10^{(14.52-6.77[C_1 P_{ijk}-9]^{0.103})}$$

where
N_{ijk} is the allowable number of axle load repetitions for axle type i and load j for foundation condition k
$C_1 = 1 - (k/2000 * 4/h)^2$ (approximately equal to 1 for granular materials and 0.9 for high-strength nonerodible materials)
P is the power of axle type i and load j for foundation condition k.

Erosion damage can be calculated using Miner's equation as follows:

$$EROSION_{ijk} = \sum_{ijk} \frac{100 C_2 n_{ijk}}{N_{ijk}}$$

where
$EROSION_{ijk}$ is the percentage of erosion damage for axle type i and load j for foundation condition k
C_2 is the 0.06 for pavements without a shoulder and 0.94 for pavements with a tied concrete shoulder
n_{ijk} is the expected number of axle load repetitions for each axle group i and load j for foundation condition k
N_{ijk} is the allowable number of repetitions for each axle group i and load j for foundation condition k to a maximum faulting

13.1.2.2.3 Yao et al. (1990) Model

Prior to determining the number of repetitions to failure, the Foundation Damage Index is first determined and is expressed as follows:

$$DI = \frac{W^2}{hk^{1.27}}$$

where
  DI is the Foundation Damage Index
  W is the corner deflection, in.
  h is the pavement thickness, in.
  k is the modulus of subgrade reaction, psi/in.

For $DI > 1.862 \times 10^{-4}$:

$$N_{ijk} = \frac{0.5064}{DI_{ijk}^{0.312}}$$

where
  N_{ijk} is the allowable number of repetitions for each axle group i and load j for foundation condition k
  DI_{ijk} is the deflection index for axle type i and load j for foundation condition k

For $DI < 1.862 \times 10^{-4}$:

$$N_{ijk} = \frac{3.75 * 10^{-22}}{DI_{ijk}^{5.96}}$$

Using Miner's damage equation, the damage at a slab corner is as follows:

$$DAM_{ijk} = \sum_{ijk} \frac{n_{ijk}}{N_{ijk}}$$

where
  DAM_{ijk} is the total damage for axle type i and load j for foundation condition k
  n_{ijk} is the expected number of axle load repetitions for each axle group i and load j for foundation condition k
  N_{ijk} is the allowable number of repetitions for each axle group i and load j for foundation condition k

The damage calculated in Yao et al.'s model may be used to assess the potential for erosion, pumping, and the loss of slab support due to void creation beneath the PCC slab. This damage index, however, was not correlated to field faulting data to predict transverse joint faulting. Yao et al.'s model shows that corner deflection could be used as a predictor for joint damage, erosion, and pumping that could ultimately induce transverse faulting.

13.1.2.2.4   NCHRP 1-26 (RIPPER Model)

This model was developed for transverse joint faulting in doweled jointed pavements. This was done since it is the belief that previously introduced joint-faulting models are not appropriate for doweled joints. Faulting in doweled joints is attributed to free vertical movement between the dowel and the contiguous or surrounding concrete. This can be attributed to excessive deformation of the concrete when its localized bearing strength is exceeded. The bearing stress between dowels and concrete has been introduced earlier and is presented here for completeness.

Dowel deflection is defined as follows:

$$D_o = \frac{P_t(2+\beta z)}{4\beta^3 E_d I_d}$$

where
  D_o is the deformation of the dowel at the face of the joint
  p_t is the shear force acting on the dowel (load on one dowel)
  z is the joint width
  E_d is the modulus of elasticity of the dowel
  $I_d = \pi d^4/64$ is the moment of inertia of the dowel

$$\beta = \sqrt[4]{\frac{Kd}{4E_d I_d}}$$

The bearing stress σ_b is proportional to the deformation:

$$\sigma_b = KD_o = \frac{KP_t(2+\beta z)}{4\beta^3 E_d I_d}$$

An estimation of the amount of load transferred by the dowel is commonly selected as 1,500,000 psi/in. (as adopted by NCHRP 1-26). For the NCHRP 1-26 study, the critical pavement response parameter, which is the maximum bearing stress at the concrete dowel interface, was directly correlated with measured transverse joint faulting.

13.1.2.2.5 Byrum et al. (1997) Model

Byrum et al. developed a mechanistic-based transverse faulting model for nondoweled jointed pavement. This model is based on the NAPCOM and PCA approaches and coupled with thermal gradients and shrinkage to accentuate the faulting due to upward slab curvature. Byrum proposed a curvature index to quantify the slab curvature due to thermal gradients and shrinkage. There was a good correlation between CI and transverse joint faulting. Hence, this approach may have a good potential for predicting joint pumping and faulting in the future (Byrum, 1999):

$$CI = 0.5(LCI + RCI)$$

where
  CI is the slab curvature index
  LCI is the left curvature index
  RCI is the right curvature index

Curvature is defined as follows:

$$Curvature = \frac{Slope_{n+1} - Slope_n}{X_{n+1} - X_n}$$

where
  Curvature is the slab curvature for a given interval or segment
  X_n is the horizontal distance along a given slab

Slope is defined as follows:

$$\text{Slope} = \frac{Y_{n+1} - Y_n}{X_{n+1} - X_n}$$

where
   Slope is the slope along the slab surface for a given interval or segment
   Y_n is the vertical elevation along a given slab
   X_n is the horizontal distance along a given slab

13.1.3   EROSION CHARACTERIZATION OF BASE/SUBBASE

The erodibility of a base is an important factor contributing to the potential pumping and faulting that may occur. Understanding, measuring, and predicting foundation material erodibility is, therefore, critical for developing accurate pumping and faulting models. Many factors may contribute to quantifying erodibility; they include soil compaction, shearing, friction, water pressure and mobility, and treatment for bound materials.

   Characterization of material erodibility was attempted by many. PIARC (the World Road Association) developed specifications for characterizing the erosion potential of base and subbase material into five classes. (PIARC, 1987). This is presented in Table 13.1.

   Other methods for developing more reliable predictors for erodibility include the *rotational shear device*, which uses water for erosion after applying a critical shear stress (this does not evaluate cohesive materials); the *jetting erosion device*, which is a modified rotational shear device that can

---

TABLE 13.1

PIARC Recommendations for Erosion Potential of Base/Subbase Material

Class	Description
A: extremely erosion resistant	Class A materials are extremely resistant to erosion. The typical material of this category is lean concrete with at least 7%–8% cement (or 6% with special addition of fines) or bituminous concrete with a bitumen content of at least 6%
B: erosion resistant	Class B materials are five times more erodible (on average) than class A materials, but they still offer good guarantees of erosion resistance because they are far from the threshold at which erodibility increases exponentially. The typical material of this category is a granular material that is cement treated in the plant and contains 5% cement
C: erosion resistant	Class C materials are five times more erodible (on average) than those in class B, and they are close to the threshold under which erodibility increases very rapidly in inverse proportion to the amount of binder. The typical material of this category is a granular material that is cement treated in the plant and contains 3.5% cement, or a bitumen-treated granular material with 3% bitumen
D: fairly erodible	Class D materials are five times more erodible (on average) than those of Class C. Their low binder content makes their erosion-resistance properties highly dependent on construction conditions and the homogeneity of the material. The typical material of this category is a granular material treated in place with 2.5% cement. Also falling within this category are fine soils treated in place, such as cement-treated silt-lime and cement-treated sand. By extension, clean, well-graded, good-quality granular materials would also fall in this category
E: very erodible	Class E materials are over five times more erodible (on average) than those of class D. Class E materials are untreated or very poorly treated mixes. The typical material of this category is an unprocessed treated material rich in fine elements, and especially untreated silt

Source: PIARC technical committee on concrete roads, World Road Association, 1987.

evaluate cohesive soils; the *brush test*, which assesses the erosion due to a standard brush and load of a soil by measuring weight loss; and the *South African erosion test*, which uses a loaded wheel on a submerged confined soil specimen, and quantifies erodibility using the Erosion Index (L).

13.1.4 CHARACTERIZING FREE WATER WITHIN A PAVEMENT STRUCTURE

The enhanced integrated climatic model (EICM) is a one-dimensional couple heat and flow program that is used for analyzing pavement soil systems with climatic conditions. EICM can model numerous climate-related parameters such as rainfall, solar radiation, cloud cover, wind speed, and air temperature, but the prediction of moisture at the interface between the PCC slab and base is essential to modeling pumping and faulting.

13.1.5 PRS M-E TRANSVERSE JOINT-FAULTING PREDICTION MODEL

The PRS M-E model estimates JPCP transverse joint faulting in two ways: estimating faulting without considering the amount of consolidation around dowels, and estimating faulting while considering the amount of consolidation around dowels.
The model is presented as follows:

$$FAULT = DAMAGE^{0.275} * [0.1741 - 0.0009911 * DAYS90 + 0.001082 * PRECIP]$$

where
  FAULT is the average transverse joint faulting per joint, in.
  DAMAGE = n/N
  n is the actual number of applied cumulative 18 kip ESALs
  N is the allowable number of applied cumulative 18 kip ESALs
  DAYS90 is the number of days per year with the maximum temperature greater than 32°C (90°F)
  PRECIP is the average annual precipitation, in.

Statistics:
  N is the 511 (JPCP sections from LTPP and FHWA RIPPER databases).
  R^2 is the 56%
  SEE is the 0.029 in. (0.74 mm)

The following equation is used to compute allowable ESALs (N):

$$Log(N) = 0.785983 - Log(EROD) - 0.92991 * (1 + 0.40 * PERM * (1 - DOWEL)) *$$

$$Log(DE * (1 - 1.432 * DOWELDIA + 0.513 * DOWELDIA^2))$$

where
  N is the allowable number of applied cumulative million 18 kip ESALs
  EROD is the base erodibility factor for the base (value between 1 and 5)
  PERM is the base permeability (0 = not permeable, 1 = permeable)
  DOWEL is the presence of dowels (1 if dowels are present, 0 if dowels are not present)
  DOWELDIA is the dowel diameter, in. (maximum allowed is 1.50 in.)
  DE is the differential energy density at a corner

DE at a corner is defined as the energy difference in the elastic base/subgrade deformation under the loaded leave slab and the unloaded approach slab. Another parameter used in the calculation of DE is that for the nondimensional aggregate interlock stiffness (AGG*) factor. When the percentage of consolidation around dowels is not considered, AGG* is computed using the following equation:

$$AGG^* = \left(\frac{AGG}{kL}\right) = 2.3 * Exp\left(-1.987 * \frac{JTSPACE}{L} + DOWELDIA^{2.2}\right)$$

where
    AGG* is the nondimensional aggregate interlock stiffness
    AGG is the aggregate load transfer stiffness, psi
    k is the dynamic modulus of subgrade reaction (dynamic k value, approximately twice the static value), psi/in.
    L is the slab's radius of relative stiffness, in.

$$= \frac{\left[E_{PCC} * h_{PCC}^3\right]}{\left(12 * (1 - v^2) * k\right)]^{0.25}}$$

E_{PCC} is the PCC modulus of elasticity, psi
h_{PCC}^3 is the slab thickness, in.
v is the PCC Poisson's ratio (assumed to be equal to 0.15)
JTSPACE is the slab length (joint spacing), ft
DOWELDIA is the dowel diameter, in. (maximum allowed is 1.50 in.)

The computation of DE involves completing a multistep process in which maximum corner deflections are computed for loaded and unloaded conditions.

13.1.6 Punchouts in Continuously Reinforced Concrete Pavements

Punchouts are the main structural distress of continuously reinforced concrete pavements (CRCP). Punchouts initiate as transverse cracks. Continued traffic loading on the pavement edge, coupled with loss of support for the slab, will eventually lead to a loss of load transfer at the transverse cracks. This scenario results in a section of the PCC slab between the transverse cracks to behave like a cantilever beam when subjected to wheel loads at the slab edge. When excessive stresses build up, micro-cracking develops, and eventually top-down cracking leads to punchouts (Figure 13.4).

FIGURE 13.4 Punchout resulting from cantilever effect at slab edge.

Load transfer efficiency (LTE) at the transverse crack is critical to potential tensile stresses that develop at the cantilever joint due to wheel edge loading. If adequate aggregate interlock exists at the transverse crack, the progression of punchouts could be retarded. Adequate load transfer is provided by aggregate at the crack interface, sufficient steel reinforcement in the CRCP, and PCC slab thickness that may increase the surface area of aggregate interlock faces.

Loss of foundation support will increase the shear at the crack faces, reduce aggregate interlock, and increase the tensile stresses at the cantilever section, resulting in increased potential for punchouts. Therefore, it is critical to maintain a nonerodible base. This can be accomplished by using cement or asphalt-treated bases.

13.1.6.1 Development of CRCP Punchout Models

Punchouts are commonly formed between transverse cracks spaced less than 2 ft apart. This, coupled with tensile stress development, LTE, and edge loading, could be used as a framework for developing CRCP punchout models. The following process is recommended:

- Assemble relevant pavement properties
- Determine transverse crack spacing
- Estimate number of crack spacing less than 2 ft
- Estimate transverse crack width and mean LTE over time
- Compute tensile stress at top of slab
- Compute tensile stress-to-strength ratio and allowable number of load repetitions
- Use Miner's damage model to compute fatigue damage for 2 ft crack spacing
- Calibrate cumulative fatigue damage with field punchout distress data to predict punchouts

Procedure for determining transverse crack development (after Abou-Ayyash, 1974):

- Determine tensile strength of CRCP slab at time t_n.
- Compute drying shrinkage Z_n, change in temperature ΔT_1 at time t_1.
- Calculate maximum tensile stress generated in CRCP.
- Compare computed CRCP stress to concrete tensile strength; cracking will occur if actual concrete slab stress exceeds concrete tensile strength.
- Increase time to t_2, and repeat the process. If failure occurs between time n and $(n+1)$, then use iterative tools to determine the exact time of cracking between intervals.
- Repeat the process at a later time and search for additional cracks, modifying the slab configuration to account for initial cracks.

Crack spacing in CRCP is affected by numerous factors such as concrete properties, age, environmental conditions, reinforcement content, placement and steel properties, slab geometry and boundary conditions, and construction factors, among others.

The bending tensile stress at the cantilever joint in the transverse direction is used to compute the critical pavement response parameter stress ratio (SR), and is calculated as follows:

$$SR = \frac{\sigma}{M_R}$$

where
σ is the top tensile stress in the transverse direction along the longitudinal crack for CRCP
M_R is the PCC flexural strength

The number of allowable load repetitions is computed using the stress ratio by the following:

$$\log N_{ij} = \forall (SR)^{\Omega}$$

where
  SR is the stress ratio
  $\Omega \forall$ are the regression constants

Fatigue damage is then determined using Miner's damage accumulation equation and can account for incremental damage due to traffic, concrete properties, age, foundation, and seasonal and climate variables. The cumulative fatigue damage due to all transverse crack spacings can be accumulated over the entire crack-spacing distribution using Miner's approach:

$$FD = \sum_{i=1}^{n} \frac{n_i}{N_i}$$

where
  FD is the accumulated fatigue damage for ith crack spacing
  n_i is the number of applied ESALs
  N_i is the number of allowable ESALs
  i is the counter for crack spacing

13.1.6.2 Punchout Distress Model

An example of an M-E-based distress model that predicts punchouts as a function of accumulated fatigue damage due to bending in the transverse direction is given by the following:

$$PO = 100e^{-(FD/\alpha)^{\beta}}$$

where
  PO is the total predicted number of punchouts per mile
  FD is the accumulated fatigue damage (due to slab bending in the transverse direction) assuming ith crack spacing
  α, β are the calibration constants for punchout prediction

It can be seen that punchouts occur due to the deterioration of transverse cracks that develop in CRCP slabs. The transverse cracks are induced by climate-related tensile stresses and built-in curling and warping stresses, and deteriorate due to traffic loading, loss of base support, and loss of load transfer through the transverse crack. Fatigue cantilever stresses due to repeated wheel edge loads will eventually fracture the slab and cause it to settle into a moderate to severe punchout. Punchouts are influenced to a certain degree by the following parameters: PCC slab thickness, PCC stiffness or elastic modulus, base/subbase support, amount and depth of reinforcing steel, coefficient of thermal expansion for steel, shrinkage of PCC, PCC creep characteristics, slab–base friction, age, temperature, and humidity during construction.

13.1.7 Smoothness Considerations

Smoothness or roughness is a rider-perceived judgment of pavement ride quality. A pavement's functionality is based more on ride quality and user comfort than on structural performance.

Road roughness does not have one standard definition. Road roughness can be described as the lack of smoothness. The American Society of Testing and Materials (ASTM) defines roughness as follows: "The deviations of a pavement surface from a true planar surface with characteristic dimensions that affect vehicle dynamics, ride quality, dynamic loads, and drainage, for example, longitudinal profile, transverse profile, and cross slope." Or it has been described as the "irregularities in the pavement surface that affect the smoothness of the ride or induce vibration in traveling vehicles."

Road roughness is caused by either longitudinal or transverse distortions in the roadway, where longitudinal is in the direction of travel and transverse is perpendicular to the direction of travel. Longitudinal distortions can occur in either long wavelengths or short wavelengths. Long wavelengths are caused by consolidation of the pavement's foundation material and usually occur with high amplitudes but low frequency. Transverse distortions are usually the result of rutting or settlement within the road. Roadways with various frequencies and amplitudes of distortions cause oscillations in vehicles traveling over the pavement, and these oscillations are referred to as *roughness.*

To determine roughness or smoothness, a pavement profile must be obtained. A profile can be defined as the elevation profile representation of a two-dimensional slice of pavement surface. To obtain a profile, one can use a rod and level, a Dipstick™ (walking elevation measuring device), or an inertial profiler.

An inertial profilometer is a vehicle that is equipped with an accelerometer mounted to an axle and provides inertial reference by measuring the vertical acceleration of the vehicle. A sensor mounted to the vehicle's body measures the distance from a set point on the vehicle to the pavement. This is the vehicle's relative elevation displacement. Different types of distance-measuring sensors are commonly used such as laser, optical, and ultrasound. Since the sensors are digital instruments, they only measure on a set interval. This interval, expressed in terms of the longitudinal vehicle motion between readings, is known as *sample spacing.*

To determine the profile, the accelerations are first integrated twice to determine the vertical displacement of the axle. The change between the height and relative displacement is the pavement profile. The longitudinal distance is measured using a distance-measuring device (DMI).

There are two different types of inertial profilers—high speed and lightweight. High-speed devices are those which are usually van mounted, and operate at highway speeds. Lightweight inertial profilers are mounted on a smaller vehicle, similar to a golf cart. The high-speed profilers are typically used for network condition assessment, while the lightweights are used for pavement construction quality acceptance.

The International Roughness Index (IRI) is an index computed from the cumulative elevation changes over a distance, as determined from a longitudinal road. This is correlated to a standard quarter-car (one wheel's suspension) model, and measures the cumulative output of the suspension travel. It was developed for the World Bank in 1982 to establish a relevant, transportable, and time-stable measure of road roughness.

The IRI is an important ride quality statistic used for M-E smoothness modeling. The M-E design procedures include performance prediction models to predict the progression of certain key pavement distresses over the design life of a pavement.

Pavement smoothness or ride quality has been used since the AASHO Road Test. The present serviceability rating (PSR) concept was a subjective assessment of ride quality, while the PSI was developed as an objective quantitative assessment. The longitudinal portion of the PSI contributed greatly to the PSI value. And, therefore, smoothness has become used as the parameter that most closely relates to serviceability.

The first model to relate serviceability and distress was developed for the AASHO Road Test (as developed by Carey and Irick, 1960). The PSR was defined as follows:

$$PSR = C + A_1 R_1 + B_1 D_1 + B_2 D_2$$

where
    C, A_1, B_1, B_2 are the regression coefficients
    R_1 is the function of longitudinal profile
    D_1 is the function of transverse profile
    D_2 is the function of surface distress

The rigid pavement model used for the AASHO Road Test was given as follows:

$$PSR = 5.41 - 1.78 \log(1 + SV) - 0.09(C + P)^{0.5}$$

where
    PSR is the present serviceability rating or panel mean rating
    SV is the slope variance
    C is the major cracking in ft per 1000 ft^2 area
    P is the bituminous patching in ft^2 per 1000 ft^2 area

Slope variance is defined as follows:

$$SV = \frac{\sum Y^2 - \frac{1}{n}\left(\sum Y\right)^2}{n - 1}$$

where
    Y is the difference between two elevations 9 in. apart
    n is the number of elevation readings

Correlation statistics show that PSR is highly correlated with smoothness and distress for rigid pavements. It was also observed that distress greatly affects smoothness or profile measurements. Other researchers have studied and developed various models to correlate smoothness to pavement distress features. Unfortunately, the models vary widely based on the type and extent of distress. Increasing quantities and severities of distresses such as cracking, faulting, joint spalling, crack and joint deterioration, and punchouts will greatly influence smoothness degradation in pavements. Hence, the development of M-E pavement distress models will be an initial step to the development of smoothness models. Therefore, input parameters for distress models, such as traffic, construction, and environment over time, will also be applicable for smoothness predictor models.

Initial smoothness was found to be highly correlated to future smoothness and degradation of smoothness. Pavements that are built with a smoother profile tend to last longer and will remain smoother longer. This probably relates to the quality of the construction, where smoother pavements are a result of good construction practices. In addition, smoother pavements will be subject to less truck-dynamic loads than a rougher pavement surface (NCHRP 1–31). This can be seen in the incorporation of smoothness specifications that are part of many state highway agencies' new pavement construction projects. Smoothness of new pavements has become so important to the owners and riding public that many new construction specifications have included a bonus/penalty clause. If the contractor exceeds expectations, he or she is paid a bonus. If the contractor has not met smoothness specifications, he or she must pay a penalty or reconstruct the pavement if it is below a certain threshold.

Any factors that will contribute to changing the pavement profile will contribute to smoothness loss. This includes foundation movement (consolidation and swelling of soils, and frost-susceptible soils), maintenance activities (such as patching and crack filling), and initial smoothness. Therefore, these conditions must be addressed when developing predictor models for pavement smoothness.

A model to predict smoothness could be structured as follows:

$$S(t) = S_0 + a_i D(t)_i + b_i M_j + c_i SF$$

where
$S(t)$ is the pavement smoothness over time (IRI, m/km)
S_0 is the initial smoothness over time (IRI, m/km)
a_i b_i c_i are the regression constants
$D(t)_i$ is the ith distress at a given time
M_j are the maintenance activities that significantly influence smoothness (e.g., patching and crack filling)
SF are the site factors, which are defined as follows:

$$SF = AGE\left(SC_1 * SC_2 * SC_3\right)$$

A model example using the proposed structure described earlier was developed under a FHWA study (Hoerner, 1998). Input parameters include initial IRI, cracking, spalling, and faulting. A site factor (SITE) was added to account for the effects of age, Freezing Index, and percentage of subgrade passing the 0.075 mm (No. 200) sieve. The model that predicts smoothness in IRI is given as follows:

$$IRI = IRI_0 + 0.013 * \%CRACKED + 0.007 * \%SPALL + 0.001 * TFAULT + 0.03 * SITE$$

where
IRI_i is the initial smoothness measured in m/km
%CRACKED is the percentage of slabs with transverse cracking and corner cracks (all severities), expressed as a number between 0 and 100
%SPALL is the percentage of joints with spalling (medium and high severities), expressed as a number between 0 and 100
TFAULT is the total joint faulting cumulated per km, mm
SITE is the site factor $= AGE (1+FI)^{1.5} (1+P_{0.075}) * 10^{-6}$
AGE is the pavement age since construction
FI is the Freezing Index, °C days
$P_{0.075}$ is the percentage of subgrade material passing the 0.075 (No. 200) sieve, expressed as a number between 0 and 100

This model can predict future smoothness in IRI as a function of initial IRI, transverse slab cracking, transverse joint spalling, transverse joint faulting, and pavement site condition.

13.2 TESTS FOR CONCRETE

13.2.1 FLEXURAL STRENGTH TESTS

The flexural strength (also called the *modulus of rupture*, or MOR) is a specified design parameter for rigid pavements. This is due to the fact that pavements are stressed in bending during traffic loading and under curling stresses. The flexural strength can be determined using two

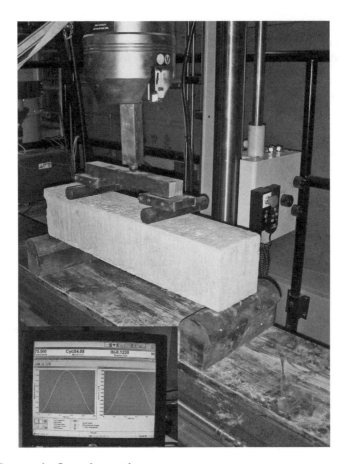

FIGURE 13.5   Center-point flexural strength test.

different tests: a center-point load (Figure 13.5) and a third-point load. The beam size is typically $6 \times 6 \times 20$ in. for an aggregate maximum size less than 2 in. The difference between the two test methods is that the third-point method provides pure moment in the middle third of the span with no shear. In the center-point load, the point of fracture is unknown and there is both moment and shear contributing to the failure. The standard flexural strength test is given by AASHTO T-97 and ASTM C-78, (Flexural Strength of Concrete [Using Simple Beam with Third-Point Loading]), and AASHTO T-177 and ASTM C-293, (Flexural Strength of Concrete [Using Simple Beam with Center-Point Loading]).

The MOR of rupture is calculated using the following equation:

$$MOR = \frac{Mc}{I}$$

where
   M is the moment
   c is the distance from neutral axis to the extreme fiber
   I is the moment of inertia

Usually, mix designs are typically tested for both flexural and compressive strength; they must meet a minimum flexural strength, which is then correlated to measured compressive strengths so that compressive strength (an easier test) can be used in field acceptance tests.

FIGURE 13.6 Compressive strength test.

13.2.2 COMPRESSIVE STRENGTH

The compressive strength test is performed using a 6 × 12 in. cylinder after 28 days of curing. The test is shown in Figure 13.6 and is conducted in accordance with AASHTO T-22 and ASTM C-39 (Compressive Strength of Cylindrical Concrete Specimens). Although the MOR is the specified strength for structural design in rigid pavements, the compressive strength is a much easier test to conduct and can be correlated reliably to the MOR. Hence, compressive strength is used by certain agencies for acceptance criteria. The compressive strength test is also less prone to testing variation than the flexural strength tests.

13.2.3 TENSILE STRENGTH

PCC tensile strength is important in pavement applications. Tensile strength may be used as a PCC performance measure for pavements because it best simulates tensile stresses at the bottom of the PCC surface course during traffic loading. Because of secondary stresses generated during gripping and pulling of a sample, a true direct tensile strength is difficult to conduct. Therefore, tensile stresses are typically measured indirectly using a splitting tension test.

A splitting tension test (AASHTO T-198 and ASTM C-496, Splitting Tensile Strength of Cylindrical Concrete Specimens) uses a standard 6 × 12 in. cylinder laid on its side. A diametral compressive load is then applied along the length of the cylinder until it fails (see Figure 13.7). Because PCC is much weaker in tension than compression, the cylinder will typically fail due to horizontal tension and not due to a vertical compression.

The modulus of elasticity for normal weight concrete (Ec) ranges between 2,000,000 and 6,000,000 psi. It can be determined by calculating the stress-to-strain ratio in the elastic range in accordance with ASTM C-469. Alternatively, it can be approximated using the ACI empirical equations by $Ec = 57,000\sqrt{f'c}$ (f'c in psi) or $Ec = 5,000\sqrt{f'c}$ (f'c in SI units).

The value of Poisson's ratio for concrete will range between 0.15 and 0.25, but is commonly selected as 0.20–0.21.

13.2.4 COEFFICIENT OF THERMAL EXPANSION TEST

In this test (AASHTO TP-60), a concrete sample is placed in a temperature-controlled chamber, and its expansion is monitored at different temperatures. The values range from 6 to 13 * $10^{-6}/°C$.

FIGURE 13.7   Tensile strength test.

13.2.5   Fatigue Testing for PCC

In a fatigue test for PCC, a repeated flexural loading is applied on a prismatic beam specimen measuring $3 \times 3 \times 15$ in. Loading is usually applied at the third points with a rate of 1–2 repetitions per second (2 Hz) for a duration of 0.1 s.

The stress (σ) in the extreme fiber of the beam is calculated using the following equation:

$$\sigma = \frac{3aP}{bh^2}$$

where
    a is the distance from the edge support to the inner support
    P is the total applied load
    b is base of the beam
    h is height of the beam

Note: when the third points are equidistant and $a = L/3$, then the stress equation reduces to PL/bh^2.

The results are plotted on a semilog plot where the number of cycles to failure are plotted on the log scale—(log N_f) is plotted versus the flexural stress ratio (the quotient of the applied cyclic stress, σ, and the modulus of rupture, S_c, as determined by ASTM C78-84).

Based on numerous published data, it has been observed that the concrete will not reach failure by fatigue if the stress ratio is less than 50% after approximately 20 million cycles or repetitions (Mindess and Young, 1981, Neville, 1981). Based on data published by others, Huang (1993) found that the relationship representing a 50% probability of failure is expressed as follows:

$$\log N_f = 17.61 - 17.61 \left(\frac{\sigma}{S_c} \right)$$

Similarly, the Portland Cement Association recommends the use of the following fatigue equations:

$$\text{For } \frac{\sigma}{S_c} \geq 0.55 : \log N_f = 11.737 - 12.077\left(\frac{\sigma}{S_c}\right)$$

$$\text{For } 0.45 < \frac{\sigma}{S_c} \geq 0.55 : N_f = \left(\frac{4.2577}{\sigma/S_c - 0.4325}\right)^{3.268}$$

$$\text{For } \frac{\sigma}{S_c} \leq 0.45 : N_f = \text{unlimited}$$

QUESTIONS

13.1 A 12 in. long concrete cylinder with a 6 in. diameter failed at a load of 122,000 lb. Determine the compressive strength and approximate modulus of elasticity.

13.2 For a concrete slab of dimension 40 ft (length) by 12 ft (width) by 6 in. (thickness), determine the stresses developed due to a temperature drop of 20°C. Use concrete property values from question 1.

14 Mix and Structural Design of Asphalt Mix Layers

Structural and mix design of asphalt pavements are interlinked. The mix design step develops the most optimum combination of materials to produce a mix, and the structural design step utilizes the properties of the designed mix to determine the most optimum thickness and number of layers. Even though it seems that the structural design step should come after mix design, in many cases preliminary structural design is done prior to mix design. This is because, for most agencies, a database of information on relevant properties of typical mixes and materials exists.

Mix design is conducted mostly on the basis of consideration of volumetric properties of asphalt mixes. The materials, aggregates and asphalts, are selected on the basis of empirical and/or mechanistic test properties. These properties are considered to ensure that the mix will be resistant against the major distresses—thermal cracking, fatigue cracking, and rutting. Empirical and, in some cases, calibrated mechanistic models are available which relate volumetric properties to levels/amounts of distress. These models are considered to determine appropriate optimum levels of the volumetric and other properties during mix design.

Structural design of asphalt pavements can be done primarily in two ways—either through the use of charts and nomographs that have been developed on the basis of experiments and/or experience (empirical method), or part through the use of principles of mechanics and part through results obtained from experiments and/or experience (mechanistic-empirical [ME]).

Note that while the tendency for structural design has been to move from purely empirical to mechanistic-empirical methods (to reduce uncertainties regarding extrapolation of data), mix design has more or less remained primarily dependent on empirical relationships between volumetric properties and distress levels in asphalt mixes (although more sophisticated models relating volumetric properties to mechanistic properties have been developed). The two major advancements made in mix design are the development and adoption of better mix compaction methods and mix evaluation methods in the laboratory. For structural design, the major advancements have been in terms of better testing methods, sophisticated data analysis techniques, and field calibration of models through instrumentation and accelerated loading and testing.

14.1 PHYSICAL AND VOLUMETRIC PROPERTIES OF ASPHALT MIX

Physical and volumetric properties are used extensively in mix design, as they help us to convert weight to volume and vice versa, and as they have been shown to have significant correlations with the performance of asphalt mixes. Correlations between volumetric properties and structural properties are also used in structural design, when testing is not possible or actual test data are not available.

Figure 14.1 shows a simplified schematic of a compacted asphalt mix (i.e., what we find in pavements). It consists of three primary components—the aggregate, the asphalt binder, and some air voids. Before the asphalt is added to the aggregates, there were only aggregates and some voids within the aggregate structure. Part of the total voids is filled with asphalt binder, and the rest remains as air voids. Of the asphalt binder, part of it is absorbed by the aggregate, and part remains as effective asphalt binder on the aggregate surface. The relative proportions of the

Definitions

$$VTM, \% = \frac{V_a}{V_{mb}} *100 \qquad P_{ba}, \% = \frac{M_{ba}}{M_s} *100$$

$$VMA, \% = \frac{V_a + V_{be}}{V_{mb}} *100 \qquad P_{be}, \% = \frac{M_{be}}{M_t} *100$$

$$VFA, \% = \frac{V_{be}}{V_a + V_{be}} *100$$

Calculation of volumetric parameters requires the determination of density or specific gravity values of the different components.

FIGURE 14.1 Conceptual block diagram.

volume of the air voids with respect to the total volume and with respect to the volume of the total voids, and the relative proportion of the volume of the effective asphalt binder with respect to the volumes of the total voids, are important volumetric parameters—those that have been shown to have significant effects on the performance of the asphalt mix. In addition, the proportion of mass of the effective asphalt with respect to the total mass is also important, since it tells us how much of the asphalt binder that we add to the aggregates is actually being effective in providing all the good properties that the asphalt binder is suppose to provide.

The different parameters discussed earlier can be separated into the following (as explained in Figure 14.1).

Voids in total mix (VTM): Total volume of pockets of air in between the *asphalt-coated aggregates* in a compacted asphalt mix, expressed as a percentage of the total volume of the mix. Neither a high nor a very low VTM is desirable in a compacted asphalt mix—a VTM of 3–5 is the optimum range for adequate stability and durability.

Absorbed asphalt (P_{ba}): The mass of asphalt binder that is absorbed by the aggregates, expressed as a percentage of the mass of the aggregate. There is no specific limit on absorption except the fact that highly absorptive aggregates (>2% absorption) may not be used for economical reasons (would need a relatively high amount of asphalt binder). However, if aggregates with relatively high absorption are used, the "aging" process during mix design in the laboratory needs to be modified to accurately simulate the expected high absorption during production in the plant.

Effective asphalt (P_{be}): The mass of the asphalt binder that is not absorbed by the aggregate (total—absorbed), expressed as a percentage of the mass of the total mass of the mix.

Voids in mineral aggregate (VMA): Total volume of the void spaces in between the *aggregates* in a compacted asphalt mix, part of which is filled with effective asphalt binder, and part with air, expressed as a percentage of the total volume of the mix. Generally, a minimum VMA is desirable to provide space for adequate asphalt binder for durability, and air voids for stability, although a very high VMA would lead to unstable mixes as well. Generally, for surface and base course paving mixes, the specified minimum VMA values range from 12 to 16, depending on the nominal maximum aggregate size (NMAS) of the mix—the higher the NMAS, the lower is the minimum VMA. This is based on the principle that a gradation with a higher NMAS has a relatively lower surface area (of aggregates) and hence would require a relatively lower amount of asphalt to provide an optimum film thickness of asphalt binder.

Voids filled with asphalt (VFA): The part of the total void spaces in between the aggregates in a compacted mix that is filled with asphalt binder, expressed as a percentage. A low VFA indicates an inadequate amount of asphalt binder or too high VTM, which are undesirable for durability, whereas a high value indicates a relatively low VTM, which can cause stability problems, Hence a range, generally 65–80, is specified on the basis of traffic levels—a higher volume of traffic would need a mix with lower asphalt content for stability, and hence a lower VFA. Note that the volumetric properties need the determination of density or specific gravity.

Next consider Figure 14.2, in addition to Figure 14.1. They show definitions of the different parameters discussed earlier, in equations, and point out the properties that we need from tests to determine these volumetric properties. These are listed next.

14.1.1 Bulk-Specific Gravity of Compacted Asphalt Mix (G_{MB})

This test determines the ratio of weight in air of a compacted asphalt mix specimen at a certain temperature to the weight of an equal volume of water at the same temperature. For dense graded asphalt paving mixes, the test can be run according to AASHTO T-166 (Bulk Specific Gravity of Compacted Hot Mix Asphalt Using Saturated Surface Dry Specimens). The test consists of measuring the weight of the sample in air, in water (after immersing for 3–5 min), and then measuring

Working formula

$$VTM, \% = \left(1 - \frac{G_{mb}}{G_{mm}}\right) * 100$$

$$VMA, \% = \left(1 - \frac{G_{mb}(1 - P_b)}{G_{sb}}\right) * 100$$

$$VFA, \% = \left(\frac{VMA - VTM}{VMA}\right) * 100$$

$$P_{ba} = 100 * \left(\frac{G_{se} - G_{sb}}{G_{sb} G_{se}}\right) * G_b$$

$$P_{be}, \% = P_b - \frac{P_{ba}}{100} P_s$$

Volume		Mass
V_a	Air	$M_v = 0$
V_{be}	Effective asphalt	M_{be}
V_{ba}	Absorbed asphalt	M_{ba}
V_{sv}	Voids not filled	$M_{sv} = 0$
V_s	Aggregate	M_s

G_{mb} = Bulk specific gravity of compacted mix

$$G_{mb} = \frac{\text{Mass of aggregate and asphalt}}{\text{Volume of aggregate, asphalt and air}} = \frac{M_t}{V_{mb}}$$

G_{mm} = Maximum (mix) specific gravity

$$G_{mm} = \frac{\text{Mass of aggregate and asphalt}}{\text{Volume of aggregate and asphalt}} = \frac{M_s + M_b}{V_{mb} - V_a}$$

P_b = Percentage of asphalt binder by weight of mix, or asphalt content, %

$$P_b = \left(\frac{M_b}{M_t}\right) * 100$$

$$G_{se} = \frac{\text{Mass of aggregate}}{\text{Volume of aggregate, impermeable and permeable voids in the aggregate − volume of absorbed asphalt}}$$

Calculated as follows:

$$G_{se} = \frac{100 - P_b}{\dfrac{100}{G_{mm}} - \dfrac{P_b}{G_b}}$$

FIGURE 14.2 Different parameters required for calculation of volumetric properties.

the saturated surface dry (SSD) weight (after patting the surface dry). The bulk-specific gravity is determined from the following formula.

$$G_{mb} = \frac{A}{B - C}$$

where
  A is the weight in grams of the specimen in air
  B is the weight in grams of the saturated surface dry specimen in air
  C is the weight in grams of the specimen in water

The bulk-specific gravity is multiplied by the density of water to obtain bulk density.

Specimens that contain open or interconnected voids or absorb more than 2% water (or both) can be tested by using either ASTM D-1188 (Bulk Specific Gravity and Density of Compacted Bituminous Mixtures Using Coated Samples) or AASHTO TP-69 (Bulk Specific Gravity and Density of Compacted Asphalt Mixtures Using Automatic Vacuum Sealing Method). Note that for testing of field cores that have been obtained by sawing or coring and that contain a significant amount of water, the procedure is the same, except that the dry mass (mass in grams of the specimen in air) is obtained last after drying the specimen at 110°C to constant mass.

14.1.2 Theoretical Maximum Density or Maximum Specific Gravity of the Mix (G_{MM})/Rice Specific Gravity (after James Rice)

This test determines the ratio of weight of a unit volume of an uncompacted asphalt mix sample to the weight of an equal volume of water at the same temperature. This test can be run according to ASTM D-2041 (Theoretical Maximum Specific Gravity and Density of Bituminous Paving Mixtures) or AASHTO T-209 (Theoretical Maximum Specific Gravity and Density of Hot-Mix Asphalt Paving Mixtures). A specified amount of mix (ranging in weight from 1.1 to 5.5 lb for a range of maximum size of aggregate of #4–1 in.) is warmed in an oven and taken out, and the particles are separated from each other. The mix is weighed in air, and then placed in a pycnometer with sufficient water to cover it. A vacuum pump is then connected to the pycnometer and turned on to remove entrapped air from the mix. The mix is agitated during this procedure to help remove the air. At the end of this step, the pycnometer is submerged in a water bath, and its weight (along with the mix inside it) is noted. The following expression is used for the calculation of G_{mm}.

$$G_{mm} = \frac{A}{A - B}$$

where
  A is the weight of sample in air, g
  B is the weight of sample in water, g

Note that there is a check for making sure that porous aggregates without an adequate amount of asphalt binder cover are not getting saturated with water during the vacuum procedure.

The different specific gravities of aggregates and asphalt binder have been discussed in Chapters 8 and 9, respectively. The importance of the three specific gravities of aggregate is discussed in the following paragraphs.

The need for the determination of the different specific gravities of aggregate arises from the fact that all aggregates absorb water (and hence asphalt as well) to some extent. Determination of the different specific gravities allows the estimation of such absorbed water, and then the estimation of absorbed asphalt. Whatever the specific gravity might be, the testing procedure always has the basic fundamental objective of determination of weight (or, more precisely, mass) divided by volume. The weight is either dry (for G_{sb}) or saturated surface dry (for $G_{sb\text{-}SSD}$), and the volume is (volume of aggregate + surface voids) for G_{sb}, and volume of aggregate only for apparent specific gravity (G_{sa}; Figure 14.3). Using G_{sb} and G_{sa}, as shown in Figure 14.2, one can determine the absorption.

Note that when the specific gravities of different size fractions are used (for a blend of aggregates), the average specific gravity value of the blend needs to be calculated and used in the determination of the volumetric properties of the mix. The procedure is explained in Figure 14.3.

Bulk specific gravity

$$G_{sb}, Dry = \frac{Mass\ of\ dry\ aggregate}{Volume\ of\ aggregate\ and\ surface\ voids}$$

$$G_{sb}, SSD = \frac{Mass\ of\ dry\ aggregate\ and\ water}{Volume\ of\ aggregate\ and\ surface\ voids}$$

Water permeable surface voids

Apparent specific gravity

$$G_{sa} = \frac{Mass\ of\ dry\ aggregate}{Volume\ of\ aggregate}$$

The following equation can be used to compute an average specific gravity of a blend of aggregates consisting of different seize fractions:

$$G = \frac{1}{\frac{P_1}{100G_1} + \frac{P_2}{100G_2} - \frac{P_n}{100G_n}}$$

where
  G is the average specific gravity
  $G_1, G_2, ...G_n$ is the appropriate specific gravity of each size fraction
  $P_1, P_2, ...P_n$ is the percentage (by mass) of each fraction in the total sample

FIGURE 14.3   Bulk and apparent specific gravity of aggregates.

The concept of effective specific gravity arises in an asphalt-coated aggregate. In this case, for the mass divided by volume parameter, the mass is still the dry aggregate, but the volume consists of the volume of the aggregate plus the volume of the surface voids that are not filled with asphalt. This definition is important since the "effective" aggregate volume is used to calculate the absorbed asphalt and hence the effective asphalt binder content. Effective specific gravity is a calculated parameter, from G_{mm}, as shown in Figure 14.2.

A worked-out example of the use of the different parameters is shown in Figure 14.4.

14.2   MIX DESIGN METHODS

In the following paragraphs, three different mix design systems are presented. Note that all of the mix design systems have the same objectives of selecting the most desirable aggregates and asphalts and combining them in the most appropriate ratio to obtain the optimum volumetric properties.

14.2.1   HVEEM METHOD (ASTM D-1560, D-1561)

The Hveem method was developed by Francis Hveem of the California Division of Highways in the 1920s. It is intended for mixes with a maximum aggregate size of 1 in. and has been used primarily for the design of dense graded mixes. Currently, it is used mainly in some of the Western states in the United States.

Example: Determine the different volumetric properties of the mix, for which the following information is provided: $G_{mb} = 2.335$, $G_{mm} = 2.530$, $G_b = 1.02$, $P_b = 5.5\%$ (by mass of total mix), G_{sb} (blend of aggregate) = 2.650

Steps
1. Determine density of the mix: Density = $G_{mb} * Y_w = 2.335*1 \ g/cm^3 = 2.335 \ g/cm^3$;
2. Determine masses: Assume 1 cm³ of compacted HMA; Total mass = 2.335 g; $M_a = 0$; $M_b = (5.5/100)*2.335 = 0.128 \ g$; $M_s = (94.5/100)*2.335 = 2.206 \ g$;
3. Determine total volume of asphalt binder: $V_b = M_b/(G_b*Y_w) = 0.128/(1.02*1.00) = 0.125 \ cm^3$
4. Determine volume of bulk aggregate: $V_{sb} = M_s/(G_{sb}*Y_w) = 2.206/(2.650*1.00) = 0.832 \ cm^3$
5. Determine volume associated with G_{mm}: $V_{mm} = M_t/(G_{mm}*Y_w) = 2.335/(2.530*1.00) = 0.923 \ cm^3$
6. Determine volume of air: $V_a = V_{mb} - V_{mm} = 1.00 - 0.923 = 0.077 \ cm^3$
7. Determine volume of effective asphalt binder: $V_{be} = V_{mm} - V_{sb} = 0.923 - 0.832 = 0.091 \ cm^3$
8. Determine volume of absorbed asphalt binder: $V_{ba} = V_b - V_{be} = 0.125 - 0.091 = 0.034 \ cm^3$
9. Determine volume of effective aggregate: $V_{se} = V_{sb} - V_{ba} = 0.832 - 0.034 = 0.798 \ cm^3$
10. Determine mass of effective asphalt binder: $M_{be} = V_{be}*G_b*Y_w = 0.091*1.02*1.00 = 0.093 \ g$
11. Determine mass of absorbed asphalt binder: $M_{ba} = V_{ba}*G_b*Y_w = 0.034*1.02*1.00 = 0.035 \ g$
12. Determine volumetric properties: VTM, % = $100\% * V_a/V_{mb} = 100\%*0.077/1.00 = 7.7 \ \%$;
    VMA, % = $100\%*(V_a + V_{be})/V_{mb} = 100\%*(0.077 + 0.091)/1.00 = 16.8\%$;
    VFA, % = $100\%*V_{be}/(V_{be} + V_a) = 100\%*0.091/(0.091 + 0.077) = 54.2\%$;
    Effectice specific gravity of aggregate, $G_{se} = Ms/(V_{se}*Y_w) = 2.206/(0.798 * 1.00) = 2.764$;
    Percent absorbed asphalt binder: P_{ba}, % = $100\% * M_{ba}/M_s = 100\% * 0.035/2.206 = 1.59\%$
    Percent effective asphalt binder: P_{ba}, % = $100\% * M_{be}/M_t = 100\% * 0.093/2.335 = 3.98\%$

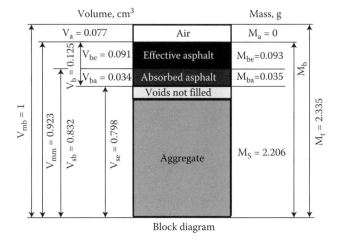

FIGURE 14.4 Worked-out example on calculation of different volumetric properties of a compacted hot mix asphalt (HMA).

The primary steps in this procedure are the following:

1. Select suitable aggregates, gradation, and asphalt binder for the project.
2. Determine the approximate asphalt (binder) content as a percentage of dry aggregate using the centrifuge kerosene equivalent (CKE) test procedure and aggregate gradation.
3. Prepare specimens, 2.5 in. high and 4 in. in diameter, for a range of asphalt contents at, below, and above the estimated asphalt content from Step 1, using a kneading compactor, at recommended temperatures (for specific asphalt grades).
4. Test using a stabilometer and a swell test apparatus.
5. Compute stabilometer value, swell, density, and air voids (using maximum specific gravity).
6. Determine the optimum asphalt content by selecting the maximum asphalt content that meets the minimum air voids and stabilometer value criteria. The other criteria that must be met are those for maximum swell and observed flushing of asphalt.

The key steps are illustrated in Figure 14.5, with the relevant charts in Figure 14.6. A kneading compactor is shown in Figure 14.7. The criteria are shown in Table 14.1.

14.2.2 Marshall Method

The Marshall method was developed by Bruce Marshall of the Mississippi Highway Department in the 1930s, and was refined by the U.S. Army Corps of Engineers in the 1940s and 1950s. Although intended for use of mixes with a maximum aggregate size of 1 in., the procedure has since been modified to accommodate large-size aggregates (up to 1 1/2 in.). This method continues to be used widely in airport pavement design jobs, and has been used for designing asphalt emulsion mixes, as well as for compaction and testing of quality control samples during construction.

Following are the main steps in this method.

1. Select aggregates using available test methods, and gradation, considering VMA; select appropriate viscosity-grade asphalt, conduct viscosity tests at different temperatures, and determine mixing and compaction temperature from temperature versus viscosity data.
2. Prepare specimens using a Marshall compactor (handheld or mechanical) with an appropriate number of blows per side (considering the traffic level of the project) at the mixing and compaction temperature determined in Step 1.
3. Determine bulk-specific gravity.
4. Condition the specimens, and run a Marshall stability and flow test.
5. Compute VTM, VMA, and VFA, using maximum specific gravity of the mix and specific gravity of aggregates, and stability and flow from test data obtained in Steps 3 and 4.
6. Determine the optimum asphalt content using criteria from the appropriate method, on the basis of VTM, VMA, VFA, stability, and flow values, for the expected traffic level in the project.

The method is explained in Figure 14.8, with the relevant charts and criteria in Figure 14.9. The fabrication of a Marshall specimen consists of placing a specified amount of mix in a mold assembly (with a protection paper disk at the bottom and top) and compacting each side of the mix inside the mold using a drop weight hammer. Generally 1200 g of HMA material are used to fabricate a Marshall specimen, which is 4 in. in diameter and 2 1/2 in. in height, or 509–522 cc in volume. The Marshall mold assembly consists of the collar (2 3/4 in. high) on the mold (3 7/26 in. high) which sits on a base (4 in. in diameter). The compaction hammer has a 10 lb sliding weight with a free fall height of 18 in., which could be raised automatically by a mechanism or manually (Figure 14.10). The mold assembly sits on a compaction pedestal which consists of an 8 in. × 8 in. × 18 in. wooden post capped with a 12 in. × 12 in. × 1 in. steel plate. The post is secured to a concrete floor by brackets, and the cap should be level. The mold assembly and compaction hammer face are heated to 300°F before compaction. The mix is transferred carefully inside the mold, with the collar on top of it, and spaded vigorously around the perimeter and at the center. After a paper disk is put on top of the smoothed surface of the mix in the mold, the mold assembly is transferred to the pedestal, and compaction starts when the temperature of the mix reaches a temperature at which the asphalt binder has a specified viscosity, for example, 280 ± 30 Cs. Each side of the specimen is compacted by the specified number of blows, the paper disks are peeled off the surface, and the specimen is left to cool down to a temperature that can be handled easily, after which the specimen is extruded out of the mold with an extruder.

The procedure for checking the stability and flow starts with conditioning the specimen in a 140°F water bath for 30–40 min. Testing is conducted using a stability and flow tester (Figure 14.11)

1. Compute equivalent surface area of the aggregate sample for the selected aggregate blend

Sieve Size (inch)	Percent Passing (Column B)	Surface Area Factor, ft²/lb (Column C)	Surface Area, ft²/lb Column B*Column C
Maximum size	100	2	
No. 4		2	
No. 8		4	
No. 16		8	
No. 30		14	
No. 50		30	
No. 100		60	
No. 200		160	
		Total surface area	Sum of above cells

2. Determine centriguge kerosene equivalent (CKE) of fine aggregates from the CKE test and percent oil retained for coarse aggregates from surface capacity test

$$CKE_{corrected} = CKE * \frac{SG}{2.65}$$

$$Percent\ oil\ retained_{corrected} = Percent\ oil\ retained * \frac{SG}{2.65}$$

$$CKE = \frac{W_w - W_D}{W} * 100$$

$$Percent\ oil\ retained = \frac{W_w - W_D}{W_D} * 100$$

SG = Specific gravity of fine aggregates

SG = Specific gravity of coarse aggregates

W_w = Weight of aggregate + absorbed kerosene

W_w = Weight of aggregate + absorbed oil

W_D = Weight of aggregate

W_D = Weight of aggregate

3. Determine K_f from Chart 1, and K_c from Chart 2

4. Determine the correction to K_f using chart 3

5. Calculate K_m as: $K_m = K_f +$ correction (correction is negative if $k_c - k_f$) is negative, otherwise positive

6. Determine the oil ratio for a cutback asphalt from chart 4

7. Determine oil ratio for the paving grade asphalt binder from chart 5 and compute asphalt content as follows:

$$Asphalt\ content = 100 * \frac{Oil\ ratio}{100 + Oil\ ratio}$$

8. Mix asphalt and aggregates using mixing temperature as follows:

Grade	Temperature Range,°F
AC 2.5, AR-1,000, or 200–300 Pen	225-250
AC-5, AR 2,000, or 120–150 Pen	250-275
AC-10, AR 4,000, or 85–100 Pen	275-300
AC-20, AR 8,000, or 60–70 Pen	300-325
AC-40, AR 16,000, or 40–50 Pen	300-325

Prepare 4 specimens - One at asphalt content determined in Step 7, 2 at 0.5% increments above, and one at 0.5% increment below - for stabilometer test; 2 at asphalt content determined in Step 7 for swell test

FIGURE 14.5   Key steps in the Hveem mix design method.

(*continued*)

9. After curing for 15 h at 140°F, reheat the mix to 230°F, and then compact using a kneading compactor

10. Conduct stabilometer test; determine stabilometer value (S) as follows:

$$S = \dfrac{22.2}{\dfrac{P_h D_2}{P_v - P_h} + 0.222}$$

D_2 is the displacement on specimen

P_v is the vertical pressure (typically 400 psi)

P_h is the horizontal pressure, from stabilometer pressure gage reading obtained at the instant P_v is 400 psi

Correct the stabilometer value for specimen height using Chart 6

11. Conduct swell test and determine swell as change in height of the specimen after 24 h of conditioning

12. Determine air voids in % in the specimen mixes

13. Determine the optimum asphalt content using Chart 7

FIGURE 14.5 (continued)　Key steps in the Hveem mix design method.

at a temperature of 70°F–100°F. The peak load at which the sample breaks and the flow are recorded. Stability correlation ratios are provided in AASHTO T-245 to adjust the values of the stability if a specimen with height different from 2 1/2 in. is used.

14.2.3　Superpave Method

The Superpave (*Superior Performing Pavement*) method was developed as a result of the Strategic Highway Research Program (SHRP) research effort in the United States during the late 1980s and early 1990s. The system includes the newly developed performance-graded (PG) asphalt binder classification system, aggregate properties identified as "consensus" properties, new mix compaction, and mix analysis procedures. The key steps in the mix design method are the following:

1. Select asphalt binder and aggregates on the basis of environmental conditions and traffic.
2. Select aggregate gradation and estimate preliminary asphalt binder contents by considering volumetric properties—VTM, VMA, and VFA—and by considering traffic levels.
3. Compact specimens at a range of asphalt contents (enveloping the estimate content from Step 3) using a Superpave gyratory compactor, with the number of gyrations commensurate with the traffic level expected.
4. Determine bulk density, and compute VTM, VMA, and VFA (using the theoretical maximum density [TMD] of mixes and bulk-specific gravities of the aggregates and asphalt).
5. Determine the optimum asphalt content on the basis of optimum VTM and allowable VMA and VFA.
6. Prepare specimens at construction voids with optimum asphalt content.
7. Subject one set of moisture-conditioned and another set of dry specimens to indirect tensile strength tests.
8. Compute the tensile strength ratio, and ensure adequate resistance against moisture susceptibility.

The method is illustrated in Figure 14.12, with the relevant plots and criteria in Figures 14.13 and 14.14, respectively.

AASHTO 312 specifies the method for using a Superpave gyratory compactor for compaction of 150 mm diameter HMA specimens for subsequent determination of bulk-specific gravity. The equipment consists of a loading system within a frame, which applies a specified load for

Chart for determining K_f from C.K.E.

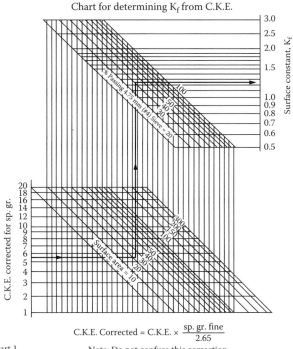

C.K.E. Corrected = C.K.E. $\times \dfrac{\text{sp. gr. fine}}{2.65}$

Chart 1

Note: Do not confuse this correction
to C.K.E. with that used in Figure 4.5

*Surface area, $\dfrac{\text{m}^2}{\text{kg}} = 0.204816 \dfrac{\text{ft}^2}{\text{lb}}$

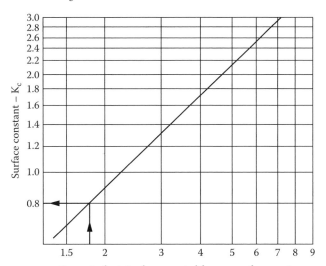

Percent oil retained – corrected for sp. gr. of aggregate

Material used: aggregate – passing 9.5 mm (3/8 in.), ret 4.75 mm (#4) sieve

Oil – SAE 10

% Oil ret. corrected = % oil ret. $\times \dfrac{\text{sp. gr. of coarse aggregate}}{2.65}$

Chart 2

FIGURE 14.6 Charts for Hveem mix design method. (Courtesy of California Department of Transportation, Sacramento, CA.)

(*continued*)

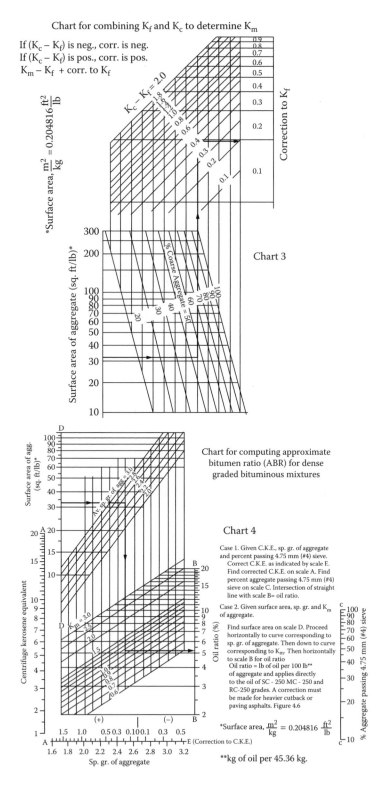

FIGURE 14.6 (continued) Charts for Hveem mix design method. (Courtesy of California Department of Transportation, Sacramento, CA.)

compaction of the mix in a mold, which sits between the loading ram and a rotating base. The loading ram applies a stress of 600 kPa, and the base rotates with a tilt of 1.25° (Figure 14.15), for a specified number of gyrations (depending on the level of compaction, which is dependent on the design traffic as well as high temperature), at 30 gyrations per minute. The final height is approximately 115 mm, and during compaction the height is continuously measured to 0.1 mm. Generally mix in the amount of approximately 5000 g is obtained following the AASHTO-specified quartering procedure from fresh samples, and is used for compaction. The mix should be compacted at a temperature at which the asphalt binder in the mix possesses a viscosity of $0.28+-0.03$ Pa·s. The mix is transferred to the preheated (to compaction temperature) steel mold (150 mm high with an inside diameter of 150 mm) with a base plate. Protective paper disks must be placed at the bottom of the mold before placing the mix, and at the top of the mix before placing it inside the gyratory compactor. The number of gyrations is then set, and the compaction started. The number of compactions could be set at N_{max} or N_{design}, depending on the agency-specific requirements. The number of gyrations versus specimen height data is generally stored to a disk via the equipment computer and/or printed out at the end of the compaction process. The specimen is then extruded out of the mold, and the paper discs are removed as quickly as possible. The specimens are then left to cool down for subsequent tests.

The height data at the final compaction are used to estimate the volume of the sample, and hence the bulk-specific gravity, which is then corrected on the basis of the actual bulk-specific gravity,

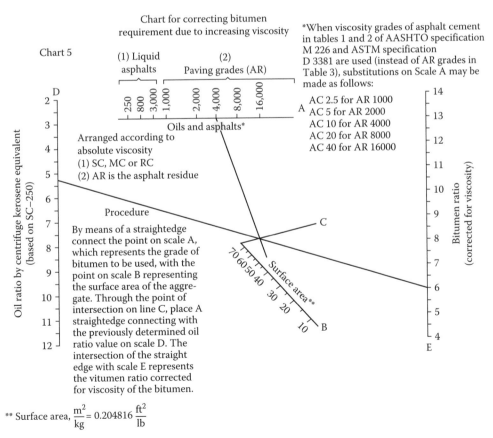

FIGURE 14.6 (continued) Charts for Hveem mix design method. (Courtesy of California Department of Transportation, Sacramento, CA.)

(continued)

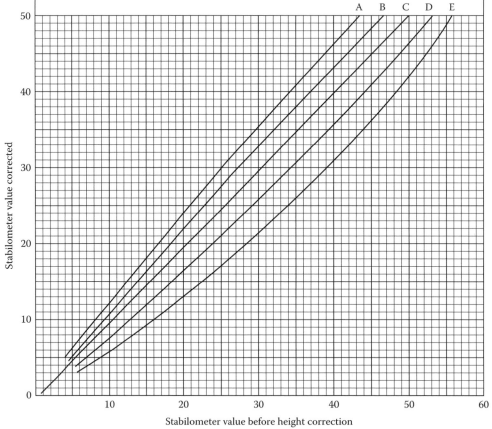

Chart 6

FIGURE 14.6 (continued) Charts for Hveem mix design method. (Courtesy of California Department of Transportation, Sacramento, CA.)

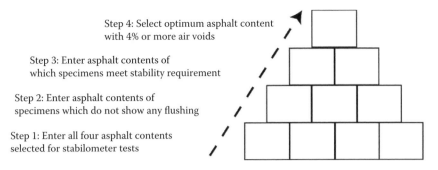

Step 4: Select optimum asphalt content
with 4% or more air voids

Step 3: Enter asphalt contents of
which specimens meet stability requirement

Step 2: Enter asphalt contents of
specimens which do not show any flushing

Step 1: Enter all four asphalt contents
selected for stabilometer tests

Chart 7

FIGURE 14.6 (continued) Charts for Hveem mix design method. (Courtesy of California Department of Transportation, Sacramento, CA.)

Compacting ram
with triangular face

Rotating sample
holder base

FIGURE 14.7 Kneading compactor.

TABLE 14.1
Requirements of Hveem Mix Design Mixes

Traffic Category	Heavy	Medium	Light
Stabilometer value	≥37	≥35	≥30
Swell	<0.030 in.		

---

which is determined by testing of the specimen. The correction factor is then used to determine the corrected bulk-specific gravities at all gyrations. The corrected bulk-specific gravity can be used along with the TMD or maximum Rice gravity (G_{mm}) to calculate the air voids. Figure 14.13 shows a complete example of all calculations. The densities at $N_{initial}$, N_{design}, and N_{max} are of primary importance, and the determined values are checked against specified values for compliance with mix design. The data can also be used for calculation and checking of voids in mineral aggregates (VMA) and voids filled with asphalt (VFA).

1. Conduct viscosity tests at different temperatures and plot viscosity versus temperature; select mixing temperature corresponding to 170 ± 20 cSt select compaction temperature corresponding to 280 ± 30 cSt

2. Mix 18 specimens - six at each of three asphalt contenst, optimum (from guidelines), 0.5 % above optimum, 0.5 % below optimum

3. Compact specimens using a marshall hammer, with numbers of blows as shown in the table in Figure 14.9, after curing mix, if required

4. Determine bulk specific gravity and compute VTM, VMA and VFA

5. Test specimens for stability and flow using the marshall stability-flow equipment

6. Plot asphalt content versus unit weight, VMA, stability, VFA, flow and VTM, as shown in Figure 14.9

7. Select the optimum asphalt content on the basis of criteria for specific method used; national Asphalt Pavement Association (NAPA) and Asphalt Institute (AI) guidelines are provided below

NAPA procedure

Select the asphalt content of the mix with the VTM equal to the median VTM in specification (for example, 4%) as the optimum asphalt content

Determine stability, flow, VMA and VFA at the optimum asphalt content and check to ensure compliance against criteria

If any one of the criteria is not met, redesign the mix

AI procedure

Determine

Asphalt content at maximum stability | Asphalt content at maximum density | Asphalt content at mid-point of allowable range of VTM

Determine the average of these three asphalt contents

Determine stability, flow, VTM, VMA and VFA at the optimum asphalt content from the plots and check to ensure compliance against criteria

If any one of the criteria is not met, redesign the mix

Note: Change aggregate gradation and redesign the mix if VMA criterion is not met

FIGURE 14.8   Key steps in the Marshall mix design method.

A test for resistance against moisture damage (AASHTO T-283) is then conducted on samples compacted with the optimum asphalt content. In this test, the resistance to moisture damage is evaluated by measuring the tensile strength of a dry set of HMA samples and another set of wet/conditioned samples, and comparing the two. The tensile strength ratio (strength of conditioned specimens divided by the strength of the unconditioned/dry specimens) is checked against a specified minimum limit (such as 80%). Note that the conditioning phase is supposed to simulate the destructive action of moisture/freezing/thawing, through which cohesive bonds inside the specimen are destroyed, and the tensile strength is lowered. This conditioning can be simply soaking and drying, or it can be freezing or thawing, depending on the environmental conditions that are expected in the location of the pavement being constructed. Specimens used in this test are compacted after conditioning the mix for 16 h in an oven at 60°C. The mix is compacted at the specified compaction

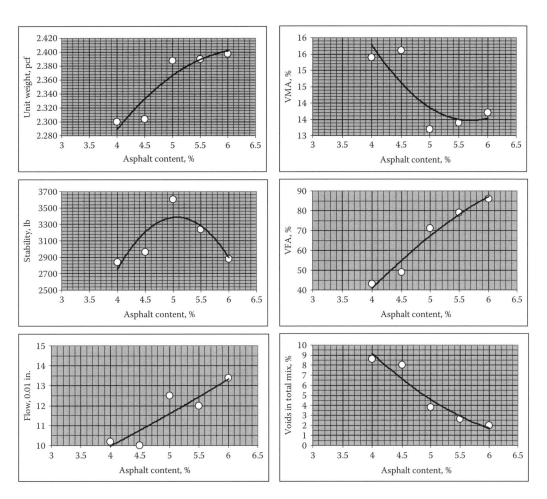

Traffic	Heavy	Medium	Light
Compaction, number of blows on each side	75	50	35
Stability, lb	Minimum: 1800	Minimum: 1200	Minimum: 750
Allowable range			
Flow, 0.01 in.	8–14	8–16	8–18
Voids in total mix, %	3–5	3–5	3–5

Nominal maximum aggregate size, inch	Minimum VMA, %		
	Voids in total mix, 3.0%	Voids in total mix, 4.0%	Voids in total mix, 5.0%
2.5	9.0	10.0	11.0
2.0	9.5	10.5	11.5
1.5	10.0	11.0	12.0
1.0	11.0	12.0	13.0
3/4	12.0	13.0	14.0
1/2	13.0	14.0	15.0
3/8	14.0	15.0	16.0
No. 4	16.0	17.0	18.0
No. 8	19.0	20.0	21.0
No. 16	21.5	22.5	23.5

FIGURE 14.9 Plots required in the Marshall mix design method.

FIGURE 14.10 Marshall hammer.

FIGURE 14.11 Marshall stability tester.

temperature such that the resulting specimens have 7% ± 1% air voids. These specimens need to be stored for 72–96 h at room temperature before the next step.

Based on the air voids (determined from bulk-specific gravity and maximum specific gravity, which are measured from an additional sample), the specimens are divided into two groups (each group with three samples) such that the average air voids of the two groups are approximately the same. Thickness and diameter measurements are made. The dry subset samples are placed inside a water bath maintained at 25°C after placing them in leak-proof plastic bags. After 2 h, the samples are removed from the bags and tested for their tensile strength. Load is applied at a rate of 50 mm/min. The peak load is recorded, and is used along with the diameter and the thickness of the sample to calculate the tensile strength. The wet subset samples are vacuum saturated to at least 55%–80% of saturation (which should be checked by measuring the SSD weight), after which they are covered with plastic wrap and placed in a plastic bag with 10 mL of water inside the bag. The bag is then placed inside a freezer at −18°C for 16 h, after which the specimen is placed in a 60°C water bath for 24 h. Then the specimen is conditioned at 25°C water, and then tested for tensile strength.

After breaking, the aggregates inside the sample could be examined to determine the degree of stripping or loss of asphalt, if any. Some agencies (such as the Georgia DOT) specify a method

1. Select asphalt binder on the basis of climate and traffic (see Chapter 9) from performance-graded classifications according to AASHTO MP1.
2. Select aggregates on the basis of consensus and source property requirement as shown in Figure 14.13.
3. Compact mixes with three trial gradations with an estimated asphalt content, using the superpave gyratory compactor, with number of gyrations recommended for specific traffic levels and average design high temperature. Use mixing and compaction temperature from the temperature versus viscosity plot obtained by testing viscosities at 135°C and 165°C. Select a mixing temperature correponding to a viscosity range of 0.15 and 0.19 Pa-s and a compaction temperature corresponding to a viscosity range of 0.25 and 0.13 Pa-s. Use 2-or-4 h aging at 135°C prior to compaction by spreading in a pan at 21–50 kg/m².
4. Select a gradation that gives a mix with the best overall match with specified VTM, VMA, and VFA.
5. Mix and compact eight specimens with the select gradation—two each at four ashalt contents enveloping the initial asphalt content.
6. Determine bulk specific gravity and compute VTM, VMA, and VFA.
7. Select the asphalt content that gives mixes with 4% VTM as the optimum asphalt content. Check whether the mix at 4% VTM meets other criteria, as shown in Figure 14.13.
8. Prepare six specimens at optimum asphalt content at construction (7%) VTM.
   Subject three to moisture conditioning. Determine tensile strengths of three unconditioned specimens. Determine tensile strength ratio (TSR), as shown in in Figure 14.13. Check if TSR is equal to or greater than minimum specified value. If not, consider adding an antistripping agent, such as lime, to the mix.

FIGURE 14.12 Key steps in the superpave mix design method.

of evaluating the degree of stripping by this method. Note that one freezing and thawing cycle is described earlier—agencies could also require multiple freeze–thaw tests, depending on the severity of the environment in which the pavement is expected to serve. Figure 14.16 shows a typical agency job mix formula, resulting from a mix design. In more and more cases, designed mixes are being "proof" tested for such properties as resistance against rutting and/or moisture effects, or compared amongst each other, with the help of simulative tests. The asphalt pavement analyzer (APA), shown in Figure 14.17, is such a test procedure, in which gyratory-compacted samples can be subjected to a pressure of 100 psi and a load of 100 lb in a temperature-controlled chamber, with or without moisture. The steel wheels apply the load through pressurized rubber hoses, and the deformation in the HMA samples after a specific number of "passes" of the wheels is used as the indicator of strength against rutting.

The asphalt vibratory compactor (AVC) can be used to compact slab or cylindrical samples at the same amplitude, frequency and relative weight of a vibratory roller. These samples could be tested for a variety of tests, including tests with the APA.

14.2.3.1 Mix Design Systems of South Africa, France, the United Kingdom and Australia

(Note that a harmonized European system of mix design has been introduced by the European Committee for Normalization, or Comité Européen de Normalisation [CEN], and readers are encouraged to view CEN pages at http://www.cen.eu/cen/pages/default.aspx.)

The South African mix design system is based on the conventional Marshall mix design system and also includes performance simulative tests such as indirect tensile strength, dynamic creep, static creep and immersion index. New test methods have been added, such as rutting tests with the model mobile load simulator (MMLS3), especially for slow moving heavy loads. Field trials are also mentioned to ensure the applicability of the mixing and the compaction procedure. The design consideration consists of preventing rutting, fatigue damage, excessive permeability, moisture susceptibility, noise and skidding, and ensuring constructability. These considerations are ranked in order to determine the specific performance tests that need to be carried out. Before conducting the performance tests however, the gradation and the target binder content are selected and volumetric analysis is conducted to finalize the binder content. The selection of the specific gradation is based on the consideration of either of the two mechanisms—substitution or filling. For substitution the space filled by the fine aggregates is replaced by an increase in the

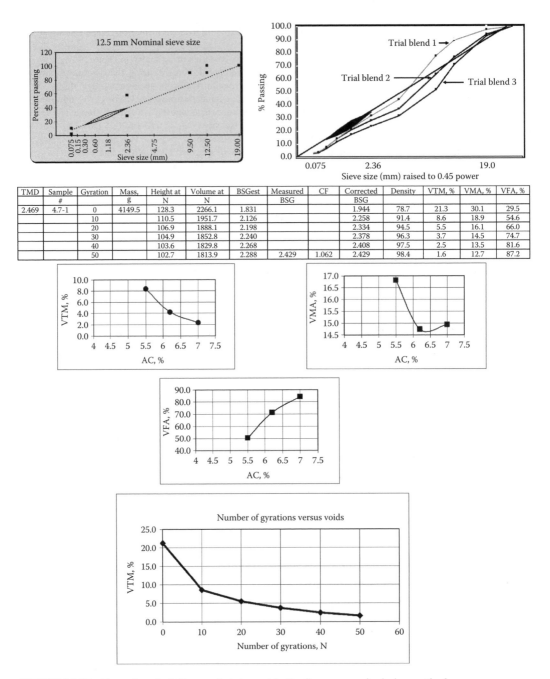

TMD	Sample #	Gyration	Mass, g	Height at N	Volume at N	BSGest	Measured BSG	CF	Corrected BSG	Density	VTM, %	VMA, %	VFA, %
2.469	4.7-1	0	4149.5	128.3	2266.1	1.831			1.944	78.7	21.3	30.1	29.5
		10		110.5	1951.7	2.126			2.258	91.4	8.6	18.9	54.6
		20		106.9	1888.1	2.198			2.334	94.5	5.5	16.1	66.0
		30		104.9	1852.8	2.240			2.378	96.3	3.7	14.5	74.7
		40		103.6	1829.8	2.268			2.408	97.5	2.5	13.5	81.6
		50		102.7	1813.9	2.288	2.429	1.062	2.429	98.4	1.6	12.7	87.2

FIGURE 14.13   Example calculations and plots used in the Superpave mix design method.

coarse aggregate fraction whereas in filling, the space between the coarse aggregate is filled by an increase in the fine aggregate concentration. The selection of one mechanism over the other depends on the priority given to the different design objectives, such as ensuring stability or durability and compactability. The various performance tests and criteria may consist of dry and wet rutting under MMLS3, CSIR wheel tracking test, gyratory refusal density (air voids at 300 gyrations of the Superpave gyratory compactor), gyrations to 93% of field density, dynamic creep test (stiffness at a specific temperature), strain in four point bending beam fatigue test, indirect

Traffic Level	Coarse Aggregate Angularity		Fine Aggregate Angularity	
	<100 mm	>100 mm	<100 mm	>100 mm
<0.3	75/ ---	50/---	40	40
0.3 to <3.0	85/80	60/---	45	40
3.0 to <30.0	95/90	80/75	45	40
>30.0	100/100	100/100	45	45

Traffic Level	Sand Equivalent, %	Flat and Elongated, %
<0.3	40	---
0.3 to <3.0	45	10
3.0 to <10.0	45	10
10 to <30.0	45	10
>30.0	50	10

$$0.6 \leq \frac{\% \text{ Weight of material passing the 0.075 mm sieve}}{\% \text{ Weight of effective asphalt binder}} \leq 1.2$$

Traffic Level, ESALs, Million	Compaction, Number of Gyrations			
	$N_{initial}$		N_{design}	$N_{maximum}$
	Gyrations	Density, % G_{mm}		
<0.3	6	<91.5	50	75
0.3–< 3	7	<90.5	75	115
3.0–<30	8	<89.0	100	160
>30.0	9	<89.0	125	250

Slow/standing traffic: increase N_{design} by 1 level

Nominal Maximum Aggregate Size, mm	Minimum VMA, %
9.5	15
12.5	14
19	13
35	12
37.5	11

Traffic (ESAL, Million)	Range of VFA, %
<0.3	70–80
1–3	65–78
>3.0	65–75

$$\text{Tensile strength ratio} = \frac{\text{Average tensile strength of conditioned specimens}}{\text{Average tensile strength of unconditioned specimens}}$$

FIGURE 14.14 Requirements in the Superpave mix design method.

tensile strength, modified Lottman test, Marshall stability and flow and filler to asphalt binder ratio. A viscosity graded binder system is used.

The French mix design system consists of four different levels, from 1 to 4, with increasing number of tests. The asphalt binder tests are penetration and Ring and Ball temperature as well as RTFOT loss, and the asphalt content is based on a minimum content (richness factor) that is similar

Controls
with disk drive

Compactor

Printer

Mix put in the mold

Filled mold with paper disc on top and bottom

Mold placed in compactor

Compacted sample

1.25°

The mold tilts at 1.25
and gyrates at the rate of
30 revolutions per minute

Compacted sample at the end of
gyrations

Schematic of mold gyration

FIGURE 14.15   Compaction using the Superpave gyratory compactor.

to the film thickness concept. The field compactability of the mix is checked by the determination of air voids after a specific number of gyrations in a gyratory shear compactor, whereas the moisture sensitivity is evaluated with the Duriez (resistance to unconfined compression for dry and water conditioned samples) test. For higher level mixes, the rutting potential is evaluated with the wheel tracking test, whereas the modulus and the fatigue properties are estimated from the complex modulus (or uniaxial tensile) test, and the two point bending tests, respectively. A slab compactor with inflated tires is used for preparation of samples for all of the tests. The rutting test is conducted with the LCPC wheel tracking rut tester, which consists of a pneumatic tire inflated to 600 kPa and a vertical load of 5 kN. A two-point bending test with a trapezoidal shaped HMA sample is utilized for the complex modulus test at a set of frequencies and temperature and a master curve is produced. Alternatively, the uniaxial direct tensile test at a constant strain is conducted on a cylindrical specimen at a range of temperatures and frequencies. The two-point bending test is conducted for fatigue tests, with trapezoidal shaped samples at different strain levels at 10°C and 25 Hz, and the fatigue curves for strain versus number of load cycles to failure are constructed. The fatigue and modulus tests are conducted for Levels 3 and 4 for which the structural design properties are required.

	Ref. No. **3566**	**HMA DESIGN** SUBMITTAL DATA	**100 GYRATIONS**

PLANT DATA	Make	Size	Type	Location
	WARREN	3 TON	BATCH	PORTLAND

BITUMEN DATA	Grade	Refiner	Supplier	Temp. Range (Aim ± 20°F [10°C])	
				Plant Aim, °F [°C]	Street Aim, °F [°C]
	PG70-28	BITUMAR CA	BOUCHARD	310 [155]	290 [143]

SUPERPAVE DESIGN DATA

ESAL's: **3 to <30**
Nominal Size: **12.5mm**

Gsb: **2.65**
Gmm: **2.440**

% Binder: **6.2**
Fines to Eff. Binder Ratio: **1.2**

Gmb, weight, g

AGGREGATE DATA

Size	Type	Original Source/Owner	Original Source/Location
12.5	QUARRY	PIKE	WELLS
9.5	QUARRY	PIKE	WELLS
DRY STONE SCREENINGS	QUARRY	PIKE	WELLS

Stockpile Gradation (Percentages Passing Sieve Sizes)

% Used	1½ in. 37.5 mm	1 in. 25.0 mm	¾ in. 19.0 mm	½ in. 12.5 mm	⅜ in. 9.5 mm	No. 4 4.75 mm	No. 8 2.36 mm	No. 16 1.18 mm	No. 30 0.600 mm	No. 50 0.300 mm	No. 100 0.150 mm	No. 200 0.075 mm
65			100	94	37	4	3	2	2	2	1	1
15				100	97	29	7	4	3	3	2	1.5
12.6					100	99	78	54	39	26	17	10.5
7.4										100	99	94
Aim			100	96	59	27	20	16	14	12	11	9
Lower			100	89	52	20	16	12	11	10	9	7
Upper			100	100	66	34	24	20	17	14	13	11
Spec.			100	90 - 100	26 - 78	20 - 28	16 - 24					8 - 10
Restricted Zone												

Status: **PENDING**
JMN: **PII-PD3-100-12**

COMMENTS:

MAINE DOT USE ONLY

JOB MIX SPECIFICATIONS

	Pb, %	%Passing No. 200 0.075mm Sieve
Aim	6.2	
Lower	5.8	
Upper	6.6	

DOT Consensus Qualities

Fractured, 1 Face (ASTM D 5821), %	99
Fractured, 2 Face (ASTM D 5821), %	99
F.A. Angularity (T 304), %	45
Flat/Elongated (ASTM D 4791), %	2
Micro-Deval (TP58 99), %	6
Sand Equivalent (T 176), %	66
Washington Degradation (MeDOT)	
Combined Agg, Gsb (PP 28)	

Authorized by: Date:

Paper Copy: *Lab File*   Electronic: *Area Supervisor; Resident; Contractor*

FIGURE 14.16 Results of a mix design in a standard format (Job Mix Formula). (Courtesy of Rick Bradbury, Maine Department of Transportation, Augusta, ME.)

The British mix design system uses a penetration graded asphalt binder, and the Marshall hammer is utilized for preparation of samples from which the volumetric and Marshall mix design properties (stability and flow) are determined. Compaction to refusal is also utilized; with either a Marshall hammer or with an electric vibrating hammer. The mix stiffness modulus is determined with the British Nottingham Asphalt Tester at 20°C. A repeated load axial test could be conducted to determine the resultant strain and creep stiffness modulus at a temperature of 30°C. A wheel tracking test

Wheels running on hose during testing

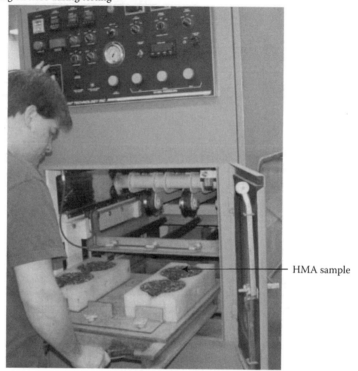

Rutted samples after testing

FIGURE 14.17 The asphalt pavement analyzer (APA).

is conducted for the evaluation of rutting potential, with either cylindrical or slab samples, at 45°C (moderate to heavily stressed) or 60°C (very heavily stressed) to check against a maximum allowable rut depth.

The Australian mix design system has three levels, 1 through 3, with increasing number of tests. The mix is compacted with a gyratory compactor and the asphalt content is determined form the criteria of volumetric properties. For heavy trafficked pavements, wheel tracking and refusal density tests are performed, with options of additional test, such as those for fatigue and moisture sensitivity. Resilient modulus and dynamic creep properties are determined for heavy trafficked pavements (or slow traffic pavements). A wheel tracking test at 60°C is also conducted at Level 3. If fatigue properties are required, tests are conducted with third point loading of beams at constant strain at 20°C. In the refusal density test, the density of sample compacted to 350 cycles in a gyratory compactor is determined, and the air voids are calculated and checked against a minimum allowable (3%) to prevent rutting.

14.3 STRUCTURAL DESIGN

The basic methods of consideration of the major distresses have been discussed in Chapter 12. Recall that some of the models are based on statistical regression with empirical data, whereas others are based on mechanics. In an empirical structural design method, only empirical models are used, whereas in a mechanistic-empirical method, both empirical and mechanistic expressions are utilized. The following paragraphs present examples of both types of design method.

14.3.1 EMPIRICAL METHODS

The empirical methods include mostly those based on the soil classification system and subgrade type (U.S. Bureau of Public Roads) and those based on road tests conducted by AASHO. In both cases the design methods are based on experience—either from in-service pavements or from controlled experiments. The advantages of such methods are that sufficient data are available and design procedures are relatively simple and less time consuming. The primary disadvantage is that the design cannot accommodate traffic, environmental conditions, and materials for which experience and/or results of experiments do not exist.

14.3.1.1 California Bearing Ratio Method

This method is based on charts relating the required thickness of pavement over the subgrade to the strength of the subgrade—the strength being represented in terms of a parameter called the *California Bearing Ratio* (CBR; ASTM D-1883-99, AASHTO T-193). The CBR concept and test method were originally developed by the California Division of Highways in the 1920s, later modified by the U.S. Army Corps of Engineers, and finally adopted by the American Society of Testing and Materials (ASTM) and the American Association of Highway and Transportation Officials (AASHTO).

The CBR test is a penetration test that uses a standard piston (3 in.[2]) which penetrates the soil at a standard rate of 0.05 in./min. A unit load is recorded at several penetration depths, typically 0.1 and 0.2 in. The CBR value is computed by dividing the recorded unit load by a standard load that is required for penetration for a high-quality crushed stone material. The CBR test is conducted on a soaked sample of soil—the soaking in water (for 96 h) is conducted to simulate the worst (saturated) condition under which the pavement would perform.

Figure 14.18 shows the original curves for CBR versus thickness for two different traffic levels, and an example to illustrate its use.

14.3.1.2 American Association of State Highway and Transportation Officials Method

The AASHTO design procedure introduced a number of important design concepts. The design procedure came out of statistical analysis of data from an extensive road test carried out between 1955 and 1960 near Ottawa, Illinois. Traffic, consisting of trucks with different loads and axles, were run to failure over different pavement sections, consisting of different materials and thickness. Two important concepts—serviceability and reliability—were utilized in the development of the design process. This design procedure has been subsequently revised to add more mechanistic components over the years. The latest version is the AASHTO *Guide for Design of Pavement Structures*, published in 1993.

The procedure started with the AASHO Road Test. The concept of Present Serviceability Index (PSI) was developed, and an equation relating traffic to thickness was developed. This equation was mapped into a nomograph, which is often used in this method. The tests were carried out in Ottawa, Illinois, between the years of 1958 and 1961. In total, there were six test tracks, with both HMA and reinforced concrete sections. For the test location, the mean temperatures in January and July were 27°F and 76°F, respectively. The annual average rainfall and frost depth

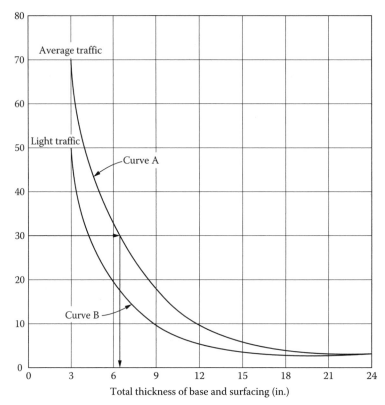

For average traffic, for a subgrade
with CBR of 30 the design thickness
of the pavement (base+surface)
is found to be approximately 6.5 in.

FIGURE 14.18 Example of the use of a CBR curve for pavement design. (From ASCE, 1950, and Monismith and Brown, 1999, reprinted with kind permission from the Association of Asphalt Paving Technologists.)

were 34 and 28 in., respectively. The subgrade soil consisted of A6 (silty clay), with 82% of the material passing the No. 200 sieve, and an average CBR of 2.9 and optimum moisture content of 13%. The subbase and base courses consisted of a sand/gravel mixture (CBR = 28–51) and crushed limestone (average = 108), respectively. The HMA was made up of 85–100 penetration grade asphalt.

There were multiple HMA sections with subbase courses varying in thickness from 0 to 16 in., base courses 0–9 in., and HMA 1–6 in. The majority of the HMA sections failed, with most failing due to the effect of spring thaw, which was found to decrease with the use of a thicker base and subbase.

The complete equation relating traffic and thickness is as follows.

$$\log W_{18} = Z_R * S_0 + 9.36 \log(SN+1) - 0.20 + \frac{\log\left(\dfrac{APSI}{4.2-1.5}\right)}{0.40 + \dfrac{1094}{(SN+1)^{5.19}}} + 2.32 \log(M_r) - 8.07$$

The use of this equation requires the selection of reliability levels in terms of Z_R and S_0, initial and terminal PSI or change in PSI, and determination of resilient modulus, M_r. The following paragraphs explain the various parameters in the equation.

14.3.1.2.1 log W18

Base 10 logarithm of the predicted number of ESALs over the lifetime of the pavement. The logarithm is taken based on the original empirical equation from the AASHO Road Test.

14.3.1.2.2 Structural Number

Structural number. An abstract number expressing the structural strength of a pavement required for given combinations of soil support (M_r), total traffic (ESALs), and allowable change in serviceability over the pavement life (ΔPSI).

The structural number is used to determine layer depths by using a parameter called *layer coefficient*, such that

$$SN = a_1 D_1 + a_2 D_2 m_2 + a_3 D_3 m_3 + \cdots$$

where
a_i is the layer structural coefficient
D_i is the layer depth, in.
m_i is the layer drainage coefficient

Typical values of layer coefficients are as follows. HMA (surface layer): 0.44; base course (crushed stone): 0.14; stabilized base course material: 0.30–0.40; subbase (crushed stone): 0.11. Values of drainage coefficients can range from 0.4 (slow-draining, saturated layers) to 1.4 (fast-draining layers that do not get saturated). The term can be neglected by using m = 1. An example chart relating the various stiffness/strength measuring for subbase materials is shown in Figure 14.19.

The concept of reliability is based on the assumption that the distribution of variables such as stress, resulting from uncontrollable factors such as loading and the environment, and strength/stiffness of materials/layers, resulting from controllable factors such as variations in construction quality and materials, can be assumed to be of the normal distribution type. *Reliability* refers to the probability that the predicted design life will exceed the required design, or, more specifically, the strength of a material will exceed the stress on the material. Two factors are considered for reliability:

Z_R is the standard normal deviate
S_0 is the combined standard error of the traffic prediction and performance prediction

Table 14.2 shows the different values of Z_R for the different levels of reliability, which could be selected on the basis of guidelines that have been provided in the AASHTO Design Guide.

14.3.1.2.3 S0

The typical values for a flexible pavement are 0.40–0.50. S_0 cannot be calculated from actual traffic or construction numbers, so it is almost always assumed to be 0.50.

14.3.1.2.4 Solution of the AASHTO Equation

The equation is solved using an iterative process. The equation is often solved assuming ESAL values. The process consists of the following steps.

1. Determine the total ESALs.
2. Determine M_r, or, using the following formula, convert existing CBR values to M_r.
   AASHTO method: M_r, psi = 1500 * (CBR)
   NCHRP 1-37 method: M_r, psi = 2555(CBR)$^{0.64}$

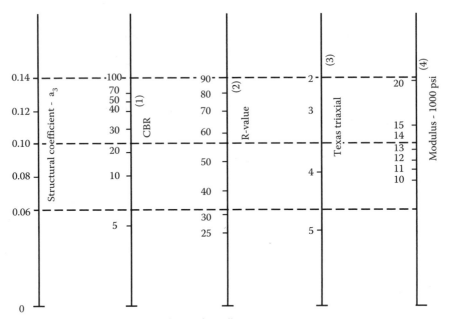

(1)   Scale derived fromp correlations from Illinois.
(2)   Scale derived fromp correlations obtained from the asphalt institute, California,
      New Mexico and Wyoimng.
(3)   Scale derived fromp correlations obtained from texas.
(4)   Scale derived on NCHRP project (3).

FIGURE 14.19   Variation in granular subbase layer coefficient.

TABLE 14.2
Z_R Values for Different Reliabilities

Reliability	Z_R
99.99	−3.750
99.9	−3.090
99	−2.327
95	−1.645
90	−1.282
80	−0.841
75	−0.674
70	−0.524
50	0

3. Select reliability (selecting a reliability means selecting a specific value of Z_R).
4. Select p_o, p_t.
5. Decide on a basic structure, and select the M_r of the materials of the layers.
6. Solve the equation for two layers (e.g., combine lower two layers and upper two layers);
   solve each layer using the M_r of the layer directly underneath.
7. Determine total SN: SN required from lower two layers (combined) and from upper two
   layers.

8. Use guidelines and experience to select thickness of the different layers to satisfy the SN requirement.

14.3.1.2.5 Correlation with Layer Coefficient

The AASHTO standard guide for design of pavement provides relationships between the resilient modulus and layer coefficients, whereas the relationship between the CBR and resilient modulus is provided by the U.K. Transportation Research Laboratories (TRL). The relationships are as follows:

$$M_r = 2555 \, (CBR)^{0.64}$$

where M_r, resilient modulus, is in psi.
And

$$M_r = 30,000 * \frac{a_i}{0.14}$$

where
M_r is in psi
a_i is the AASHTO layer coefficient

The nomograph used in the AASHTO design method is shown in Figure 14.20.

Example 14.1

Determine the thicknesses for two layers in an asphalt pavement, where a total structural number of 4.5 is required. Layer 1 is a HMA layer with a layer coefficient of 0.44, and layer 2 is a granular base course with a layer coefficient of 0.13. Consider the drainage coefficient or the base as 0.9, and the minimum thickness of the HMA layer as 2 in.
  Consider the minimum thickness of the HMA first, say, 2 in.

$$SN = a_1 D_1 + a_2 D_2 m_2$$

where
  a_i is the layer structural coefficient
  D_i is the layer depth, in.
  m_i is the layer drainage coefficient

$$SN_{required} = 4.5 = (0.44)(2) + (0.13)(D_2)(0.95)$$

where D_2 is the thickness of the base = 29.3 in., approximately 29.5 in.
  If a 4-in. HMA layer is used instead, the required base thickness is approximately 22 in.

14.3.2 Mechanistic-Empirical Methods

A typical mechanistic-empirical design procedure basically consists of two parts—a mechanistic part or an analytical part which helps in the determination of pavement response under load, and an empirical part that relates the response to pavement performance or (more specifically)

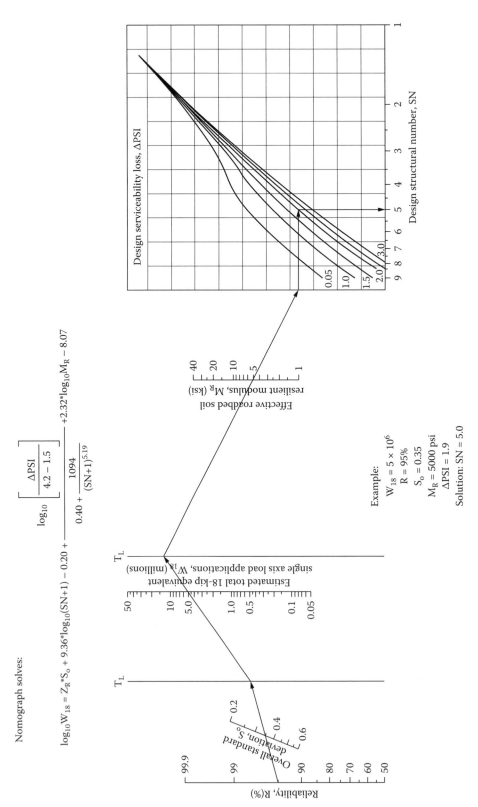

FIGURE 14.20 Nomograph for solving the AASHTO equation. (From American Association of State Highway and Transportation Officials (AASHTO), *AASHTO Guide for Design of Pavement Structures*, AASHTO, Washington, DC, 1993. With permission.)

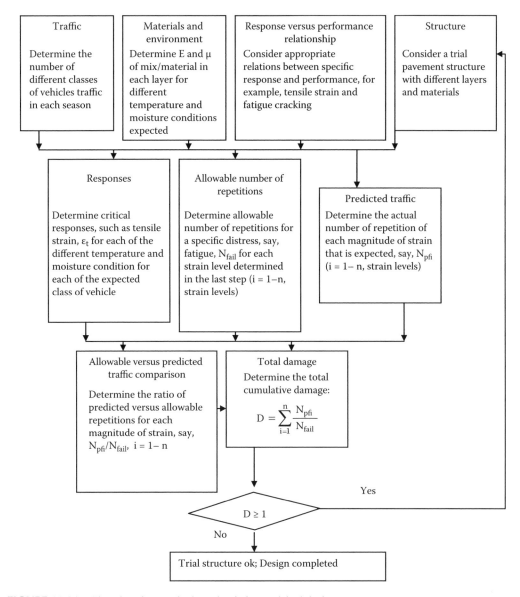

FIGURE 14.21 Flowchart in a typical mechanistic-empirical design.

to distress, such as rutting. The complete flow chart for a mechanistic-empirical design is shown in Figure 14.21.

A typical schematic of a layered pavement structure with material properties relevant for ME design is shown in Figure 14.22. The layers are shown with typical values for a specific location. The layer thicknesses as well as the material properties are then changed within reasonable ranges. The effects of these changes on a critical response, the tensile strain at the bottom of the HMA surface layer, are also shown in Figure 14.22.

Once the critical responses are determined, they are used in *relationships* between such responses and critical performance conditions or distresses, to predict the number of load repetitions that the pavement can sustain before reaching the threshold/critical values of the distress conditions. Alternatively, if the number of load repetitions is given for a specific design period, the cumulative "damage" to the pavement for the given number of load repetitions is determined.

FIGURE 14.22 Schematic of pavement layers with parameters required for mechanistic–empirical design and example of sensitivity of response to change in parameters.

This cumulative "damage" is then checked against the appropriate criteria, and if found unsuitable, the layer thicknesses and/or material properties are changed, and the process is repeated until an acceptable "damage" is obtained—and this represents the end of the "structural design" step.

In the design methods, the damage from each group of traffic (for each season) is considered, and the damage factor for the ith loading is defined as the number of repetitions (N_{actual}) of a given response parameter divided by the "allowable" repetitions ($N_{allowable}$) of the response parameter that would cause failure. The cumulative damage factor (CDF) is obtained by summing the damage factors over all the loadings in the traffic spectrum. If the CDF reaches 1.0, then the pavement is presumed to have reached the end of its design life. Hence, the objective of the design process is to change the different factors until the computed CDF is found to be close to but less than 1.0.

A simple application of the two steps is illustrated in the following example.

Example 14.2

Check the design of a pavement with the following information.

Traffic
One million repetitions of a standard 18-kip dual-tire axle.

Load
Considering one side of the axle only, there are two tires, each supporting a load of 4500 lb, with a tire pressure of 100 psi, 13 in. apart, center to center.

Assumed Pavement Structure
The pavement is made up of four layers: HMA surface and binder (3 in.), stabilized base (6 in.), crushed gravel subbase (10 in.), and subgrade.

Pavement Materials
Properties

$$E_{HMA} = 3500 \text{ MPa} = 507 \text{ ksi}; \quad \upsilon_{HMA} = 0.35$$

$$E_{stabilized\,base} = 1500 \text{ MPa} = 217 \text{ ksi}; \quad \upsilon_{stabilized\,base} = 0.35$$

$$E_{subbase} = 150 \text{ MPa} = 21.7 \text{ ksi}; \quad \upsilon_{subbase} = 0.35$$

$$E_{subgrade} = 50 \text{ MPa} = 7.2 \text{ ksi}; \quad \upsilon_{subbase} = 0.4$$

Performance Models

$$N_{HMA} = \left[\frac{0.005889}{\text{Tensile strain}}\right]^5$$

$$N_{subgrade} = \left[\frac{0.003900}{\text{Vertical strain}}\right]^{7.10}$$

Results
Maximum horizontal tensile strain at the bottom of the HMA = 75 µε.

$$N_{HMA\,allowable} = \left[\frac{0.005889}{0.000075}\right]^5 = 2984 * 10^6$$

$$CDF_{HMA} = \sum_{i=1}^{N} \frac{N_{actual}}{N_{allowable}} = \frac{10^6}{2984 * 10^6} = 0.00033$$

Maximum vertical compressive stress on top of the subgrade = 390 µε.

$$N_{subgrade} = \left[\frac{0.003900}{0.000390}\right]^{7.10} = 12.59 * 10^6$$

$$CDF_{subgrade} = \sum_{i=1}^{N} \frac{N_{actual}}{N_{allowable}} = \frac{10^6}{12.59 * 10^6} = 0.0079$$

Since the CDFs are both less than 1, the pavement is sufficiently strong to withstand the expected design traffic. In fact, since the CDFs are very low compared to the limiting value of 1, there is scope for redesigning the pavement for the design traffic.

Note that in this case fatigue cracking and rutting, and relatively simple models relating tensile strains and compressive strains to them (respectively), have been used. A design could involve more sophisticated models and/or other distresses such as thermal cracking, as discussed in Chapter 12.

14.3.2.1 Example of Structural Design Procedure Using Mechanistic Principles

The Asphalt Institute procedure (Asphalt Institute, 1991) contains multilayer elastic theory; is built on the basis of data from the AASHTO Road Test, the Western Association of State Highway Organizations (WASHO) Road Test, and other different state and local road tests; and considers rutting and fatigue cracking as the distresses. For rutting the vertical compressive strain (ε_c) at the top of the subgrade is considered, whereas for fatigue cracking the horizontal tensile strain (ε_t) at the bottom of the lowest asphalt mix layer is considered. Rutting in the asphalt mix layer only is considered to be due to consolidation due to improper compaction of the layer during construction. Design charts are the basis of this design procedure. The design charts were generated with the use of the computer program DAMA, originally developed in 1983 (Asphalt Institute, 1983), which determined the largest of the two strains, and recommended the thickness of the layer to meet the higher (or more critical) of the two strains, on the basis of inputs which include resilient modulus of the subgrade, selection of type of surface and base layers, traffic information, and mean annual air temperature.

The procedure is illustrated with an example in Figure 14.23. In the AI method, a minimum thickness of asphalt mix surface layers is provided for different types of bases—emulsion treated and untreated. Charts are provided for different mean annual air temperatures—45°F, 60°F, and 75°F. Limits of CBR, Atterberg Limits, and percentage passing the No. 200 sieve for subbase/base are also provided.

14.3.2.1.1 Staged Construction

When adequate funds are not available for constructing a pavement to meet the demands of a full design period, the pavement can be constructed in stages. The initial thickness can accommodate a certain number of years of traffic, and the second construction adds the thickness required for sustaining the remaining traffic for the full design period.

The principle of design in staged construction is that the second construction will take place (say, as an overlay) before the first pavement has reached the end of its life. Based on the concept of cumulative damage (discussed earlier), it is often assumed that the second overlay will be applied when the first pavement has at least 40% of its life left (i.e., damage is a maximum of 60%).

If

n_1 is the actual accumulated ESAL for the first stage

N_1 is the allowable number of ESAL for the stage 1 pavement thickness

D_1 is the percentage of life of the pavement expended right before the second stage

$$\frac{n_1}{N_1} = D_1 = 0.6$$

$$N_1 = \frac{n_1}{0.6} = 1.67 n_1$$

Determine h_1 on the basis of N_1.

If

n_2 is the actual accumulated ESAL for the second stage

N_2 is the allowable number of ESAL for stage 2 pavement thickness

$$\frac{n_2}{N_2} = (1 - D_1) = 0.4$$

$$N_2 = \frac{n_2}{0.4} = 2.5 n_2$$

Determine h_2 on the basis of N_2, then thickness to be added in stage $2 = (h_2 - h_1)$.

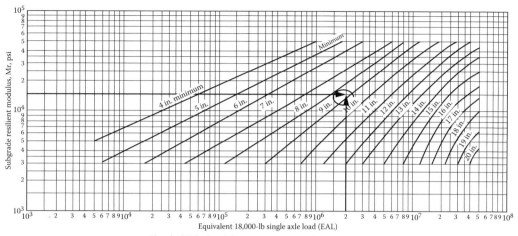

Chart for full depth asphalt pavement, mean annual air temperature, 60°F

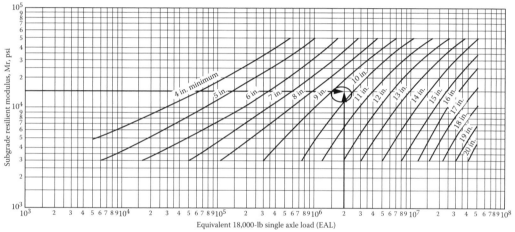

Chart for type I emulsified asphalt mix base, mean annual air temperature, 60°F

Chart for type III emulsified asphalt mix base, mean annual air temperature, 60°F

FIGURE 14.23 Example of use of the Asphalt Institute (AI) method. (Reprinted from Asphalt Institute, *Thickness Design: Asphalt Pavements for Highways and Streets*, Manual Series No. 1 (MS-1), 9th edn., AI, Lexington, KY, 1999. With permission. Copyright 1999, the Asphalt Institute, Inc.)

14.3.2.2   NCHRP 1-37A Mechanistic-Empirical Design Guide

The NCHRP 1-37A Mechanistic Empirical Design Method (commonly referred to as the *2002 Design Guide*) was developed in NCHRP Project 1-37A (Development of the 2002 Guide for the Design of New and Rehabilitated Pavement Structures). The *Guide* presents a very comprehensive method, with considerations of appropriate sophisticated models, and uses state-of-the-art testing procedures. The software and detailed instructions of the design guide can be obtained from www.trb.org/mepdg/(NCHRP, n.d.).

The *Guide* suggests three different levels of design: Level 1, which requires very basic information and the use of correlations; Level 2, which requires some test data; and finally Level 3, for which most of the data needed to be input are obtained from testing. The basic premises of this *Guide* are the same as the mechanistic principles discussed in Chapter 12. Note that the models have been calibrated against national data (in the United States and Canada)—it is extremely important that those models be calibrated against locally available data before the *Guide* is adopted by any specific state or local transportation department. In the meantime, the *Guide/* software serves as a very useful tool for checking the sensitivity of the performance/distress of the pavements to the different control variables. The most powerful feature of the guide is the ability to cycle through variations in material properties as a function of changes in seasonal conditions, and compute the incremental damage due to cumulative traffic loading through the different seasons for long design periods. This concept is illustrated with an example output in Figure 14.24. Note that it shows three relevant factors—cumulative traffic, changes in modulus of asphalt mix, and total rutting during the entire 20-year design period. Such plots can also be generated for other types of distresses.

The different parameters that need to be input are shown in Table 14.3, and the steps in running the software (for a Level 1 design) for the guide are presented thereafter. A brief description of the different steps is provided in the following paragraphs.

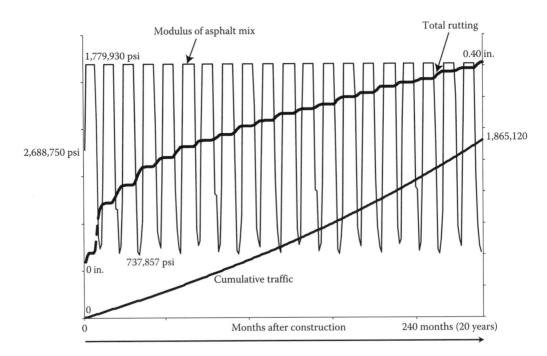

FIGURE 14.24   Example of information obtained from M-E design.

TABLE 14.3

Inputs from Test Results for NCHRP 1-37A Design Guide

	Inputs from Tests		
	Levels		
Material	**1**	**2**	**3**
Soil	Resilient modulus from AASHTO T-307; Plasticity index, AASHTO T-90; Gradation, AASHTO T-27; Specific gravity, Maximum dry unit weight, Optimum moisture content, AASHTO T-99, T-180; Saturated hydraulic conductivity, AASHTO T-215	Resilient modulus from correlation with CBR, AASHTO T-193; Layer coefficient, AASHTO 1993; PI/gradation, AASHTO T-27/T-90, DCP, ASTM D-6951	Resilient modulus from CBR from soil classification, AASHTO M-145/ASTM D-2487 (default values available)
Asphalt binder	Performance grading tests, AASHTO T-315; or viscosity, AASHTO T-201; penetration, AASHTO T-49; specific gravity, ASTM D-70; softening point, ASTM D-36	Performance grading tests, AASHTO T315; or viscosity, AASHTO T201; penetration, AASHTO T49; specific gravity, ASTM D70; softening point, ASTM D36	A and VTS
Aggregates	Coefficient of thermal contraction, ASTM D-4535	Gradation: retained on 3/4 in., 3/8 in., and No. 4 sieve, and passing No. 200 sieve, AASHTO T-30	Gradation: retained on 3/4 in., 3/8 in., and No. 4 sieve, and passing No. 200 sieve, AASHTO T-30
Asphalt mix	Dynamic modulus, AASHTO T-P62; volume of air voids, AASHTO T-269; unit weight, AASHTO T-166; indirect tension test at 3 temperatures, AASHTO T-322; voids in mineral aggregates, AASHTO R-35	Volume of air voids, AASHTO T-269; unit weight, AASHTO T-166; indirect tension test at 1 temperature, AASHTO T-322; voids in mineral aggregates, AASHTO R-35	Volume of air voids, AASHTO T-269; unit weight, AASHTO T-166; voids in mineral aggregates, AASHTO R-35
Existing pavement for rehabilitation concrete	Unbound layers: resilient modulus from falling weight deflectometer (FWD) tests, rutting from trench	Unbound layers: resilient modulus from correlations with other tests, rutting from user	Unbound layers: resilient modulus from soil classification, rutting from user
	Stabilized base layers: maximum modulus, existing modulus from FWD	Stabilized base layers: maximum modulus and % fatigue cracking	Stabilized base layers: typical maximum modulus and pavement condition
	Fractured concrete: modulus from FWD		Fractured concrete: crack spacing
	Asphalt mix layer: existing modulus from FWD, volumetric, and binder properties	% alligator cracking, volumetric, and binder properties	Pavement rating, estimates of mix properties

14.3.2.2.1 Overview of the Software

The NCHRP 1-37A Mechanistic Empirical Pavement Design (commonly referred to as the Design Guide 2002, or DG2002) software is a Windows-based application for the simulation of pavement structures. It can be used to simulate pavement structures under different traffic and climatic conditions. The software has the ability to predict the amount of damage that a structure will display at the end of its simulated design life. In addition to using laboratory test results, predicting properties for pavement mixes can be done by using correlations. Values (either default or entered by the user) that represent structural, climatic, or traffic properties are used in calculating distresses. The following example demonstrates how to provide the different inputs for a design project. The project is based in Guilford, Maine, and is designed for a 20-year life.

14.3.2.2.2   Creating a New Project

To create a new project, first open the Design Guide 2002 software. Click on *File*, then *New*; a window, as shown in Figure 14.25, will pop up. Here you can specify the name of the project, the folder to create it in, and the measurement system (currently the software can only handle U.S. Customary). After this is complete, clicking *OK* will bring you into the main software console (shown in Figure 14.26). When running the software, the red boxes next to each section indicate that information needs to be entered; yellow boxes indicate that there are default values for that section which need to be either accepted or changed.

FIGURE 14.25   Creating a new project.

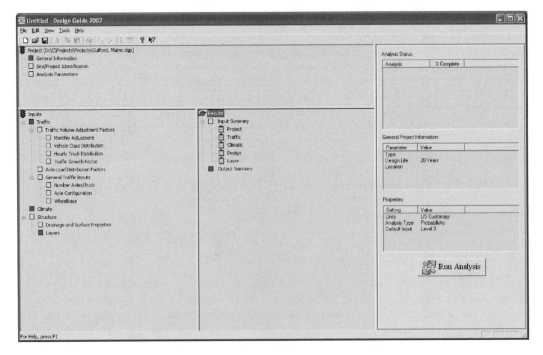

FIGURE 14.26   Main software console.

14.3.2.2.3   Specifying Project Information

On the top left corner of the console, double click on *General Information* (see Figure 14.27). This will bring up a window where you can specify the Design Life, Base/Subgrade Construction Month, Pavement Construction Month, and Traffic Open Month (see Figure 14.28). Then choose the Type of Pavement.

In the main console, double click on *Site/Project Information*. In the window that pops up, enter the Location, and the Project ID shown in Figure 14.29 (for the date and traffic direction, defaults are used here and can be changed if needed). Click OK.

Double click on Analysis Parameters in the main console. These are the values that will determine at the end of the design life whether that aspect of the pavement structure has failed; these are shown in Figure 14.30. For example, the limit for AC Thermal Fracture is 1000 ft/mile. In the final summary, if the simulation shows that the structure has thermal cracking in excess of 1000 ft/mile, then it has failed. Figure 14.30 shows the default values; accept them by clicking *OK*.

14.3.2.2.4   Specifying Traffic Inputs

On the main console under *Inputs*, double click on *Traffic*. Figure 14.31 shows the traffic window where the specific inputs are shown for Guilford; enter these values as shown. After these are finished, click on the *Edit* button next to Traffic Volume Adjustment. This will pull up a window like the one in Figure 14.32. Going through each tab,

- *Monthly adjustment*: these are default values.
- *Vehicle class distribution*: enter 100 for *Class 9*, and set all others to 0.

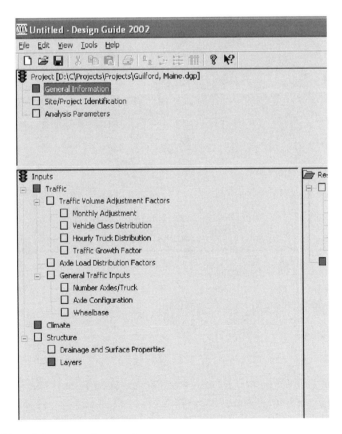

FIGURE 14.27   Upper left corner of a console.

FIGURE 14.28   General information.

FIGURE 14.29   Site/project identification.

- *Hourly distribution*: these are default values.
- *Traffic growth factors*: the compound growth should be set to 4%. All other parameters have been set already.

Press *OK*.

Next, click on the *Edit* button next to Axle Load Distribution Factor. The window that pops up (see Figure 14.33) will contain default values; press *OK*.

General Traffic Inputs is the last parameter set that needs to be specified; click *Edit*. All the parameters in this window (see Figure 14.34) under each tab are default values; click on each tab

FIGURE 14.30   Analysis parameters.

FIGURE 14.31   Traffic parameters.

FIGURE 14.32   Traffic volume adjustment factors.

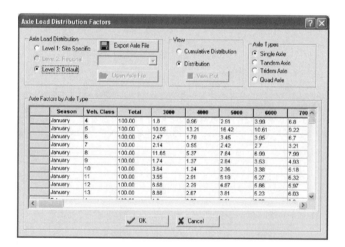

FIGURE 14.33   Axle load distribution factors.

until all are green, and then press *OK*. Now the original Traffic window is displayed; click *OK* to return to the main console.

14.3.2.2.5 *Specifying Climate*

In the main console, double click on *Climate*. A window will pop up where you can designate climate and water table information (see Figure 14.35). For the *Depth of Water Table*, enter 10. Next, click on *Generate*. Figure 14.36 shows the window that pops up; click on *Interpolate Climatic Data for Given Station*. Enter the latitude, longitude, and elevation values for Guilford that are shown in Figure 14.36. Select the first three options (Millinocket, Greenville, and Bangor) to interpolate weather data for Guilford. Click *Generate*. Save the generated weather file; this will make it available for use in future projects.

FIGURE 14.34 General traffic inputs.

FIGURE 14.35 Environment/climatic window.

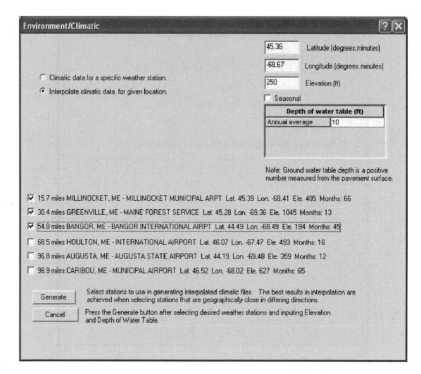

FIGURE 14.36   Environment/climatic.

FIGURE 14.37   Structure.

14.3.2.2.6 Defining Pavement Structure

14.3.2.2.6.1 Hot Mix Asphalt Layer On the main console, double click on *Structure*. Figure 14.37 shows the window that pops up. Highlight the Asphalt Concrete layer, and click on *Edit*. Here you can enter and select the properties of the HMA layer.

- Under the Asphalt Mix tab, enter the gradations shown in Figure 14.38. The layer thickness should be 3 in., and be sure that Level 3 is selected.
- Under the Asphalt Binder tab, select Superpave binder grading. Choose 64–28 for the binder (high and low temperature, respectively; shown in Figure 14.39).
- Under the Asphalt General tab, change the values for Effective Binder Content and Air Voids, as shown in Figure 14.40.

Click *OK* to return to the Structure window.

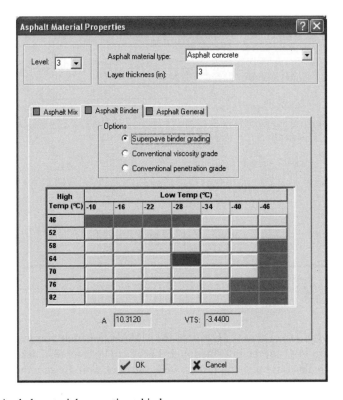

FIGURE 14.38 Asphalt material properties—gradation.

FIGURE 14.39 Asphalt material properties—binder.

FIGURE 14.40   Asphalt material properties—volumetric properties.

FIGURE 14.41   Inserting a new layer.

14.3.2.2.6.2 Granular Base Layers   Highlight the HMA Layer that you just created, and click *Insert*. Refer to Figure 14.41 for the information that needs to be entered. Click *OK*.

Highlight the Granular Base that you just created, and click *Edit*. Be sure that the Modulus (psi) button is selected, and enter 51,428 as shown in Figure 14.42. Note: the software is able to calculate modulus values based on other material properties if you are using a level 1 or 2 design.

FIGURE 14.42 Granular layer strength properties.

FIGURE 14.43 Inserting a second granular layer.

Now, repeat the process for a second granular base layer beneath the one you just created. Follow Figures 14.43 and 14.44 for relevant inputs.

14.3.2.2.6.3 Subgrade To complete the structure, highlight the second granular layer and click *Insert*. Refer to Figures 14.45 and 14.46 for subgrade inputs. Once finished, return to the main console.

FIGURE 14.44   Second granular layer strength properties.

FIGURE 14.45   Inserting a subgrade layer.

FIGURE 14.46   Subgrade layer strength properties.

Thermal cracking: on the main console, double click on *Thermal Cracking*. Figure 14.47 shows the window that pops up; the values are calculated by the software and are default Click *OK*.

Distress potential: on the main console, double click on *Distress Potential*. Figure 14.48 shows the window that appears. Click *OK* since these are default values.

14.3.2.2.7   Running the Analysis

Before running the analysis, there are two simple steps that need to be taken care of:

- Check all values that you entered. This ensures that there will be no surprises or inaccuracies on your part when the simulation is finished.
- Save your work before running the simulation. If anything happens to the computer or the software, you will not have to reenter all of your work.

Once you have verified your inputs and saved the program, run the simulation by clicking *Run Analysis*. Depending on the speed of your computer and the design life that you have entered, the software can take anywhere from 30 to 90 min or more (based on runs on a laptop computer, Dell Inspiron 8100, with Microsoft Windows XP 2002 Professional, 1200 MHz Intel Pentium III, 256 MB of RAM, and 20 GB hard drive).

14.3.2.2.8   Practical Issues

While using the DG2002, some glitches and problems could be encountered, since the software is still relatively new and is most likely prone to compatibility problems. The user is strongly advised to participate in the Highway Community Exchange Online Forum (Highway Community Exchange, n.d.) to discuss any problems.

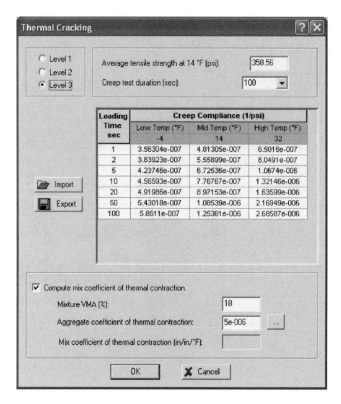

FIGURE 14.47   Thermal cracking values.

FIGURE 14.48   Distress potential window.

QUESTIONS

14.1 Determine the VTM, VMA, and VFA of a compacted HMA sample with the following test data: bulk-specific gravity = 2.351, theoretical maximum density = 2.460, and aggregate bulk-specific gravity = 2.565, asphalt content = 5.6%, and specific gravity of asphalt binder = 1.03.

14.2 Determine the optimum asphalt content for a mix, on the basis of VTM, from the following data:

AC (%)	VTM
4.5	6.7
5	5.2
5.5	3.4
6	2.4

14.3 Using the AASHTO procedure, determine the thickness required for a base and a surface course, over a subgrade. The structural number (SN) required for the pavement is 4.5. The available materials are a granular base with a layer coefficient of 0.13 and an HMA surface with a layer coefficient of 0.40. The drainage coefficient (m_2) of the base course can be considered to be 0.9.

14.4 Design a four-lane pavement for a 15-year design period, with the following traffic, layers, and material properties, using the mechanistic-empirical design method. Use rutting and fatigue failure criteria. Calculate the tensile strain at the bottom of the HMA surface and vertical strain on top of the subgrade layer using any layered analysis program. Use Asphalt Institute models for rutting and fatigue failure.

Layer	Modulus (MPa)	Poisson's Ratio
HMA surface	3500	0.35
Recycled base course	1500	0.35
Gravel subbase	300	0.4
Subgrade	40	0.4

Note: Traffic: initial two-way annual average daily truck traffic (AADTT) = 345.% of trucks in design direction: 50% of trucks in design lane: 95%; traffic growth rate: 4%.

Assume 80 kN tandem axles for all trucks, and tire pressure of 690 kPa.

15 Mix Design and Structural Design for Concrete Pavements

15.1 MIX DESIGN

The absolute volume method detailed in ACI 211 (American Concrete Institute, ACI Committee 211, 1991) is commonly used to proportion the local concrete materials to achieve the desired fresh and ultimate hardened concrete properties. This includes selecting the water–cement ratio, coarse and fine aggregate, cement and supplementary cementitious materials, water, and admixture requirements. The effects of all of these volumetric properties are interrelated. Their selection depends on many factors, including the compressive strength, which is the simplest and easiest to measure.

15.1.1 CONCRETE STRENGTH

According to ACI 318 (American Concrete Institute, ACI Committee 318, 2002), a concrete strength test is defined by the average compressive strength of two concrete cylinders. The running average of three consecutive tests must be equal to or greater than the specified compressive strength $\left(f_c'\right)$ at 28 days. ACI 318 also requires the f_c' to be at least 2500 psi (17.5 MPa), and no individual test (average of two cylinders) can be more than 500 psi (3.5 MPa) below the specified strength (ACI 318). To allow for variability in materials, batching, mixing, placing, finishing, curing, and testing the concrete, the ACI recommends a higher target strength that would assure the designer to achieve the minimum specified strength, which is called f_{cr}' (see Tables 15.1 through 15.3).

15.1.2 WATER-TO-CEMENTITIOUS MATERIALS RATIO

The water-to-cementitious materials ratio (or water–cementitious ratio; W/CM), is the mass of water divided by the mass of all cements, blended cements, and pozzolanic materials such as fly ash, slag, silica fume, and natural pozzolans. In most literature, *water–cement ratio* (W/C) is used synonymously with W/CM.

Assuming that a concrete is made with clean sound aggregates, and that the cement hydration has progressed normally, the strength gain is inversely proportional to the water–cementitious ratio by mass. The paste strength is proportional to the solids volume or the cement density per unit volume (Abrams, 1918) and is shown schematically in Figure 15.1. It should be mentioned, however, that the cement hydration process is more complex in its microstructural development of calcium-silicate-hydrates and pore structure. But even at a given water–cementitious ratio, differences in strength may result based on the influence of aggregate gradation, shape, particle size, surface texture, strength, and stiffness, or factors associated with cement materials (e.g., different sources, chemical composition, and physical attributes), the amount of entrained air, and the effects of admixtures and curing. A typical relationship between W/C, compressive strength, and air-entrainment is shown in Figure 15.2.

TABLE 15.1

Modification Factor for Standard Deviation When Less Than 30 Tests Are Available

Number of Tests[a]	Modification Factor for Standard Deviation[b]
Less than 15	Use Table 15.2/Table 15.3
15	1.16
20	1.08
25	1.03
30 or more	1.00

Source: Reprinted from Portland Cement Association (PCA), *Design and Control of Concrete Mixtures*, Engineering Bulletin 001, 14th edn., PCA, Skokie, IL, 2002.

[a] Interpolate for intermediate numbers of tests.

[b] Modified standard deviation to be used to determine required average strength, f'_{cr}.

TABLE 15.2

(Metric) Required Average Compressive Strength When Data Are Not Available to Establish Standard Deviation

Specified Compressive Strength, f'_c (MPa)	Required Average Compressive Strength, f'_{cr} (MPa)
Less than 21	$f'_c + 7.0$
21–35	$f'_c + 8.5$
Over 35	$1.10 f'_c + 5.0$

Source: Reprinted from Portland Cement Association (PCA), *Design and Control of Concrete Mixtures*, Engineering Bulletin 001, 14th edn., PCA, Skokie, IL, 2002.

15.1.3 SELECTION OF THE WATER-TO-CEMENTITIOUS MATERIALS RATIO

The selection of the W/CM is influenced by strength requirements and exposure conditions. The W/CM will govern the permeability or watertightness that is necessary for preventing aggressive salts and chemicals from entering the paste pore system and causing deleterious effects in the concrete or reinforcing steel. For different exposure conditions, ACI recommendations for a maximum W/CM and minimum design compressive strengths are provided in Tables 15.4 and 15.5. For pavement design, flexural strength is recognized as the more appropriate strength characteristic, and is specified for pavement design. However, the ACI uses only compressive strength in its specifications because of the lower variability associated with compressive strength testing.

When durability does not govern, the W/CM should be selected based on compressive strength. In this case, adequate knowledge about strength and mix proportions should be obtained through trial mixes or field data. For example, Figure 15.2 and Tables 15.6 and 15.7 show the relationship

TABLE 15.3

(Inch-Pound) Required Average Compressive Strength When Data Are Not Available to Establish Standard Deviation

Specified Compressive Strength, f'_{cr} (psi)	Compressive Strength, f'_{cr} (psi)
Less than 3000	$f'_c + 1000$
3000–5000	$f'_c + 1200$
Over 5000	$1.10 \, f'_c + 700$

Source: Reprinted from Portland Cement Association (PCA), *Design and Control of Concrete Mixtures*, Engineering Bulletin 001, 14th edn., PCA, Skokie, IL, 2002.

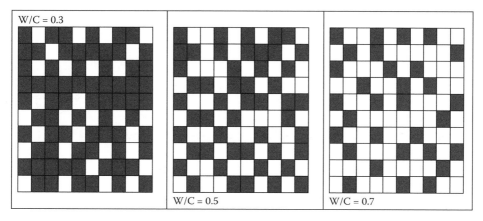

FIGURE 15.1 Schematic showing the effect of the water-to-cementitious materials ratio (W/CM) on solids density per unit volume. (dark = solid; white = voids.)

between the W/CM and compressive strength for a given concrete mix and curing conditions. It should be noted that the ACI-recommended values given in subsequent tables and figures should be a starting point from which to initiate the mix design process. Variations in mix ingredients may require a few iterations to optimize the desired properties.

15.1.4 AGGREGATES

Aggregates have an important role in proportioning concrete. Aggregates have a great effect on the workability of fresh concrete. The aggregate particle size and gradation, shape, and surface texture will influence the amount of concrete that is produced with a given amount of paste (cement plus water). The selection of the maximum size aggregate is governed by the thickness of the slab and by the closeness of the reinforcing steel. The maximum size aggregates should not be obstructed and should flow easily during placement and consolidation.

Guidelines provided by ACI state that the maximum size coarse aggregate should not exceed one-fifth of the narrowest dimension between the sides of forms, or three-fourths of the clear space between individual reinforcing bars or the clear spacing between reinforcing bars and form walls.

FIGURE 15.2 Typical relationship between compressive strength and W/CM for concrete using 19–25 mm nominal maximum aggregate size coarse aggregates. (Reprinted with kind permission from Portland Cement Association (PCA), *Design and Control of Concrete Mixtures*, Engineering Bulletin 001, 14th edn., PCA, Skokie, IL, 2002.)

TABLE 15.4
Maximum Water–Cement Ratios and Minimum Design Strengths for Various Exposure Conditions

Exposure Condition	Maximum Water-to-Cementitious Materials Ratio by Mass for Concrete	Minimum Design Compressive Strength, f_c', MPa (psi)
Concrete protected from exposure to freezing and thawing, application of deicing chemicals, or aggressive substances	Select water-to-cementitious materials ratio on the basis of strength, workability, and finishing needs	Select strength based on structural requirements
Concrete intended to have low permeability when exposed to water	0.50	28 (4000)
Concrete exposed to freezing and thawing in a moist condition, or deicers	0.45	31(4500)
For corrosion protection for reinforced concrete exposed to chlorides from deicing salts, saltwater, brackish water, seawater, or spray from these sources	0.40	35 (5000)

Source: Reprinted from Portland Cement Association (PCA), *Design and Control of Concrete Mixtures*, Engineering Bulletin 001, 14th edn., PCA, Skokie, IL, 2002.

For unreinforced slabs on grade, the maximum aggregate size should not exceed one-third of the slab thickness. For pavements, typically 1–1.5 in. maximum size aggregate is used.

The most desirable gradation for fine aggregate will depend on the properties of the concrete the designer is trying to seek and will depend on coarse aggregate volume and maximum size, and amount of paste. Figure 15.3 and Table 15.8 provide the recommended bulk volume fraction

TABLE 15.5
Requirements for Concrete Exposed to Sulfates in Soil or Water

Sulfate Exposure	Water-Soluble Sulfate (SO₄) in Soil,% by Mass[a]	Sulfate (SO₄) in Water (ppm)	Cement Type[b]	Maximum Water-to-Cementitious Materials Ratio, by Mass	Minimum Design Compressive Strength, f_c', MPa (psi)
Negligible	Less than 0.10	Less than 150	No special type required	—	—
Moderate[c]	0.10–0.20	150–1,500	II, MS, IP(MS), IS(MS), P(MS), I(PM)(MS), I(SM)(MS)	0.50	28 (4000)
Severe	0.20–2.00	1,500–10,000	V, HS	0.45	31 (4500)
Very severe	Over 2.00	Over 10,000	V, HS	0.40	35 (5000)

Source: Reprinted from Portland Cement Association (PCA), *Design and Control of Concrete Mixtures*, Engineering Bulletin 001, 14th edn., PCA, Skokie, IL, 2002.

[a] Tested in accordance with the *Method for Determining the Quantity of Soluble Sulfate in Solid (Soil and Rock) and Water Samples* (Bureau of Reclamation, 1977).

[b] Cement types II and V are in ASTM C-150 (AASHTO M-85), types MS and HS in ASTM C-1157, and the remaining types are in ASTM C-595 (AASHTO M-240). Pozzolans or slags that have been determined by test or service record to improve sulfate resistance may also be used.

[c] Seawater.

TABLE 15.6
(Metric) Relationships between the Water-to-Cementitious Materials Ratio and the Compressive Strength of Concrete

Compressive Strength at 28 Days (MPa)	Water-to-Cementitious Materials Ratio by Mass	
	Non-Air-Entrained Concrete	Air-Entrained Concrete
45	0.38	0.30
40	0.42	0.34
35	0.47	0.39
30	0.54	0.45
25	0.61	0.52
20	0.69	0.60
15	0.79	0.70

Source: Reprinted from Portland Cement Association (PCA), *Design and Control of Concrete Mixtures*, Engineering Bulletin 001, 14th edn., PCA, Skokie, IL, 2002.

Note: Strength is based on cylinders moist-cured 28 days in accordance with ASTM C-31 (AASHTO T-23). Relationship assumes a nominal maximum size aggregate (NMSA) of about 19–25 mm.

TABLE 15.7

(Inch-Pound Units) Relationships between Water-to-Cementitious Material Ratio and Compressive Strength of Concrete

Compressive Strength at 28 Days (psi)	Water-to-Cementitious Materials Ratio by Mass	
	Non-Air-Entrained Concrete	Air-Entrained Concrete
7000	0.33	—
6000	0.41	0.32
5000	0.48	0.40
4000	0.57	0.48
3000	0.68	0.59
2000	0.82	0.74

Source: Reprinted from Portland Cement Association (PCA), *Design and Control of Concrete Mixtures*, Engineering Bulletin 001, 14th edn., PCA, Skokie, IL, 2002.

Note: Strength is based on cylinders moist-cured 28 days in accordance with ASTM C-31 (AASHTO T-23). Relationship assumes a nominal maximum size aggregate (NMSA) of about ¾ to 1 in.

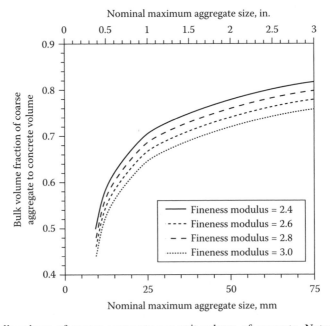

FIGURE 15.3 Bulk volume of coarse aggregate per unit volume of concrete. Note: Bulk volumes are based on aggregates in a dry-rodded condition. (Reprinted with kind permission from Portland Cement Association (PCA), *Design and Control of Concrete Mixtures*, Engineering Bulletin 001, 14th edn., PCA, Skokie, IL, 2002.)

TABLE 15.8
Bulk Volume of Coarse Aggregate per Unit Volume of Concrete

Nominal Maximum Size of Aggregate, mm (in.)	Bulk Volume of Dry-Rodded Coarse Aggregate per Unit Volume of Concrete for Different Fineness Moduli of Fine Aggregate[a]			
	2.40	2.60	2.80	3.00
9.5 (3/8)	0.50	0.48	0.46	0.44
12.5 (1/2)	0.59	0.57	0.55	0.53
19 (3/4)	0.66	0.64	0.62	0.60
25 (1)	0.71	0.69	0.67	0.65
37.5 (11/2)	0.75	0.73	0.71	0.69
50 (2)	0.78	0.76	0.74	0.72
75 (3)	0.82	0.80	0.78	0.76
150 (6)	0.87	0.85	0.83	0.81

Source: Reprinted from Portland Cement Association (PCA), *Design and Control of Concrete Mixtures*, Engineering Bulletin 001, 14th edn., PCA, Skokie, IL, 2002.

[a] Bulk volumes are based on aggregates in a dry-rodded condition as described in ASTM C-29 (AASHTO T-19).

of coarse aggregate per unit volume of concrete based on the maximum nominal size of the coarse aggregate and the fineness modulus of the sand or fine aggregate. PCC for pavements may have a stiffer consistency and therefore coarse aggregate volumes may be increased by 10%.

15.1.5 Air Content in Concrete

The purposeful entrainment of air in concrete provides tremendous protection against freezing and thawing action and against deicing salts. Hydraulic pressure is generated when water in the paste pore structure freezes and pushes against the unfrozen water. The tiny entrained air bubbles act as relief valves for this developed hydraulic pressure. Figure 15.4 shows the recommended relationship between the target percentage air content, the maximum nominal aggregate size, and the severity of exposure.

15.1.6 Slump

Fresh concrete must have the appropriate workability, consistency, and plasticity suitable for construction conditions. Workability is a measure of the ease of placement, consolidation, and finishing of the concrete. Consistency is the ability of freshly mixed concrete to flow, and plasticity assesses the concrete's ease of molding. If the concrete is too dry and crumbly, or too wet and soupy, then it lacks plasticity. The slump test is a measure of consistency and indicates when the characteristics of the fresh mix have been changed or altered. However, the slump is indicative of workability when assessing similar mixtures. Different slumps are needed for different construction projects. Table 15.9 provides recommendations for slump for different types of construction. For pavements, state DOTs may use different slump specifications for fixed-form and slip-form construction.

15.1.7 Water Content

The water demands for a concrete mix depend on many variables, including mix proportions, air content, aggregate gradation, angularity and texture, and climate conditions. Tables 15.10 and 15.11 and Figure 15.5 present the relationship between mix water, different slumps, and nominal maximum

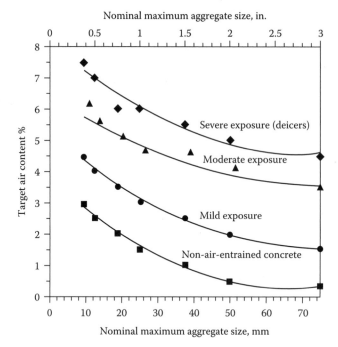

FIGURE 15.4 Target total air content requirements for concrete using different sizes of aggregate. Note: The air content in job specifications should be specified to be delivered within −1 to +2 percentage points of the target value for moderate and severe exposure. (Reprinted with kind permission from Portland Cement Association (PCA), *Design and Control of Concrete Mixtures*, Engineering Bulletin 001, 14th edn., PCA, Skokie, IL, 2002.)

TABLE 15.9
Recommended Slumps for Various Types of Construction

Concrete Construction Reinforced Foundation	Slump, mm (in.)	
	Maximum[a]	Minimum
Walls and footings	75 (3)	25 (1)
Plain footings, caissons, and substructure walls	75 (3)	25 (1)
Beams and reinforced walls	100 (4)	25 (1)
Building columns	100 (4)	25 (1)
Pavements and slabs	75 (3)	25 (1)
Mass concrete	75 (3)	25 (1)

Source: Reprinted from Portland Cement Association (PCA), *Design and Control of Concrete Mixtures*, Engineering Bulletin 001, 14th edn., PCA, Skokie, IL, 2002.

Note: Plasticizers can safely provide higher slumps.

[a] May be increased 25 mm (1 in.) for consolidation by hand methods, such as rodding and spading.

TABLE 15.10

(Metric) Approximate Mixing Water and Target Air Content Requirements for Different Slumps and Nominal Maximum Sizes of Aggregate

Slump (mm)	Water, Kilograms per Cubic Meter of Concrete, for Indicated Sizes of Aggregate[a]							
	9.5 mm	12.5 mm	19 mm	25 mm	37.5 mm	50 mm[b]	75 mm[a]	150 mm[b]
Non-air-entrained concrete								
25–50	207	199	190	179	166	154	130	113
75–100	228	216	205	193	181	169	145	124
150–175	243	228	216	202	190	178	160	—
Approximate amount of entrapped air in non-air-entrained concrete (%)	3	2.5	2	1.5	1	0.5	0.3	0.2
Air-entrained concrete								
25–50	181	175	68	160	150	142	122	107
75–100	202	193	184	175	165	157	133	119
150–175	216	205	197	184	174	166	154	—
Recommended average total air content (%) for level of exposure[c]								
Mild exposure	4.5	4.0	3.5	3.0	2.5	2.0	1.5	1.0
Moderate exposure	6.0	5.5	5.0	4.5	4.5	4.0	3.5	3.0
Severe exposure	7.5	7.0	6.0	6.0	5.5	5.0	4.5	4.0

Source: Reprinted from Portland Cement Association (PCA), *Design and Control of Concrete Mixtures*, Engineering Bulletin 001, 14th edn., PCA, Skokie, IL, 2002.

[a] These quantities of mixing water are for use in computing cementitious material contents for trial batches. They are maximums for reasonably well-shaped angular coarse aggregates graded within the limits of accepted specifications.

[b] The slump values for concrete containing aggregates larger than 37.5 mm are based on slump tests made after the removal of particles larger than 37.5 mm by wet screening.

[c] The air content in job specifications should be specified to be delivered within –1 to +2 percentage points of the table target value for moderate and severe exposures.

size aggregates (NMSAs). The recommendations given in Tables 15.10 and 15.11 and Figure 15.5 are for angular aggregates, and the water estimates may be reduced by 20–45 lb for subangular to rounded aggregates, respectively.

15.1.8 Cementing Materials Content and Type

The cementing materials content is usually determined based on the specified water-to-cementitious materials ratio. However, usually a minimum amount of cement is also specified to ensure satisfactory durability, finishability, and wear resistance of slabs even though strength needs may be satisfied at lower cement contents, as shown in Table 15.12. However, economic mixes are obtained with the least amount of cement. To accomplish this, the concrete should be as stiff as possible, should use the largest practical maximum coarse aggregate, and should use the optimum ratio of coarse to fine aggregate.

For durability, the type of cement used is also important. When the concrete will be exposed to significant amounts of sulfates, cements with low C_3A contents should be used, as presented in Table 15.5. When the potential for corrosion of reinforcing bars exists due to a favorable chloride condition, then blended cements and a low W/CM should be considered. ACI 357R should be consulted for designing for corrosion protection. Table 15.13 shows limits on the amounts of pozzolans and

TABLE 15.11

(Inch-Pound Units) Approximate Mixing Water and Target Air Content Requirements for Different Slumps and Nominal Maximum Sizes of Aggregate[a]

Slump (in.)	Water, Pounds per Cubic Yard of Concrete, for Indicated Sizes of Aggregate[a]							
	⅜ in.	½ in.	¾ in.	1 in.	1¼ in.	2 in.[b]	3 in.[b]	6 in.[b]
Non-air-entrained concrete								
1–2	350	335	315	300	275	260	220	190
3–4	385	365	340	325	300	285	245	210
6–7	410	385	360	340	315	300	270	—
Approximate amount of entrapped air in non-air-entrained concrete (%)	3	2.5	2	1.5	1	0.5	0.3	0.2
Air-entrained concrete								
1–2	305	295	280	270	250	240	205	180
3–4	340	325	305	295	275	265	225	200
6–7	365	345	325	310	290	280	260	—
Recommended average total air content (%) for level of exposure[c]								
Mild exposure	4.5	4.0	3.5	3.0	2.5	2.0	1.5	1.0
Moderate exposure	6.0	5.5	5.0	4.5	4.5	3.5	3.5	3.0
Severe exposure	7.5	7.0	6.0	6.0	5.5	5.0	4.5	4.0

Source: Reprinted from Portland Cement Association (PCA), *Design and Control of Concrete Mixtures*, Engineering Bulletin 001, 14th edn., PCA, Skokie, IL, 2002.

[a] These quantities of mixing water are for use in computing cement factors for trial batches. They are maximums for reasonably well-shaped angular coarse aggregates graded within the limits of accepted specifications.

[b] The slump values for concrete containing aggregates larger than 1½ in. are based on slump tests made after the removal of particles larger than 1½ in. by wet screening.

[c] The air content in job specifications should be specified to be delivered within –1 to +2 percentage points of the table target value for moderate and severe exposures.

mineral admixtures to be used when designing concrete exposed to deicing chemicals. It should be noted that these ACI recommendations should be used as a starting point for trial mixes. However, local experience should be consulted whenever it exists.

15.1.9 Admixtures

Admixtures are used in concrete to enhance the desirable characteristics in the fresh and hardened concrete. This includes reducing the W/CM, entraining air, retarding or accelerating the set time, increasing the slump without increasing the W/CM, inhibiting corrosion, and self-leveling. Some admixtures are calcium chloride based and may affect steel corrosion. ACI 318 places limits on the maximum chloride-ion content for corrosion protection, as presented in Table 15.14.

The design of a concrete mix using the ACI method is explained with the following example.

15.1.10 Example of Mix Design

Concrete is specified for a 10-in. pavement slab that will be exposed to severe freeze–thaw conditions and a moist environment. The required 28 day compressive strength is 3500 psi.
Cement: Type II, Sp. Gr. 3.15.

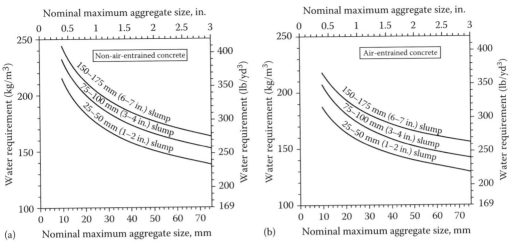

FIGURE 15.5 Approximate water requirement for various slumps and crushed aggregate sizes for (a) non-air-entrained concrete and (b) air-entrained concrete. (Reprinted with kind permission from Portland Cement Association (PCA), *Design and Control of Concrete Mixtures*, Engineering Bulletin 001, 14th edn., PCA, Skokie, IL, 2002.)

TABLE 15.12

Minimum Requirements of Cementing Materials for Concrete Used in Flatwork

Nominal Maximum Size of Aggregate, mm (in.)	Cementing Materials, kg/m³ (lb/yd³)ᵃ
37.5 (11/2)	280 (470)
25 (1)	310 (520)
19 (34)	320 (540)
12.5 (1/2)	350 (590)
9.5 (3/8)	360 (610)

Source: Reprinted from Portland Cement Association (PCA), *Design and Control of Concrete Mixtures*, Engineering Bulletin 001, 14th edn., PCA, Skokie, IL, 2002.

ᵃ Cementing materials quantities may need to be greater for severe exposure. For example, for deicer exposures, concrete should contain at least 335 kg/m³ (564 lb/yd³) of cementing materials.

Coarse aggregate: well-graded gravel containing some crushed particles. The maximum nominal size is 3/4 in., with a specific gravity of 2.65 and an oven dry-rodded unit weight of 100 pcf; the moisture content for the trial batch is 3%; and the absorption is 0.7%.

Fine aggregate: a natural sand with a specific gravity of 2.60; moisture content for the trial batch is 1.2%, the absorption is 2.8%, and the fineness modulus is 2.50.

The required W/CM is determined by strength (or structural design) and durability requirements. Our strength requires a 3500 psi concrete. Since no statistical data exist, Table 15.3 is used to establish a target strength, $f_{cr}' = f_c' + 1200$ psi. Our new target compressive strength is (3500 + 1200 = 4700 psi). For an environment with moist freezing and thawing, the maximum W/CM is 0.45 (Table 15.4). From Figure 15.2, the recommended W/CM = 0.42 for a compressive strength of 4700 psi. Or it could be

TABLE 15.13

Cementitious Materials Requirements for Concrete Exposed to Deicing Chemicals

Cementitious Materials[a]	Maximum % of Total Cementitious Materials by Mass[b]
Fly ash and natural pozzolans	25
Slag	50
Silica fume	10
Total of fly ash, slag, silica fume, and natural pozzolans	50[c]
Total of natural pozzolans and silica fume	35[c]

Source: Reprinted from Portland Cement Association (PCA), *Design and Control of Concrete Mixtures*, Engineering Bulletin 001, 14th edn., PCA, Skokie, IL, 2002.

[a] Includes portion of supplementary cementing materials in blended cements.

[b] Total cementitious materials include the summation of Portland cements, blended cements, fly ash, slag, silica fume, and other pozzolans.

[c] Silica fume should not constitute more than 10% of total cementitious materials, and fly ash or other pozzolans shall not constitute more than 25% of cementitious materials.

TABLE 15.14

Maximum Chloride-Ion Content for Corrosion Protection

Type of Member	Maximum Water-Soluble Chloride Ion (Cl^-) in Concrete, % by Mass of Cement[a]
Prestressed concrete	0.06
Reinforced concrete exposed to chloride in service	0.15
Reinforced concrete that will be dry or protected from moisture in service	1.00
Other reinforced concrete construction	0.30

Source: Reprinted from Portland Cement Association (PCA), *Design and Control of Concrete Mixtures*, Engineering Bulletin 001, 14th edn., PCA, Skokie, IL, 2002.

[a] ASTM C-1218.

interpolated from Table 15.7. Table 15.4 suggests a W/CM of 0.45 to protect against moisture and freeze–thaw conditions. However, since the lower W/CM governs, the W/CM = 0.42 is based on strength.

Select slump when it is not specified according to the desired application from Table 15.9. For a pavement slab, choose a slump of 3 in. Note that you can add 1 in. to the maximum values when placing by hand or rodding. For slip forming construction, the slump could be as low as zero.

Specified maximum coarse aggregate size: 3/4 in. This could be dictated by availability and the ACI, which states that the maximum coarse aggregate size should be less than one-third of the slab thickness or not more than three-fourths of the distance between reinforcements, and not more than one-fifth of form dimensions.

The water content requirement depends upon the coarse aggregate maximum size, particle shape and grading of aggregates, and amount of air-entrainment. From Table 15.11 for a 3-in. slump, a 3/4-in. maximum coarse aggregate, and an air-entrained concrete, we need 305 lb of water per yd^3. Since our aggregate is a gravel with some crushed particles, the water content may be reduced by 35 lb. Therefore, $(305 - 35 = 270)$.

Air content: for a severe freeze–thaw exposure, Table 15.11 recommends a target air content of 6%. Since it is difficult to achieve a target value for air, a range of +2% and −1% is acceptable, hence a range of 5%–8%. Use 7% air.

Cement content: calculate based on the maximum W/CM.

$$\text{Cement content} = \frac{(\text{Water Requirement})}{(\text{W/CM})} = \frac{(270)}{(0.42)} = 643 \text{ lb/yd}^3$$

Coarse aggregate content: estimate the quantity of 3/4-in. maximum size coarse aggregate with a sand fineness modulus of 2.5.

From Table 15.8 and Figure 15.3, an amount of 0.65 yd^3 may be used. Since our coarse aggregate weighs 100 pcf, the required dry weight per yd^3 is as follows:

$$100 \text{ pcf} * 27 \text{ ft}^3/\text{yd}^3 * 0.65 = 1755 \text{ lb (dry)}$$

Admixture content: for a 7% air content, the air-entraining admixture manufacturer recommends a dosage of 0.9 fl oz per 100 lb of cement.

$$0.9 * \frac{643}{100} = 5.8 \text{ fl oz/yd}^3$$

Fine aggregate content: at this point, all ingredients of the mix have been estimated except for the fine aggregate. We shall use the method of absolute volume to compute the fine aggregate.

$$\text{Volume} = \frac{\text{mass}}{(\text{Sp. Gr.} * \text{density of water})}$$

$$\text{Water volume} = \frac{(270)}{(1 * 62.4)} = 4.33 \text{ ft}^3/\text{yd}^3$$

$$\text{Cement volume} = \frac{(643)}{(3.15 * 62.4)} = 3.27 \text{ ft}^3/\text{yd}^3$$

$$\text{CA volume} = \frac{(1755)}{(2.65 * 62.4)} = 10.61 \text{ ft}^3/\text{yd}^3$$

$$\text{Air volume} = 0.07 * 27 = 1.89 \text{ ft}^3$$

$$\text{Total volume of known ingredients} = 20.1 \text{ ft}^3$$

(Note that the admixture volume is insignificant to add to the water volume.)

Solid volume of "fine aggregate" required is the difference between the unit volume and total volume of known ingredients.

$$27 - 20.1 = 6.9 \text{ ft}^3$$

Weight of dry "fine aggregate"

$$6.9 * 2.60 * 62.4 = 1119 \text{ lb}$$

So far, the mixture proportions are as follows (per 1 yd³):

Water	270 lb
Cement	643 lb
Coarse aggregate (dry)	1755 lb
Fine aggregate (dry)	1119 lb
Air-entraining admixture	5.8 fl oz
Total weight	3787 lb
Slump	3 in. (±3/4 in. for trial batch)
Air content	7% (±0.5% for trial batch)
Estimated density	270 + 643 + (1755 * 1.007) +
(using SSD aggregates)	(1119 * 1.012)/27 = 3812/27 = 141 lb/ft³

Moisture corrections: since the mix water is based on oven-dried aggregate, the batch weights must be corrected for absorbed and free moisture in aggregates. For the aggregates with the indicated moisture content (MC), the wet batch weights are as follows:

$$\text{Coarse aggregate (MC} = 3\%): \ 1755 * 1.03 = 1808 \text{ lb/yd}^3$$

$$\text{Fine aggregate (MC} = 1.2\%) \ 1119 * 1.012 = 1132 \text{ lb/yd}^3$$

Note: absorbed water should be excluded from mix water, and surface moisture should be included in mix water.

$$\text{Coarse aggregate absorption} = 0.7\%$$

$$\text{Fine aggregate absorption} = 2.8\%$$

$$\text{Coarse aggregate contribution} = 3\% - 0.7\% = +2.3 \text{ moisture (subtract from mix water)}$$

$$\text{Fine aggregate contribution} = 1.2\% - 2.8\% = -1.6\% \text{ moisture (add to mix water)}$$

Corrected mix water:

$$270 - 1755 * (0.023) + 1119 * (0.016) = 247 \text{ lb/yd}^3$$

The revised batch weights for 1 yd³ are as follows:

Water to be added to mix	247 lb
Cement	643 lb
Coarse aggregate (wet with MC = 3%)	1808 lb
Fine aggregate (wet with MC = 1.2%)	1132 lb
Total	3830

Note: Air-entraining admixture = 5.8 fl oz.

Trial batch: at this point, trial batches are used to check if the target parameters such as slump, air, and workability are met. In addition, concrete cylinders for strength testing at 28 days are made at this point. For a laboratory batch, mixing a smaller volume is adequate. For this example, a 2 ft^3 or 2/27 yd^3 batch will be made.

Water to be added to mix	247 * (2/27) = 18.30 lb
Cement	643 * (2/27) = 47.63 lb
Coarse aggregate (wet with MC = 3%)	1808 * (2/27): = 133.93 lb
Fine aggregate (wet with MC = 1.2%)	1132 * (2/27): = 83.85 lb
Total	283.6

Air-entraining admixture: 5.8 * (2/27) = 0.43 fl oz (Note: usually admixtures are batched in milliliters). Therefore, 5.8 * 29.573 mL/oz = 171.52 mL * (2/27) = 12.71 mL.

After mixing the trial batch, the measured slump was 4 in., the air content was 8%, and the fresh concrete density was 140.3 pcf. During mixing, not all of the mix water was used. The net water used was 18.30 − 0.5 = 17.8 lb. Therefore, the updated batch weights are as follows:

Water to be added to mix	17.8 lb
Cement	47.63 lb
Coarse aggregate (wet with MC = 3%)	133.93 lb
Fine aggregate (wet with MC = 1.2%)	83.85 lb
Total	283.21 lb

The yield for the trial batch is as follows:

$$\frac{283.21}{140.3} = 2.018 \text{ ft}^3$$

The mixing water content is determined from the added water plus the free water on the aggregates, and is determined as follows:

Water added	17.8 lb
Free water on coarse aggregate: 133.93/1.03 * (3% − 0.7%)/100	2.998
Free water on fine aggregate: 83.85/1.012 * (1.2% − 2.8%)/100	−1.325
Total	19.47 lb

(The negative sign indicates that the fine aggregate is drier than the absorption moisture content and therefore additional water should be added to the mix to compensate for it.)

The mix water required for a cubic yard of the same slump concrete as the trial batch is as follows:

$$19.47 * \frac{27}{2.018} = 260.5 \text{ lb}$$

Batch adjustments: the measured slump of the trial mix was 4 in. This is higher than the target slump of 3 in. and is unacceptable. The measured air content was also too high (more than target of 7% ± 0.5%), and the yield was higher too. Therefore, we shall adjust the yield, air-entrainment, and water content to obtain a 3 in. slump.

When adjusting the mix design from the trial batch to the mix proportions for the large batch, use the following rule of thumb: increase the water by 5 lb for each 1% air reduction to maintain a similar slump, and reduce the water by 10 lb for each reduction in slump by 1 in. Therefore, the adjusted mix water for the reduced slump and air is as follows:

$$(5*1) - (10*1) + 247 = 242 \text{ lb/yd}^3$$

Now that the water is reduced, the cement content should decrease for a constant W/CM:

$$\text{Cement} = \frac{242}{0.42} = 576 \text{ lb}$$

Since the workability is adequate, the coarse aggregate content will not be readjusted.
The new adjusted batch weights are as follows:

Water to be added to mix	242/(1 * 62.4)=3.87 ft³
Cement	643/(3.15 * 62.4)=3.27 ft³
Coarse aggregate (wet with MC=3%)	1755/(2.65 * 62.4)=10.61 ft³
Air content	7% * 27=1.89 ft³
Total	19.61 ft³
Fine aggregate volume	27 – 19.61=7.39 ft³

Fine aggregate (dry weight basis)=7.39 * 2.60 * 62.4=1199 lb
  Air-entrainment admixture: use 0.8 fluid oz per 100 lb cement to obtain a 7% air content.

$$\text{AEA} = 0.8 * \frac{643}{100} = 5.14 \text{ fl oz}$$

Adjusted batch weights per cubic yard of concrete are as follows:

Water	242 lb	240 lb
Cement	643 lb	643 lb
Coarse aggregate (wet with MC=3%)	1755 lb (dry)	1767 lb (SSD)
Fine aggregate (wet with MC=1.2%)	1199 lb (dry)	1233 lb (SSD)
Total	5606 lb	5652 lb

15.2 STRUCTURAL DESIGN

15.2.1 AASHTO Method (AASHTO, 1993)

The rigid pavement design guide was developed similarly to the flexible pavement design guide based on the AASHO Road Test's results and knowledge gained.
  The basic equations developed for rigid pavements are similar to and of the same form as the flexible pavements but with different regression equation constants.

$$G_t = \beta(\log W_t - \log \rho)$$

$$\beta = 100 + \frac{3.63(L_1 + L_2)^{5.20}}{(D+1)^{8.46} L_2^{3.52}}$$

$$\log \rho = 5.85 + 7.35 \log(D+1) - 4.62 \log(L_1 + L_2)$$

G_t is the logarithm of the ratio of the loss in serviceability at time t to the potential loss in serviceability taken at a point in time where $p_t = 1.5$; p_t is the serviceability at any time t and therefore $G_t = \log[4.5 - p_t/4.5 - 1.5]$
β is the function of design and load variables that dictates the shape of the ρ versus W_t curve
ρ is the function of design and load variables that relates the expected number of load applications to a p_t of 1.5 and $\rho = W_t$
W_t is the axle load application at the end of time t

L_1 is the load on one single axle or a set of tandem axles (in kips); for example 18 for an equivalent 18 kip single axle load; 32 kip for a 32 kip equivalent tandem axle load

L_2 is the axle code; 1 used for single axle; 2 used for tandem axle, and 3 used for tridem

In the AASHO Road Test, the initial serviceability for rigid pavements was assigned a value of 4.5. This value is slightly higher than for flexible pavements which was taken as 4.2; D is the slab thickness which replaces the term SN for flexible pavements. The equation simplifies greatly when using an equivalent 18 kip single axle load with $L_1 = 18$; $L_2 = 1$.

$$\log W_{t18} = 7.35 \log(D+1) - 0.06 + \frac{\log\left[\dfrac{4.5 - p_t}{4.5 - 1.5}\right]}{1 + \dfrac{1.624 * 10^7}{(D+1)^{8.46}}}$$

W_{t18} predicts the number of 18 kip single axle load applications to time t. The original equation is only applicable to the conditions present at the AASHO Road Test. The conditions at the road test included: the concrete modulus of elasticity $E_c = 4.2 \times 10^6$; the modulus of rupture of concrete $S_c = 690$ psi, modulus of subgrade reaction k = 60 pci, the joint load transfer coefficient J = 3.2 and the drainage coefficient $C_d = 1.0$.

Over the years, other modifications were made to the original equation to take into consideration factors such as different subgrades and conditions other than those that existed during the AASHO Road Test. In 1972, AASHTO adopted a modification developed using the Spangler (1942) equation to consider corner loading and to extend the original AASHO Road Test equation to other conditions. Other modifications include the drainage coefficient C_d, $Z = E_c/k$, and the reliability term $Z_R S_o$, replacing the term (4.5-p_t) by ΔPSI.

15.2.2 DESIGN INPUT PARAMETERS

15.2.2.1 Reliability

The original AASHO equations were based on the mean values, which means that there is a 50% chance that the actual pavement would not reach its design life. This is not an acceptable risk for large capacity highways that carry a significant traffic volume and are essential to the economy. Therefore a higher reliability should be expected.

The reliability term in the AASHTO equation is the product of the $Z_R S_o$. The overall standard deviation S_o should be based on the variability of the local conditions. AASHTO recommends a value range for S_o between 0.3 and 0.4. When variability in traffic is considered, then $S_o = 0.39$, otherwise an $S_o = 3.4$ is typically used. The recommended level of reliability and the standard normal deviate Z_R are provided by AASHTO and reproduced in Table 14.2. For interstate highways, the reliability R% = (85%–99.9%) corresponding to $Z_R = (-1.037$ to $-3.75)$ (AASHTO, 1993).

15.2.2.2 Serviceability

The initial serviceability p_o is an estimate of the perceived smoothness condition of the pavement by the user. For the AASHO Road Test conditions, a value of $p_o = 4.5$ was established for rigid pavements and $p_o = 4.2$ for flexible pavements. These values may vary based on local conditions and SHA recommendations. The terminal serviceability index p_t is the lowest acceptable level of perceived ride and comfort that a SHA will tolerate before major rehabilitation is performed. Values of p_t of 2.0, 2.5, or 3.0 are commonly used based on the highway function. The ΔPSI term is the total loss in serviceability. This could be due to traffic or foundation support deterioration based on environmental conditions.

TABLE 15.15

Recommended Values of Drainage Coefficient, C_d, for Rigid Pavement Design

	Percent of Time Pavement Structure Exposed to Moisture Levels Approaching Saturation			
Quality of Drainage	Less Than 1%	1%–5%	5%–25%	Greater Than 25%
Excellent	1.25–1.20	1.20–1.15	1.15–1.10	1.10
Good	1.20–1.15	1.15–1.10	1.10–1.00	1.00
Fair	1.15–1.10	1.10–1.00	1.00–0.90	0.90
Poor	1.10–1.00	1.00–0.90	0.90–0.80	0.80
Very poor	1.00–0.90	0.90–0.80	0.80–0.70	0.70

15.2.2.3 Drainage Coefficient (C_d)

The effect of drainage is critically important to the performance of rigid pavements. The quality of drainage could affect the severity and extent of pumping distress, bearing capacity loss, frost damage, and D-cracking among others. Table 15.15 (reproduced from AASHTO, 1993) shows the recommended drainage coefficients based on the quality of drainage and the percent of time that the pavement moisture levels approach saturation.

15.2.2.4 Load Transfer Coefficient (J)

The load transfer coefficient J accounts for the ability of the rigid pavement joint to distribute the load across the joint. The value of J depends on the pavement type (JPCP, JRCP, or CRCP); if transfer devices such as dowels are used; and the type of shoulder-tied concrete or asphalt. Table 15.16 provides the load transfer coefficient values with the lowest J value for a CRCP with tied concrete shoulders. In general the J value for a given set of conditions will increase with an increase in traffic volume since aggregate interlock decreases with load repetitions. As a guide, higher J values should be used with lower k-values, high values for CTE, and large temperature variations. The most common value $J = 3.2$ is for a JPCP or JRCP that is doweled but not tied to a concrete shoulder (AASHTO, 1993).

15.2.2.4.1 Modulus of Subgrade Reaction

The effective modulus of subgrade reaction k-value is a composite value based on a seasonally adjusted annual average k-value that is modified for subbase type and thickness, loss of support due to erosion and proximity to rigid foundation and bedrock. AASHTO recommends the following

TABLE 15.16

Recommended Load Transfer Coefficient for Various Pavement Types and Design Conditions

Shoulder	Asphalt		Tied P.C.C.	
Load Transfer Devices	Yes	No	Yes	No
Pavement type				
1. Plain jointed and jointed reinforced	3.2	3.8–4.4	2.5–3.1	3.6–4.2
2. CRCP	2.9–3.2	N/A	2.3–2.9	N/A

process for determining the composite k-value. The AASHTO 1993 Guide for determining composite k-value is shown in the next example.

15.2.2.4.2   Concrete Flexural Strength (S_c)

The modulus of rupture (S_c) or flexural strength for concrete is required for pavement design. The modulus of rupture is determined by using the third point (ASTM C78) bend test for a 28 day cured sample. AASHTO recommends that if a SHA commonly uses the center point bend test for determining the modulus of rupture, then a correlation between the two methods should be performed.

15.2.2.4.3   Concrete Modulus of Elasticity (E_c)

The modulus of elasticity E_c for Portland cement concrete is estimated using the ACI equation:

$$E_c = 57,000\sqrt{f_c'} \text{ psi} = 4,730\sqrt{f_c'} \text{ MPa}$$

The value for E_c does not vary substantially over the range of concrete strengths used for slab construction and does not commonly have a large influence over pavement thickness design.

15.2.2.4.4   Traffic, ESALs

AASHTO provides methods for converting single, tandem and tridem axle loads to ESALs using the axle equivalency factor which is covered in Chapter 5. As presented in Chapter 5, the LEF can be approximated by the fourth power rule. Appendix D of the AASHTO 1993 Design Guide, provides tables of the axle equivalency factors which depend on the type of axle (single, tandem, and tridem), the axle load magnitude, the slab thickness and the terminal serviceability, p_t. Ultimately, the AASHTO equation predicts W_{18} the number of 18 kip ESAL that will be carried by a pavement over its design life.

When the AASHO equations were developed in the 1960s and the use of computers was not yet common, nomographs were developed to aid in solving the complex equations. Figure 15.6 shows the nomograph for solving the following AASHTO equation:

$$\log W_{18} = Z_R * S_0 + 7.35\log(D+1) - 0.06 + \frac{\log\left[\dfrac{(\Delta PSI)}{4.5-1.5}\right]}{1+\dfrac{1.624*10^7}{(D+1)^{8.46}}}$$

$$+ (4.22 - 0.32 p_t)\log\left\{\frac{S_c C_d (D^{0.75} - 1.132)}{215.63 J\left[D^{0.75} - \dfrac{18.42}{\left(\dfrac{E_c}{k}\right)^{0.25}}\right]}\right\}$$

where

  S_c is the flexural strength of concrete
  D is the slab thickness
  E_c is the modulus of elasticity of concrete
  k is the modulus of subgrade reaction
  J is the joint transfer coefficient: 3.2 for plain jointed and jointed reinforced pavements, 2.9–3.2 for CRCP for doweled but not tied pavements, and 3.6–4.2 for doweled and tied pavements

The rest of the parameters are the same as those used in the asphalt pavement thickness design. Examples to demonstrate the use of the AASHTO equation and input parameters are presented next. These are the same examples presented in the AASHTO 1993 Design Guide.

Example 15.1

Determine the thickness D given the input parameters.
Given
$k = 72$ pci (19.5 MN/m³)
$E_c = 5 * 10^6$ psi (34.5 GPa)
$S_c = 650$ psi (4.5 MPa)
$J = 3.2$
$C_d = 1.0$
$\Delta PSI = 4.2–2.5 = 1.7$
$R = 95\%$
$S_o = 0.29$
$W_t = 5.1 * 10^6$

Nomograph solves:

$$\log_{10} W_{18} = Z_R {}^*S_O + 7.35{}^*\log_{10}(D + 1) - 0.06 + \frac{\log_{10}\left[\dfrac{\Delta\,PSI}{4.5 - 1.5}\right]}{1 + \dfrac{1.624{}^*10^7}{(D + 1)^{8.46}}} + (4.22 - 0.32p_t){}^*\log_{10}\left[\frac{S'_C {}^* C_d\left[D^{0.75} - 1.132\right]}{215.63{}^*J\left[D^{0.75} - \dfrac{18.42}{(E_C/k)^{0.25}}\right]}\right]$$

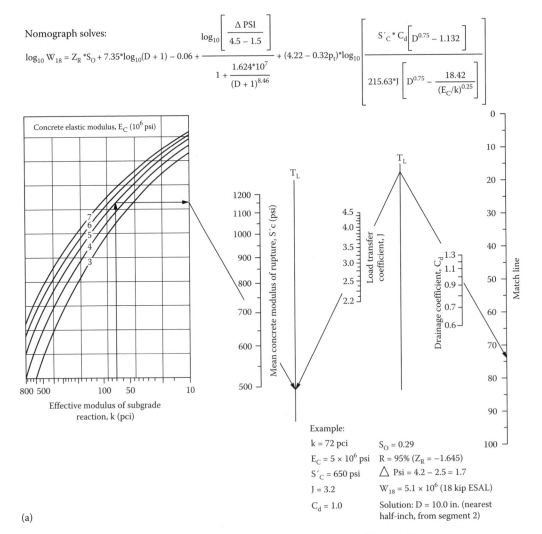

(a)

FIGURE 15.6   (a) Design chart for rigid pavements. (Segment 2 refers to Figure 15.6b.)

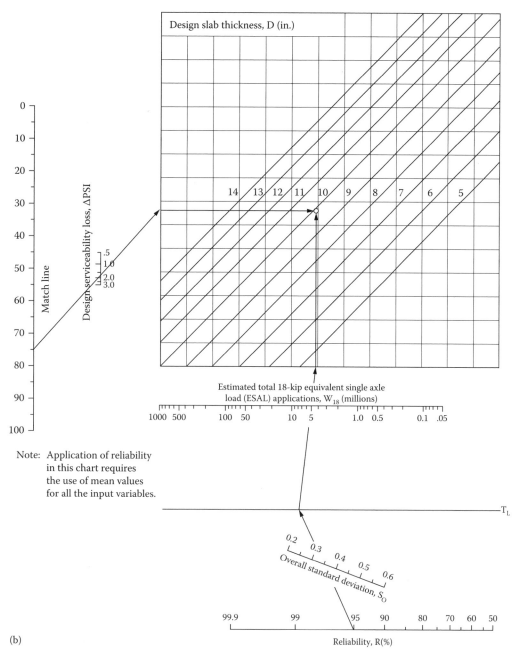

FIGURE 15.6 (continued) (b) Design chart for rigid pavements. (From American Association of State Highway and Transportation Officials (AASHTO), *AASHTO Guide for Design of Pavement Structures*, AASHTO, Washington, DC, © 1993. Used with permission.)

Solution

1. Starting with $k = 72$ pci (19.5 MN/m³) in Figure 15.6a, draw a vertical line until it connects with $E_c = 5 * 10^6$ psi (34.5 GPa). Draw a horizontal line until the end of the k versus E_c figure is reached. Draw a line from this end point that goes through $S_c = 650$ psi (4.5 MPa) and that ends at the turning line (T_L). Similar lines are drawn through $J = 3.2$ and $C_d = 1.0$ until the end line hits the value of 74 on the Match Line.

Nomograph Solves

$$\log_{10} = W_{18} = Z_R{}^*S_O + 7.35^*\log_{10}(D+1) - 0.06 + \frac{\log_{10}\left[\dfrac{DPSI}{4.5 - 1.5}\right]}{1 + \dfrac{1.624 * 10^7}{(D+1)^{8.46}}}$$

$$+ (4.22 - 0.32p_t)^*\log_{10}\left[\frac{S_C'{}^*C_d\left[D^{0.75} - 1.132\right]}{215.63^*J\left[D^{0.75} - \dfrac{18.42}{(E_C/k)^{0.25}}\right]}\right]$$

2. Starting at 74 on the Match Line in Figure 15.6b, a line is drawn through ΔPSI = 1.7 until it reaches the vertical axis.
3. Using the reliability axis, start with R = 95% and draw a line through $S_o = 0.29$, and then draw a line through $W_t = 5.1 * 10^6$ until it intersects the horizontal axis.
4. A horizontal line is drawn from the last step in Step 2. A vertical line is drawn from the last step in Step 3. The intersection of these two lines gives D = 9.75 in. (246 mm).

The nomograph can also be used to determine the amount of traffic that will produce a certain degradation in the Present Serviceability Index (PSI).

The property of the subgrade used for rigid pavement design is the modulus of subgrade reaction k instead of the resilient modulus M_r. Therefore, it is necessary to convert M_r to k. Values of k are also season dependent and will vary over the course of the year. Therefore, the relative damage parameter that includes the effect of seasonal change in k-value needs to be evaluated. If the pcc slab is placed directly on the subgrade without the use of a base or subbase, AASHTO recommends the use of the following theoretical relationship between k-values from a plate load bearing test and the resilient modulus, M_r:

$$k = \frac{M_r}{19.4}$$

where
   k-values are in pci
   M_r values are in psi

If a subbase exists between the slab and the subgrade, the composite modulus of subgrade reaction can be determined from Figure 15.7 (AASHTO, 1993). The composite modulus value is based on a subgrade of infinite depth, k_∞. The chart was developed using the same method as for a homogeneous half-space except that the 30-in. plate is applied on a two-layer system. Hence, the k-values obtained from the chart are too large and are not representative of values obtained in the field.

Example 15.2

Given a subbase thickness D_{SB} of 6 in. (152 mm), a subbase resilient modulus of E_{SB} of 20,000 psi (138 psi), and a roadbed soil resilient modulus M_r of 7000 psi (48 MPa), determine the composite modulus of subgrade reaction k_∞.

Solution

The composite of the subgrade reaction can be determined as follows.

Using Figure 15.7, draw a vertical line from $D_{SB} = 6$ in. to the curve of $E_{SB} = 20,000$ psi. The same line is drawn downward until it intersects with $M_r = 7000$ psi, and then the line is turned horizontally until it intersects with the turning line.

A horizontal line is drawn from the point in step 1, and a vertical line from the point on the turning line in Step 2. The intersection of these two lines gives a k_∞ of 400 pci.

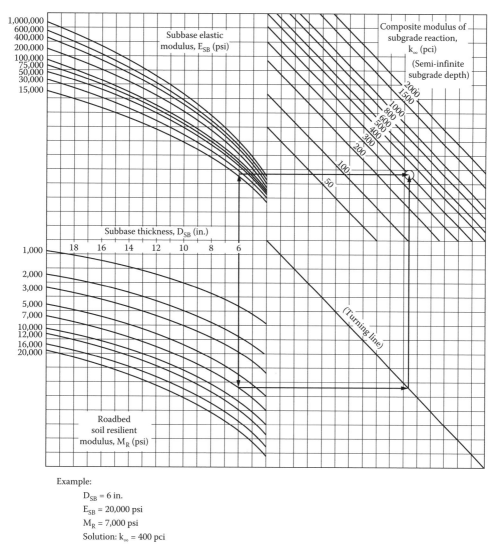

Example:

    D_{SB} = 6 in.

    E_{SB} = 20,000 psi

    M_R = 7,000 psi

    Solution: k_∞ = 400 pci

FIGURE 15.7 Estimating the composite modulus of subgrade reaction. (From American Association of State Highway and Transportation Officials (AASHTO), *AASHTO Guide for Design of Pavement Structures*, AASHTO, Washington, DC, © 1993. Used with permission.)

15.2.3 Rigid Foundation at Shallow Depth

If the rigid foundation lies below the subgrade and the subgrade depth to rigid foundation D_{SB} is smaller than 10 ft (3 m), then the modulus of subgrade reaction must be modified using the chart in Figure 15.8 (AASHTO, 1993). This chart can be used to modify PCC slabs with or without a subbase.

Example 15.3

Determine k

Given

    M_r = 4000 psi

    D_{SG} = 5 ft

    k_∞ = 230 pci

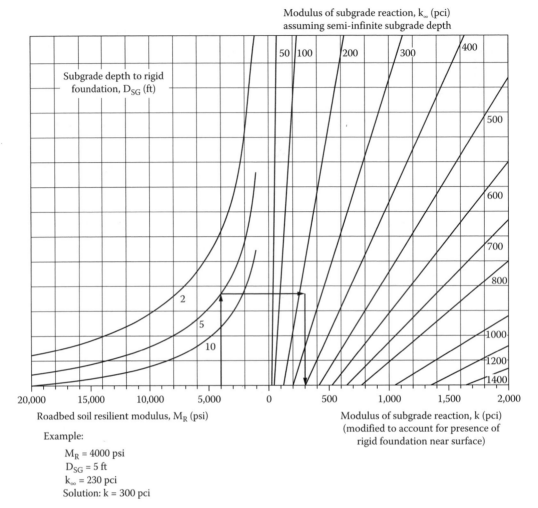

FIGURE 15.8 Modifying the modulus of subgrade reaction due to a rigid foundation near the surface. (From American Association of State Highway and Transportation Officials (AASHTO), *AASHTO Guide for Design of Pavement Structures*, AASHTO, Washington, DC, © 1993. Used with permission.)

Solution

Using Figure 15.8, a vertical line is drawn from the horizontal scale with a $M_r = 4000$ psi until it intersects the curve with a $D_{SG} = 5$ ft. The line is then drawn horizontally until it reaches a point with $k_\infty = 230$ pci, and then vertically until a k of 300 pci is obtained.

15.2.4 EFFECTIVE MODULUS OF SUBGRADE REACTION

The effective modulus of subgrade reaction is an equivalent modulus that would result in the same damage if seasonal values were used throughout the year. The relative damage to rigid pavements u_r is given by the following equation and Figure 15.9:

$$u_r = (D^{0.75} - 0.3k^{0.725})^{3.42}$$

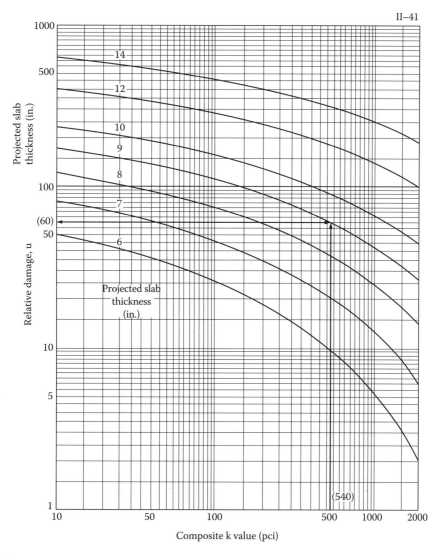

FIGURE 15.9 Estimating relative damage to rigid pavements. (From American Association of State Highway and Transportation Officials (AASHTO), *AASHTO Guide for Design of Pavement Structures*, AASHTO, Washington, DC, © 1993. Used with permission.)

Example 15.4

Given D = 9 in. and k = 540 pci, determine the u_r using the aforementioned equation and Figure 15.9.

Solution

From the equation $u_r = [9^{0.75} - 0.3(540)^{0.725}]^{3.42} = 60.3$, which is similar to the value obtained using Figure 15.9.

To account for a potential loss of support by foundation erosion or differential vertical soil movement, the effective modulus of subgrade reaction must be reduced by the loss of subgrade support (LS) factor as shown in Figure 15.10. Table 15.17 provides recommendations for LS values for different types of subbases and subgrades.

FIGURE 15.10 Correction of the effective modulus of subgrade reaction due to a loss of foundation contact. (From American Association of State Highway and Transportation Officials (AASHTO), *AASHTO Guide for Design of Pavement Structures*, AASHTO, Washington, DC, © 1993. Used with permission.)

TABLE 15.17

Recommended Loss of Subgrade Support Factors

Type of Material	Loss of Support (LS)
Cement-treated granular base (E = 1,000,000–2,000,000 psi)	0.0–1.0
Cement aggregate mixtures (E = 500,000–1,000,000 psi)	0.0–1.0
Asphalt-treated base (E = 350,000–1,000,000 psi)	0.0–1.0
Bituminous stabilized mixtures (E = 40,000–300,000 psi)	0.0–1.0
Lime stabilized (E = 20,000–70,000 psi)	1.0–3.0
Unbound granular materials (E = 15,000–45,000 psi)	1.0–3.0
Fine-grained or natural subgrade materials (E = 3000–40,000 psi)	2.0–3.0

Source: Reprinted from Portland Cement Association (PCA), *Design and Control of Concrete Mixtures*, Engineering Bulletin 001, 14th edn., PCA, Skokie, IL, 2002.

Note: E in this table refers to the general symbol for the elastic or resilient modulus of the material.

TABLE 15.18

Computation of Effective Modulus of Subgrade Reaction

Trial Subbase: Type Granular Thickness (in.) 6 Loss of Support (LS) 1.0			Depth to Rigid Foundation (ft) 5 Projected Slab Thickness (in.) 9		
(1)	(2)	(3)	(4)	(5)	(6)
Month	Roadbed Modulus, E_{SG} (psi)	Subbase Modulus, E_{SB} (psi)	Composite k-Value (pci) (Figure 15.7)	k-Value (pci) on Rigid Foundation (Figure 15.8)	Relative Damage, u_r (Figure 15.9)
January	20,000	50,000	1100	1350	0.35
February	20,000	50,000	1100	1350	0.35
March	2,500	15,000	160	230	0.86
April	4,000	15,000	230	300	0.78
May	4,000	15,000	230	300	0.78
June	7,000	20,000	410	540	0.60
July	7,000	20,000	410	540	0.60
August	7,000	20,000	410	540	0.60
September	7,000	20,000	410	540	0.60
October	7,000	20,000	410	540	0.60
November	4,000	15,000	230	300	0.78
December	20,000	50,000	1100	1350	0.35
				Summation: $\Sigma u_r = 7.25$	

Source: Reprinted from Portland Cement Association (PCA), *Design and Control of Concrete Mixtures*, Engineering Bulletin 001, 14th edn., PCA, Skokie, IL, 2002.

Average: $\bar{u}_r = \dfrac{\Sigma u_r}{n} = \dfrac{7.25}{12} = 0.60$

Effective modulus of subgrade reaction, k (pci) = 540.
Corrected for loss of support: k (pci) = 170.

Table 15.18 shows an example for determining the effective modulus of a subgrade reaction for a slab thickness of 9 in. The slab is to be placed directly onto the subgrade with the monthly resilient moduli shown in Table 15.18. Note that the year is divided into 12 months, each with different moduli. The normal summer modulus is 7000 psi, and the maximum of modulus of 20,000 psi occurs in the winter months, when the subgrade is frozen (December–February). The k-values are obtained from the equation $K = M_r/19.4$.

The relative damage can be obtained from Figure 15.9 or the following equation:

$$u_r = (D^{0.75} - 0.3k^{0.725})^{3.42}$$

The sum of the relative damage = 7.25, and the average over the 12 months is 0.6, which is equivalent to an effective modulus = 540 pci.

15.2.4.1 Software Solutions

The AASHTO design equation may be solved using the nomographs or using a spreadsheet or equation solver software. Two commercial software packages are also available to design pavements using the AASHTO 1993 Design Guide. DarWIN is an AASHTO product that is no longer available through AASHTO. The AASHTO bookstore currently sells only the new DarWIN-ME software to support the new MEP-Design Guide (http://darwin.aashtoware.org/overview.htm). WinPAS is available through ACPA (http://www.acpa.org/Concrete_Pavement/Technical/Downloads/Software.asp).

Both software are versatile and can solve for pavement thickness and other unknowns and can conduct a reliability analysis.

It should be noted, that the new DarWIN-ME software that includes the new MEPDG Design Guide has replaced the old DarWIN software. The MEPDG Design Guide and software were adopted as the Interim AASHTO pavement design standard in 2007 (Von Quintus et al., 2007)

15.2.4.2 AASHTO 1998 Supplement to Design Guide

The AASHTO Design Guide was developed over the years after the AASHO Road Test was completed in 1960 and the Interim Guides published in 1961 and 1962 (HRB Special Report 61E). Various enhanced versions of the AASHTO Design Guide were published in 1972, 1986, and 1993. The Design Guidelines for Rigid Pavements were similar from 1986 to 1993. Development of the AASHTO equations over the years is provided by Huang (2004).

The original AASHTO equations were developed to predict the number of axle loads (18 kip EASLs) for a given slab thickness and loss in serviceability (ΔPSI). These empirical regression equations were valid for the specific pavement and environmental conditions at the AASHO Road test; namely a concrete modulus $E_c = 4.2 \times 10^6$ psi, a modulus of rupture of the concrete $S_c = 690$ psi, a modulus of subgrade reaction $k = 60$ pci, a load transfer coefficient $J = 3.2$, and a drainage coefficient $C_d = 1.0$. Extension of the AASHTO equation to predict traffic loading for a given serviceability beyond those conditions was in many cases not adequate. Over the years, the equation was modified to take into consideration variation in strength, drainage and environmental conditions, load transfer and reliability.

In 1998, AASHTO published a supplement to the 1993 Guide; entitled Part II—Rigid Pavement Design & Rigid Pavement Joint Design. These new modifications were based on the study conducted under NCHRP Project 1–30 that used the data from GPS-3, GPS-4, and GPS-5 (JPCP, JRCP, and CRCP) sections in the Long-Term Pavement Performance (LTPP) studies for model calibration. The supplement and design procedure was incorporated in a spreadsheet software funded by FHWA, and is now a user-friendly online software which can be accessed through LTPP Products Online (http://www.ltpp-products.com/Rigid/Rigid.aspx).

The Rigid Pavement Design Software features four new or modified design inputs: subgrade support, joint spacing, slab/subbase friction, and temperature differential throughout the slab. In the 1998 guidelines, the subgrade support is characterized by the effective elastic k-value (or seasonally adjusted k-value) of the subgrade, compared to the "composite" k-value of the subgrade/base system used in the 1993 Guide.

Figure 15.11 shows a screen shot of the input page. The user inputs concrete properties, base properties, reliability and standard deviation, climatic properties, subgrade k-value, design equivalent single-axle loads (ESALs), pavement type, joint spacing, and edge support through the input screen. A sensitivity analysis feature can allow the designer to change input conditions based on design constraints.

A design example using the AASHTO 1998 Supplement procedure for a pavement in Worcester, Massachusetts, having a mean annual temperature of 46.8°F, a precipitation of 47.6 in. per year, and an average wind speed 12.4 mph is presented in Figures 15.11 through 15.16.

The thickness design for the JPCP with a 15 ft joint spacing, an adjusted seasonal composite k-value of 120 pci, a base stiffness of 20,000 psi, a thickness of 6 in. and a friction factor of 1.4, with an equivalent traffic loading of 20.3 million ESALs and a 95% reliability is determined to be 11.68 in. as shown in Figure 15.11. The next step is to determine the dowel diameter that will limit the faulting check to 0.06 in. This requires input of a project duration, a temperature range of 60°F, CTE (α of 5×10^{-6} in./in./°F), a shrinkage strain of −0.00015 in./in., a freezing index of 500°F degree days, and a drainage coefficient C_d of 1.0. Figure 15.12 shows an example of a pavement with a dowel diameter of 1.25 in. that fails and Figure 15.13 shows an example of a pavement with a diameter of 1.75 in. that just passes. Improving the drainage coefficient C_d to 1.25, will allow a dowel diameter size of 1.25 in. to also pass the faulting criteria. Sensitivity analysis may also be performed using the Rigid Pavement Online Software. Figures 15.14 and 15.15 shows sensitivity analysis plots for effective subgrade support using the composite k-value with traffic demands and joint spacing, respectively.

FIGURE 15.11   Rigid pavement design software input page.

15.2.4.2.1   PCA 1984 Design Procedure

The PCA design method for concrete roads and streets evaluates a pavement design with respect to two potential failure modes: fatigue and erosion. For fatigue failure, the concrete will fail due to cumulative damage when the stress to strength ratio exceeds a threshold. For erosion, the pavement fails by pumping and erosion of the foundation and joint faulting. The procedure was developed using the results of the finite element computer program JSLAB (Tayabji and Colley, 1986) to determine the critical stresses and deflections induced in concrete pavements by joint, edge, and corner loading. The PCA method considers the degree of load transfer provided by dowels or aggregate interlock and the degree of edge support provided by a concrete shoulder.

The PCA procedure uses the "composite-k" concept in which the design k is a function of the subgrade soil k, base thickness, and base type (granular or cement treated), similar to the 1986 and 1993 AASHTO design procedure.

Fatigue analysis in the PCA method is based on the edge stress midway between transverse joints. The fatigue analysis assumes that approximately 6% of all truck loads will pass sufficiently close to the slab edge to produce a significant tensile stress. However, this critical tensile stress is reduced considerably when the main traffic lane and the concrete shoulder are tied. Stresses due to warping and curling are not considered in the PCA method.

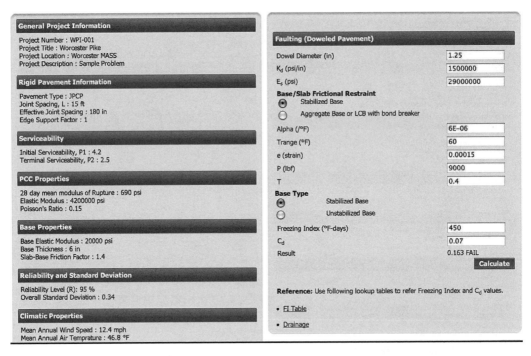

FIGURE 15.12   Rigid pavement design software example showing faulting of pavement.

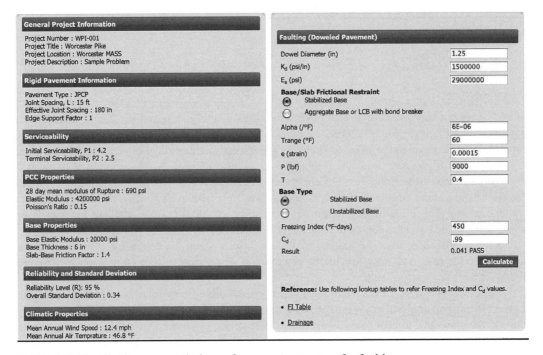

FIGURE 15.13   Rigid pavement design software output screen for faulting.

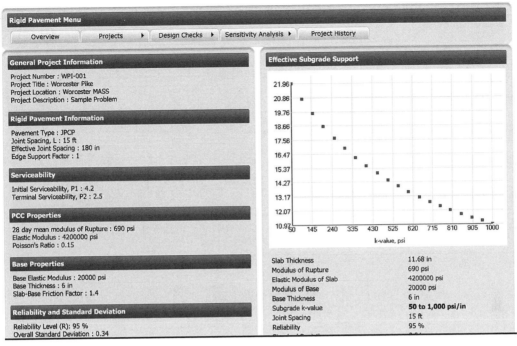

FIGURE 15.14 Rigid pavement design software output showing effective subgrade support.

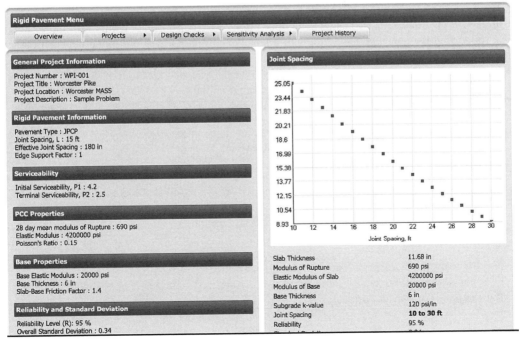

FIGURE 15.15 Rigid pavement design software showing output joint spacing.

The fatigue model relates the log of allowable load repetitions versus the stress-to-strength ratio and is given by

$$\log N_f = f_1 - f_2\left(\frac{\sigma}{S_c}\right)$$

N_f is the allowable number of repetitions
σ is the flexural stress in the slab
S_c is the modulus of rupture of the concrete

$$\text{For}\left(\frac{\sigma}{S_c}\right) \geq 0.55 \quad \log N_f = 11.737 - 12.077\left(\frac{\sigma}{S_c}\right)$$

$$\text{For } 0.45 < \left(\frac{\sigma}{S_c}\right) < 0.55 \quad \log N_f = \left(\frac{4.2577}{\dfrac{\sigma}{S_c} - 0.4325}\right)^{3.268}$$

$$\text{For} < \left(\frac{\sigma}{S_c}\right) \leq 0.45 \quad N_f = \text{unlimited}$$

Ultimately the damage ratio Dr accumulated over the design period is

$$D_r = \sum_{i=1}^{m} \frac{n_i}{N_i}$$

where
n_i is the predicted number of repetitions
N_i is the allowable number of repetitions

In the AASHTO Road Test, the principal mode of failure was the erosion or pumping of the granular base supporting the concrete slab. Although a positive correlation was not found between corner slab deflections and performance, a better correlation was found when relating the power or rate of work with which a slab corner is deflected by a wheel load as a function of the slab thickness, foundation k-value, and estimated pressure at the slab–foundation interface (Huang, 2004).

Therefore, the cumulative percent erosion damage should be less than 100% and is given by

$$\% \text{ Erosion damage} = 100 \sum_{i=1}^{m} \frac{C_2 n_i}{N_i}$$

where $C_2 = 0.06$ for pavements with concrete shoulders, 0.94 for tied shoulders

$$\log N = 14.524 - 6.777(C_1 P - 9.0)^{0.103}$$

where
N is the allowable number of load repetitions (PSI = 3.0)
C_1 is the adjustment factor (1-treated subbase; 0.9-stabilized subbase)
P is the power or rate of work

$$P = 268.7\frac{p^2}{hk^{0.73}}$$

The solution for the PCA method was originally developed for hand solutions with charts and tables. The procedure was developed using the results of the finite element computer program JSLAB (Tayabji and Colley, 1986) to determine the critical stresses and deflections induced in concrete pavements by joint, edge, and corner loading. The PCA method considers the degree of load transfer provided by dowels or aggregate interlock and the degree of edge support provided by a concrete shoulder.

The PCA, 1984 design procedure is presented in depth by Huang (2004) and others, with complete tables and charts and worksheets to use with the tedious hand calculation method. The PCA design procedure and charts will not be provided here. The PCA software PCAPAV or StreetPave was developed and can be used to quicken the process. Although the PCA design procedure is not as commonly used for highways compared to the AASHTO method, it still is used for lighter duty pavements, and used to develop design tables for parking lots (ACI Committee 330) and for streets and local roads (ACI Committee 325), Datelle, 2008.

An example from the PCA 1984 Design Guide will be summarized here (see Figure 12.15 Huang) (Table 15.19).

TABLE 15.19
Rigid Pavement Design Example Using the PCA Method

Fatigue analysis

A doweled pavement has a thickness = 9.5 in.

k for subgrade = 130 pci

Equivalent stress for single axle = 206 psi

Equivalent stress for tandem axle = 192 psi

Concrete modulus of rupture S_c = 650 psi

Stress ratio single axle = 206/650 = 317

Stress ratio tandem axle = 192/650 = 295

Projected 30 kip single axle loads = 6310

Load safety factors (LSF) = 1.2 for interstate highway

30 k × 1.2 = 36 kip single axle

Using PCA (1984) Nomograph (Huang Figure 12.12)

For 36 kip single axle; stress ratio = 0.317; allowable load repetitions = 27,000

% fatigue life consumed = 6,310/27,000 = 23.3%

Repeat process for all axle loads

For 28 kip; % fatigue life = 14,690/77,000 = 19.1%

For 26 kip; % fatigue life = 30,140/230,000 = 13.1%

For 24 kip; % fatigue life = 64,410/1,200,000 = 5.4%

For 22 kip; % fatigue life = unlimited

Tandem axle

52 k × 1.2 = 62.4 kip tandem axle

For 52 kip; % fatigue = 21,320/1,100,000 = 1.9%

For 48 kip; % fatigue = unlimited

Total fatigue life used = 23.3 + 19.1 + 13 + 5 + 1.9 = 62.8% < 100% for fatigue; design acceptable

Erosion analysis

Use chart for doweled joints, no concrete shoulders (Huang Table 12.8)

For a slab thickness = 9.5 in.; k = 130 pci; erosion factors 2.59 single; 2.79 tandem

For a 36 k single axle; erosion factor 2.59; allowable load = 1,500,000

For 36 kip; % erosion life used = 6,310/1,500,000 = 0.4%

Repeat for all axle loads (Huang, Figure 12.15)

Total erosion life used = 38.9% < 100% for erosion; design is acceptable

---

TABLE 15.20

JPCP Design Simulation Using MEPDG

Pavement type: JPCP

Thickness (in.): 10; 11; 12; 13

Dowel size (in.): case1-no dowels; case1a-1.25; case1b-1.5

Joint spacing (ft): case 1c-12, case1–15, case1d-20

Shoulder system: case1—not tied; case1e—tied;

Base course layer: case1–6 in. clean gravel; case1f- asphalt permeable

Subgrade: A-7–6 unbound; Mr = 13,000 psi

Traffic: case1—medium; case1g—heavy

Climate: Missouri

CTE: (5.5; 6.2; and 7) $\times 10^{-6}$/°F

Mix properties

Cement type:	Type I
Cementitious material content (lb/yd³):	500
Water/cement ratio:	0.42
Aggregate type:	Limestone
PCC zero-stress temperature (°F)	Derived
Ultimate shrinkage at 40% R.H (microstrain)	Derived
Reversible shrinkage (% of ultimate shrinkage):	50
Time to develop 50% of ultimate shrinkage (days):	35
Curing method:	Curing compound

Strength properties

Input level:	Level 3
28-day PCC modulus of rupture (psi):	620

15.2.4.3 MEPDG Design Guide

The MEPDG Design Guide was developed under NCHRP Project 1-37A. The guidelines were introduced in Chapter 14. Similarly for rigid pavements, the MEPDG Design Guide evaluates the effects of various pavement materials, traffic loading conditions, design features, and construction practices. It considers both long-term (age) and short-term (temperature and moisture) changes in material properties, and provides more accurate performance predictions so that the frequency of premature failure is reduced.

A rigid pavement design example showing the use of the MEPDG Design Guide is presented in Table 15.20 and Figures 15.16 through 15.22.

Figure 15.16a through f shows the input pages in the MEPDG Software including traffic, materials, base layer, temperature and environment, strength, physical characteristics, etc. The output results for predicted faulting, predicted distress, LTE, IRI, and cumulative damage are shown in Figures 15.17 through 15.22 respectively. Table 15.21 shows the results of changing the various parameters for the selected JPCP.

15.2.4.4 Continuously Reinforced Concrete Pavements

Continuously reinforced concrete pavements (CRCP) are heavily reinforced concrete slabs as shown in Figure 15.23 that contain continuous longitudinal reinforcement, and do not have transverse joints except when needed for construction purposes. These include end-of-day construction header joints or joints at bridge approaches or transitions to other pavement structures (see Figure 15.24). The concrete in CRCP will have similar behavior to concretes used for JPCP or JRCP. Meaning that that the concrete matrix will undergo similar volume and dimensional changes with changes in temperature and moisture in addition to traffic load stresses. And since concrete is weak in tension, even new concrete pavements will crack if the appropriate induced stress conditions exist. Transverse cracks will develop to relieve these tensile stresses that may build up due to normal cement hydration, chemical shrinkage, and variations in temperature and moisture gradients. These transverse hairline cracks can develop through the entire depth of the slab. The continuous steel reinforcement in CRCP is used to keep the hairline cracks held tightly together to ensure adequate shear transfer through aggregate interlock between contiguous cracked slabs, as well as reduce the applied bending stresses due to traffic loads and due to curling and warping stresses.

Following construction of a CRCP, a number of mechanisms influence development of stresses in the slab and thus the formation of cracks. Figure 15.25 provides a schematic representation of some of the factors influencing CRCP behavior. During the early age, temperature and moisture fluctuations induce volume changes in the concrete that are restrained by reinforcement and base friction, leading to the development of stresses and closely spaced cracks. The addition of repeated traffic loading and possible undermining of the foundation support can result in the typical CRCP distress called punchouts.

Project: New_JPCP-tek-1.dgp

General Information

Design Life	20 years
Pavement construction:	September, 2006
Traffic open:	October, 2006
Type of design	JPCP

Description:
JPCP design in Missouri, 20 year project duration, gravel base, tie shoulder

Analysis Parameters

Performance Criteria

	Limit	Reliability
Initial IRI (in/mi)	63	
Terminal IRI (in/mi)	172	90
Transverse Cracking (% slabs cracked)	15	90
Mean Joint Faulting (in)	0.12	90

Location:
Project ID:
Section ID:

Date: 5/27/2009

Station/milepost format:
Station/milepost begin:
Station/milepost end:
Traffic direction: East bound

Default Input Level

Default input level   Level 3, Default and historical agency values.

Traffic

Initial two-way AADTT:	4000
Number of lanes in design direction:	2
Percent of trucks in design direction (%):	50
Percent of trucks in design lane (%):	95
Operational speed (mph):	60

Traffic -- Volume Adjustment Factors

Monthly Adjustment Factors   (Level 3, Default MAF)

Month	Vehicle Class									
	Class 4	Class 5	Class 6	Class 7	Class 8	Class 9	Class 10	Class 11	Class 12	Class 13
January	1.00	1.00	1.00	1.00	1.00	1.00	1.00	1.00	1.00	1.00
February	1.00	1.00	1.00	1.00	1.00	1.00	1.00	1.00	1.00	1.00
March	1.00	1.00	1.00	1.00	1.00	1.00	1.00	1.00	1.00	1.00
April	1.00	1.00	1.00	1.00	1.00	1.00	1.00	1.00	1.00	1.00
May	1.00	1.00	1.00	1.00	1.00	1.00	1.00	1.00	1.00	1.00

(a)

FIGURE 15.16 Example of JPCP using MEPDG.

(continued)

Vehicle Class Distribution
(Level 3, Default Distribution)

AADTT distribution by vehicle class

Class 4	1.8%
Class 5	24.6%
Class 6	7.6%
Class 7	0.5%
Class 8	5.0%
Class 9	31.3%
Class 10	9.8%
Class 11	0.8%
Class 12	3.3%
Class 13	15.3%

Hourly truck traffic distribution
by period beginning:

Midnight	2.3%	Noon	5.9%
1:00 am	2.3%	1:00 pm	5.9%
2:00 am	2.3%	2:00 pm	5.9%
3:00 am	2.3%	3:00 pm	5.9%
4:00 am	2.3%	4:00 pm	4.6%
5:00 am	2.3%	5:00 pm	4.6%
6:00 am	5.0%	6:00 pm	4.6%
7:00 am	5.0%	7:00 pm	4.6%
8:00 am	5.0%	8:00 pm	3.1%
9:00 am	5.0%	9:00 pm	3.1%
10:00 am	5.9%	10:00 pm	3.1%
11:00 am	5.9%	11:00 pm	3.1%

Traffic Growth Factor

Vehicle Class	Growth Rate	Growth Function
Class 4	4.0%	Compound
Class 5	4.0%	Compound
Class 6	4.0%	Compound
Class 7	4.0%	Compound
Class 8	4.0%	Compound
Class 9	4.0%	Compound
Class 10	4.0%	Compound
Class 11	4.0%	Compound
Class 12	4.0%	Compound
Class 13	4.0%	Compound

Traffic -- Axle Load Distribution Factors
Level 3:   Default

Traffic -- General Traffic Inputs

Mean wheel location (inches from the lane marking):	18
Traffic wander standard deviation (in):	10
Design lane width (ft):	12

Number of Axles per Truck

Vehicle Class	Single Axle	Tandem Axle	Tridem Axle	Quad Axle
Class 4	1.62	0.39	0.00	0.00
Class 5	2.00	0.00	0.00	0.00
Class 6	1.02	0.99	0.00	0.00
Class 7	1.00	0.26	0.83	0.00
Class 8	2.38	0.67	0.00	0.00

(b)

FIGURE 15.16 (continued)   Example of JPCP using MEPDG.

Axle Configuration
Average axle width (edge-to-edge) outside dimensions,ft):   8.5
Dual tire spacing (in):   12

Axle Configuration
Tire Pressure (psi) :   120

Average Axle Spacing
Tandem axle(psi):   51.6
Tridem axle(psi):   49.2
Quad axle(psi):   49.2

Wheelbase Truck Tractor

	Short	Medium	Long
Average Axle Spacing (ft)	12	15	18
Percent of trucks	33%	33%	34%

Climate
icm file: C:\DG2002\Projects\StLouis.icm
Latitude (degrees.minutes)   41.81
Longitude (degrees.minutes)   -90.47
Elevation (ft)   686
Depth of water table (ft)   12

Structure--Design Features
Permanent curl/warp effective temperature difference (°F):   -10

Joint Design
Joint spacing (ft):   15
Sealant type: Liquid
Dowel diameter (in):   1.25
Dowel bar spacing (in):   12

Edge Support
    None
Long-term LTE(%):   n/a
Widened Slab (ft):   n/a

Base Properties
Base type: Granular
Erodibility index: Very Erodable (5)
PCC-Base Interface   Zero friction contact
Loss of full friction (age in months):   n/a

Structure--ICM Properties
Surface shortwave absorptivity: 0.85

(c)

FIGURE 15.16 (continued) Example of JPCP using MEPDG.

(continued)

Structure--ICM Properties
  Surface shortwave absorptivity: 0.85

Structure--Layers

Layer 1 -- JPCP
  General Properties
    PCC material   JPCP
    Layer thickness (in):   10
    Unit weight (pcf):   150
    Poisson's ratio   0.2

  Thermal Properties
    Coefficient of thermal expansion (per F° x 10- 6):   5.5
    Thermal conductivity (BTU/hr-ft-F°) :   1.25
    Heat capacity (BTU/lb-F°):   0.28

  Mix Properties
    Cement type: Type I
    Cementitious material content (lb/yd^3):   500
    Water/cement ratio: 0.42
    Aggregate type: Limestone
    PCC zero-stress temperature (F°)   Derived
    Ultimate shrinkage at 40% R.H (microstrain)   Derived
    Reversible shrinkage (% of ultimate shrinkage):   50
    Time to develop 50% of ultimate shrinkage (days):   35
    Curing method: Curing compound

  Strength Properties
    Input level: Level 3
    28-day PCC modulus of rupture (psi):   620
    28-day PCC compressive strength (psi):   n/a

Layer 2 -- A-1-a
  Unbound Material: A-1-a
  Thickness(in):   6

  Strength Properties
    Input Level: Level 3
    Analysis Type: ICM inputs (ICM Calculated Modulus)
    Poisson's ratio: 0.35
    Coefficient of lateral pressure,Ko: 0.5
    Modulus (input) (psi):   40000

  ICM Inputs
    Gradation and Plasticity Index
    Plasticity Index, PI: 1
    Liquid Limit (LL)   6
    Compacted Layer   Yes
    Passing #200 sieve (%):   8.7
    Passing #40   20

(d)

FIGURE 15.16 (continued)   Example of JPCP using MEPDG.

Calculated/Derived Parameters
Maximum dry unit weight (pcf):   97.7 (derived)
Specific gravity of solids, Gs: 2.70 (derived)
Saturated hydraulic conductivity (ft/hr):   8.946e-006 (derived)
Optimum gravimetric water content (%):   22.2 (derived)
Calculated degree of saturation (%):   82.7 (calculated)

Soil water characteristic curve parameters: Default values

Parameters	Value
a	136.42
b	0.51828
c	0.03238
Hr.	500

Distress Model Calibration Settings - Rigid (new)

Faulting
  Faulting Coefficients
    C1   1.0184
    C2   0.91656
    C3   0.00218
    C4   0.00088
    C5   250
    C6   0.4
    C7   1.83312
    C8   400

  Reliability (FAULT)
    Std. Dev.   POWER(0.0097*FAULT,0.5178)+0.014

Cracking
  Fatigue Coefficients
    C1   2
    C2   1.22

  Cracking Coefficients
    C4   1
    C5   -1.98

  Reliability (CRACK)
    Std. Dev.   POWER(5.3116*CRACK,0.3903) + 2.99

IRI(jpcp)
    C1   0.8203
    C2   0.4417
    C3   20.37
    C4   1.4929
    C5   25.24

(e)

FIGURE 15.16 (continued)   Example of JPCP using MEPDG.

(*continued*)

ICM Inputs

Gradation and Plasticity Index

Plasticity Index, PI:	1
Liquid Limit (LL)	6
Compacted Layer	Yes
Passing #200 sieve (%):	8.7
Passing #40	20
Passing #4 sieve (%):	44.7
D10(mm)	0.1035
D20(mm)	0.425
D30(mm)	1.306
D60(mm)	10.82
D90(mm)	46.19

Sieve	Percent Passing
0.001mm	
0.002mm	
0.020mm	
#200	8.7
#100	
#80	12.9
#60	
#50	
#40	20
#30	
#20	
#16	
#10	33.8
#8	
#4	44.7
3/8"	57.2
1/2"	63.1
3/4"	72.7
1"	78.8
1 1/2"	85.8
2"	91.6
2 1/2"	
3"	
3 1/2"	97.6
4"	97.6

Calculated/Derived Parameters

Maximum dry unit weight (pcf):	127.7 (derived)
Specific gravity of solids, Gs:	2.70 (derived)
Saturated hydraulic conductivity (ft/hr):	0.05054 (derived)
Optimum gravimetric water content (%):	7.4 (derived)
Calculated degree of saturation (%):	62.2 (calculated)
Soil water characteristic curve parameters:	Default values

(f)

FIGURE 15.16 (continued)   Example of JPCP using MEPDG.

Predicted faulting: Project New_JPCP-tek-1.dgp

Pavement age mo	yr	Month	Epcc Mpsi	Ebase ksi	Dyn. k psi/in	Ave. R.H. %	Ave. ΔT °F	Joint open. in	LTE %	18-kip single deflection (in)		36-kip tandem deflection (in)		D.E. in-lb	Faulting in	Faulting at specified reliability
										Loaded	Unloaded	Loaded	Unloaded			
1	0.08	October	4.04	54.2	173	71.9	-19.1	0.049	94.4	0.023	0.023	0.028	0.028	1.6	0	0.02
2	0.17	November	4.15	53.37	173	74.3	-17.3	0.064	93.1	0.022	0.022	0.026	0.026	2	0	0.02
3	0.25	December	4.22	119.5	178	79	-17.4	0.074	95	0.022	0.022	0.026	0.026	1.2	0.001	0.021
4	0.33	January	4.26	395.5	208	78.1	-16.8	0.08	95	0.022	0.022	0.026	0.026	1.4	0.001	0.022
5	0.42	February	4.3	243.6	188	76.9	-15	0.075	95	0.021	0.021	0.025	0.024	1.2	0.001	0.022
6	0.5	March	4.33	34.96	158	70.5	-15.9	0.07	91.6	0.02	0.02	0.025	0.024	2.3	0.001	0.023
7	0.58	April	4.35	35.03	158	66	-16.9	0.059	91.6	0.021	0.021	0.025	0.025	2.4	0.002	0.023
8	0.67	May	4.37	43.31	164	69.8	-17.8	0.049	92.1	0.022	0.021	0.026	0.026	2.3	0.002	0.024
9	0.75	June	4.39	50.59	169	73	-18.8	0.041	93.7	0.023	0.022	0.027	0.027	1.8	0.002	0.025
10	0.83	July	4.41	53.98	172	75.2	-20.5	0.037	94.9	0.024	0.024	0.029	0.029	1.4	0.002	0.025
11	0.92	August	4.42	54.18	173	78.4	-21.2	0.04	93.8	0.025	0.024	0.03	0.029	2	0.002	0.025
12	1	September	4.43	53.99	174	73.2	-21.2	0.047	91.6	0.025	0.024	0.03	0.029	3.1	0.003	0.026
13	1.08	October	4.44	54.2	173	71.9	-19	0.059	89.8	0.023	0.022	0.028	0.027	3.9	0.003	0.027
14	1.17	November	4.46	53.37	173	74.3	-17.3	0.07	88.8	0.022	0.021	0.026	0.025	4.2	0.004	0.028
15	1.25	December	4.46	119.5	178	79	-17.5	0.079	95	0.022	0.022	0.026	0.026	1.3	0.004	0.029
16	1.33	January	4.47	395.5	208	78.1	-16.9	0.084	95	0.022	0.022	0.026	0.026	1.4	0.004	0.029
17	1.42	February	4.48	243.6	188	76.9	-15.1	0.078	95	0.021	0.02	0.024	0.024	1.2	0.004	0.029
18	1.5	March	4.49	34.96	158	70.5	-15.9	0.072	88	0.02	0.02	0.025	0.024	3.9	0.005	0.03
19	1.58	April	4.5	35.03	158	66	-16.8	0.061	88.2	0.021	0.02	0.025	0.024	4	0.005	0.031
20	1.67	May	4.51	43.31	164	69.8	-17.8	0.051	89	0.022	0.021	0.026	0.025	3.9	0.006	0.032
21	1.75	June	4.51	50.59	169	73	-18.8	0.043	91	0.023	0.022	0.027	0.026	3.2	0.006	0.032
22	1.83	July	4.52	53.98	172	75.2	-20.6	0.038	92.6	0.024	0.024	0.029	0.028	2.6	0.006	0.033

FIGURE 15.17   MEPDG example output predicted faulting.

Predicted distress: Project New_JPCP-tek-1.dgp

Pavement age mo	yr	Month	Epcc Mpsi	Ebase ksi	Dyn. k psi/in	Faulting in	Percent slabs cracked	IRI in/mile	Heavy Trucks (cumulative)	IRI at specified reliability
1	0.08	October	4.04	54.2	173	0	0	63	57831	87.6
2	0.17	November	4.15	53.37	173	0	0	63.4	115663	88.5
3	0.25	December	4.22	119.5	178	0.001	0	63.6	173494	89
4	0.33	January	4.26	395.5	208	0.001	0	63.8	231325	89.4
5	0.42	February	4.3	243.6	188	0.001	0	64	289156	89.8
6	0.5	March	4.33	34.96	158	0.001	0	64.1	346988	90
7	0.58	April	4.35	35.03	158	0.002	0	64.4	404819	90.5
8	0.67	May	4.37	43.31	164	0.002	0	64.6	462650	91
9	0.75	June	4.39	50.59	169	0.002	0	64.9	520481	91.5
10	0.83	July	4.41	53.98	172	0.002	0.1	65.1	578313	91.8
11	0.92	August	4.42	54.18	173	0.002	0.1	65.3	636144	92.2
12	1	September	4.43	53.99	174	0.003	0.1	65.5	693975	92.6
13	1.08	October	4.44	54.2	173	0.003	0.1	65.8	754120	93.1
14	1.17	November	4.46	53.37	173	0.004	0.1	66.2	814264	93.8
15	1.25	December	4.46	119.5	178	0.004	0.1	66.5	874409	94.4
16	1.33	January	4.47	395.5	208	0.004	0.1	66.7	934553	94.8
17	1.42	February	4.48	243.6	188	0.004	0.1	66.8	994698	95
18	1.5	March	4.49	34.96	158	0.005	0.1	67	1054840	95.3
19	1.58	April	4.5	35.03	158	0.005	0.1	67.4	1114990	95.8
20	1.67	May	4.51	43.31	164	0.006	0.2	67.7	1175130	96.5
21	1.75	June	4.51	50.59	169	0.006	0.2	68.1	1235280	97.1
22	1.83	July	4.52	53.98	172	0.006	0.2	68.4	1295420	97.7

FIGURE 15.18   MEPDG example output for JPCP.

An important factor in preventing punchouts is the use of a non-erodible pavement base material to minimize loss of foundation support. Other factors that contribute to reducing the risk of punchouts include using an adequate amount of steel reinforcement to maintain tight cracks, reducing load stresses and enhancing load transfer across cracks; a minimum slab thickness to reduce traffic load stresses; the use of hard angular and large size coarse aggregate with a low CTE to enhance load transfer; the use of a bond breaker between slab and base such as plastic sheeting; well controlled concrete mix design and concrete curing to avoid over heating of the PCC slab and subsequent cooling and shrinkage stresses that develop; and finally tied shoulders and widened lanes (Rasmusen, Rogers and Farragut [CRCP Manual, 2012]).

FIGURE 15.19 MEPDG example output for predicted IRI.

FIGURE 15.20 MEPDG example output for LTE.

Keeping the hairline cracks tightly held together is beneficial since this will prevent water infiltration and prevent the intrusion of incompressibles from entering the cracks and resulting in crack joint damage. Crack spacing that develops will vary with concrete properties, environmental conditions, wheel loads, and age of the pavement. It's observed that the crack spacing decreases rapidly during the early age of the pavement, up until about 1 or 2 years. After this stage, the transverse cracking pattern remains constant until the slab reaches the end of its fatigue life. The addition of transverse reinforcement is commonly used to tie any longitudinal cracks that develop. However, longitudinal joints are typically used when pavement widths exceed 14 ft (CRCP Manual, 2012).

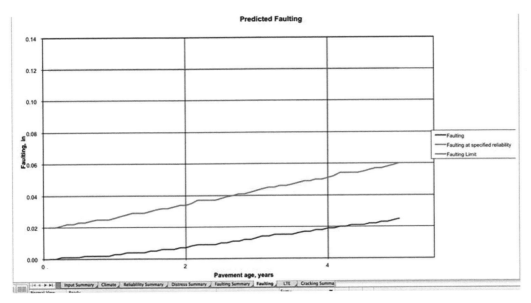

FIGURE 15.21 MEPDG example output for predicted faulting.

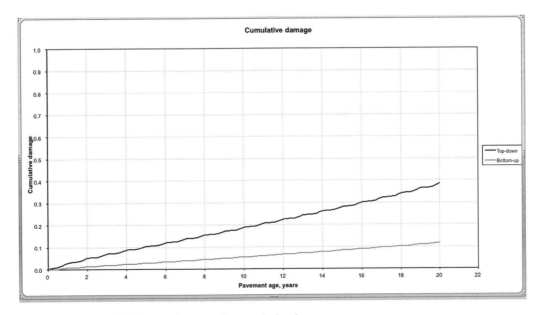

FIGURE 15.22 MEPDG example output for cumulative damage.

A recent survey on CRCP design practices in the United States (Rasmusen, Rogers and Farragut [CRCP Manual, 2012]) has shown that most states that commonly build and use this type of pavement, use the AASHTO design procedure published in 1986 (and later in 1993). One exception is the State of Illinois, which uses a modified version of this method. Therefore, the guidelines referred to in the CRCP Guidelines (2012) and summarized here refer to this method as the "AASHTO-86/93" design procedure.

The CRCP Guide Rasmusen, Rogers and Farragut (CRCP Manual, 2012) reports that since the development of the AASHTO-86/93 design procedures, a number of new findings have been reported based on research conducted worldwide. These studies on CRCP sections were conducted based on a FHWA research studies, LTPP database studies, and others.

TABLE 15.21
JPCP Performance Based on Simulation Parameters

Case	Performance Criteria	Distress Target	Distress Predicted	Reliability Target	Reliability Predicted	Acceptable
1	Terminal IRI (in./mile)	172	82.4	90	99.89	Pass
	Transverse cracking (% slabs cracked)	15	14.3	90	53.36	Fail
	Mean joint faulting (in.)	0.12	0.025	90	99.98	Pass
1a	Terminal IRI (in./mile)	172	168.7	90	52.46	Fail
	Transverse cracking (% slabs cracked)	15	14.3	90	53.22	Fail
	Mean joint faulting (in.)	0.12	0.177	90	13.15	Fail
1b	Terminal IRI (in./mile)	172	168.7	90	52.46	Fail
	Transverse cracking (% slabs cracked)	15	14.3	90	53.22	Fail
	Mean joint faulting (in.)	0.12	0.177	90	13.15	Fail
1c	Terminal IRI (in./mile)	172	76.7	90	99.97	Pass
	Transverse cracking (% slabs cracked)	15	0.3	90	99.98	Pass
	Mean joint faulting (in.)	0.12	0.012	90	99.999	Pass
1d	Terminal IRI (in./mile)	172	165.9	90	55.05	Fail
	Transverse cracking (% slabs cracked)	15	99.7	90	0	Fail
	Mean joint faulting (in.)	0.12	0.06	90	95.76	Pass
1e	Terminal IRI (in./mile)	172	77.3	90	99.98	Pass
	Transverse cracking (% slabs cracked)	15	3	90	97.87	Pass
	Mean joint faulting (in.)	0.12	0.017	90	99.999	Pass
1f	Terminal IRI (in./mile)	172	76.3	90	99.99	Pass
	Transverse cracking (% slabs cracked)	15	15.2	90	48.97	Fail
	Mean joint faulting (in.)	0.12	0.014	90	99.999	Pass
1g	Terminal IRI (in./mile)	172	97.5	90	98.38	Pass
	Transverse cracking (% slabs cracked)	15	40.6	90	1.07	Fail
	Mean joint faulting (in.)	0.12	0.048	90	98.6	Pass
1h	Terminal IRI (in./mile)	172	88.8	90	99.57	Pass
	Transverse cracking (% slabs cracked)	15	54.9	90	0.05	Fail
	Mean joint faulting (in.)	0.12	0.031	90	99.9	Pass
1j	Terminal IRI (in./mile)	172	117.2	90	91.46	Pass
	Transverse cracking (% slabs cracked)	15	88.5	90	0	Fail
	Mean joint faulting (in.)	0.12	0.049	90	98.47	Pass

The Design and Construction Guidelines provided in CRCP Manual (2012) (Rasmusen, Rogers and Farragut [CRCP Manual, 2012]) summarizes some key factors that ensure successful CRCP design and construction practices that build upon successful projects in states like Texas and Illinois:

- Decisions related to mix design, steel reinforcement design and construction practices should maximize load transfer efficiency and minimize flexural stresses in the slab.
- Cracks should be closely spaced together but not more than approximately 3–4 ft apart; and cracks that are held tightly together with a crack width of 0.02 in. at the depth of the reinforcement will result in a maximized load transfer efficiency and minimize flexural stresses, while maintaining steel stress below the yield strength.
- Closely spaced cracks with crack widths within 0.02 in. will result when CRCP has adequate longitudinal steel content of approximately 0.6%–0.8% of the slab cross-section area; optimum reinforcement diameter; adequate lapping of reinforcement splices; reinforcement placed at an appropriate depth and adequate consolidation of concrete around the reinforcement bars.

FIGURE 15.23 CRCP under construction.

FIGURE 15.24 CRCP under construction.

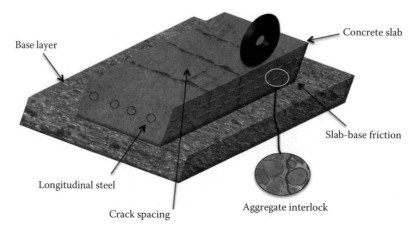

FIGURE 15.25 Factors affecting CRCP behavior.

- Reinforcement design should consider possible fracture and/or excessive plastic deformation. Stresses developed in the reinforcement are usually conservative and limited to a stress level usually 75% of the ultimate tensile strength to avoid fracture and limit the amount of plastic deformation.
- Large, hard, and abrasion-resistant aggregates will force tortuous crack patterns around the coarse aggregates and promote good aggregate interlock and thus enhance load transfer efficiency.
- Adequate minimum slab thickness is required to accommodate transverse tensile and longitudinal stresses due to truck traffic and curling and warping stresses.
- The supporting foundation must be uniform and stable, provide adequate drainage, should extend beyond the slab edge through the shoulder area and through special transitions at bridge approaches, cuts, and fills.
- As with JPCP, JRCP, the integration of pavement design, mix design, and construction practices used for CRCP should maximize load transfer efficiency and minimize bending and flexural stresses.
- Concrete curing should be appropriate and adequate to each CRCP application, construction requirements, temperature, moisture, and wind conditions.
- Longitudinal construction joints must be tied to adjacent pavement at the centerline or shoulder.
- Longitudinal contraction joints at shoulders should be sawed directly over the transverse reinforcement.

Similar to JPCP or JRCP the design of CRCP also involves the dimensioning of a pavement structure that includes slab thickness and mix design, construction joints, reinforcement details, slab width, shoulders, pavement transitions, and foundation support, that can handle the required traffic loads, climate and environment conditions, and provide the expected level of performance at the desired cost. Figure 15.26 shows the cross-section of a typical CRCP pavement that conforms to good practices observed in the field.

15.2.4.5 CRCP Design Criteria
Limiting Criteria on Crack Spacing, Crack Width, and Steel Stress
  The AASHTO-86/93 Guide recommends controlling crack spacing to fall within 3.5–8 ft (1.1–2.4 m). However, the new CRCP design procedure described in the AASHTO MEPDG does not provide recommendations on the control of minimum crack spacing. It does however recommend

FIGURE 15.26 Typical CRCP cross-section. (From Rasmussen, R.O. et al., *Continuously Reinforced Concrete Pavement (CRCP) Design and Construction Guidelines*, Federal Highway Administration and Concrete Reinforcing Steel Institute, 2012.)

TABLE 15.22
Allowable Steel Working Stress, ksi (MPa)

Indirect Tensile Strength of Concrete at 28 Days, psi (MPa)	Reinforcing Bar Diameter, in. (mm)		
	0.5 (12.7)	0.625 (15.9)	0.75 (19.1)
300 (2.1) or less	65 (448)	57 (308)	54 (372)
400 (2.8)	67 (462)	60 (414)	55 (379)
500 (3.4)	67 (462)	61 (421)	56 (386)
600 (4.1)	67 (462)	63 (434)	58 (400)
700 (4.8)	67 (462)	65 (448)	59 (407)
800 (5.5) or greater	67 (462)	67 (462)	60 (414)

Source: American Association of State Highway and Transportation Officials, AASHTO Guide for Design of Pavement Structures, Washington, DC, 1993.

a maximum average crack spacing of 6 ft (1.8 m) and advocates for small crack widths to ensure long-term performance (Rasmusen, Rogers and Farragut [CRCP Manual, 2012]).

The AASHTO-86/93 Guide limits the crack widths to 0.04 in. (1 mm) to avoid spalling. Additional studies and observations since the AASHTO-86/93 Guide have shown that crack widths of 0.024 in. (0.6 mm) have been found to be more effective in reducing water penetration, minimizing corrosion of the steel, maintaining the integrity of the support layers, and ensuring high load transfer efficiency. However, the AASHTO Interim MEPDG predicts and requires a maximum crack width of 0.020 in. (0.5 mm) over the entire design period (Rasmusen, Rogers and Farragut [CRCP Manual, 2012]).

Table 15.22 shows the maximum allowable working stress for steel with yield strength of 60 ksi (420 MPa) that is recommended by the AASHTO-86/93 Guide. It can be noted that in some cases, the AASHTO-86/93 Guide allows a working stress above the yield strength that could result in a possibility of some plastic deformation. This could result in somewhat wider crack openings when permanent deformation is allowed (CRCP Guide, 2012).

15.2.4.6 Structural Performance

In mechanistic-empirical design procedures such as those developed for the AASHTO Interim MEPDG, structural performance for CRCP is typically expressed in terms of allowable punchouts per unit of distance (i.e., punchouts/mile) before rehabilitation is needed.

Recommended threshold values for the allowable number of punchouts are commonly expressed as a function of the functional highway classification or traffic level. The limit that is selected is also a function of the design reliability (risk). The AASHTO Interim MEPDG recommends a maximum of 10 medium- and high-severity punchouts per mile (6 punchouts/km) for interstates and freeways, 15 punchouts per mile (9 punchouts/km) for primary highways, and 20 (12) for secondary highways. The American Concrete Pavement Association (ACPA) recommends a maximum of 10 punchouts per mile (6 punchouts/km) for Average Daily Traffic (ADT) greater than 10,000 vehicles/day, 24 punchouts per mile (15 punchouts/km) for ADT between 3,000 and 10,000 vehicles/day, and 39 punchouts per mile (24 punchouts/km) for ADT below 3000 vehicles/day (Rasmusen, Rogers and Farragut [CRCP Manual, 2012]).

15.2.4.7 Functional Performance

Like structural performance, functional performance threshold levels in terms of international roughness index (IRI) as a function of time or load applications are commonly defined based on the functional highway classification or traffic level. The AASHTO Interim MEPDG recommends a maximum IRI of 175 in./mile (2.7 m/km) for interstates and freeways, 200 in./mile (3.2 m/km) for primary highways, and 250 in./mile (4 m/km) for secondary highways. The ACPA recommends a maximum IRI of 158 in./mile (2.5 m/km) for ADT greater than 10,000 vehicles/day, 190 in./mile (3.0 m/km) for ADT between 3,000 and 10,000 vehicles/day, and 220 in./mile (3.5 m/km) for ADT below 3,000 vehicles/day. In the AASHTO Interim MEPDG, the threshold value is selected based on the design reliability (risk) (Rasmusen, Rogers and Farragut [CRCP Manual, 2012]).

15.2.4.8 Reinforcement for CRCP

The most common type of reinforcement in CRCP is deformed steel bars. However, other new materials including solid stainless steel, stainless steel clad, and other proprietary materials such as fiber reinforced polymer (FRP) bars have also been used on a smaller and experimental scale but mostly for dowel bars. Stresses generated in the CRCP due to volume changes from the steel to concrete depend on the steel surface area and the shape of the surface deformations on the reinforcing bar (rebar). Therefore, it is critical that the rebar used comply with requirements specified in AASHTO M 31, M 42, or M 53 for billet-steel, rail-steel, or axle-steel deformed bars respectively. Alternatively, ASTM A 615 for billet-steel, and ASTM A 996 for rail- and axle-steel deformed bars, may be used (Rasmusen, Rogers and Farragut [CRCP Manual, 2012]).

Bar designations as well as requirements for deformations and steel tensile strength or steel grade are provided in both the AASHTO and ASTM specifications. Table 15.23 shows weight and dimensions of ASTM standard reinforcing steel bars. The required yield strength of reinforcing steel for use in CRCP is typically 60,000 psi (420 MPa), designated as English Grade 60 (metric Grade 420). Higher strength steels may be used, but this does not imply that the concrete cracks will be held tighter together. This function of controlling crack widths still depends on the MOE or stiffness of the steel which is still around 29 million psi.

Another property of interest for CRCP reinforcement design is the coefficient of thermal expansion of the steel. Depending on the difference in the steel and concrete CTE, varying restraint will result, leading to different crack patterns. The AASHTO 86/93 Guide recommends the use of a steel CTE of 5×10^{-6} in./in./°F (9×10^{-6} m/m/°C) for design. The CTE values from the MEPDG Guide range from 6.1 to 6.7×10^{-6} in./in./°F and 11 to 12×10^{-6} m/m/°C.

15.2.4.9 Design Methods for CRCP

In the past, it was common practice by some states to design CRCP thickness based on jointed concrete pavement methodology, and then reduce the thickness by as much as 20% to account for the effect of increased load transfer efficiency at the cracks. In some cases, this resulted in an under-design with subsequent failure. This practice is no longer being applied. Today, typical CRCP thicknesses vary from 7 to 15 in. (178 to 381 mm) depending on the level of traffic and

TABLE 15.23
Weight and Dimension of ASTM Standard Reinforcing Steel Bars

	Nominal Dimensions		
Bar Size US (SI)	Diameter, in. (mm)	Cross-Sectional Area, in.^{2} (mm^2)	Weight, lb/ft (kg/m)
#3 (#10)	0.375 (9.5)	0.11 (71)	0.376 (0.560)
#4 (#13)	0.500 (12.7)	0.20 (129)	0.668 (0.994)
#5 (#16)	0.625 (15.9)	0.31 (199)	1.043 (1.552)
#6 (#19)	0.750 (19.1)	0.44 (284)	1.502 (2.235)
#7 (#22)	0.875 (22.2)	0.60 (387)	2.044 (3.042)
#8 (#25)	1.000 (25.4)	0.79 (510)	2.670 (3.973)
#9 (#29)	1.128 (28.7)	1.00 (645)	3.400 (5.060)
#10 (#32)	1.270 (32.3)	1.27 (819)	4.303 (6.404)
#11 (#36)	1.410 (35.8)	1.56 (1006)	5.313 (7.907)

environmental conditions, although most common practice is between 10 and 12 in. (254–305 mm) (Rasmusen, Rogers and Farragut [CRCP Manual, 2012]).

The two most common design methods for determination of slab thickness and the amount of reinforcement required in CRCP are the (1) ASSHTO-86/93 Guide for Design of Pavement Structures and the (2) AASHTO Interim MEPDG.

15.2.4.10 AASHTO Interim MEPDG

The National Cooperative Highway Research Program (NCHRP) has taken the lead on a major effort to develop the next generation of pavement design procedure based on mechanistic-empirical methods. This has been conducted under the funded research project 1-37A, and has resulted in the proposed AASHTO Interim MEPDG. In this design procedure, specific mechanistic-empirical models for prediction of CRCP performance have been developed. Figure 15.27 shows the flow diagram of the AASHTO Interim MEPDG CRCP design process. The process starts with the selection of a trial design including layer thicknesses, materials, reinforcement, shoulder characteristics, and construction information. Site-specific conditions including environment, foundation, and traffic are also considered. Performance criteria in terms of punchouts and IRI are then specified, along with the reliability level for each criterion.

The MEPDG also has limiting design criteria on crack width (over design period), crack spacing, and crack load transfer efficiency (over design period as well). The procedure predicts punchout failure development as a function of the crack width and load transfer efficiency due to aggregate interlock at transverse cracks. Stresses due to loading are predicted as a function of load transfer efficiency at cracks, and continuously evaluated and modified throughout the design period. Fatigue damage as a function of the stress level and concrete strength is evaluated and accumulated, and punchout development is subsequently predicted. Pavement roughness using the summary statistic in IRI values is also predicted throughout the design period as a function of the initial smoothness conditions, punchout development, and site-specific conditions. Once the trial design is evaluated, its predicted performance is checked against design criteria at the specified reliability level. If the design requirements are satisfied, the trial design is considered as a viable alternative that can later be evaluated in terms of life-cycle costing. Otherwise, a new trial design is evaluated (Rasmusen, Rogers and Farragut [CRCP Manual, 2012]).

15.2.4.11 AASHTO-86/93 Design Procedure

The AASHTO-86/93 Guide procedure for CRCP uses the same consideration as for jointed concrete pavements. It does not however consider punchout failure which is a major CRCP distress.

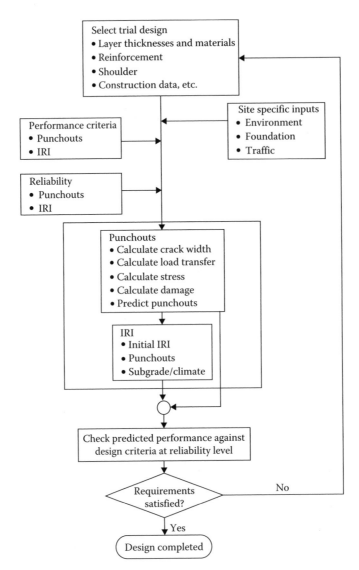

FIGURE 15.27 Flow diagram of the AASHTO MEPDG CRCP design process. (AASHTO Interim Report, NCHRP 1-37A.)

In addition, due to the heavy reinforcement which keeps the cracks tight in CRCP, an improved load transfer coefficient is commonly used, which typically results in a moderate reduction in thickness for this pavement type under similar traffic and environmental conditions.

The AASHTO-86/93 method also includes design procedures for the selection of steel reinforcement in CRCP. These procedures are based on a desired range of crack spacing, maximum crack width, and maximum steel stress. It should be noted that it has been reported that this design procedure tends to underestimate the required steel.

15.2.4.12 Reinforcement Design

Reinforcement design for CRCP involves selecting the proper percentage, bar size, and bar configuration for optimum CRCP performance. Proper reinforcement design provides the minimum reinforcement necessary to develop the desired crack spacing and desired crack widths, while maintaining the steel stresses at an acceptable level.

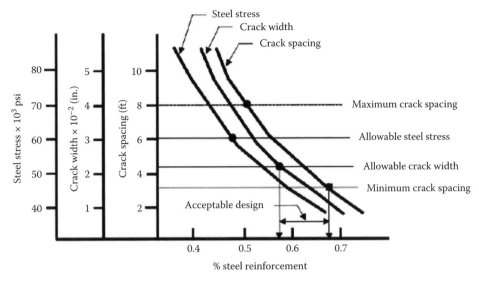

FIGURE 15.28 Conceptual design approach for CRCP. (From Rasmussen, R.O. et al., *Continuously Reinforced Concrete Pavement (CRCP) Design and Construction Guidelines*, Federal Highway Administration and Concrete Reinforcing Steel Institute, 2012.)

Longitudinal steel reinforcement content is defined as the ratio of the area of longitudinal steel to the area of concrete (As/Ac) across a transverse section, commonly expressed as a percentage. Figure 15.28 shows a conceptual design approach for CRCP showing higher amounts of steel reinforcement will result in shorter crack spacings, smaller crack widths, and lower steel stresses. An increase in the percent of longitudinal reinforcement will result in an increase in restraint. As the level of restraint increases, so does the number of cracks that develop, resulting in shorter crack spacings. In addition, as the amount of reinforcement increases, the average steel stresses are reduced, producing less reinforcement elongation (Rasmusen, Rogers and Farragut [CRCP Manual, 2012]).

15.2.4.13   Design Variables for Longitudinal Reinforcement: AASHTO Method

The input variables required for the design of longitudinal reinforcement in CRC pavements are presented next.

15.2.4.13.1   Concrete Tensile Strength

The 28-day concrete indirect tensile strength as determined using ASTM C 496 or AASHTO T 198 splitting tensile should be used to determine the longitudinal steel reinforcement. The indirect tensile strength may be approximated from the modulus of rupture or the flexural strength used for thickness design from the third-point loading test. The indirect tensile strength can be approximated as 86% of the modulus of rupture value at 28 days.

15.2.4.13.2   Concrete Coefficient of Thermal Expansion

The coefficient of thermal expansion for concrete depends on a number of variables such as the mix design, the water–cement ratio, the duration of curing, the relative humidity and moisture content, and coarse aggregate type. But the most influential factor is the type of coarse aggregate since it occupies a significant volume of the concrete. Values of the coefficient of thermal expansion for Portland cement concrete as a function of aggregate type is provided in Table 15.24.

15.2.4.13.3   Reinforcing Bar Diameter

Typical bar sizes used in CRCP range from # 4 (0.5 in.) to # 7 (0.875 in.) (#13 [13 mm] to # 22 [22 mm]) (Rasmusen, Rogers and Farragut [CRCP Manual, 2012]).

TABLE 15.24

Typical CTE Values for Concrete Made with Common Aggregate Types

Aggregate Type	Concrete CTE × 10^{-6} in./in./°F (m/m/°C)
Quartz	6.6 (11.9)
Sandstone	6.5 (11.7)
Gravel	6 (10.8)
Granite	5.3 (9.5)
Basalt	4.8 (8.6)
Limestone	3.8 (6.8)

15.2.4.13.4 Steel Coefficient of Thermal Expansion

A steel CTE of 5 ± 10^{-6} in./in./°F (9 ± 10^{-6} m/m/°C) is recommended for longitudinal reinforcement design unless specific knowledge of the CTE for the steel to be used in the project is known.

15.2.4.13.5 Design Temperature Drop

The design temperature drop is determined based on the difference between the average concrete curing temperature after placement and a design minimum temperature and can be expressed as:

$$\text{Design temperature drop, } \Delta TD = TH - TL$$

TH is the average daily high temperature during the month the pavement is constructed
TL is the average daily low temperature during the coldest month of the year where the CRC pavement will be constructed

15.2.4.13.6 Concrete Drying Shrinkage

Concrete drying shrinkage is an important design parameter for design of the reinforcement for CRCP. Concrete shrinkage depends on many factors, but ultimately it depends on the properties of the paste phase and the moisture and temperature environment of the concrete. The water content, the water–cement ratio, and the strength will affect the shrinkage greatly. A correlation between the indirect tensile strength of the concrete and shrinkage is presented in Table 15.25.

15.2.4.13.7 Wheel Load Tensile Stress

The wheel load tensile stresses that develop in the concrete pavement due to construction traffic or wheel loads during its early age may impact the crack spacing pattern and should be accounted

TABLE 15.25

Approximate Relationship between Shrinkage and Indirect Tensile Strength of Concrete

Indirect Tensile Strength psi (MPa)	Shrinkage in./in. (m/m)
300 (2.07) or less	0.00080
400 (2.76)	0.00060
500 (3.45)	0.00045
600 (4.14)	0.00030
700 (4.83) or greater	0.00020

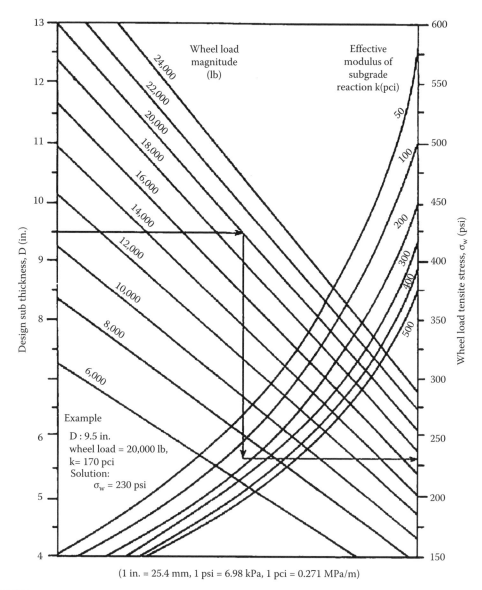

(1 in. = 25.4 mm, 1 psi = 6.98 kPa, 1 pci = 0.271 MPa/m)

FIGURE 15.29 Example showing wheel load stress (AASHTO).

for in longitudinal reinforcement design. Figure 15.29 shows an example of wheel load stress as a function of the design slab thickness, magnitude of wheel load, and effective k-value.

15.2.4.14 Limiting Criteria

There are three limiting criteria that are considered for the design of longitudinal steel in CRCP. These include crack spacing, crack width, and stress developed in steel.

1. Limiting crack spacing between 3.5 and 8 ft (1.2 and 2.4 m) for the purpose of reducing the risk of punchout failure.
2. Limiting the width of cracks to 0.04 in. (1 mm) for reducing the risk of water infiltration and concrete spalling. According to Huang (2004), this can be accomplished by the selection of a higher steel percentage or smaller diameter reinforcing bars.
3. Limiting the steel stress developed to not more than 75% of the yield stress.

The design procedure adopted by AASHTO was developed by McCullough and Elkins (1979) and McCullough and Cawley (1981). The design equations and nomographs are provided in the AASHTO Design Guide. All three criteria must be checked to determine the required percent of longitudinal steel. Huang (2004) presents the AASHTO equations for determining the percent steel, P.

15.2.4.14.1   Crack Spacing

For a given crack spacing, the percentage of steel P can be determined by the following equation.

$$P = \frac{1.062\left(1+\dfrac{f_t}{1000}\right)^{1.457}\left(1+\dfrac{\alpha_s}{2\alpha_c}\right)^{0.25}(1+\phi)^{0.476}}{(\overline{X})\left(1+\dfrac{\sigma_w}{1000}\right)^{1.13}(1+1000Z)^{0.389}} - 1$$

P is the amount of longitudinal steel in percent
f_t is the concrete indirect tensile strength, psi
α_s/α_c is the thermal coefficient ratio
ϕ is the reinforcing bar diameter
σ_w is the tensile stress due to wheel load, psi
Z is the concrete shrinkage at 28 days, in./in.

The example shown in Figure 15.30 and presented by AASHTO can be solved for the percent steel as follows and checked with the value from the chart:

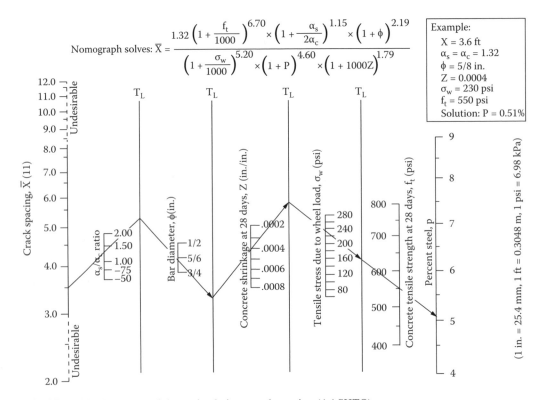

FIGURE 15.30   Percent steel determined given crack spacing (AASHTO).

$$P = \frac{1.062\left(1+\dfrac{550}{1000}\right)^{1.457}\left(1+\dfrac{1.32}{2}\right)^{0.25}(1+0.625)^{0.476}}{(3.5)^{0.217}\left(1+\dfrac{230}{1000}\right)^{1.13}(1+1000*0.0004)^{0.389}}-1=0.52\%$$

15.2.4.14.2   Crack Width

For a given crack width, the percentage of steel P can be determined by the following equation:

$$P = \frac{0.358\left(1+\dfrac{f_t}{1000}\right)^{1.435}(1+\phi)^{0.484}}{(CW)^{0.220}\left(1+\dfrac{\sigma_w}{1000}\right)^{1.079}}-1$$

The example shown in Figure 15.31 and presented by AASHTO can be solved for the percent steel as follows and check with the value from the chart:

$$P = \frac{0.358\left(1+\dfrac{550}{1000}\right)^{1.437}(1+0.75)^{0.484}}{(0.04)^{0.220}\left(1+\dfrac{230}{1000}\right)^{1.079}}-1=0.43\%$$

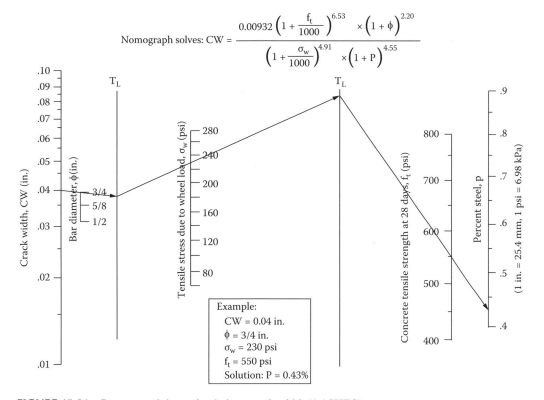

$$\text{Nomograph solves: } CW = \frac{0.00932\left(1+\dfrac{f_t}{1000}\right)^{6.53}\times\left(1+\phi\right)^{2.20}}{\left(1+\dfrac{\sigma_w}{1000}\right)^{4.91}\times\left(1+P\right)^{4.55}}$$

Example:
CW = 0.04 in.
ϕ = 3/4 in.
σ_w = 230 psi
f_t = 550 psi
Solution: P = 0.43%

FIGURE 15.31   Percent steel determined given crack width (AASHTO).

15.2.4.14.3 Steel Stress

For a given steel stress, the percentage of steel P can be determined by the following equation:

$$P = \frac{50.834\left(1+\dfrac{DT_D}{100}\right)^{0.155}\left(1+\dfrac{f_t}{1000}\right)^{1.493}}{\sigma_s^{0.365}\left(1+\dfrac{\sigma_w}{1000}\right)\left(1+1000Z\right)^{0.180}} - 1$$

$$DT_D = T_H - T_L$$

DT_D is the design temperature drop, °F
T_H is the average daily high temperature during the month which the pavement was constructed
T_L is the average daily low temperature during the coldest month of the year
σ_s is the allowable steel stress

The example shown in Figure 15.32 and presented by AASHTO can be solved for the percent steel as follows and checked with the value from the chart:

$$P = \frac{50.8334\left(1+\dfrac{55}{100}\right)^{0.155}\left(1+\dfrac{550}{1000}\right)^{1.493}}{(57)^{0.365}\left(1+\dfrac{230}{1000}\right)\left(1+1000(0.0004)\right)^{0.180}} - 1 = 0.47\%$$

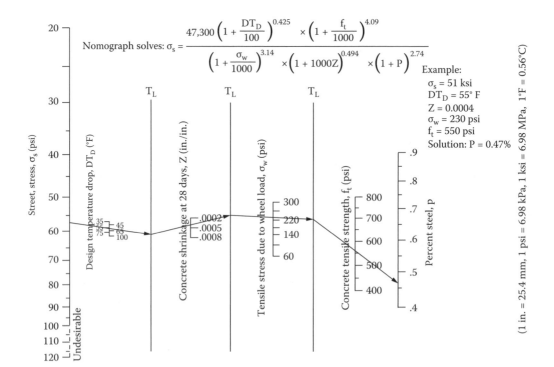

FIGURE 15.32 Percent steel determined given steel stress (AASHTO).

15.2.4.15   Longitudinal Reinforcement Design Procedure

The following procedure should be followed to determine the amount of longitudinal reinforcement needed for CRCP.

If P_{max} is greater than or equal to P_{min}, continue with the next step in the design process. When P_{max} is less than P_{min}, the design is considered unsatisfactory and the input variables should be modified until P_{max} is greater than P_{min}.

The following equations can be used to determine the range in the number of reinforcing bars or wires required for design:

$$N_{min} = 0.01273 P_{min} \frac{W_s D}{\phi^2}$$

$$N_{max} = 0.01273 P_{max} \frac{W_s D}{\phi^2}$$

where

N_{min} is the minimum number of reinforcing bars = reinforcing bar or wire diameter (in or required, mm)

N_{max} is the maximum number of reinforcing bars required

P_{min} is the minimum required steel percentage

P_{max} is the maximum required steel percentage

W_s is the total width of pavement section, in. or mm

D is the slab thickness (in or mm)

ϕ is the reinforcing bar or wire diameter, in. (may be increased if potential for corrosion exists)

The final steel design is determined by selecting the total number of bars or wires N; where the value of N is a whole number between N_{min} and N_{max}. To check if the final design is appropriate, the number of bars or wires may be converted to the percentage of steel P and then used in the equations and nomographs to back calculate the approximate values for crack spacing, crack width, and steel stress.

QUESTIONS

15.1 Determine the corrected batch weights for a 14 ft³ mix volume and for the following conditions.

W/CM = 0.55.

Air = 7%.

Bulk volume of coarse aggregate = 59%.

Coarse aggregate dry-rodded unit weight = 95 pcf.

Estimated water = 324 lb/yd³.

Cement type = 1.

Sand (moisture content = 5%; absorption = 3%) G_s = 2.60.

Coarse aggregate (moisture content = 0%; absorption = 0.9%) G_s = 2.70.

After mixing, the unit weight of the fresh concrete is 142 pcf.

Calculate the yield, the gravimetric percentage of air content, the relative yield, and the cement factor (bags/yd³).

15.2   Determine the thickness for a rigid pavement given the following:

$K = 100$ pci

$E_c = 4 * 10^6$ psi

$S_c = 650$ psi

$J = 3.2$

$C_d = 1.0$

$\Delta PSI = 4.2 - 2.5 = 1.7$

$R = 95\%$

$S_o = 0.29$

$W_t = 7 * 10^6$

16 Construction of Asphalt Pavements

16.1 OVERVIEW

The essential steps in the construction of asphalt pavements consist of selecting proper materials and conducting a proper mix design, ensuring drying and mixing of aggregates from properly maintained stockpiles, mixing aggregates with asphalt binder, delivering the mix in insulated trucks in a smooth way so as to match with paver speed, laying down with a paver, and using a properly selected combination of rollers of desirable weight and type, with correct vibration frequency and/or tire pressure, in a correct rolling pattern. The rollers should be close to the paver and operating with desirable speed and manner of reversing and turning. All of the equipment should be in good operating condition. And, very importantly, good quality control must be ensured at every step in the plant and during laydown and rolling operations. There must be clear communication of any problems between the plant and the site, and if required, adjustments should be made in mix design, plant production, transportation, and laydown and rolling operations. Provisions should be made to use additives in the materials, such as natural rounded sand to reduce the harshness of a mix, and/or employ special equipment, such as a material transfer vehicle to reduce temperature-related segregation, if required.

16.1.1 PRODUCTION

The production of loose hot mix asphalt (HMA) starts in the HMA plant—which can be of either batch (Figure 16.1) or drum type. In both plants, the steps consist of heating and drying aggregates and mixing them with heated liquid asphalt binder. In a batch plant, the production happens in "batches"—one batch of aggregates of different sizes is dried in a drum, moved up in a hot elevator, and mixed with asphalt in a pugmill. In the case of a drum plant, the production is a continuous process—aggregates of different sizes are dried in the front part of the dryer drum and then mixed with the liquid asphalt binder in the back part of the drum. In a batch plant, aggregates are collected on cold feed conveyors, heated in a dryer drum, moved up a hot elevator, rescreened over a screen deck and separated into hot bins, and then mixed with asphalt in the pugmill. In the drum plant, the tower, consisting of the hot elevator, hot bins, and pugmill, is eliminated, and mixing is done in the drum. Hybrid plants, using some features of both batch and drum plants, are also being used.

Drum plants are continuous as opposed to batch plants where every batch of aggregate should be weighed and mixed individually. However, batch plants are also more efficient in cases of short production runs or delays in hauling, or when multiple types of mixes are required (such as paving for highways, parking lots, and driveways) in urban areas.

The production of HMA is a complex process and requires preparation and continuous supervision by experienced and trained people. Once the HMA is produced, it is either stored in a silo or put in trucks for transportation to the job site. The silo must be properly insulated to prevent oxidation and heat loss from the HMA. The trucks also need to be insulated (sometimes heated) to minimize heat loss from the HMA during transportation.

FIGURE 16.1 Hot mix asphalt plant. (Courtesy of Matthew Teto, All States Asphalt, Sunderland, MA.)

16.1.2 TRANSPORTATION AND LAYDOWN

Once the trucks arrive at the job site, the HMA is transferred either directly to the paver or to a paver through a material transfer device (MTD). The role of the MTD is to store HMA such that it can be used when the paver hopper becomes empty and before more trucks arrive with new HMA. The MTD also helps keep the HMA heated and mixes it to maintain its uniformity and hence reduce the potential of segregation.

The paver lays down the HMA at a specific thickness with the help of the screed. It also provides some compaction to the mix. The remaining compaction is provided by the roller(s) following the paver. Generally, rollers with these distinct features are used. A steel drum roller with one vibratory roller is used for breakdown rolling, a static steel drum or pneumatic roller is used for intermediate rolling, and usually a static steel drum roller is used for finishing.

Generally, a mix will get compacted by 20%–25%. Generally used thicker lifts are 2–4 in., and thinner lifts are 1–2 in. Thicker lifts are easier to compact since they retain heat longer and also help avoid the bridging tendency of thinner lifts. Thinner lifts, though, provide good results through better smoothness and help avoid shoving, which could happen in thicker lifts.

Tack coats are used to provide a bond between the old surface and the new asphalt mix layer. This is applied on the existing surface before the application of the HMA. The right kind and amount should be used, and adequate time should be given for curing before the HMA is laid down. The typical application rate is 0.05 gal/yard^2. Typically, it is applied in double lap with a distributor truck.

16.2 DESCRIPTION AND REQUIREMENTS OF COMPONENTS IN HOT MIX ASPHALT–PRODUCING PLANTS

16.2.1 AGGREGATE STOCKPILES

Aggregate stockpiles (Figure 16.2) should have uniform gradation of aggregates in all parts and must be free from foreign materials and excessive amounts of moisture. The moisture content should be fairly constant throughout the depth of the stockpile.

Stockpiles should be constructed in the proper sequence using power equipment. Proper stockpiling should include remixing of materials, layering of remixed materials, and dispersion of newly delivered materials into existing piles. Front-end loaders are generally used. If trucks are used, then the pile should be built in layers, with each pile being one truck load of material.

FIGURE 16.2   Aggregate stockpiles. (Courtesy of Matthew Teto, All States Asphalt, Sunderland, MA.)

Manipulation by dozers should be limited. The stockpiles should be separated from each other and built on a surface with good drainage, and all parts of the stockpiles should be accessible and reachable from all sides.

16.2.2   COLD FEED BINS

The purpose of the cold feed bin system, which consists of multiple bins, is to provide a controlled distribution of aggregate in the conveyor system. The cold feed system (Figure 16.3) is controlled by manipulating the relative amounts of the different constituents as well as by the rate at which the material is fed into the drum. Cold bins contain the aggregates and direct them toward the cold feed through their sloping sides. The cold feed can be of the belt, vibratory, or apron type. The rate of flow from a bin into the cold feed is controlled by the adjustable gate at the bottom of the bins as well as by the rpm of the belt feeders or the frequency/amplitude of the vibratory feeders. The actual flow rate has to be calibrated for different aggregates, with difference in gradation, shape, and moisture content.

From the cold feeder the aggregate drops into a moving belt, which collects aggregates from multiple bins, which deliver the materials directly or through a scalping screen to the drum, as well as to sampling chutes. The final drop into the drum happens through an inclined belt, with scales for measuring the amount of aggregate, which dictates the amount of asphalt that should be added to the aggregates in the drum, if it is a drum plant.

(a)   (b)

FIGURE 16.3   (a) Cold feed bins and (b) aggregate feeder system. (Courtesy of Matthew Teto, All States Asphalt, Sunderland, MA.)

The aggregate feeders need to be calibrated to determine the speed required to feed aggregates at a desired rate during production. The following steps are conducted to calibrate the feeders:

1. Close all feeders except the one being calibrated.
2. Run materials into the belt.
3. Determine the amount of material in a given length of the belt.
4. Adjust the feeder, and repeat steps 2 and 3.
5. Calibrate the next feeder using steps 1–4.
6. Determine the speed of the belt.

The relationship between the different parameters can be expressed as follows (NAPA, 1996):

$$R = \frac{1.8ws}{B}$$

where
R is the rate of production in ton/h
w is the weight of sample on B length (ft) of belt in lb
s is the speed of belt in ft/s

16.2.3 DRYER DRUM

In a batch plant the drum acts as a dryer. The drum is tilted, and the aggregates enter the upper end of the drum and flow toward the lower end, where the burner provides the heat for drying the aggregates (Figure 16.4). The burner works on oil and/or gas with a regulated flow of air from a blower. The system for supplying fuel and air needs to be checked to ensure complete combustion of the fuel. The flow of hot air is further increased by an exhaust fan. The long thin flame of the burner as well as the hot gasses leaving the drum near the entry point of the aggregates dry off their moisture and heat them to a high temperature. The aggregates tumble through the flame through a set of flights inside the drum as it turns during the drying process. The time for which the aggregate remains in the dryer drum is known as the *dwell time*. It is important to use a proper dwell time, taking into consideration the amount of moisture in the aggregates, and the temperature to which it should be heated to mix with the asphalt binder. This temperature is equal to the temperature of the asphalt binder at which it has the optimum viscosity for mixing. For the proper dwell time, the moisture content of aggregates leaving the drum must not exceed 0.5%. Dwell time can be increased by slowing the drum rotation, lowering the tilt angle of the drum, rearranging the flights inside the

(a)   (b)

FIGURE 16.4 (a) Dryer drum with burner and (b) inside of drum. (Courtesy of Matthew Teto, All States Asphalt, Sunderland, MA.)

FIGURE 16.5   Hot elevator. (Courtesy of Matthew Teto, All States Asphalt, Sunderland, MA.)

drum, or increasing the total length of the dryer drum. If required, the amount of aggregate that is fed into the dryer, per unit time, should be decreased. Note that aggregates with higher moisture content will require more heat and hence more fuel to drive the moisture, and it is always advisable to store aggregates in enclosed storage areas.

16.2.4   HOT ELEVATOR AND BINS

The hot aggregates leave the dryer drum in a chute and are picked up by buckets in the hot elevator (Figure 16.5), which delivers it to the screen deck in the tower. The screen deck has a number of vibrating screen decks, arranged in a way such that the screen with the largest opening is at the top and the smallest opening is at the bottom. The screen sizes may range from 1½ to 5/32 in. The screened aggregates land in hot bins, which are placed beneath the screens. The screens help in separating the composite aggregate sample batch into different portions, depending on their size. Note that it is the final screening in this step that determines the blend of the aggregate in the mix, and hence the screens must operate efficiently without getting blinded due to processes such as high rates of loading or loading of improperly sized aggregates or due to blinding of the screens themselves due to poor maintenance. The malfunctioning of screens causes the dumping of aggregates into inappropriate bins ("carryover"), and this affects the quality of the asphalt mix.

The function of the hot bins is to provide adequate temporary storage for the separate hot aggregates before mixing with the asphalt binder in the weigh hopper below them. Narrow hot bin gates open into the weigh hopper. Quality control samples should be taken at this point from hot bins to ensure proper functioning. To ensure a smooth flow of materials from the screen decks to the bins and into the hopper and avoid segregation, it is important that the aggregates are supplied through the cold feed at a rate that is consistent with the demand for production. The plant operator could check on the relative amounts of materials in the different bins and, if needed, adjust the cold feed. A predetermined mass of aggregate from each hot bin (based on the mix design of the asphalt mix being produced) is dropped from the hot bins, one by one, to a hopper on a scale (weigh hopper). The batch is complete when all of the constituent aggregates have been dropped on the weigh hopper, and the cumulative mass is printed on a batch ticket, which is used as supporting documentation for verification of compliance with mix design.

16.2.5   PUGMILL

In the next step, the weigh hopper empties the aggregates into the pugmill (Figure 16.6), where they are mixed with each other and with the asphalt binder with two mixing shafts with paddles, which

FIGURE 16.6 Pugmill, view from the bottom with discharge door removed. (Courtesy of Matthew Teto, All States Asphalt, Sunderland, MA.)

rotate in opposite directions. It is important to make sure that the paddles are in good condition (and not worn out) and that the pugmill is filled to its optimum height (bottom to top of the paddles—the "live zone") to ensure good mixing. There are two stages of mixing in the pugmill: dry, when the aggregates are mixed with each other, typically 2 s or less, and wet, when the asphalt binder is introduced either by gravity or through a pressurized spray, typically 25–35 s. The total mixing time, which includes dry mixing time, wet mixing time, opening time of the pugmill gates below, discharge time of the mix, and closing time of the pugmill gates, affects the quality of the mix as well as the rate of production.

16.2.6 Hauling and Storage

The mix from the pugmill is either carried away by hauling trucks to a job site or taken to a storage silo through a chute in a drag slat conveyor. The truck must be positioned underneath the pugmill to avoid spillage and one-sided drops, and ensure proper discharge, which is generally made by loading the front end of the truck first, then the back end, and finally the middle portion, to avoid segregation.

Drag slat conveyors (Figure 16.7) move the asphalt mix from the pugmill to either surge silos (Figure 16.7) for temporary storage (a few hours) or storage silos for long-term storage (overnight to several days). Drag slat conveyors have steel plates set at a 90° angle to the drag chain and are typically enclosed to prevent any loss of heat from the mix.

(a)   (b)

FIGURE 16.7   (a) Slat conveyor and (b) silos. (Courtesy of Matthew Teto, All States Asphalt, Sunderland, MA.)

Whenever there is movement of HMA mix, there is a tendency of coarser aggregate particles to roll down and get separated from the finer aggregates. This not only separates the aggregates of different sizes, but also creates an uneven distribution of asphalt binder in the mix, since it is the finer particle/matrix that contains more asphalt than the coarse aggregates. To minimize movement, sometimes a batcher is used for loading the silos from the slat conveyor. The batcher has a small hopper, which gets filled through a chute at the center, with a batch of mix. This batch of mix is then released to fall into the silo through gates at the bottom of the batcher. This way the silo is filled up in batches, and not in a continuous stream of HMA (thus limiting movement and hence segregation).

To minimize segregation from loading out from the silo, the trucks must be positioned properly beneath it, and loaded front first, then back, and finally the middle. Mix from the silo should not be used unless there is adequate mix inside the silo above the bottom cone portion of the silo. Also, if required, heated oil–carrying pipes throughout the shell of the silo could keep the mix warm, and inert gases could be used to reduce the oxidation of the mix inside.

16.2.7 DRUM PLANT

In a drum plant, the aggregates are dried, heated, and mixed with the asphalt binder in the same drum, which is usually longer than the dryer drum used in a batch plant, and which is generally of the parallel flow type, in which the aggregates enter the drum in the same side as the burner flame. The drum has three distinct zones—the drying, heating, and mixing zones. The flame in the drum is shorter and wider, and the aggregates are kept in flights along the perimeter of the drum in the first zone, to prevent them from coming in direct contact with the flame. In the second zone, the flights cascade the aggregates through the center of the drum and through the hot burner gases, and as a result the aggregates get heated up to a high temperature. The configuration of the flights as well as the geometry of the drum could be varied in this zone to achieve this objective. Finally, in the mixing zone, flights are used to coat the aggregates with asphalt binder which is injected into the drum.

Although a typical drum is as it is described earlier, numerous variations, as a result of continuous development processes, have resulted over the years, specifically to improve the efficiency of the heating process, and often to accommodate the use of recycled asphalt pavement (RAP) materials (Figure 16.8). Two such developments are the concepts of the double-barrel drum and the triple-barrel drum, which have outer "shells" that improve the heating process of aggregates and at the same time cut down the potential of excessive oxidation of the asphalt binder–RAP materials.

FIGURE 16.8 RAP stockpile and RAP feeding system in a drum plant. (Courtesy of Robert Frank.)

Another important development, in the context of paving jobs in remote areas, is the availability of portable drum plants, which can be transported from one site to another and readied for operation in a relatively short period of time. Note that counterflow drums are indeed used in many of these newly developed drum mix plants.

16.2.8 Dust Collection from HMA Plants

Dust is created during the drying, heating, and mixing of aggregates and asphalt. The dust primarily results from the existing fine portions of the aggregate; is created by abrasion during mixing, blown out by gases, or created by the air used for the burner; and comes from steam from the evaporation of water from the aggregates. The dust is actually captured through a collecting mechanism and prevented from polluting the environment. Air quality codes from regulatory agencies direct the plants to test for particulate emissions in the exhaust gases from the plant to ensure the efficiency of the dust-collecting system.

The dust-collecting system consists of a primary and secondary mechanism. In the primary system, the relatively bigger size particles are separated from the exhaust gases. This is accomplished by either expanding the gases, reducing their velocity, and making the heavier particles drop out in the "knockout box" (expansion chamber process) or by speeding up the gases by directing them in a spiral flow through a chamber and separating the bigger particles by centrifugal force (cyclone method). The particles collected in the knockout box or the cyclone chamber are directed to a mineral filler silo, from where they could be reused for producing HMA.

In the secondary system of dust collection, either a baghouse (Figure 16.9) or a wet scrubber method is employed. In the baghouse method, bags stacked in vertical rows capture the particles as exhaust gases flow through them. The bags need to be cleaned periodically (generally by flexing, shaking, and blowing air pulses), and the recovered particles could be reused.

16.2.9 Asphalt Storage Tanks

Asphalt tanks (Figure 16.10) maintain asphalts in a plant at a required temperature, such that they can be supplied steadily to the pugmill or the drum at the desirable viscosity. Coiled pipes carrying hot oils around the periphery of the tanks maintain the high temperature. Delivery pipes are insulated to prevent loss of heat. It is important to note the grade and level of the binder before pumping new asphalt from a tanker to a storage tank.

FIGURE 16.9 Baghouse. (Courtesy of Matthew Teto, All States Asphalt, Sunderland, MA.)

FIGURE 16.10 Asphalt storage tanks. (Courtesy of Matthew Teto, All States Asphalt, Sunderland, MA.)

16.3 EQUIPMENT USED FOR TRANSPORTATION, LAYDOWN, AND COMPACTION

16.3.1 TRUCKS

For long-distance hauling, trucks should be insulated. A canvas should be used to hold heat, keep light rain showers off the HMA, and keep away dust. Truck beds should be clean, and trucks should be loaded properly to avoid segregation. Typically, trucks should be loaded in the front first, then in the back, and finally in the middle. Sometimes a material transfer vehicle is used to receive the mix from the truck, store it, remix it with augers, and then transfer it to the paver. This helps in maintaining the uniformity of the mix and also in maintaining constant paving operations. Note that prior to the placement of an HMA layer, a tack coat of emulsion (at the rate of 0.01–0.2 gal/square yard, depending on the type of surface) is generally applied on the existing surface to help bond it with the new layer (Figure 16.11). The surface must be cleaned prior to the application of the tack coat. Figure 16.12 shows a view of an HMA paving site with the different equipment.

FIGURE 16.11 Application of tack coat. (Courtesy of Carolina Carbo, Road Recycling Council.)

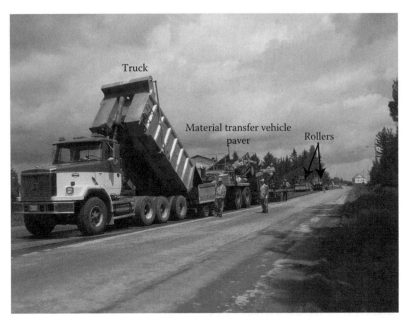

FIGURE 16.12   Different equipment used in hot mix asphalt paving. (Courtesy of Carolina Carbo, Road Recycling Council.)

16.3.2 PAVERS

The HMA is placed and compacted to a certain degree with the paver (Figure 16.13). The paver is wheel, rubber, or steel-track mounted and consists of a hopper to receive the material from the truck or a material transfer vehicle, conveyors to send the material at the back, augers (with available extensions) to distribute the mix transversely across the width of the pavement, and a screed (often heated and capable of vibrating, and with extensions), which ensures a specific depth of the material and preliminary compaction. Paving widths with available equipment can range from 8 to 48 ft. Grade control is usually done with a stringline/mobile ski, sonic/laser, as well as GPS system. Recent developments in paver technology include the use of variable speed engine to save fuel and reduce noise whenever possible, specially designed pipes to remove heat and exhaust from the crew, better turning capability, design of paver components to reduce the chance of segregation of the mix, and redesign of the paver to improve operator visibility.

FIGURE 16.13   Paver. (Courtesy of Ed Kearney.)

(a) (b)

FIGURE 16.14 (a) Steel-wheel and (b) rubber-tired roller. (Courtesy of Ed Kearney and Mike Marshall, Wirtgen GmbH, Windhagen, Germany.)

16.3.3 ROLLERS

Rollers are used for breakdown and intermediate rolling, compaction, and finish rolling to remove roller marks. Available drum widths (rolling widths) range from 3 to 7 ft, approximately. Generally, a vibratory or static steel-wheel roller, with the first drum as the driver, is used for breakdown. The contact pressure that ensures compaction can be altered by changing the ballast load or dynamic force (in the case of a vibratory roller) or the tire pressure (for a rubber-tired roller). Rollers (Figure 16.14) can be rubber tired (these rollers are usually 15–20 ton in weight, and their tire pressure is generally around 90 psi, with 4500 lb per tire), a vibratory steel-wheel roller (usually 10 ton in weight, with high-frequency and low-amplitude vibrations), and a static steel-wheel roller (10–12 ton). Recent developments in rollers include the provision for oscillation (for compaction of thin layers and sensitive areas such as bridge decks) or vibration in both drums and sensors for continuous monitoring of compaction of the mat. Another type of roller is one that consists of three wheels of the same diameter and with equal compressive force per linear inch—all of which are powered. Such rollers are suitable for compaction of thin lifts and city pavements built of fragile structures where vibrations are not appropriate. Such rollers reduce shoving during compaction and also provide the option of increasing the forces using water ballast. Finally, short radius high vibration with smaller drums (e.g., 35.4 in., Sakai CR 270, http://www.sakaiamerica.com/rollers_cr270_series.dws) are available for compaction around utilities as well as for compaction of driveways, bike paths and road maintenance areas.

16.4 IMPORTANT FACTORS

Several factors influence the asphalt pavement construction process, and hence their proper consideration is absolutely necessary to ensure good construction. The factors, with discussion, are listed as follows:

1. Meet specifications for aggregates and binders.
2. Control quality with properly trained and equipped personnel.
3. Maintain the flow of materials to the paver at a speed so as to limit the time the paver is kept waiting.
4. Ensure the smooth transfer of materials from the truck to the paver, while the paver is moving.

5. All the mechanisms, such as the conveyor, hopper gate, and auger, in the paver should be working properly so as to maintain a good and consistent amount of HMA at the correct temperature in front of the screed.
6. Ensure paving with a minimum number of transverse joints.
7. The reference point for screed operation should not be readjusted too much—avoid constant/frequent readjustments of the paver's automatic controls.
8. Use proper guidelines to select the correct rolling techniques—the types and number of rollers, number of passes, and rolling pattern.
9. Make sure that the equipment, such as the paver and rollers, is in good operating condition.
10. During laydown or rolling, if a problem is identified as a material- or mix design–related problem, then it must be reported to the laboratory/plant as quickly as possible to ensure readjustments/corrective actions.

The different factors, along with their associated problems, are listed in Table 16.1.

16.5 SPECIFICATIONS

Specifications include description of construction materials, methods, and/or end product. There are primarily three types of specifications: performance-based or end product–based method specification and proprietary specification.

There are advantages and disadvantages of different kinds of specifications. For example, a performance- or end result–based specification could be shorter, but relatively difficult to prepare, and while it is easy to write a method specification, the results may not correlate well with performance, and while writing a proprietary specification is relatively easy, it does not allow the use of equal or better materials, and does not encourage innovations and competition.

An example of method-based specification, for compaction of an HMA pavement, is as follows:
Use two passes with a vibratory roller weighing at least 10 ton. Vibrations per minute must exceed 1000, and the roller shall not travel faster than 4 miles/h.

For the same purpose, a performance-based specification could be written as follows: compact the HMA to at least 98% of laboratory density. Most specifications combine aspects of both method- and performance-based specification.

Contracts for construction of asphalt pavements can be made through the low-bid system, best-bid system, or negotiated contract. In the low-bid system, generally the lowest bidder is selected. Although this system provides a fair and rational method of decision, it could lead to cutting corners. In the best-bid system, more emphasis is placed on the performance and experience of the contractor. In a negotiated contract, the owner negotiates the work with a specific contractor.

In the construction process, standards and specifications from ASTM and AASHTO are also followed. Most agencies would prepare a project specification from a list of guideline specifications from ASTM or AASHTO. Specifications and standards are particularly important with respect to sampling and testing for quality control and quality assurance.

A good specification must include a description of the following:

1. Lot size is the amount of materials that needs to be evaluated for acceptance or rejection. For example, 1 day of production of HMA.
2. Test properties indicate the properties that need to be tested. For example, density, aggregate gradation, asphalt content, and smoothness for an asphalt pavement.
3. Number of tests for each lot.
4. Point of sampling for tests. For example, aggregates from stockpiles and belt.
5. Method of sampling. For example, random sampling or representative sampling.

TABLE 16.1
Steps, Potential Problems, and Solutions

Construction Step	Potential Problems/Suggested Solutions
Materials selection	Bleeding: check type of filler, grade of asphalt binder
	Blisters: ensure steel slag, if used, is cured; avoid aggregates with inorganic nitrogen or phosphorus, and quicklime, roots, or trapped volatiles from cutback asphalts
	Brown streaks: avoid highly absorptive aggregates; check quality of asphalt
	Checking under roller, tender mix: check quality of asphalt and its temperature susceptibility; check if source of asphalt has remained the same
	Tearing of mat: check maximum aggregate size
Mix design	Bleeding, poor-quality longitudinal joints: check amount of filler, VMA and asphalt content, voids, and laboratory compaction procedures
	Tender mix and checking under roller: check filler and asphalt content, asphalt binder temperature–viscosity characteristics, and residual moisture
	Segregation: check gradation
	Checking under roller, or tearing of mat: check asphalt content, VMA, asphalt content variation during production, and filler-metering system
	Transverse cracks: check asphalt content, filler content and asphalt binder temperature–viscosity characteristics, and residual moisture
	Poor-quality longitudinal joint: check filler and asphalt content, asphalt binder temperature–viscosity characteristics, residual moisture, and VMA
	Lean, brown, dull appearance of surface: check filler content
	Shoving of mat ahead of roller, or roller marks: check amounts of smooth, rounded aggregates, mix design temperature, asphalt content, and VMA
	Poor compaction: check for asphalt content and VMA
Storage	Bleeding: check stockpiling and drying techniques
	Segregation: check stockpiling
	Blisters: check drying techniques
	Shoving of mat ahead of roller, or lean, brown, dull appearance of surface: check for moisture in stockpiles
Plant production	Bleeding, or lean, brown, dull appearance of surface: check fuel/combustion/burner/nozzles
	Segregation: check storage bins
	Checking under roller: check mix temperature
	Poor-quality longitudinal/transverse joint: check filler-metering system and mix temperature
	Shoving of mat ahead of roller; lean, brown, dull appearance of surface; tearing of mat; or uneven thickness and quality: check mix temperature, asphalt cement type, and content; check filler-metering system, drying of aggregates, dwell time in dryer drum, and for proper quality control
	Poor compaction, or wavy surface: check mix temperature; asphalt viscosity–temperature characteristics; asphalt properties, if overheated; filler-metering system; and for proper quality control at the plant
Tack/prime coating	Bleeding, transverse cracks, or shoving of mat ahead of roller: check type and application rates, and time available for curing of emulsion and/or penetration of prime coat into granular layers
Lift thickness	Checking under roller, shoving of mat ahead of roller, tearing of mat, or poor compaction: check proper lift thickness and number of lifts

(continued)

TABLE 16.1 (continued)
Steps, Potential Problems, and Solutions

Construction Step	Potential Problems/Suggested Solutions
Existing surface and environmental conditions	Bleeding: check for moisture in underlying layers; check if porous layer with water
	Blisters: check drainage
	Transverse cracks, checking under roller, shoving of mat ahead of roller, tearing of mat, or uneven thickness and quality: check for grading, drainage, density and weak areas, air temperature, and dust on existing surface
	Depression, or shoving of mat ahead of roller: check base preparation—density, drainage, and weak areas
	Poor-quality longitudinal joint: check air temperature
	Poor compaction: check preparation, ruts (if any), crack filling (if any), drainage and density, and existing air temperature
Transportation	Bleeding: check release agents on truck beds
	Poor-quality longitudinal/transverse joint; lean, brown, dull appearance of surface; tearing of mat; uneven thickness and quality; or poor compaction: check insulation in trucks
	Wavy surface: check for brakes in trucks
Transfer	Segregation, or poor-quality longitudinal/transverse joint: check for temperature variation across mat—consider using materials transfer vehicle
Paving	Bleeding, or segregation: check paver speed, and different parts such as auger and screed and mix in hopper
	Checking under roller, depression, or poor-quality longitudinal joint: check screed
	Poor-quality longitudinal/transverse joint: check screed, hopper and feed, and sensor, and check raking, bumping, and luting
	Tearing of mat: check paver speed and screed, feeder gate, kicker screws, and end plates
	Uneven thickness and quality, or wavy surface: check hopper gates, conveyor and spreading screw, sensor and auger, and screed pull point
	Poor compaction, or wavy surface: check for proper type of screed and hydraulics of screed operation
Rolling	Bleeding: check rolling pattern and number of passes
	Tender mix: change type of roller
	Checking under roller: check rolling pattern, speed, and number of rollers
	Depression, shoving of mat ahead of roller, roller marks, tearing of mat: check speed, reverse, turns, rolling pattern, parking of roller, tire pressure, roller weight, time of starting rolling, amplitude and frequency of vibration for vibratory roller, and pattern of rolling in superelevation
	Poor-quality longitudinal/transverse joint: check number and speed of rollers, and rolling direction
	Poor compaction: check for number, type and speed of rollers, frequency of vibrations, and pressure in tires of rubber-tired roller
Compacted surface	Brown streak: check for spilled gas or oil

Source: National Asphalt Pavement Association (NAPA), *Constructing Quality Hot Mix Asphalt Pavements: A Troubleshooting Guide and Construction Reference*, QIP 112, NAPA, Lanham, MD, 2003.

6. Number of tests to be reported. For example, an average of five tests should be reported.
7. Test method to be used for each property. For example, the density of the HMA layer should be measured with a nuclear density gage test.
8. The target value for each test property. For example, the target density is 95% of the laboratory density.
9. Tolerance of test results around the target value.
10. Actions to be taken if specification requirements are not met.

A good specification should be free of vague and confusing words (such as "aggregate should be clean or to the satisfaction of the engineer"). Some other examples of words that should be avoided are as follows:

1. "... promptly and uniformly compacted ..."—should state time of compaction, or that before the temperature drops to a certain value compaction should be completed.
2. "... heavy equipment or rollers ..."—should state minimum weight of the roller.

An example of a specification for density of the HMA, including all of these features, can be written as follows.

Example 16.1

The lot size of the construction for density evaluation is 1 day of production. The property to be tested is density. The lot will be divided into four equal sublots, and one random sample should be taken from each sublot. The in-place density should be measured. A 4 in. diameter core will be taken at each location. Use one test for each sample. Use ASTM D-2000 for determination of density. The average of the four core samples shall be 98%–100% of laboratory density. When the density is below 98%, the contractor will be paid at the percentage shown in Table 16.2.

16.5.1 Variability of Materials

Test results are always variable because of errors of sampling and testing as well as material variability. Material variability will always be there, but the errors due to sampling and testing can be minimized by following specified methods of sampling, and using competent and qualified testing personnel, and appropriate testing equipment.

The variability of a number of test properties, such as density, can be approximated by a normal distribution. Using the different parameters, as listed in the following, allows the determination of variability of materials and making decisions regarding acceptance or rejection in quality control and assurance.

TABLE 16.2
Density Pay Factors

% of Density	% of Pay
Above 100	100
97.9–100	99
96.0–97.8	97
Below 96.0	Choice of paid 60%, or remove and replace

The relevant parameters are as follows:

$$\text{Variance: } \quad s^2 = \frac{\sum_{i=1}^{n} (x_i - \bar{x})^2}{n-1}$$

$$\text{Standard deviation: } \quad s = \sqrt{\frac{\sum_{i=1}^{n} (x_i - \bar{x})^2}{n-1}}$$

$$\text{Coefficient of variation: } \quad s = \frac{s}{x} * 100\%$$

$$\text{Standard error of mean: } \quad \bar{s} = \frac{s}{\sqrt{n}}$$

where
  n is the number of samples averaged to give each test result
  $\bar{x}$ is the average
  x_i is the individual test results

Typical standard deviation values have been determined as shown in Table 16.3.

In many specifications, the evaluation of a job for acceptance or rejection is based on the *percent within limits* (PWL) of results for each lot of the pavement. PWL (or *percent conforming*) is defined as the percentage of the lot falling above the lower specification limit (LSL), beneath the upper specification limit (USL), or between the USL and LSL. Although tests can be applied to verify it, in general, the population of most test results is assumed to be normally distributed, and the use of this procedure ensures the consideration of both average and variability of the test results for evaluation of the "quality" of the product. The PWL concept is based on the use of the area under a standard normal distribution.

A Z value, where $Z = y - \mu/\sigma$, can be used to determine the percentage of population that is within or outside a certain limit, using Table 16.4. Note that Z is the statistic that is to be used with the table, Y is the point within which the area is determined from the table, and μ and σ are the mean and standard deviation of the population, respectively. For example, consider the following test results for an in-place density measurement (as a percentage of theoretical maximum density).

Number of test results = 10; mean value = 94.0, standard deviation = 1.15. What percentage of the test results are expected to be below 92%?

$$Z = \frac{92-94}{1.15} = -1.74$$

From Table 16.4, (1−0.5−0.4591) * 100 = 4.1% of the test results are expected to be below 92%.

TABLE 16.3

Examples of Typical Standard Deviations

Property	Standard Deviation	Coefficient of Variation (%)
Subbase density	3.5	3.7
Asphalt content	0.2	0.03
Density of asphalt mix	1.02	1.0
Base course density	2.5	2.5

TABLE 16.4
Areas under the Standard Normal Distribution

Z	0.00	0.01	0.02	0.03	0.04	0.05	0.06	0.07	0.08	0.09
0.0	0.0000	0.0040	0.0080	0.0120	0.0160	0.0199	0.0239	0.0279	0.0319	0.0359
0.1	0.0398	0.0438	0.0478	0.0517	0.0557	0.0596	0.0636	0.0675	0.0714	0.0753
0.2	0.0793	0.0832	0.0871	0.0910	0.0948	0.0987	0.1026	0.1064	0.1103	0.1141
0.3	0.1179	0.1217	0.1255	0.1293	0.1331	0.1368	0.1406	0.1443	0.1480	0.1517
0.4	0.1554	0.1591	0.1628	0.1664	0.1700	0.1736	0.1772	0.1808	0.1844	0.1879
0.5	0.1915	0.1950	0.1985	0.2019	0.2054	0.2088	0.2123	0.2157	0.2190	0.2224
0.6	0.2257	0.2291	0.2324	0.2357	0.2389	0.2422	0.2454	0.2486	0.2517	0.2549
0.7	0.2580	0.2611	0.2642	0.2673	0.2704	0.2734	0.2764	0.2794	0.2823	0.2852
0.8	0.2881	0.2910	0.2939	0.2967	0.2995	0.3023	0.3051	0.3078	0.3106	0.3183
0.9	0.3159	0.3186	0.3212	0.3238	0.3264	0.3289	0.3315	0.3340	0.3365	0.3389
1.0	0.3413	0.3438	0.3461	0.3485	0.3508	0.3531	0.3554	0.3577	0.3599	0.3621
1.1	0.3643	0.3665	0.3686	0.3708	0.3729	0.3749	0.3770	0.3790	0.3810	0.3830
1.2	0.3849	0.3869	0.3888	0.3907	0.3925	0.3944	0.3962	0.3980	0.3997	0.4015
1.3	0.4032	0.4049	0.4066	0.4082	0.4099	0.4115	0.4131	0.4147	0.4162	0.4177
1.4	0.4192	0.4207	0.4222	0.4236	0.4251	0.4265	0.4279	0.4292	0.4306	0.4319
1.5	0.4332	0.4345	0.4357	0.4370	0.4382	0.4394	0.4406	0.4418	0.4429	0.4441
1.6	0.4452	0.4463	0.4474	0.4484	0.4495	0.4505	0.4515	0.4525	0.4535	0.4545
1.7	0.4554	0.4564	0.4573	0.4582	0.4591	0.4599	0.4608	0.4616	0.4625	0.4633
1.8	0.4641	0.4649	0.4656	0.4664	0.4671	0.4678	0.4686	0.4693	0.4699	0.4706
1.9	0.4713	0.4719	0.4726	0.4732	0.4738	0.4744	0.4750	0.4756	0.4761	0.4767
2.0	0.4772	0.4778	0.4783	0.4788	0.4793	0.4798	0.4803	0.4808	0.4812	0.4817
2.1	0.4821	0.4826	0.4830	0.4834	0.4838	0.4842	0.4846	0.4850	0.4854	0.4857
2.2	0.4861	0.4864	0.4868	0.4871	0.4875	0.4878	0.4881	0.4884	0.4887	0.4890
2.3	0.4893	0.4896	0.4898	0.4901	0.4904	0.4906	0.4909	0.4911	0.4913	0.4916
2.4	0.4918	0.4920	0.4922	0.4925	0.4927	0.4929	0.4931	0.4932	0.4934	0.4936
2.5	0.4938	0.4940	0.4941	0.4943	0.4945	0.4946	0.4948	0.4949	0.4951	0.4952
2.6	0.4953	0.4955	0.4956	0.4957	0.4959	0.4960	0.4961	0.4962	0.4963	0.4964
2.7	0.4965	0.4966	0.4967	0.4968	0.4969	0.4970	0.4971	0.4972	0.4973	0.4974
2.8	0.4974	0.4975	0.4976	0.4977	0.4977	0.4978	0.4979	0.4979	0.4980	0.4981
2.9	0.4981	0.4982	0.4982	0.4983	0.4984	0.4984	0.4985	0.4985	0.4986	0.4986
3.0	0.4987	0.4987	0.4987	0.4988	0.4988	0.4989	0.4989	0.4989	0.4990	0.4980
3.1	0.4990	0.4991	0.4991	0.4991	0.4992	0.4992	0.4992	0.4992	0.4993	0.4993
3.2	0.4993	0.4993	0.4994	0.4994	0.4994	0.4994	0.4994	0.4995	0.4995	0.4995
3.3	0.4995	0.4995	0.4995	0.4996	0.4996	0.4996	0.4996	0.4996	0.4996	0.4997
3.4	0.4997	0.4997	0.4997	0.4997	0.4997	0.4997	0.4997	0.4997	0.4997	0.4998

Source: Burati, J.L. et al., *Optimal Procedures for Quality Assurance Specifications*, FHWA-RD-02-095, Federal Highway Administration, Washington, DC, 2003.

Now, in the case of the PWL procedure, the quality index value, Q, is used in instead of Z, and the sample mean, $\bar{x}$, and sample standard deviation, s, are used instead of the population mean and standard deviations. Also, instead of the "Y" value used for the calculation of Z, the USL and LSL values are used, as follows:

$$Q_L = \frac{\bar{x} - LSL}{s}$$

$$Q_U = \frac{USL - \bar{x}}{s}$$

where

Q_L is the quality index for lower specification limit, used when there is a one-sided lower specification limit

Q_U is the quality index for upper specification limit, used when there is a one-sided upper specification limit

LSL is the lower specification limit

USL is the upper specification limit

$\bar{x}$ is the sample mean

s is the sample standard deviation

The use of the quality index is illustrated through the following example. Consider the in-place density data as follows:

n = number of samples = 10

$\bar{x}$ = sample mean = 97.5% of density of samples compacted by the Marshall procedure in the laboratory

S = sample standard deviation = 1.25

The specification states that in order to get 100% paid, 90% of the test results should be above 96.3% of density of the samples compacted by the Marshall procedure in the laboratory. That is, a 90% PWL is required for 100% pay. In this case, can the contractor expect 100% pay?

$$Q = \frac{97.5 - 96.3}{1.25} = 0.96$$

Referring to the quality index table (Table 16.5), for n = 10, a Q value of 0.96 indicates a PWL of 83%, which is less than 90%. Hence, the contractor cannot expect to get paid 100% of the pay for the lot being tested.

Note that when two-sided specifications are used, the PWL is calculated as follows:

$$PWL_T = PWL_U + PWL_L - 100$$

where

PWL_U is the percentage below the upper specification limit (based on Q_U)

PWL_L is the percentage above the upper specification limit (based on Q_L)

PWL_T is the percentage within the upper and lower specification limits

Generally, pay adjustments, on the basis of pay factors, are made on the basis of the PWL of a lot. It is desirable that the pay factors are tied to the quality and resulting performance of the pavement on a rational basis. Generally, payments are reduced for substandard work and increased (above 100%, as a bonus) for superior-quality work. Both stepped and continuous pay factors are used, as shown in an example in Table 16.6.

TABLE 16.5
Example of Quality Index Value Table for Estimating PWL

PWL	n = 3	n = 4	n = 5	n = 6	n = 7	n = 8	n = 9	n = 10–11
100	1.16	1.50	1.79	2.03	2.23	2.39	2.53	2.65
99	—	1.47	1.67	1.80	1.89	1.95	2.00	2.04
98	1.15	1.44	1.60	1.70	1.76	1.81	1.84	1.86
97	—	1.41	1.54	1.62	1.67	1.70	1.72	1.74
96	1.14	1.38	1.49	1.55	1.59	1.61	1.63	1.65
95	—	1.35	1.44	1.49	1.52	1.54	1.55	1.56
94	1.13	1.32	1.39	1.43	1.46	1.47	1.48	1.49
93	—	1.29	1.35	1.38	1.40	1.41	1.42	1.43
92	1.12	1.26	1.31	1.33	1.35	1.36	1.36	1.37
91	1.11	1.23	1.27	1.29	1.30	1.30	1.31	1.31
90	1.10	1.20	1.23	1.24	1.25	1.25	1.26	1.26
89	1.09	1.17	1.19	1.20	1.20	1.21	1.21	1.21
88	1.07	1.14	1.15	1.16	1.16	1.16	1.16	1.17
87	1.06	1.11	1.12	1.12	1.12	1.12	1.12	1.12
86	1.04	1.08	1.08	1.08	1.08	1.08	1.08	1.08
85	1.03	1.05	1.05	1.04	1.04	1.04	1.04	1.04
84	1.01	1.02	1.01	1.01	1.00	1.00	1.00	1.00
83	1.00	0.99	0.98	0.97	0.97	0.96	0.96	0.96
82	0.97	0.96	0.95	0.94	0.93	0.93	0.93	0.92
81	0.96	0.93	0.91	0.90	0.90	0.89	0.89	0.89
80	0.93	0.90	0.88	0.87	0.86	0.86	0.86	0.85
79	0.91	0.87	0.85	0.84	0.83	0.82	0.82	0.82
78	0.89	0.84	0.82	0.80	0.80	0.79	0.79	0.79
77	0.87	0.81	0.78	0.77	0.76	0.76	0.76	0.75
76	0.84	0.78	0.75	0.74	0.73	0.73	0.72	0.72
75	0.82	0.75	0.72	0.71	0.70	0.70	0.69	0.69
74	0.79	0.72	0.69	0.68	0.67	0.66	0.66	0.66
73	0.76	0.69	0.66	0.65	0.64	0.63	0.63	0.63
72	0.74	0.66	0.63	0.62	0.61	0.60	0.60	0.60
71	0.71	0.63	0.60	0.59	0.58	0.57	0.57	0.57
70	0.68	0.60	0.57	0.56	0.55	0.55	0.54	0.54
69	0.65	0.57	0.54	0.53	0.52	0.52	0.51	0.51
68	0.62	0.54	0.51	0.50	0.49	0.49	0.48	0.48
67	0.59	0.51	0.47	0.47	0.46	0.46	0.46	0.45
66	0.56	0.48	0.45	0.44	0.44	0.43	0.43	0.43
65	0.52	0.45	0.43	0.41	0.41	0.40	0.40	0.40
64	0.49	0.42	0.40	0.39	0.38	0.38	0.37	0.37
63	0.46	0.39	0.37	0.36	0.35	0.35	0.35	0.34
62	0.43	0.36	0.34	0.33	0.32	0.32	0.32	0.32
61	0.39	0.33	0.31	0.30	0.30	0.29	0.29	0.29
60	0.36	0.30	0.28	0.27	0.27	0.27	0.26	0.26
59	0.32	0.27	0.25	0.25	0.24	0.24	0.24	0.24
58	0.29	0.24	0.23	0.22	0.21	0.21	0.21	0.21
57	0.25	0.21	0.20	0.19	0.19	0.19	0.18	0.18
56	0.22	0.18	0.17	0.16	0.16	0.16	0.16	0.16
55	0.18	0.15	0.14	0.14	0.13	0.13	0.13	0.13
54	0.14	0.12	0.11	0.11	0.11	0.11	0.10	0.10

(continued)

TABLE 16.5 (continued)
Example of Quality Index Value Table for Estimating PWL

PWL	n = 3	n = 4	n = 5	n = 6	n = 7	n = 8	n = 9	n = 10–11
53	0.11	0.09	0.08	0.08	0.08	0.08	0.08	0.08
52	0.07	0.06	0.06	0.05	0.05	0.05	0.05	0.05
51	0.04	0.03	0.03	0.03	0.03	0.03	0.03	0.03
50	0.00	0.00	0.00	0.00	0.00	0.00	0.00	0.00

Source: Burati, J.L. et al., *Optimal Procedures for Quality Assurance Specifications*, FHWA-RD-02-095, Federal Highway Administration, Washington, DC, 2003.

TABLE 16.6
Example of Pay Factors

Estimated PWL	Payment Factor (%)
95.0–100.0	102
85.0–94.9	100
50.0–84.9	90
0.0–49.9	70

Source: Burati, J.L. et al., *Optimal Procedures for Quality Assurance Specifications*, FHWA-RD-02-095, Federal Highway Administration, Washington, DC, 2003.

Continuous payment factors:

$$PF = 55 + 0.5 * PWL$$

where
   PF is the payment factor as a percentage of contract price
   PWL is the estimated percentage within limits

16.5.2 Use of Quality Control Charts

During construction, it is important to summarize and analyze the test results. A quality control chart is a plot of test results and numbers. The control chart can be used to summarize the data, identify trends, show specification limits and percentages within limits, and evaluate changes in any process.

Different data sets can be shown in quality control charts:

1. *Plot of individual values*: The advantage is that it shows all the data, and the disadvantage is that it is difficult to identify trends.
2. *Plot of running average*: It shows trends but does not show actual data.
3. *Plot of running variability*: It shows the variability of the mix.

Typical test properties that are plotted include the density and water content for the subgrade; the density, water content, and gradation for the subbase and base; and the gradation, asphalt content, in-place density, laboratory density, and voids for the HMA.

An example of a quality control chart for individual data values is shown in Figure 16.15.

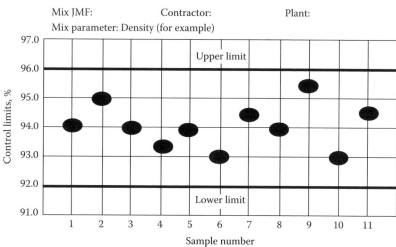

FIGURE 16.15 Example of a control chart.

16.6 PREPARATION OF SUBGRADE AND CONSTRUCTION OF BASE AND SUBBASE LAYERS

The subgrade can consist of relatively high-quality materials such as sand and gravel or poor-quality materials such as silt and clay. The pavement structure is designed to protect the subgrade from excessive deformation. However, the subgrade may need improvement before pavement construction begins. The improvement is done with compaction and/or stabilization.

Compaction can be measured by the nuclear density test, using a sand cone or a water balloon, or by testing samples recovered with cylinders. Density of the subgrade is specified as a percentage of Proctor or modified Proctor density or as a percentage of density of a control strip. The laboratory test data on optimum moisture content and maximum dry density are used.

Low density in subgrade can be due to the following factors:

1. Inadequate weight of rollers
2. Unsatisfactory roller patterns
3. Improper moisture content
4. Incorrect laboratory density due to wrong compactive effort and/or use of nonrepresentative samples
5. Poor testing, because of uncalibrated equipment or untrained personnel

The following tests could be run for characterizing the subgrade for performance.

1. California bearing ratio (CBR)
2. Plate bearing test
3. Unconfined compression test
4. Resilient modulus

Subbase and base materials generally consist of higher quality materials, such as clean, well-graded aggregates with fractured faces. Mixing is generally done in a portable pugmill, with close control on water content. Subbase and base materials are placed using a grader, spreader, or paver. Compaction of the subbase and base can be accomplished using a static and/or vibratory steel-wheel roller as well as rubber-tired rollers. Potential problems with the base and subbase include low density, segregation, grade control, and gradation of aggregates.

16.7 QUALITY CONTROL AND QUALITY ASSURANCE

An important part of the production process is the testing of plant-produced material as part of the quality control (QC) methods. QC sampling and testing are done on a regular basis to continuously evaluate the produced material, by the material producer. The results of the QC process can be fed back to the plant to correct or modify any setting, as required. Quality assurance (QA) tests are generally conducted by the owner or a consultant agency (or sometimes even by the producer) at longer time intervals to make sure that the produced material meets the specification.

The important tests and methods are presented in the following:

1. *Standard test method for random sampling (ASTM 3665)*: This standard outlines the process that can be used to eliminate bias during sampling. Sampling can be done either by random procedure or stratified random procedure. In the latter process, the lot is divided into sublots and then the random method is used to pick samples from each sublot. Tables for picking random numbers, by selecting any two numbers and rows and columns, are provided in ASTM 3665. Sampling should be done from different points, for example, aggregates from a belt or truck or from in-place paving material.

2. *Checking temperature of asphalt, HMA, and air*: HMA and asphalt temperatures are measured using dial- or digital-type thermometers (two thermometers) made with metal stems (which are inserted into the material) and armored glass. AASHTO T-245 specifies the requirements of the thermometers, which must be calibrated before use. Infrared thermometers, when calibrated against conventional thermometers, could be used. HMA temperatures are generally taken for each sample as well as at the start of the day, by pushing the stem to its full depth into the material, after waiting for 3 min or when there is no appreciable temperature difference between two readings at 1 min intervals. The same type of thermometer could be used to check the temperature of the asphalt binder from sampling valves from the asphalt tank. For measuring air temperature, the thermometer should be kept out of direct sunlight.

3. *Preparing Marshall hammer compacted specimens*: AASHTO T-245 specifies the procedure for compacting specimens with a Marshall hammer and testing the specimens for stability and flow, after running bulk-specific gravity (AASHTO T-166). The bulk-specific gravity could be used along with the theoretical maximum density to determine the percentage of density and air voids.

4. *Preparing a specimen with a Superpave gyratory compactor*: AASHTO 312 specifies the method for using a Superpave gyratory compactor for the compaction of 150 mm diameter HMA specimens for subsequent determination of bulk-specific gravity.

5. *Determination of bulk-specific gravity of a compacted HMA specimen*: AASHTO T-166 specifies the procedure for the determination of bulk-specific gravity of a laboratory-compacted HMA specimen and in-place cores from pavement. Note that cores should be dried at 110°C to constant mass before running this test.

6. *Maximum specific gravity of HMA*: AASHTO T-209 specifies the method for determination of maximum specific gravity of HMA specimens, by determining the ratio of the weight of a void-less than the mass of HMA and the mass of an equal volume of water.

7. *Determination of thickness of HMA specimen*: ASTM D-3549 specifies the method of determination of thickness of a compacted HMA specimen. The devices that can be used are metal tape, a rule, a measurement jig, and a set of calipers. Required accuracy for measurements is specified in the standard. Measurements are made between the horizontal planes of the upper and lower surfaces of the specimens or between lines of demarcation (as in a thickness of a layer inside a full-depth core). The measurements are taken at four

different points around the periphery, and the average of the four numbers is reported as the thickness of the specimen.

8. *Determination of asphalt content of HMA by ignition method (AASHTO T-308)*: In this test, an HMA sample (generally ranging in weight from 1.2 to 4 kg, for a range of nominal maximum aggregate size [NMAS] of 4.75–37.5 mm) is weighed, it is placed inside an oven maintained at 650°C, the asphalt binder is burnt off, the remaining aggregates are weighed, and the asphalt content is determined on the basis of the two weights. Aggregate calibration information, which accounts for loss of aggregates by burning, if any, can be input in the ignition oven computer. The ignition oven reports the weight of the sample to the nearest 0.1 g, taking into consideration the temperature correction, and prints out the asphalt content (using an onboard computer program) from the pre- and postignition weights, taking the aggregate calibration factor into consideration. Important items are provision of methods for reducing furnace emissions, availability of a self-locking oven door (which will not open until the end of the test), an audible alarm and lights for indication of completion of the test, as well as safety equipment such as a face shield and thermal gloves for the operator. The asphalt-free aggregates can be used for running a sieve analysis and checking gradation, although care must be taken to consider the breakdown of some aggregates and the creation of some fines (if any) during the ignition test of some types of aggregates.

9. *Determination of asphalt content by the nuclear method (AASHTO T-287)*: In this test the amount of asphalt is indicated by the amount of hydrogen in a mix as measured with a neutron source in the nuclear gage. The gage needs to be calibrated with both aggregate and mix samples before using. A pan filled with consolidated HMA is placed inside the nuclear gage, and the asphalt content is read off the equipment.

10. *Quantitative extraction of asphalt from HMA (AASHTO T-164)*: In this test, a solvent is used to dissolve the asphalt binder out from an HMA sample, and the aggregates are recovered. The weight of the mix sample before and after the washing out of the asphalt is used to calculate the asphalt content, and the recovered aggregates can be used for sieve analysis. A special bowl is used to soak a specified amount of mix (ranging in weight from 0.5 to 4 kg, for an NMAS of 4.75 to 37.5 mm) in a solvent. A filter ring is placed on the base of the bowl, and the lid is clamped down. A centrifuge is used to spin the sample in the bowl at a speed of 3600 rpm (slowly increasing the speed to the maximum). The extraction fluid is run off the bowl, and the process is continued with fresh charges of solvent until all of the asphalt has been washed away. At the end of this step, the fines adhering to the filter paper can be scraped out on the remaining aggregate sample. The filter paper and the sample are then dried at 110°C to constant mass.

11. Resistance of a compacted HMA specimen to moisture-induced damage (AASHTO T283) as discussed in Chapter 14.

16.8 CONSTRUCTION OF LONGITUDINAL JOINTS

A longitudinal construction joint occurs when a lane of HMA is constructed adjacent to previously placed HMA. Longitudinal joints are inevitable in both highway and airfield pavements, unless paving is done in the echelon formation (which is generally not the case). Damaged longitudinal joints are of very serious concern in airfield pavements. Loose materials from such areas can cause foreign object damage (FOD) to aircrafts, leading to loss of life and equipment. Potential sharp edges along open longitudinal joints can also endanger aircraft. In addition, such joints can lead to ingress of moisture and undesirable materials and lead to premature failures in the subsurface and ultimately the entire pavement, leading to a cycle of costly and time-consuming repairs.

For these reasons, engineers, consultants, and contractors have continuously tried to develop methods for constructing better performing longitudinal joints in pavements. Such methods include overlapping and luting operations; different types of rolling patterns for compacting the joint; and the use of special joint construction techniques and equipment, such as cutting wheels, restrained edge devices, notched wedge joints, and joint heaters. At the same time, many agencies have started using specifications that are written specifically for the construction of better joints, such as density requirements at joints.

There are several steps in the construction of a good-quality joint. These include paving the first lane in a uniform, unwavering line; compacting the unsupported edge of the first lane (cold lane) properly; controlling the height of the uncompacted HMA in the hot lane; and proper overlapping during the paving operation. Raking or luting at the longitudinal joint can be eliminated if the minimal overlapping is done. An excessive overlap will require the removal of extra material from the cold lane onto the hot lane; otherwise, the aggregate in the mix remaining on the compacted lane will get crushed, resulting in raveling. When that happens, the excessive overlapped material on the cold lane may be "bumped" with a lute onto the hot mat *just* across the joint. The bump should lie just above the natural slope or the wedge at the edge of the cold lane. Sometimes, there is a tendency to broadcast the raked material onto the HMA in the hot lane. This not only is undesirable for obtaining a good longitudinal joint but also affects the surface texture of the mat adversely.

Obtaining adequate compaction at the joint is the final key in obtaining a durable longitudinal joint. Joints with high densities generally show better performance than those with relatively low densities. By paying attention to construction details, it is possible to obtain a joint density within 1.5% of the mainline density.

Using proper specifications is another key requirement for obtaining good-quality joints. Joint density is best measured by obtaining a 6 in. (150 mm) diameter core centered on top of the visible line between the two lanes. It is recommended to make the compaction level at both the joint and mat based on theoretical maximum density (TMD) rather than the bulk-specific gravity of a daily compacted Marshall specimen, which is more variable.

Section 401-5.2 of Engineering Brief EB59A of the FAA, dated May 12, 2006, provides a good example of a specification of constructing longitudinal joints. The acceptance criteria are as follows.

Evaluation for acceptance of each lot of in-place pavement for joint density and mat density shall be based on PWL. The contractor shall target production quality to achieve 90 PWL or higher.

The percentage of material within specification limits (PWL) shall be determined in accordance with procedures specified in Section 110 of the General Provisions. The acceptance limits shall be as follows:

> Mat density,% of TMD   92.8 minimum
> Joint density,% of TMD   90.5 minimum

16.8.1   TECHNIQUES OF CONSTRUCTING GOOD LONGITUDINAL JOINTS

It is preferable to produce hot longitudinal joints by operating two or more pavers in echelon. If echelon paving is not possible, then any of the following good techniques can be used (Figure 16.16).

16.8.1.1   Combination of Notched Wedge Joint, Rubberized Asphalt Tack Coat, and Minimum Joint Density Requirements

Construct a notched wedge longitudinal joint. The unconfined edge of the first paved lane has a vertical notch at the edge generally ranging from ½ in. (13 mm) to 3/4 in. (19 mm) in height depending upon the nominal maximum aggregate size of the HMA mixture. Generally, a vertical notch of about ½ in. (13 mm) height is considered adequate for most surface course mixtures. It is recommended to end the taper with a minimal height such as 3/8 in. (9.5 mm) to avoid dragging of the material. Usually a loaded wheel, which is attached to the paver, is used to compact the taper. Typically, the roller weighs 100–200 lb (45–91 kg) and is approximately 14 in. (356 mm) wide by 12 in. (305 mm) in diameter. There is no need to compact the taper with a conventional steel- or pneumatic-tired roller because it will simply destroy the vertical notch. The overlap layer of the adjacent paving lane is required to

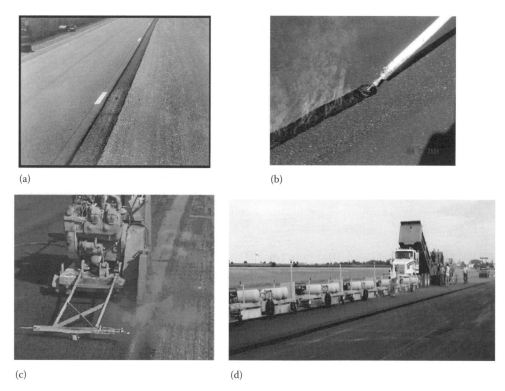

(a)

(b)

(c)

(d)

FIGURE 16.16 Different longitudinal joint construction techniques. (a) Notched wedge joint. (b) Application of rubberized asphalt tack coat at joint. (c) Joint cutting. (d) Infrared heating of joints. (Courtesy of Prithvi S. Kandhal, the U.S. Air Force, Arlington, VA; Tom Allen, Ray-Tech Infrared Corp., Charlestown, NH.)

be placed and compacted within 24 h unless delayed by inclement weather. The vertical notch and taper are tack-coated with rubberized asphalt binder prior to placing the overlap wedge, as described later. The notched wedge joint can be formed by using a homemade sloping steel plate or commercial devices attached to the inside corner of the paver screed extension. The notched wedge joint does not always work well for thinner HMA lifts. Ideally, it gives the best results with a minimum lift thickness of 1½–2 in. (37–51 mm). On the other hand, excessively thick lifts produce a long taper, which may not be desirable. In those cases, the length of the taper is generally restricted to 12 in. (305 mm). The top-course taper shall overlap and slope in the opposite direction of the lower-course taper.

After the first lane (cold lane) is paved with a notched wedge and compacted, a rubberized asphalt tack coat is applied on the face of the unconfined edge of the cold lane. The thickness of the tack coat is about 1/8 in. (3 mm) on the slope of the HMA edge. The rubberized asphalt tack coat need not be applied on the entire taper. It is considered adequate to apply it on the vertical notch and the top 3–4 in. (76–102 mm) wide band of the taper. Application excesses should not exceed more than ½ in. (13 mm) at the top of the joint. The sealant should preferably be applied within 4 h of the time that the adjacent HMA lane is placed. The heat from the HMA in the adjacent lane and the roller pressure causes the sealant to adhere strongly along the joint face, resulting in a strong bond between the two lanes and providing a built-in sealer at the joint. After the rubberized asphalt tack coat is applied, the adjacent lane (hot lane) is placed. The height of the uncompacted HMA should be about 1¼ in. (32 mm) for each 1 in. (25 mm) of the compacted lift thickness in the cold lane. The end gate of the paver should extend over the top surface of the previously placed HMA by a distance of approximately 1–1½ in. (25–38 mm). The most efficient joint compaction method is to roll the longitudinal joint from the hot side, overlapping the cold lane by approximately 6 in. (150 mm). The steel-wheel roller can be operated in vibratory or static mode, preferably the vibratory mode to obtain better compaction.

For the following procedures, the best practices for paving and compacting the first lane, paving the second lane, and overlapping, raking and luting, and compacting the longitudinal joint, as well as minimum joint density and mat density requirements, are similar to those mentioned earlier.

16.8.1.2 Rubberized Asphalt Tack Coat and Minimum Joint Density Requirements

This practice is similar to that in Section 16.8.1.1 except that no notched wedge joint is used. The first lane (cold lane) is paved as usual with the normal, unconfined edge slope. A rubberized asphalt tack coat is applied on the entire face of the unconfined edge of the cold lane using the procedure described in Item 1.

16.8.1.3 Notched Wedge Joint and Minimum Joint Density Requirements

This practice is similar to that in Section 16.8.1.1 except that a conventional tack coat material (which is used on the main line) is applied to the entire face of the notched wedge joint in lieu of rubberized asphalt material.

16.8.1.4 Cutting Wheel and Minimum Joint Density Requirements

The cutting wheel technique involves cutting 1.5–2.0 in. (38–51 mm) of the unconfined, low-density edge of the first paved lane after compaction, while the mix is plastic. The cutting wheel is usually 10 in. (254 mm) in diameter, with the cutting angle about 10° from the vertical toward the mat to be cut and about 45° on the open side to push the trimmings away. The cutting wheel can be mounted on an intermediate roller or a motor grader. The HMA trimmings can be collected and recycled. A reasonably vertical face at the edge is obtained by this process, which is then tack-coated before placing the abutting HMA. It is important to restrict the overlap to about ½ in. (13 mm) while placing the adjacent lane. It is important to remove all low-density material at the edge of the first paved lane. Some contractors remove as much as a 3 in. (75 mm) strip to meet and exceed joint density requirements. It is very important to have a skilled cutting wheel operator, who must cut straight without wavering, and a skilled paver operator, who must closely match the cut line with minimal overlap.

16.8.1.5 Infrared Joint Heating and Minimum Joint Density Requirements

The objective of the infrared joint heating system is to obtain a hot joint during conventional longitudinal joint construction. The infrared joint heating system consists of two or three preheaters and one paver-mounted heater. The preheaters, which are connected in series, are towed with a small tractor over the joint approximately 100 ft (30 m) ahead of the paver. Propane cylinders are used to feed the infrared heaters. Both the preheaters and paver-mounted heaters are placed about 2–3 in. (51–76 mm) above the pavement surface and straddle the joint. The infrared heaters usually target a final surface temperature of 340°F (171°C) behind the paver-mounted heater. The target surface temperature can be achieved by changing the number of preheaters, the distance between the preheaters and paver, and the height of the heaters above the pavement surface. All these variables need to be adjusted, taking into account prevailing ambient conditions such as air temperature and wind velocity. The preheaters run continuously as long as the towing vehicle is operated at a speed high enough to prevent overheating the HMA at the joint. If the towing vehicle slows down too much, the preheater is designed to shut down automatically. The paver-mounted heater has also been designed to shut down once the HMA at the joint exceeds a specified temperature.

QUESTIONS

16.1 What are the different steps in the production of hot mix asphalt?

16.2 A contractor notices tender mix/shoving of the mix during compaction. What could be the probable reasons?

16.3 Ten tests were conducted to determine the density of a sample section of a new pavement. The densities were reported as a percentage of theoretical maximum density. Using a standard z distribution table, estimate the percentage of results that could be expected to fall below 92%. If the specification states that the PWL must be 90% to get full payment, and the lower limit for density is 91.0%, can the contractor expect to get paid 100%?

Test No.	Density (% of Theoretical Maximum)
1	95.0
2	92.1
3	91.2
4	93.3
5	92.5
6	95.0
7	95.5
8	92.2
9	92.6
10	93.2

16.4 Review a specification that is commonly used by your local or state department of transportation for pavement construction. Can you suggest some improvements?

16.5 What are some of the better techniques that could be utilized for constructing good-quality longitudinal joints?

17 Construction of Concrete Pavements

17.1 OVERVIEW

The most important aspect of constructing high-quality pavements is producing a quality concrete mix that uses good quality materials. The construction process is just as critical for producing high-quality pavements. The importance of having skilled contractors, operators, and construction personnel cannot be overemphasized. The properties of the fresh or plastic concrete will affect the construction operation including mixing, transporting, placing, paving, texturing, curing, and jointing. Early-age properties will affect plastic shrinkage and early-age behavior and should be considered as seriously as strength requirements. The hardened properties of the concrete will eventually affect long-term pavement performance including strength, dimensional stability, and durability. In addition, the use of good performing construction equipment and quality skilled operators and contractors is critical. Poor materials and poor construction quality will never produce a high-quality pavement. The importance of having skilled personnel in all aspects of the production, construction, and quality control operations is worth emphasizing. A quality control plan must be implemented to recognize problems if they arise. Skilled contractors in the production or the paving process must recognize and/or anticipate problems arising and being able to make quick changes to still yield a positive outcome and an acceptable product.

Concrete pavements are commonly produced using two types of paving technology: slipform paving and fixed form paving. Producing quality pavements that are long lasting requires quality concrete, skilled paving contractors, proper placement, consolidation, finishing, curing, and jointing. Preplanning and coordination between all stakeholders and project delivery team is critical. This will ensure that concrete production, continuous delivery, and paving operations are performed as planned and that pavement quality is not compromised.

Once a concrete mix has been optimized in the lab using trial mixes, field verification should be conducted. Depending on the type of paving operation selected, slipform or fixed form paving, field verification of concrete mixtures should use the production equipment anticipated for the job and, if possible, should also include the construction of a test strip.

The concrete plant can be either a permanent stationary facility or a portable facility. The decision will depend on many factors such as access to the volume of concrete production needed, the distance from the job, and economical factors. Plant location and setup will depend on site factors such as zoning, access to utilities, availability of materials, and public traffic (urban or rural).

A detailed presentation of concrete paving operations is provided by the AASHTO-FHWA-Industry Joint Training Participants Manual, Construction of Portland Cement Concrete Pavements (ACPA, 1996a). The Integrated Materials and Construction Practices for Concrete Pavement: A State-of-the-Practice Manual, FHWA HIF-07-004; and other agency specifications such as the FAA P-501 (Standards for Specifying Construction of Airports, AC 150/5370-10B (FAA, 2005); Texas Design & Construction Manual; http://onlinemanuals.txdot.gov/txdotmanuals/pdm/steel_placement.htm).

17.2 CONCRETE PRODUCTION

The production of concrete is a manufacturing process that incorporates different materials with inherent variability. Therefore, producing a concrete mix with consistent properties is very important. A quality control process should be in place to manage all aspects of material stockpiles, transport and delivery, batching, mixing, and concrete delivery. Aggregate stockpiles should maintain uniform gradation and moisture content and good coordination between the mixing plant and materials supply chain.

All Portland cement concrete used for pavement applications is ready-mix concrete. ASTM C-94—AASHTO M-157 provides the standard specifications for the manufacture and delivery of freshly mixed concrete. Usually customers have a few options regarding the manner in which they can order the concrete. A performance-based option is where the customer specifies the strength, slump, and air, and the producer selects the mix proportions needed to achieve the desired performance. A prescribed option is where the customer specifies the mix proportions, whereas a combination option allows the producer to select the mix proportions with the minimum amount of cement and strength specified.

Batching is the process of combining the mix ingredients to produce the desired concrete. Batching is done by either volume or mass. Batching by weight, as specified in ASTM C-94—AASHTO M-157, is desirable and yields more accurate measurements since fine sands usually change apparent volume with a change in moisture content. Volumetric batching (ASTM C-685—AASHTO M-241) is used primarily for concrete mixed in continuous mixers.

Mixing is the process of mixing all the ingredients thoroughly without overloading the equipment above its capacity. When a concrete is adequately mixed, it should have a uniform appearance, with all ingredients uniformly distributed. And when samples are taken from different portions of a batch, they should have the same slump, density, air content, and coarse aggregate content. In a central mix concrete plant, the ingredients are batched into a stationary mixer. The mixer then completely mixes the components before discharging the concrete into a delivery vehicle for transporting to the paving site. The delivery vehicle can be a truck mixer operating at low mixing speed, a truck agitator, or a non-agitating flatbed truck. A shrink-mixed concrete is partially mixed in a stationary mixer, transported, then mixing is completed during delivery in a truck mixer. A truck-mixed concrete is mixed entirely in the truck mixer during the delivery to the site.

The challenging aspect about concrete hauling and delivery is negotiating the traffic in congested areas. Any delays in delivery could adversely affect the concrete properties and the paving process. For example, a constant rate of supply of concrete is needed slipform paving; operations should be continuous and moving at a constant rate to produce a quality smooth pavement. Another example is that concrete delivery trucks should not be backed up or waiting to deliver the concrete. This could affect concrete setting, laydown operations, and compromise or affect the quality of the final slab. In any event, the complete cycle of mixing, discharging, transporting, and depositing concrete must be coordinated for the capacity of the mixing plant, hauling and transport distance, and spreader and slip-paver capabilities. Figure 17.1 shows a typical concrete ready-mix plant.

17.3 PREPARATION OF SUBGRADE AND BASE

Prior to concrete placement and the construction of the PCC slab on grade, preparation of the base and subgrade is essential. This includes compacting, trimming, and wetting the subgrade. It is assumed that the subgrade has adequate bearing capacity, is suitable for structural loads, and is readied for construction. The subgrade should be moistened in dry conditions so that the dry subgrade soil does not draw too much moisture from the concrete. For base placing, the appropriate

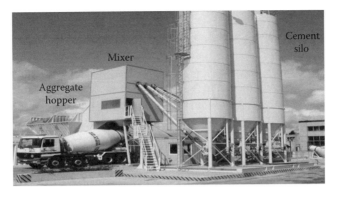

FIGURE 17.1 Concrete plant with mixer, aggregate hopper, and cement silos. (Courtesy of Kim Franklin, Northeast Cement Shippers Association, Castleton, NY.)

FIGURE 17.2 Preparations of subgrade and base: (a) subgrade stabilization and compaction; (b) compacted base; (c) stabilized base; (d) asphalt base. (Courtesy of Wouter Gulden, ACPA-SE Chapter, Duluth, GA.)

materials and compaction are important. For bound base, the material could be placed with a paver and compacted with rollers. For concrete pavements, it is more critical to have a uniform and consistent subbase and subgrade than to have higher bearing capacity. The consistency is extremely important in paving operations. Figure 17.2 shows different stages of subgrade and base preparation.

17.4 PRESETTING REINFORCEMENTS SUCH AS DOWEL BARS, TIEBARS, AND CONTINUOUS REINFORCEMENT

17.4.1 DOWEL BARS

Dowel bars and tiebars are used to provide support at transverse and longitudinal joints. The use of dowel bars for proper load transfer at PCC pavement joints is highly dependent on proper placement. Dowels can be placed either before PCC placement by using dowel baskets or after PCC placement by using an automatic dowel bar inserter. Incorrect placement of dowels resulting in skewed alignment, shallow positioning, or excessively corroded dowels can cause failure and lead to faulting and cracking at the joints. The placement of dowel bars can be evaluated with nondestructive testing (magnetic induction tomography equipment, discussed in Chapter 20). To maintain long-term functionality and durability for PCC pavements, dowel bars must be kept in good functioning condition, including protection from corrosion. This can be accomplished by sealing the joints and minimizing the amount of water that can seep through the joint. However, this is impossible to achieve since water will seep through the concrete over time and, combined with deicing salts, may corrode unprotected steel bars. Typically dowel bars are protected from corrosion by the application of epoxy coating or stainless steel cladding. However, the best approach is to have a free draining base so water does not collect at the joints.

Prior to placing the concrete, dowels should be lightly coated with a lubricant such as grease or oil to prevent bonding with the PCC. Traditionally only half the bar was greased. The FHWA recommends greasing the entire bar, since partial greasing may still cause a tight bond in one slab and prevent free movement (FHWA, 1990).

Dowel baskets are used to hold dowel bars at the appropriate location and elevation before concrete placement. The baskets are simple truss structures fabricated from thick gage wire and are left to remain as part of the PCC pavement structure. The FHWA recommends that a minimum of eight stakes be used to secure the baskets, and that the steel stakes must have a minimum diameter of 0.3 in. (8 mm) embedded at least 4 in. (100 mm) in stabilized bases, 6 in. (150 mm) in treated permeable bases, and 10 in. (250 mm) for untreated bases or subgrade (FHWA, 1990). Figure 17.3 shows dowel bars preset for PCC pavement construction.

Dowel bar inserters are automated attachments to slipform pavers that allow the paver to insert transverse joint dowel bars as part of the paving process. Figure 17.4 shows a dowel bar and tiebar inserter. Dowel bar insertion usually occurs after the vibrator but before the tamper bar. Dowel bars are placed on the fresh PCC surface, then pushed down to the correct elevation by forked rods. The rods are usually vibrated while they insert the dowel bar to facilitate insertion and consolidation around the dowels. Regardless of the technique used to position dowel bars or tiebars, the locations must be accurately marked for sawing joints later (ACPA, 1995).

17.4.2 TIEBARS

Tie bars are placed along the longitudinal joint to tie the two slabs together. Tie bars are not considered as load transfer devices and should not be designed as such. Typically, tie bars are deformed bars about 0.5 in. (12.5 mm) in diameter and between 24 and 40 in. (0.6 and 1.0 m) long.

Tiebars can be placed into position using chairs ahead of paving or inserted after concrete placement either by hand or using a tie bar inserter attachment in the slipforming process. When one lane at a time is paved, tie bars are inserted at midslab depth and bent back until the adjacent lane is ready to be paved. When using slipform pavers, tie bars are inserted through slab edges that will become future longitudinal joints. If two lanes are being paved in one paver operation, then the tie bars are inserted or pushed into a midslab area that will be cut as a future longitudinal joint similar to dowel bar construction. In fixed-form construction, standard tiebars or two-piece bars with a threaded coupling are inserted through holes in side forms for longitudinal

(a)

(b)

(c)

(d)

FIGURE 17.3   Dowels in PCC pavements: (a) dowels on asphalt base; (b) dowels with caps; (c) greased dowels; (d) greased dowels on chairs. (Courtesy of Wouter Gulden, ACPA-SE Chapter, Duluth, GA.)

construction joints. Proper consolidation of the concrete around these bars is needed. And similar to dowel bars, tie bars should be protected from corrosion. Figure 17.4 shows slipform paving with a tie bar inserter.

17.4.2.1   Reinforcing Steel (CRCP)

Proper placement of reinforcing steel is critical to the performance of CRCP. Failures in CRCP and inadequate performance are commonly associated with reinforcing steel issues and inadequacies. These include insufficient reinforcement bar lapping, inadequate consolidation of the fresh concrete around the steel, improper position of the steel in the slab, and extreme hot weather during construction. Placement of reinforcing steel for CRCP can be achieved using the manual method or the mechanical method.

The manual method involves hand-placing the reinforcing steel before the fresh concrete is placed and finished. The rebar is placed in the proper location using supporting plastic or metal anchors called *chairs*. The chairs must be well anchored with adequate metal stakes to withstand the concrete placement and consolidation. Once the chairs are secured, transverse bars are placed on the chairs as support for the longitudinal rebar. The longitudinal rebar is placed on the transverse bars and secured firmly using appropriate ties, typically every 4–6 ft (1.2–1.8 m). The advantage of the hand placement method is that it allows for easy verification of rebar placement, including elevation and lap distances. However, the manual method is slower and more labor-intensive than mechanical methods. Figure 17.5 shows reinforcing steel before placement of concrete in a CRCP pavement.

(a)

Tie bar
inserter

(b)

(c)

FIGURE 17.4 Slipform paving with tie bar and dowel bar inserters: (a) dowel bar inserter; (b) tie bar inserter; (c) tie bar inserter. (Courtesy of Wouter Gulden, ACPA-SE Chapter, Duluth, GA.)

Reinforcing steel can also be placed using the mechanical method or using tube feeders. This involves placing rebar in a prepositioned location while the paver places and consolidates the fresh concrete around it. Great care should be taken to assure proper placement of the longitudinal reinforcement, since the precision of the mechanical systems could deviate by inches from intended locations. A double-lift construction method would help in placing the reinforcing steel more precisely; however, this method could be costly. Reinforcing mesh or bars for JPCP can also be placed using the double-lift method or pushed into the plastic concrete using a mesh depressor (ACPA, 1995).

When connecting two longitudinal rebar together using a splice, these splices should not occur at the same longitudinal location. Transverse cracks that occur at the surface of the concrete at the location of the splice could cause steel bonding failure, resulting in wide cracks and performance problems. To make sure that all the splices do not occur at the same longitudinal location, staggering splices to avoid having more than 1/3 of the splices within a 2 ft longitudinal length of each lane of the pavement is recommended (Texas Design & Construction Manual; http://onlinemanuals.txdot. gov/txdotmanuals/pdm/steel_placement.htm). Figure 17.5 shows rebar prepared for skewed splicing.

17.5 PCC SLAB CONSTRUCTION

17.5.1 SLIPFORM PAVING

There are two methods of constructing PCC pavement slabs: using a slipform paver and fixed form paving. In slipform paving, the paving train is a term that refers to the combination of mechanical systems that place, consolidate, and finish the concrete slab. For highway applications, a typical

(a)

(b)

(c)

FIGURE 17.5 Steel reinforcement in CRCP pavement: (a) continuous reinforcement; (b) skewed reinforcement splicing; (c) checking rebar elevation. (Courtesy of Wouter Gulden, ACPA-SE Chapter, Duluth, GA.)

paving train includes a spreader with belt placer; a slipform paver, a texturing machine, and a curing cart that is usually combined with the texturing unit (IMCP Manual, 2007). Figure 17.6 shows a typical paving train. The major functions of a slipform paver are to receive and manage the concrete head. An auger distributes the concrete uniformly, a striker bar is used to remove excess concrete ahead of the vibrators that consolidate and fluidize the concrete for the slipform and the profile pan.

The paving machine rides on treads (or pad line) over the area to be paved. The paver is guided using previously set stringlines and paver sensing wands (or new laser- and GPS-guided sensors). Fresh concrete is deposited in front of the paving machine, which then spreads, shapes, consolidates, screeds, and float-finishes the concrete in one continuous operation. Great coordination between concrete production, delivery, and placement is needed to maintain adequate forward progress of the slipform paver, which is required for producing smooth pavements. Slipform paving is an extrusion process that uses very low slump concrete, and allows for high-production paving on the order of 1 mile (1.6 km) per day.

Depositing the correct amount of concrete in front of the paving machine (termed the head) must be controlled to ensure that it does not get too high or too low. If the head gets too high, it creates a pressure surge under the paving machine and could result in a surface bump. If there is not enough material in front of the paving machine, then the concrete head may run out or the grout box may run empty, creating a low spot and voids or pockets in the pavement surface. Experienced pavers can manage the head appropriately by using a placer/spreader machine or by carefully depositing concrete from the haul trucks.

(a)   (b)

(c)   (d)

FIGURE 17.6 Typical paving train showing (a) slipform paver, (b) auger, (c) finishing and texturing unit, and (d) striker bar. (Courtesy of Wouter Gulden, ACPA-SE Chapter, Duluth, GA.)

Slipform pavers use mechanical vibration to consolidate the concrete. A series of vibrators fluidize the concrete and remove large air voids. For normal paving operations and common concrete mixtures, a frequency from 5000 to 8000 vibrations per minute, at paver speeds greater than 3 ft/min, can adequately fluidize and consolidate the concrete without loss of entrained air or segregation of large particles. Figure 17.7 shows a typical slipform vibration system (ACPA).

17.5.2 STRINGLINES FOR SLIPFORM PAVER

Stringlines are used to guide and provide a reference for the slipform paver and the profile pan. The profile pan is the part of a slipform paver that controls the pavement surface. String lines are placed along the grade and on both sides of the paver as shown in Figure 17.8. The paver's elevation sensing wand rides beneath the string, and the alignment sensing wand rides against the inside of the string The string line material may be wire, cable, woven nylon, polyethylene rope, or another similar material. The stakes should be no more than 25 ft apart and closer for horizontal and vertical curves. The resulting smoothness and quality of the resulting pavement surface depends greatly on the tolerances and the stability of the stringline construction. Ultimately, the paver follows the elevation input from the sensor wand that's guided by the stringline position. Alternatives to stringline guiding systems include laser and GPS systems. Also, another system referred to as being "locked to grade" uses a ski system, which rides off of a previously paved surface (IMCP Manual, 2007).

FIGURE 17.7 Typical slipform vibration system. (Courtesy of Robert Rasmussen, Transtec Group Inc, Austin, TX.)

(a)   (b)

FIGURE 17.8 Paver with stringlines used for wand sensors to adjust paver elevation. (Courtesy of Robert Rasmussen, Transtec Group Inc, Austin, TX.)

17.5.3 FIXED FORM PAVING

In fixed-form paving, stationary metal forms are set, aligned, and staked rigidly on a solid foundation. Final compaction, preparation, and shaping of the subgrade or subbase are completed after the forms are set. Forms are cleaned and oiled to ensure easy release after the concrete hardens. A variety of fixed-form paving machines are available, from simple hand-operated and self-propelled vibratory screeds, single-tube finishers, and revolving triple tubes. Other larger and more complex form-riding machines can place and consolidate the concrete between forms in one pass. These machines ride on the forms or on pipes laid outside the forms (IMCP Manual, 2007). Figure 17.9 shows typical fixed form construction operations.

The ACPA (1996a) recommends the use of metal forms for pavement applications. The forms should be stable and may not warp or bulge under the concrete pressure. The metal forms should be ¼ in. thick and 10 ft long with adequate flange bracing. The forms should lock together

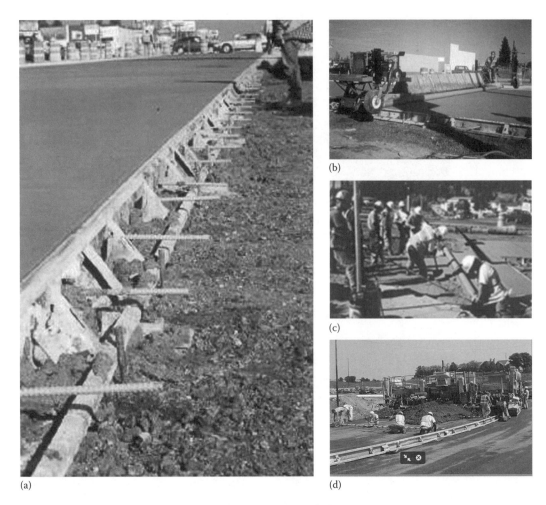

FIGURE 17.9 Typical fixed form construction operations. (ACPA)

tightly and be flush when assembled contiguously. The depth of the forms should be the depth of the slab. The tops of the forms should have adequate flange strength to support paving equipment.

17.5.4 CONCRETE PLACEMENT

Concrete should be placed as close to its final location as possible. This holds true for both paving types: slipform and fixed form. This is done using concrete-placing equipment. Concrete should not be moved excessively once deposited to minimize the potential for segregation. Concrete should not be placed in large piles and moved horizontally over long distances since the mortar fraction tends to flow ahead of the coarse aggregate and would increase segregation. Figure 17.6 shows concrete being placed ahead of the paver using a transfer device.

17.5.5 CONSOLIDATION

Consolidation is the process of compacting the fresh concrete by removing the entrapped air and causing it to mold without gaps and honeycomb around reinforcement and forms. In a slipform paver, this occurs during the vibration phase after the auger and screed operation (Figure 17.7). In fixed form paving, usually the vibration that a vibratory or roller screed produces is adequate to consolidate the surface of most pavement slabs (Figure 17.9b). Additional internal vibration with hand-operated spud vibrators is usually necessary for adequate consolidation of unreinforced concrete slabs thicker than 3 in. The *spud* or *poker vibrators* are a flexible shaft vibrator probe, which consists of a vibrating head connected to a driving motor by a flexible shaft. An unbalanced weight inside the head rotates at high speed, causing the head to rotate in a circular orbit. During vibration, the friction between aggregate particles is temporarily destroyed and the concrete behaves like a liquid; solids settle under the influence of gravity, and large entrapped air bubbles rise. The vibrated sphere of influence depends on the vibrator size and frequency of operation, and ranges between 3 and 14 in. for a vibrator head diameter between 0.75 and 3.0 in., respectively. An insertion time of 5–15 s will usually provide adequate consolidation. Vibrator head should not be dragged when inserted in the concrete, since it will cause segregation.

Improper vibration may be harmful to long-term durability of the concrete by removing entrained air required for freeze–thaw durability. Some effects of under-vibration include honeycombing; an excessive amount of entrapped air voids that are called *bugholes*; sand streaks (which result when heavy bleeding washes mortar out from along the form, or due to segregation from striking reinforcement steel without adequate vibration); cold joints (from delay in placing adjacent concrete lifts or insufficient vibration of the initially placed concrete before the second lift is added); and placement lines or "pour" lines (darker lines between adjacent placement of concrete batches that are not knit together due to insufficient vibration).

17.5.6 Screeding

Screeding or strikeoff is the process of cutting off excess concrete to bring the surface in line with the desired slab elevation and grade. This is usually done using a template called a *straightedge*. This is moved across the concrete in a sawing motion with approximately a 1 in. surcharge ahead of the straightedge to fill in low areas. There are also a number of mechanical systems that use a straight edge combined with a vibrator (called a *vibratory screed*) to ease the consolidation and strikeoff effort (Figure 17.6d).

17.6 FINISHING

Finishing operations include floating and texturing. However, hand finishing of the pavement surface using bull floats is necessary only where the surface left from the paving equipment contains voids or imperfections. Mechanical longitudinal floats should not be overused behind the screed or slipform equipment. Commonly, the slip forming operation is optimized to include finishing. If consistently excessive additional hand finishing is required, then the paving contractor should make concrete mix and equipment modifications. Figure 17.10 shows finishing operations (IMCP Manual, 2007).

A hand-operated straightedge 10–20 ft (3–6 m) should be used for checking the surface behind the paving equipment. Successive straightedge checks should overlap by one-half the length of the straight edge to help ensure that the tool detects high and low spots in the surface.

A slight hardening or stiffening of the concrete surface is necessary before finishing can take place. Usually finishing can proceed when the bleed water sheen has evaporated and the concrete surface can withstand foot pressure with no more than a ¼ in. depression or indent.

17.6.1 Bullfloating

Bullfloating is the process of smoothing the surface, leveling off high and low spots, and embedding large aggregate particles just below the slab surface (Figure 17.10). A long-handle bullfloat is typically used in pavement construction since the slabs are too wide to fully reach by a short-handle darby. For nonair-entrained concrete, a wood bullfloat is used. However, for air-entrained concrete, an aluminum or magnesium alloy tool should be used.

Consolidation, screeding, and bullfloating must be completed before excess bleed water collects on the surface of the concrete. Care should be taken to avoid overworking the concrete surface and preventing undesirable segregation and increasing the W/C ratio of the concrete surface by incorporating

(a)   (b)

(c)   (d)

FIGURE 17.10   Finishing operations in PCC pavement construction: (a) bullfloating; (b) texturing; (c) tining comb; (d) tining grooves. (Courtesy of Wouter Gulden, ACPA-SE Chapter, Duluth, GA.)

the bleed water. Increasing the surface W/C ratio has significant consequences on durability since it may reduce the strength and watertightness and increase the potential for crazing, dusting, and scaling.

17.6.2   Texturing

Texturing is the process of increasing the roughness of the concrete surface to improve traction, skid resistance, and reduce hydroplaning. The pavement texture also affects tire-pavement noise. Texturing of PCC pavements can be achieved using different methods. Dragging a rough textured burlap or cloth along the surface produces striations that are 1/16–1/8 in. (1.5–3 mm) deep (Figure 17.10). A tining machine drags a metal wire comb-like device or metal rake along the transverse or longitudinal direction of the pavement (Figure 17.10). Tining practices will vary by state agency; however, many require transverse grooves that are 3–5 mm (0.12–0.20 in.) deep, 3 mm (0.12 in.) wide, and spaced 12–20 mm (0.47–0.79 in.) apart. Artificial turf drag is produced by dragging an inverted section of artificial turf and produces 1.5–3 mm (1/16–1/8 in.) deep striations. Brooming in the transverse or longitudinal direction uses either a hand broom or mechanical broom device that lightly drags the stiff bristles across the surface and produces striations that are 1/16–1/8 in. (1.5–3 mm) deep. Other methods such as exposing aggregate by washing the surface or diamond grinding and grooving have also been used occasionally (IMCP Manual, 2007).

Texturing should be provided uniformly on the concrete surface and is affected by numerous factors such as concrete consistency and workability, amount of bleed water present on the fresh concrete surface, pressure on the texturing tools, tool variability, burlap variability, and cleanliness. At times, the area over the future joint locations is not textured in order to provide a good sawing and sealing surface (ACPA, 1995).

17.6.3 Protection of Pavement Surface from Rain

Rain protection of the concrete should be anticipated, and the contractor should be well prepared when it occurs during the concrete-placing operation. When rain occurs, all batching and placing operations should stop. The fresh or plastic concrete should be protected by plastic sheeting so that the rain does not wash the cement from the surface or indent the surface. Resurfacing should proceed as needed after the rain stops and operations resume.

17.6.4 Construction Headers

The construction and finishing of headers (transverse construction joints) are consistently a potential problem area contributing to the roughness of concrete pavements. Headers occur at the end of a work day or at an interruption for a bridge, intersection, or leave-out. Figure 17.11 shows the finishing of a construction header at the end of the day.

The paving machine must stop at these locations, and the most common practice is to place a wooden form to create the joint. The forming of a header in this manner increases the chance of a bump in the surface due to the hand work necessary to blend the mechanized paving surface with the

(a)

(b)

(c)

FIGURE 17.11 Finishing construction header in PCC pavement construction. (Courtesy of Wouter Gulden, ACPA-SE Chapter, Duluth, GA.)

hand-placed area. A "cut back" method can be used to create a joint and avoid hand-forming headers. The paver operator continues paving until all of the concrete is used. The following day a transverse saw cut is made about 5 ft (1.6 m) from the end of the hardened concrete slab and dowels are grouted into holes drilled into the smooth saw face. This method, when done correctly, produces a smoother riding construction joint than is typically achieved with hand-forming. (IMCP Manual, 2007).

17.7 CURING

Curing is the process of providing a satisfactory moisture and temperature condition that allows the concrete to develop its "timely" desirable properties. Curing allows the hydration process of Portland cement and other pozzolanic materials to proceed in a timely manner, avoid plastic shrinkage, drying shrinkage, thermal stress cracking, and continue with strength gain properties. The amount and rate of curing will vary based on the mix properties (specifically the paste and cement fineness), the moisture and temperature environment at the surface of the concrete, base drainage, the wind velocity, use of admixtures and pozzolanic materials, and any other parameters that will affect the cement hydration process. Curing in a timely manner is critical for pavement slabs since they have a large surface to volume ratio. Improper curing will negatively affect the following properties: strength gain, water tightness, abrasion resistance, volume stability due to temperature and moisture, and ultimate durability such as resistance to cycles of freezing and thawing and scaling due to deicing chemicals.

Curing in a timely manner will depend on the bleeding and the evaporation characteristics at the surface layer of the concrete slab. If the evaporation rate exceeds the bleed rate, then the potential for plastic shrinkage increases. The average rate of bleeding in kilograms per square meter per hour (kg/m²/h) can be estimated by an empirical equation proposed by T. Poole (2005).

$$\text{Bleed Rate (kg/m}^2/\text{h)} = [0.051 * \text{w/cm} - 0.015] * h$$

where

w/cm is the water to cementitious ratio
h is the pavement thickness in centimeters

The evaporation rate in concrete is a function of air temperature, concrete temperature, relative humidity, and wind speed. To approximate the evaporation rate from the surface of a concrete slab, the PCA and ACI 308 nomograph shown in Figure 17.12 can be used. This however does not predict plastic shrinkage, which is more complex and depends on the environment at the surface of the concrete and the behavior of a particular fresh cement paste.

17.7.1 EVAPORATION RATE

T. Poole (2005) presents a numerical equation that can be used to provide the rate of evaporation, which provides the same information as presented in the nomograph shown in Figure 17.12. The evaporation rate can be easily computed using a spreadsheet or an equation computing device.

$$\text{ER} = 4.88 \left[0.1113 + 0.04224 \frac{\text{WS}}{0.447} \right] (0.0443) \left(e^{0.0302(\text{CT}*1.8)+32} \right) - \left[\left(\frac{\text{RH}}{100} \right) \left(e^{0.0302(\text{AT}*1.8)+32} \right) \right]$$

where

ER is the evaporation rate, kg/m²/h
WS is the wind speed, m/s
CT is the concrete temperature, °C
AT is the air temperature, °C
RH is the relative humidity, %

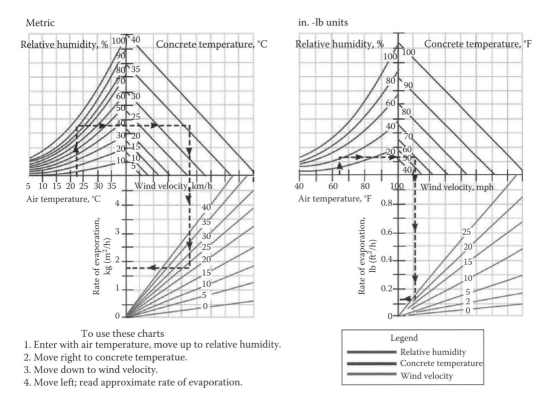

FIGURE 17.12 PCA and ACI 308 nomograph to approximate the evaporation rate from the surface of a concrete slab. (Courtesy of PCA.)

The most common curing method is the application of a liquid membrane. The objective of using liquid membranes is to seal the surface and minimize moisture migration into the air. This in turn reduces the potential for plastic shrinkage, drying shrinkage, slab curling and warping. A liquid membrane–forming compound that meets ASTM C 309/AASHTO M 148 material requirements and application rate of 200 ft^2/gal (5.0 m^2/L) is adequate for most normal paving situations (ACPA, 2005). Application timing of curing compound is also critical and should be applied as soon as possible after texturing and just as the water sheen on the surface is disappearing.

The two common methods used for applying liquid membranes are the sprayer bar and the handheld wand. Figure 17.13 shows the application of curing material using the handheld wand and the sprayer bar. The sprayer bar traverses the width of the finished concrete slab and applies the membrane through sprayer jets evenly spaced across the pavement width. The handheld wand is person controlled and should only be applied for small surfaces. Due to coloring pigmentation, the applied liquid membrane is designed to leave a white residue on the pavement to assist the operator to be assured of complete surface coverage. The white color can also enhance solar reflections and minimize slab overheating. It is important that the entire surface is covered, since even a small area could be a source of concrete weakness and damage. A single coverage for either method does not guarantee complete coverage. However, a double or triple application of liquid membrane can lead to an even-covered surface and lower the potential for moisture loss–related stresses and plastic shrinkage cracking.

Plastic sheeting used in addition to curing compounds is an effective method in retaining and trapping the moisture in PCC slabs and protecting the concrete from cold temperatures and rain. Plastic sheeting can also act as a thermal insulator due to the thin air layer trapped between the plastic and the concrete. This property can be beneficial if the construction site is subjected to rapid

(a)   (b)

(c)

FIGURE 17.13 Application of curing compound using the (a) handheld wand, (b) sprayer bar, and (c) sprayer bar.

cooling following construction. However, it can be detrimental if used improperly. The insulation, in concert with the excess heat produced by fast-track PCC mixes, can result in too great a curing temperature, causing cracking later when the slab cools. The plastic sheeting must be well secured; otherwise, strong winds may cause the sheeting to become insecure and increase the potential for slab cooling, temperature shock, and early-age cracking.

Curing blankets can also be used for curing. Caution should be used when removing curing sheeting or blankets from a recently placed concrete pavement. If the concrete is still warm when the blankets are removed, and the ambient temperature is low, thermal shock can occur, which may cause cracking (Huang et al., 2001).

17.8 PAVING IN HOT AND COLD WEATHER CONDITIONS

Hot or cold weather paving should be planned in advance. This can be achieved by making trial batches in the laboratory under similar field temperature conditions to be aware of anticipated changes in concrete properties and anticipate modifications in paving, finishing, and curing operations. Problems that could arise during hot weather concreting include rapid slump loss, reduced air content, premature stiffening, plastic shrinkage cracking, and thermal cracking. For cold weather paving, the primary concern is to keep the temperature of the concrete above freezing so that the hydration reaction and strength gain continues, and to control cracking through proper joint placement.

Planning for hot weather paving conditions should begin as soon as the contractor is aware of upcoming hot weather conditions. The contractor should take precautions to limit the evaporation rate from the concrete surface. Plastic shrinkage cracking results when the loss of moisture from the concrete exceeds the rate of bleed water reaching the surface as described previously.

If the evaporation rate exceeds 0.2 lb/ft^2/h (1.0 kg/m^2/h), it is recommended to provide a more effective curing application, such as fog spraying, or to apply an approved evaporation reducer such as a lithium-based compound. Other precautions include keeping aggregates moist, keeping the concrete temperature low by cooling aggregates and mixing water with ice; erecting temporary windbreaks to reduce wind velocity over the concrete surface; erecting temporary sunshades to reduce concrete surface temperatures. In cases where surface evaporation conditions are extreme and plastic shrinkage mitigation efforts are not successful, paving operations should be stopped until weather conditions improve. For additional measures to mitigate plastic shrinkage, drying shrinkage, and hot weather concreting, refer to ACI 305, Hot Weather Concreting.

Preplanning for low-temperature concreting should be anticipated in advance as forecasts for freezing temperatures are known. Some recommendations for cold weather concreting include the use of higher Portland cement content, the use of less pozzolanic materials such as ground, granulated blast-furnace slag or fly ash unless they are required for durability; use accelerating admixture but not chloride based due to potential for corrosion of reinforcement, admixtures will react differently in cold temperatures and mix trial batches should confirm this. Since strength gain is slow, the potential for shrinkage cracking increases. Use heated water and materials if possible for the size of the pour. The use of insulating blankets is highly recommended for curing the concrete in freezing temperatures, although concrete should not be placed if the air temperature is less than 40°F. Strength gain will be delayed due to the slow hydration process in cold temperatures. Joint sawing and opening to traffic should be monitored closely to prevent premature failure. For additional information on cold weather concreting, refer to ACI 306.

17.8.1 Edge Slump

Edge slump is the term referred to when the top edge of a freshly placed, slipformed concrete pavement sags down after the slab is extruded from behind the paver. Figure 17.14 shows the two common types of edge slump. A small amount of edge slump is more tolerated at the absolute free edge of the pavement than along the longitudinal joint, which will have some traffic moving across it. The most common form of edge slump is when the top edge slumps down. When the bottom edge slumps out, this usually indicates a more serious problem with the mix design and is often associated with higher slump mixtures that are not appropriate for slipform paving. The factors that affect concrete workability, consistency, and rheology will affect edge slip behavior. When this type of edge slump occurs, paving should be suspended until the concrete mixture has been modified to work with slipform paving (IMCP Manual, 2007).

17.8.2 Smoothness

Pavement smoothness is important to the traveling public. Smoothness is a measure of pavement quality and should be considered seriously in pavement construction. Factors affecting pavement smoothness include such things as the construction of a level base and subbase. Horizontal alignment, curves, and cross-slopes must be well aligned to minimize roughness. Preconstruction surveying and staking must be accurate. Embedded fixtures such as manholes and reinforcement bars must be well positioned and aligned. The concrete mixture should have the right consistency to flow smoothly through the slipform paver. Headers, block-outs, and leave-outs all add to roughness and should be minimized if possible. Figure 17.15 shows a smoothness measuring device and the output with a simulated straight edge and blanking band.

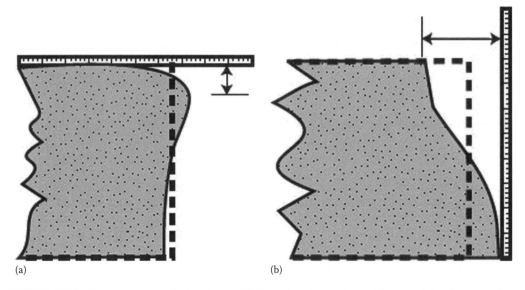

(a)   (b)

FIGURE 17.14 Common types of edge slump. (a) Top edge slumps down. Place straightedge on surface. Measure vertical distance between edge of slab at its greatest point of slump and straightedge. (b) Bottom edge slumps out. Place straightedge along slab edge at greatest point of slump. Measure horizontal distance between top edge of slab and straightedge.

(a)   (b)

FIGURE 17.15 Pavement smoothness measuring process. (a) Measuring device for elevation profile and (b) elevation profile output with a simulated straight edge and blanking band. (Courtesy of Wouter Gulden, ACPA-SE Chapter, Duluth, GA.)

17.9 JOINTING

After placing and finishing the concrete slab, joints are constructed to control cracking and to provide relief for concrete expansion and contraction caused by temperature and moisture changes. Joints are usually constructed using special saw-cutting equipment. If designed and constructed properly, joints in concrete pavements create vertical weakening planes in the concrete pavement to induce controlled cracks just below the saw cut. Properly constructed joints will be easier to maintain than uncontrolled random cracks. Once PCC joints have cured sufficiently, they should be cleaned and sealed with jointing compound to prevent noncompressible foreign debris from harming the concrete's contraction and expansion cycle. The FHWA HIPERPAVE software (FHWA, 2005) provides guidelines for timing joint construction based on concrete properties and environmental conditions.

FHWA (1994) recommends that joint sawing should be timed correctly before stresses develop in the pavement that are large enough to cause cracking. These stresses are the result of restrained volumetric changes from both temperature and moisture changes in the new pavement. Joint construction should not begin until the new PCC pavement has gained enough strength to support the weight of the sawing equipment and operator, and to minimize excessive raveling due to the forces introduced by the cutting blade. This provides a short window where the concrete conditions are right for joint sawing.

Selecting the appropriate time to saw a joint is influenced by many factors, including mix properties, weather, subbase, and the type of sawing and saw blade. Concrete mixtures that increase the potential for plastic shrinkage will shorten the potential window for joint sawing. This includes mixes with a high water demand, more fine materials, a small maximum coarse aggregate size that reduces the amount of course aggregate in the mix. Large weather changes that may increase the potential for rapid moisture evaporation from the surface of the concrete will also increase the potential for plastic shrinkage cracking and therefore shorten the window for joint sawing. This includes sudden temperature drop or rain shower, sudden temperature rise, high winds and low humidity, cool temperatures and cloud cover, hot temperatures and sunny conditions.

The coarse aggregate type influences the temperature sensitivity of the concrete. Concrete that is more temperature sensitive with high CTE will expand or contract more with temperature change, increasing cracking potential. A dry and porous base will have similar effects of drawing moisture from the concrete. The friction level between the concrete slab and the subbase has an effect on the plastic shrinkage potential. A large friction level and a strong bond between concrete slab and subbase will shorten the window for successful joint operations. Figure 17.16 shows joint sawing operations in PCC pavements.

17.10   HIPERPAV SOFTWARE

HIPERPAV (HIgh PERformance concrete PAVing) is a simulation tool of early-age concrete pavement behavior (Ruiz, 2001). The software was funded by the Federal Highway Administration (FHWA) and developed by The Transtec Group, Inc. (Qinwu Xu, J. Mauricio Ruiz, 2009) Version III of HIPERPAV can be downloaded free of charge from http:/www.hiperpav.com/

The HIPERPAV software simulates how pavement design features, materials, climate, and construction procedures impact strength and stress development, and predicts the potential for cracking. The predicted concrete temperatures drive a number of key behavior models in HIPERPAV including strength and stiffness gain (based on the Arrhenius maturity concept), creep, and thermal expansion properties. The software can be used to manage pavement early-opening to traffic, optimal mix characteristics, saw-cutting windows, and curing methods for specific conditions.

The HIPERPAV program models the behavior of concrete given numerous inputs about the concrete materials, construction information, climate information, and slab-base restraint.

The HIPERPAV program estimates the allowable stress, which is the predicted strength development over the initial 72 h; and the applied stresses developed due to climate and restraint conditions. If the predicted stress exceeds the allowable stress, then cracking conditions are favorable and are indicated as likely to occur. Therefore, it is recommended that changes to the concrete mix, construction practices, and/or protection from the environment must be made.

HIPERPAV III includes several improvements over versions I and II and include the following (Qinwu Xu, J. Mauricio Ruiz, 2009):

Simple versus advanced strategy view: The user can toggle between Simple View (less inputs) and Advanced View (more inputs). In the Simple View, a minimum number of inputs are required for an analysis and all other inputs are estimated by the software that reduces the time for analysis. In the Advanced View, more input parameters are needed for a more detailed analysis. For example, some of the input parameters required include axial restraint, PCC stiffness, PCC drying shrinkage, maturity, and heat of hydration.

(a)   (b)

(c)   (d)

FIGURE 17.16 (a–d) Joint sawing operations in PCC pavements. (Courtesy of Wouter Gulden, ACPA-SE Chapter, Duluth, GA.)

Multiple strategy batch mode: Multiple strategies can run at once by selecting "batch mode" from the Strategy Menu.

Quick compare: The Quick Compare screen offers the user options to quickly make comparisons of up to four strategies simultaneously. Once the strategies of interest are selected, the user can compare differences in input data, strength, stress, and cracking risk plots.

Sensitivity comparisons: Multiple sensitivity analysis can be performed varying one input at a time for a given base case strategy.

Enhanced moisture characterization: HIPERPAV III uses a one-dimensional finite-difference model that relates materials, environmental effects, and curing procedures to predict the potential for PCCP early-age cracking due to moisture changes. This is a significant enhancement over HIPERPAV versions 1 and 2. Changes in moisture content in the concrete are the result of drying due to hydration, evaporation, or moisture transport into the base or subbase. Moisture removal from the concrete results in concrete shrinkage. Concrete shrinkage, coupled with temperature gradients, may cause axial, curling, and warping stresses in the concrete and the potential for early-age cracking.

Heat evolution characterization: HIPERPAV III characterizes the heat of hydration of concrete mixtures from semi-adiabatic calorimetry testing. This is an improvement over previous versions and improves the software accuracy in predicting thermal stress and strength development.

A brief demonstration of the software is presented here by showing an example and selected output windows, starting with an example for an Early Age JPCP using default values in the software.

Strategy information window: Information and background is logged in this window.

Design window: Reliability can be a set or defaults at 90%. Information on slab geometry, joint spacing, and slab support is provided (see Figure 17.17). The window shows a 10 in. slab, with 15 ft joints and a smooth asphalt base.

Mix design window: Information on the concrete mix design, type of cement, binders and admixture, strength at 28 days is provided (see Figure 17.18).

Construction window: Information on concrete initial mix temperature, layer support temperature, curing method, saw-cutting, and strength gained for opening to traffic is provided (see Figure 17.19).

FIGURE 17.17   Hiperpave III design window output.

FIGURE 17.18   Hiperpave III mix design window output.

FIGURE 17.19 Hiperpave III construction window output.

FIGURE 17.20 Hiperpave III climate window output.

Climate window: From the location of the project, climate data are provided from a database and maybe updated hourly for analysis (see Figure 17.20).

Analysis window: The analysis is performed based on all previous inputs. Figure 17.21 shows strength gain development in the concrete (top curve) and the tensile stress built in the concrete (bottom curve) for the first 72 h after construction. The evaporation rate analysis during the plastic state is also presented on the same window. The two strength and stress curves are intersecting at 17 h after construction and the stress exceeds the concrete strength. This indicates a high risk for cracking and merits investigating a change in mix design and/or construction process. Changing the base conditions from a rough asphalt-base to a smoother base changes the stress versus strength prediction curve and the risk for cracking is reduced. The design engineer can change numerous variables

FIGURE 17.21   Hiperpave III stress analysis output showing concrete strength gain and tensile strength builtup.

and evaluate the parameters that will significantly increase the risk for cracking. HIPERPAV III is a powerful tool but should be used with caution and should be coupled with local experience if possible. Numerous SHA are using this tool in the design and construction process, including Texas and Ohio DOT.

17.11   JOINT SEALING

The purpose of sealing joints is to minimize infiltration of surface water and incompressible material into the pavement layers (ACPA, 1993). Excess water contributes to subgrade or base softening, erosion, and pumping of subgrade or base fines over time. This degradation and erosion results in a loss of structural support, pavement settlement, and/or faulting.

Sealing of joints prevents hard or incompressible objects from entering joint reservoirs. During hot weather when slabs need to expand and joints need to move, stresses develop when hard incompressibles restrict this movement. Incompressibles contribute to spalling and, in extreme cases, may induce blow-ups. In either case, excessive pressure along the joint faces results as incompressibles obstruct pavement expansion in hot weather.

There are many types of materials that are acceptable for sealing joints in concrete pavements. Sealants are either placed in a liquid form or are preformed and inserted into the joint reservoir. Sealants installed in a liquid form depend on long-term adhesion to the joint face for successful sealing. Preformed compression seals depend on lateral rebound for long-term success.

The selection of an appropriate sealant material depends on the environment of use, performance, joint type, and spacing and cost. Figure 17.22 shows different forms of joint sealant.

The following factors affect the function and the performance of some joint sealants.

- Concretes with a large CTE and containing harder coarse aggregates, such as gravel or granite, will expand and contract more with a given temperature change than a concrete containing limestone. Therefore, the sealant material will be strained more and stretched farther for a given joint spacing. An appropriate shape factor (width/depth ratio) for the sealant is recommended.

Notes:
A - Initial cut to a depth of T/4 or T/3 as required for conventional sawing
B - Initial cut to a depth of 1¼ in. (32 mm) minimum for early-entry sawing
C - As required to accomodate sealant and backer rod
D - As required by the manufacturer
E - A single-cut or double-cut process may be used to saw joints.
    The field-poured sealant and performed seal above illustrate a double
    cut, in which a first, narrow cut is followed by a widening cut.
    A single wide cut is also acceptable.

FIGURE 17.22   Different forms of joint sealants. (Adapted from ACPA.)

- Silicone sealants are known to have poor adhesion to concrete containing dolomitic lime-stone. Use of a chemical primer application to the sealant reservoir walls will ensure that the silicone adheres and functions.
- A reservoir shape factor (width/depth ratio) of one is recommended for hot-poured asphalt-based sealants.
- Silicone and two-component cold-poured sealants typically need a reservoir shape factor of 0.5.
- Compression sealants are selected such that the maximum compression of the seal is 50% and the minimum is 20% through the anticipated ambient temperature cycles in the area.

For more specific information on joint sealing materials and required shape factors and sizes, consult ACPA (1993).

17.11.1   QUALITY ASSURANCE/QUALITY CONTROL (QC/QA)

As discussed in Chapter 14, quality assurance (QA) is the set of tests and activities conducted by the owner or highway agency to confirm that the product delivered is in compliance with the specifications. AASHTO defines quality assurance as "All those planned and systematic actions necessary to provide adequate confidence that a product or service will satisfy given requirements for quality."

Quality control (QC) is the set of tests and activities conducted by the contractor or service provider to monitor the process of material delivery, batching, placing, and finishing the concrete to be sure that the pavement will meet the minimum QA test criteria.

QA testing can be done for acceptance by the owner, or it can also be done by the contractor and verified by the owner. As presented in Chapter 16, statistical acceptance procedures, such as percent within limits (PWL) specifications, should be used to monitor the QC process

TABLE 17.1
Phase I Initial Suite of Tests Matrix

Concrete Property	Test Name (Standard Test Method)	Test Location
Workability	Differential scanning calorimetry	Central lab
	X-ray diffraction	Central lab
	Blaine fineness (ASTM C204/AASHTO T153)	Central lab
	Combined gradation	Mobile lab
	Shilstone coarseness and workability factors	Mobile lab
	Time of setting—premature stiffening (ASTM C359/AASHTO T 185)	Mobile lab
	Cementitious materials temperature profile: "coffee cup test"	Mobile lab
	Water/Cementitious materials ratio (microwave oven) test	Mobile lab
	(AASHTO T 318)	Mobile lab
	Unit weight (ASTM C 138/AASHTO/T121M & T121)	Mobile lab
	Slump	
	Heat signature	
	Concrete temperature, subgrade temperature, and weather data	
Strength development	Concrete maturity (ASTM C 1074/AASHTO T325)	Mobile lab
	Flexural and compressive	Mobile lab
	Strength (ASTM C78/ASTM C39/AASHTO T97/AASHTO T22)	Mobile lab
Air content	Air-void analyzer	Mobile lab
	Air content (pressure meter) (ASTM C231/AASHTO T152)	Mobile lab
	Air content from hardened concrete (ASTM C457)	Central lab
Permeability	Chloride ion penetration (ASTM C1202/AASHTO T277)	Central lab
Shrinkage and	Coefficient of thermal expansion (ASTM C531/AASHTO TP 60)	Central lab
thermal movement	Free shrinkage test	

continuously. For example, a target air content can be specified at 6%; however, a ±95% range could be between 4.9% and 7.3%. The QC/QA operators should keep good records that will track the test results and are able to flag the problem areas as they become evident. Typically the tests used are those that can be easily verified such as air content, strength, thickness, and smoothness. Others that are equally important such as distribution of voids and curing tests are less emphasized.

A suite of tests for use during the planning and construction of concrete pavements has been identified in the TPF-5(066) project (Fisk, 2008) and some of the tests are presented in Table 17.1. The recommended test methods should be used during the three phases of a mix's evolution during a project: material selection and mix design stage; preconstruction verification stage; and QC/QA stage. The tests are categorized according to five concrete properties: workability; strength development; air content, permeability; and thermal movement. Some of these tests are briefly described here. The final tests selected for a project will depend on the magnitude of the project and the needs for the QC/QA program.

17.11.2 DIFFERENTIAL SCANNING CALORIMETER

The DSC is used to determine the form and content of calcium sulfate in the cement. Setting problems associated with false set and flash set are caused by unbalanced amounts of gypsum (calcium sulfate dihydrate) and plaster (calcium sulfate hemi-hydrate). Monitoring the level of calcium sulfate and adjusting the mix appropriately will prevent setting problems during construction, strength development, and long-term durability.

17.11.3 Blaine Fineness

The fineness of cement affects the rate of reaction and cement hydration, water demand, workability, strength gain, and permeability. Monitoring the cement fineness will provide for better quality control in adjusting for mix water and potential strength gain. Monitoring the cement fineness and adjusting the mix and curing conditions accordingly will reduce the variability in concrete produced and resulting pavement quality.

17.11.4 Gradation

As mentioned earlier, the gradation of the aggregates will have a big influence on the plastic properties of the concrete. The gradation will influence the workability, water demand, cement paste requirements, segregation, bleeding, and ultimate shrinkage. A uniform gradation with all particles coated with a paste layer is a desirable outcome for a rigid pavement concrete.

17.11.5 Penetration Resistance

Some Portland cements and cement blends can experience false set. False set is the condition when the concrete loses workability without strength gain. Remixing of the concrete without the addition of water can restore workability. Performing the penetration resistance test can provide valuable information on whether a particular cement may be prone to false set.

17.11.6 Cement Materials Temperature Profile (the "Coffee Cup Test")

When Portland cement hydrates, it generates heat. The coffee cup test is a simple test to monitor the chemical reaction during the hydration process. Variations in the temperature profiles may indicate problems with cement hydration, workability and uniformity, and compatibility with admixtures.

17.11.7 Water–Cement Ratio (Microwave Oven Test)

The water–cement ratio of the resulting concrete mix has a big influence on the ultimate strength and durability of the hardened concrete. The microwave test method provides a quick way (within a couple hours) of determining the water–cement ratio. A concrete sample is dried in the microwave oven. From the moisture content and assumptions that the sample tested is representative of particular concrete mix, then the w/cm can be determined.

17.11.8 Concrete and Subgrade Temperature and Environmental Conditions

Temperature and environmental conditions have a big effect on early-age concrete behavior and ultimate strength gain. Monitoring temperature and project temperature and weather conditions will allow the pavement construction decisions makers to respond appropriately and minimize problems that would be detrimental to the concrete's strength gain and long-term durability. For example, knowing the temperature, relative humidity, wind velocity, solar radiation, and temperature changes will affect the rate of drying, setting time, and strength gain of the concrete, changing the saw-cutting window will reduce the chances of cracking and curling in the slab.

17.11.9 Concrete Strength (Early Age)

Assessing the concrete strength at an early age is a good indicator for concrete quality and the potential for carrying stress. Concrete strength however should not be taken as a guarantee for durability and long-term performance. A concrete with adequate strength, for example, may not have the correct entrained air volume and air bubble size distribution, and would perform poorly under freeze–thaw conditions. Details of concrete compression and flexural testing have been presented in Chapter 13.

17.11.10 AIR VOID ANALYZER

The air void analyzer (AVA) is a sophisticated testing apparatus that can determine the air void system in fresh mortar. The AVA determines the air void distribution by taking the mortar fraction of the concrete and stirring it gently in a solution. The air bubbles are released and rise to the top where they are captured by a submerged bowl. A very sensitive scale monitors the change in mass of the bowl as its buoyancy changes. According to Stokes Law, the larger bubbles will rise faster than the smaller bubbles and this theory is used to estimate the air size distribution of the fresh concrete. The test equipment was developed under the SHRP program in the 1990s. Since the freeze–thaw resistance of concrete depends largely on the spacing factor of the air void system, knowing this while the concrete is still plastic would provide the decision makers during the construction project of a powerful tool to reject inferior concrete that could become a big problem later on.

17.11.11 MATURITY TEST

The maturity test is a nondestructive test method for estimating in-place concrete strength. The maturity method can be used as criterion for opening a pavement to traffic, for estimating strength gain and for quality control purposes. The principle behind the maturity test is that strength gain of concrete is a function of time and temperature that follows the extent of cement hydration in a given mix. Therefore, samples of the same concrete will have equal strength if they have the same maturity index independent of their actual time histories.

The maturity curve is developed by casting, curing, and testing standard strength specimens while measuring and recording the temperature of those specimens over time (ASTM C, 1074). The maturity index is computed by the summation of the time interval times the temperature. Only temperatures above 14°F are used in the maturity index since hydration is virtually inactive at very low temperatures.

In the field, a maturity test is conducted by inserting a temperature probe in the concrete slab as shown in Figure 17.23. A computerized data acquisition system is used to download the slab temperature data, which is then compared to the maturity curve. The maturity correlation curves are developed for a particular mix design that estimate concrete strength based on its maturity.

The procedure for developing the maturity correlation curve for a specific concrete involves the following:

- Cast at least 15 strength specimens from a mixture that will be the same as used for pavement construction.
- Embed temperature sensors in at least two of the specimens. These specimens are used only for recording the temperature over time and will not be broken.
- Cure all the strength specimens in the same location with exposure to a constant temperature and similar relative humidity.
- Test the strength of the specimens at 1, 3, 5, and 7 days. Break and average three specimens at each age.
- From the temperature data, determine the maturity index for each set of strength specimens.
- Plot the strength and Maturity Index data for the strength specimens on a graph, with Maturity Index on the x-axis and concrete strength on the y-axis.
- Fit a smooth curve through the plotted points. This can now be used to estimate the strength from the field pavement Maturity Index.

FIGURE 17.23 Field testing of concrete temperature for using the maturity test. (Courtesy of Wouter Gulden, ACPA-SE Chapter, Duluth, GA.)

17.11.11.1 Consistency

The slump test, ASTM C-143 (AASHTO T-119), is the most commonly accepted method used to measure the consistency of concrete (see Figure 17.24). The test is conducted by filling a metal conical mold 12 in. high with an 8 in. diameter base and a 4 in. diameter top. The slump cone is placed upright on a flat rigid surface, dampened, and filled in three layers of approximately equal volume. Each layer is rodded 25 times using a steel rod that is 5/8 in. in diameter and 24 in. long with a hemi-spherically shaped tip. Following the rodding, the last layer is screeded off, and the cone is slowly raised in

FIGURE 17.24 Field testing of concrete consistency using the slump test. (Courtesy of James Cable.)

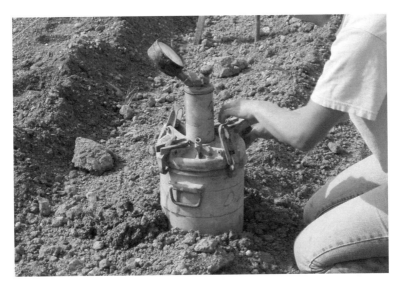

FIGURE 17.25 Field testing of air content in concrete using the air pressure meter. (Courtesy of James Cable.)

approximately 5 ± 2 s. As the concrete subsides or settles to a lower elevation, the empty slump cone is inverted and gently placed next to the settled concrete (note: any additional vibration at this stage will induce additional subsidence in the slumped concrete). The "slump" is measured as the vertical distance of the concrete settles, measured to the nearest 1/4 in. Since concrete loses slump with time, the entire slump test should be completed in 2.5 min. Note that if a portion of the molded concrete falls away or shears off another test should be run on a different portion of the sampled concrete.

17.11.11.2 Air Content

A number of methods for measuring the air content of freshly mixed concrete are available. ASTM standards include the pressure method (C-231; AASHTO T-152), the volumetric method (C-173; AASHTO T-196), and the gravimetric method (C-138; AASHTO T-121). The pressure method (Figure 17.25) is based on Boyle's law, which relates pressure to volume. The air pressure meters are calibrated to read air content directly when a predetermined pressure is applied. The pressure meter method is the most commonly used method for determining air content in fresh concrete. The test requires less time than other methods, and the specific gravities for the concrete ingredients need not be known compared to the gravimetric method. The volumetric method is outlined in ASTM C-173 (AASHTO T-196) and requires removal of air from a known volume of concrete by agitating the concrete in an excess of water. The addition of alcohol accelerates the removal of air. The gravimetric method is based on determining the density (unit weight) of concrete. The measured density of concrete is subtracted from the theoretical density as determined from the absolute volumes of the ingredients, assuming no air is present (ASTM C-138 or AASHTO T-121).

17.11.11.3 Density and Yield

The unit weight of the concrete is a good indicator of uniformity of a concrete mix from mix to mix. The density (unit weight) and yield of freshly mixed concrete are determined in accordance with ASTM C-138 (AASHTO T-121). A calibrated known-volume bucket (usually $\frac{1}{2}$ ft^2) is filled with the fresh concrete in approximately three equal layers and rodded 25 times per layer. The quotient of weight (or mass in SI units) and volume is the unit weight or density. The volumetric quantity of concrete produced per batch is called the *yield*, and is determined by summing the weights of all ingredients in the batch and dividing by the density of the fresh concrete. An approximate calculation of the volumetric air in concrete can be conducted by using the density and yield values provided the relative densities of the ingredients are known.

17.11.11.4 Sampling Fresh Concrete

It is critically important to obtain truly representative samples of freshly mixed concrete for quality control tests; otherwise, the test results will be misleading. Samples should be obtained and handled in accordance with ASTM C-172 (AASHTO T-141). It should be noted that a sample composed of two or more portions should not be taken from the very first or last portion of the batch discharge and should be protected from sunlight, wind, and other sources of rapid evaporation during sampling and testing.

QUESTIONS

17.1 What are the different stages of concrete pavement construction?

17.2 What are the two methods of paving concrete, and explain the differences?

17.3 What is the importance of jointing?

17.4 What are the concrete properties and environment conditions that affect the saw-cutting and how?

17.5 What are the different types of steel provided for concrete pavement construction, and why?

17.6 Why is curing important?

17.7 Describe the various curing technologies commonly used.

17.8 What are the various finishing and texturing applications and describe the differences between them.

17.9 What are the different tests performed for QC/QA operations?

17.10 What concrete tests are used to monitor concrete performance at an early age versus a later age?

18 Maintenance and Rehabilitation of Pavements
Pavement Management Systems

18.1 OVERVIEW

Maintenance and rehabilitation (M&R) of pavements should ideally be conducted through the use of pavement management, which can ensure the optimum use of tax dollars through the selection and use of the most cost-effective design, construction, and rehabilitation strategy. Pavement management involves all activities regarding the planning, design, construction, maintenance, and rehabilitation of pavements. It is based on the pavement management system (PMS), which consists of a set of tools or methods that help pavement managers to plan for constructing and maintaining highway or airport pavements in a serviceable condition over a given period of time.

Pavement information management systems (PIMS) or PMSs were mentioned in the 1986 *AASHTO Design Guide*. This guide defined PMS, indicated its importance, and presented the steps needed to adopt and implement it. The guide also discussed the information needed, the importance of quantification of pavement performance and optimization in selecting strategies, different levels of PMS, and steps for improving PMS. The *Pavement Management Guide*, published by AASHTO in 2001, is the latest guide on PMS from AASHTO.

Pavement management is often considered as part of a bigger management scheme—such as asset management (by the U.S. Federal Highway Administration System)—and requires the consideration of a more holistic approach in the consideration, analysis, and implementation of highways. In the PMS, the implementation activities can be separated into three distinct parts: corrective maintenance, preventive maintenance, and rehabilitation. Corrective maintenance is provided in response to an existing problem, such as filling existing cracks; preventive maintenance is implemented in anticipation of a drop in the quality of the pavement, such as providing a thin overlay to extend the life of the pavement by another 5 years; and rehabilitation is adopted when a major structural improvement of the pavement is required, such as by recycling, placement of a new overlay, or total reconstruction.

PMS is implemented in two different levels—network and project. In the network level, PMS is utilized to select the best strategies for design, construction, and rehabilitation of all the pavements within an agency, to result in the best benefit-to-cost ratio within a given analysis period. In the project level, PMS is used to select the best design, construction, or rehabilitation alternative for a specific project within the network such that the project results in the maximum benefit-to-cost ratio over the given analysis period.

18.2 STEPS IN PMS

The information on the traffic, existing condition, environmental data, and construction history of the pavement network provides the basis of all subsequent activities in a PMS. Particular attention needs to be given to accurate but practical testing of existing pavements—the use of automated and nondestructive testing is gaining more and more favor. This information/data can be used to develop models for predicting the future condition of the pavements. This condition can be with

515

respect to a conditioning index or residual life. Linked to this information is user input information on available M&R techniques, and their relative benefits and costs. The benefits and costs could be expressed simply in terms of extension of life or improvement in condition and money, or in terms of those factors as well as other fuel consumption of vehicles, user delays, and tire wear. The information collected and obtained by analysis up to this point can then be used to select the most appropriate M&R activity for any specific pavement at the most appropriate time so as to make the best use of the available budget. This step is carried out, in increasing order of sophistication, through ranking, prioritization, and optimization. This involves consideration of the "consequence" of adaptation of several alternative strategies on the condition of the different pavements. The optimization can be based on different criteria, for example, either on the concept of minimizing the total cost while keeping all pavements at or above a minimum condition, or on maximizing the total benefit, with the available budget in the PMS. Finally, mathematical modeling is used to select the most optimum strategy.

Therefore, the steps in PMS can be broadly divided into the following:

1. Collect information on pavements (distress survey).
2. Set up criteria for making decisions.
3. Identify alternative strategies.
4. Predict cost and performance of alternative strategies, and compare them.
5. Select and implement the most cost-effective strategy.

18.3 DIFFERENT PMS APPROACHES

Although the same in principle, there are different types of PMSs software, with appropriate models available to the user. For example, the World Bank has developed the Highway Development and Management System (HD-4) as a tool to evaluate pavement construction and maintenance strategies using technical and economical considerations. Its successful use depends on the calibration of pavement performance prediction models to local conditions for each agency.

MicroPaver, based on the work by the U.S. Army Corps of Engineers (n.d.), is a pavement management software that can be used for selection of cost-effective maintenance and repair (M&R) strategies for highways and airfield pavements. It allows the user to prepare a network inventory with condition ratings, develop pavement condition deterioration models, predict the condition of a pavement at a specific time and determine M&R needs, and evaluate the effect of different budget strategies. MicroPaver (also referred to as *Paver*) can be obtained from the University of Illinois at Urbana-Champaign, Technical Assistance Center (phone: 800-895-9345; e-mail: techctr@uiuc. edu). The key steps in the PMS are illustrated next, with reference to the MicroPaver approach.

A complete PMS consists of a series of steps, which starts with setting up a pavement inventory for the network that needs to be analyzed (e.g., all of the pavements in a city or county or in an airport). The network is made up of several branches, each of which consists of several pavement sections. Each section should have specific information regarding its inventory, maintenance, and inspection information. Images of distresses could be stored and retrieved for each section. Input of work, traffic, and test (nondestructive and destructive) data could also be stored for each section. The results of a pavement inspection are then input for each "sample unit," several of which make up one section. Using inbuilt tables of distresses (which are also coded with two-digit numbers for speedy entry) specific to the type of the pavement being inspected, the type, severity, and quantity of the distresses observed during inspection could be input. Video inspection data could also be input. An example of a pavement survey form as used in MicroPaver is presented in Table 18.1.

The next step is the determination of the Present Condition Index (PCI) of the entire section, using the inbuilt condition calculator in the software (for an example, see Table 18.2). PCI is a numerical indicator that rates the surface condition of the pavement, and is used as a measure of the present condition of the pavement. It can be used as a basis for establishing maintenance and

TABLE 18.1

Example of Pavement Survey Form

Distress	Description	Severity	Quantity	Units
1	Alligator cracking	L	6	ft^2
1	Alligator cracking	M	30	ft^2
4	Bumps/sags	L	28	ft
7	Edge cracking	L	142	ft
7	Edge cracking	M	15	ft
10	Longitudinal/transverse cracking	L	26	ft
10	Longitudinal/transverse cracking	M	27	ft
10	Longitudinal/transverse cracking	H	28	ft

TABLE 18.2

Present Condition Index of Several Sections in a Network

Age at Inspection	PCI	Model	Difference	Status	Network ID	Branch ID	Section ID	Surface	Rank	Inspection Date
9	61	55	6		1	IFARB	1	AAC	S	10/16/1992
1	100	95	5		1	IFARB	1	AAC	S	3/24/1984
10	64	49	15		1	IFARB	1	AAC	S	10/12/1993
8	72	60	12		1	IFARB	1	AAC	S	10/16/1991
7	80	65	15		1	IFARB	1	AAC	S	5/23/1990
6	79	70	9		1	IFARB	1	AAC	S	6/6/1989
4	89	80	9		1	IFARB	1	AAC	S	5/6/1987
3	94	85	9		1	IFARB	1	AAC	S	10/13/1986
2	95	90	5		1	IFARB	1	AAC	S	9/28/1985
1	100	95	5		1	IFARB	1	AAC	S	9/30/1984
11	50	44	6		1	IFARB	1	AAC	S	11/17/1994
1	94	95	1		1	IINTE	1	AAC	P	9/30/1984
8	44	60	16		1	IINTE	1	AAC	P	10/16/1991

repair needs. The procedure for determination of PCI is explained in ASTM D-6433-03 for roads and in ASTM D-5340-04 for airfields. The procedure consists of the following: conducting a survey of distress (quantity and density); determining a "deduct value" for each distress and computing the total corrected damage value, according to guidelines given in the standard; and then subtracting the total corrected damage value from 100, to obtain the PCI at any time. The PCI can range from 0 to 100, corresponding to a condition of "failed" to "excellent," respectively.

MicroPaver uses a hierarchical structure with networks, branches, and sections in its inventory management. The results of an inspection can be input through an interface, and the online distress guide can be consulted to aid in the inspection and input. The distress type, quantity, and severity are combined to determine the PCI and then the condition of the pavement.

The results give the condition index and values along with distress summaries (and projected distresses). Based on minimum condition criteria or projected deterioration rates, an inspection schedule report can be generated. Finally, summary charts can be used to develop a graph and compare any two attributes of the pavement network. The software allows the user to query the whole database or for specific sections to generate specific reports. Geographical Information Systems (GIS) tools have been integrated into MicroPaver from its 5.1 version, which allows the user to link individual pavement sections to available GIS databases and display summaries and PCI reports in graphical format.

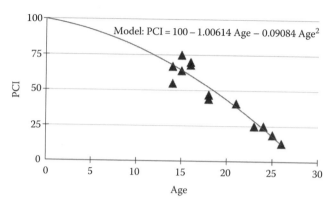

FIGURE 18.1 Change in PCI with age of pavement.

For the purpose of building predictive models relating age with condition/performance, Paver identifies and groups families of pavements that are constructed in the same way and are subjected to similar conditions such as traffic and environment. This process is called *family modeling*. Each pavement section is assigned to a family, and the family model is used to predict the section's future condition. Options of using unique knowledge about specific pavement sections, and viewing (such as outliers, equations, and coefficients of regression) and selecting the statistics (such as constraints) used to generate the models, are available. An existing model can be modified or updated with new data. The data are used to form predictive equations/models relating age with PCI, and then view the condition of the entire pavement network or any specific portion of it. The models (see an example in Figure 18.1) can be used to view the condition of any pavement section (condition analysis) at any given time, as well as to view the deterioration of the condition of the pavement with age—the results are provided in the condition analysis report.

Several forms of information can be viewed in this report, including the average condition of pavement for each year in the reporting period, and a histogram for each year with sections grouped into several PCI groups (which range from excellent to failed). *Critical PCI* is defined as the PCI value at which the rate of PCI loss increases with time, or the cost of applying localized preventive maintenance increases significantly.

18.3.1 CRITERIA FOR MAKING DECISIONS

MicroPaver provides a "work plan" tool that helps the user in planning, scheduling, budgeting, and analyzing alternative M&R activities, for specific sections and analysis periods. There are three ways in which the analysis can be conducted: (1) critical PCI methods, which optimize the M&R activity against a specific available budget or determine the budget needed to maintain a specified condition level; (2) minimum condition, which works by rationing M&R by pavement condition; and (3) consequence model, which works by measuring the impact of localized M&R action over the first year.

Inbuilt in the Paver database is a list of global maintenance work types, along with application interval, improvement in the condition every year (delta), and cost (which can be edited as required). Cost of major M&R is determined by the PCI at the time the work is performed. *Global M&R* is defined as activities applied to entire pavement sections with the primary objective of slowing the rate of deterioration. This policy is applied to pavements above the critical PCI. The work quantity is counted using "multipliers" listed for specific distress types and quantities. Paver allows the user to evaluate the effect of a selected budget, and compare it with that for unfunded M&R on the condition of a pavement (Figure 18.2). Options are also available to overlay the section condition plot with family condition plots, view the difference between the predicted condition and that with the M&R option, and the difference in condition from year to year.

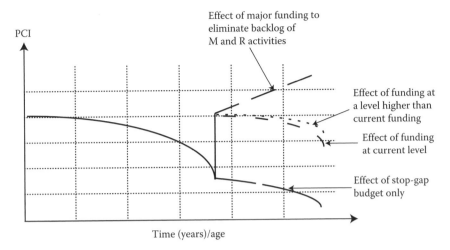

FIGURE 18.2   Conceptual change in PCI with time for different budget scenarios.

18.4 DISTRESS SURVEY

Distress surveys involve the measurement and evaluation of type, extent, and intensity of various types of distresses—surface defects, rutting or distortion, cracking, and patching. Such surveys can be conducted by an individual walking along a section of the pavement or by using multiple cameras from a moving specialized vehicle. An example of a vehicle used for such surveys is shown in Figure 18.3. The vehicle is equipped with sensors and video cameras for continuous recording of different types of distresses such as cracks, which can be analyzed by pavement managers. Guidelines for identifying and quantifying different types of distresses are provided in ASTM D-6433 for roads and ASTM D-5340 for airfields.

Structural evaluation of pavements can be done in a number of different ways. In general, the deflection or the curvature of a pavement when subjected to a specific load is used for such measurement. Such evaluation can be first done with large spacing, followed by testing at closer locations to quantify the structural capacity, taking into consideration the variability of the results. The falling weight deflectometer can be used according to ASTM D-4694; the equipment is shown

FIGURE 18.3   Automatic road analyzer.

FIGURE 18.4   Falling weight deflectometer.

in Figure 18.4. The basic principle is the use of impulse load to measure the resulting deflection in the pavement—the higher the deflection for a specific load, the worse is the structural condition of the pavement.

Skid resistance of pavements is measured as an evaluation of pavement safety. Skid resistance is usually measured in terms of friction factor, f, defined as F = F/L, where F is frictional resistance to motion in the plane of the interface, and L is the load acting perpendicular to the interface. Since the friction factor is dependent on many parameters such as tire type, speed, temperature, and water film on pavement thickness, standards are followed to measure it. Measurements can be done according to ASTM E-274 (using a full scale tire). In this test, a locked tire is towed by a vehicle on a wet pavement (water is sprayed in front of the tire), and friction force between the tire and the pavement surface and the speed of the vehicle are recorded (Figure 18.5). The testing results in the determination of a skid number (SN), which is calculated as follows:

FIGURE 18.5   Skid tester. (Courtesy of David Morrow, Dynatest Consulting Inc., Ventura, CA.)

$$SN = \frac{F}{W} * 100$$

where
   F is the tractive horizontal force applied to the test tire at the tire–pavement contact patch, lbf
   W is the dynamic vertical load on the test wheel, lbf

Note that the type of tire used in the test (ribbed or smooth) must be noted in the results by following the appropriate designation in the standard. In general, instrumentation in modern equipment automatically and continuously measures the SN for a specific test time period and averages it for reporting.

For pavements such as those in airports, under icy conditions, or in winter, spot measurement of frictional properties can also be conducted with the use of a specialized test vehicle, according to ASTM E-2101. This is important to detect otherwise hidden effects of contaminants on the pavement surface that can lead to unsafe conditions. In this test (valid only for icy conditions where the depth of loose snow does not exceed 1 in.), the steps consist of accelerating a vehicle over the pavement and applying brakes to lock all the wheels, and measuring the deceleration (which is related to the friction between the tire and the pavement surface) of the vehicle during braking.

For airports, the International Runway Friction Index (IRFI) is reported as a harmonized value of the friction characteristics of the pavement (ASTM E-2100). The parameter is calculated from friction numbers determined by local friction devices through established correlations, and can be used to monitor winter frictional characteristics of runways for maintenance.

Surface roughness can be measured using different types of profilometers, such as through a multiwheel >23-ft-long profilometer (ASTM E-1274). The test consists of moving the profilometer over the pavement at less than 3 mph, and determining the roughness and rate of roughness of the pavement as the height of each continuous "scallop" or excursion of the surface records above and below the base reference level. Roughness can also be measured by vehicular response to pavement roughness (ASTM E-1082). In this type of device, the response of a spring mass system supported by a vehicle suspension system in response to pavement roughness is measured using sensors.

18.5 MAINTENANCE AND REHABILITATION OF ASPHALT PAVEMENTS

The choice between maintenance and rehabilitation is generally made on the basis of the existing surface and structural condition of the pavement. For example, if the surface condition is above a certain level, maintenance can be selected, whereas if it is below that level, both maintenance and rehabilitation can be considered. For the latter case, depending on the structural condition of the pavement, either a significant maintenance activity (such as a thin overlay) or rehabilitation (reconstruction) can be selected. Table 18.3 shows a general guideline for the selection of different techniques.

18.5.1 MAINTENANCE

Proper maintenance of pavements is vital for the safety and comfort of the traveling public, and the overall economy of a nation. Furthermore, appropriate maintenance work results may result in significant enhancement of pavement life and/or lowering of rehabilitation cost in the future. The importance of proper maintenance has been recognized worldwide, and steps are being taken to develop better techniques for selecting appropriate techniques and new techniques. For example, in the United States since 1997, the FHWA, along with the Foundation of Pavement Preservation (FPP) and AASHTO, have vigorously started the adoption of good practices of pavement maintenance under the common theme of "pavement preservation" (www.fhwa.dot.gov/preservation, FHWA, n.d.;

TABLE 18.3
Maintenance and Rehabilitation Methods

Distress	Maintenance	Rehabilitation	Appropriate Recycling Process
Fatigue/alligator cracking		Reconstruction, thick hot mix asphalt (HMA) layer	Cold in-place recycling (CIR), full-depth reclamation (FDR) with overlay
Bleeding	Chip seal		
Block cracking	Slurry seal, chip seal, sealing	Overlay	CIR, FDR with overlay
Corrugations	Thin overlay	Thick overlay	CIR with thin overlay
Joint cracks	Seal		
Polished aggregate	Chip seal, slurry seal, open-graded friction course		
Potholes	Patching, full-depth repair		CIR
Slippage cracks	Thin overlay on milled surface	Thick overlay	CIR with thin overlay
Thermal cracks	Thin overlay		Hot in-place recycling (HIR) with/without thin overlay
Rutting	Thin overlay	Thick overlay	CIR with thin overlay
Raveling	Chip seal, slurry seal, fog seal, sand seal, thin overlay		HIR with/without thin overlay

Note: For overlays on pavements with surface defects such as corrugations, the pavement must be milled to the depth of good mix prior to the application of the overlay.

www.fp2.org, FPP, n.d.). A wealth of information can be accessed from the reference library section of the National Center for Pavement Preservation (NCPP, n.d.) at http://www.pavementpreservation. org/reference/. Some of the more important corrective and preventive maintenance activities are discussed later.

18.5.1.1 Primary Corrective Maintenance Activities

Crack *sealing* is the placement of specialized materials into working cracks (horizontal movement >2 mm) to reduce the intrusion of incompressible materials and prevent intrusion of water into the underlying pavement. The sealant material must be an agency-approved product, selected on the basis of temperature and traffic, and applied properly (e.g., with a backer-rod) on a clean and dry surface. The preparation equipment (such as a melter for hot applied sealant materials) must be in good working condition, and the operation must be completed in dry weather conditions with minimum temperature (typically 4°C and rising). Crack *filling* is the placement of materials into nonworking cracks (horizontal movement <2 mm) to substantially reduce infiltration of water and reinforce adjacent pavement.

For pothole repairs, *pothole patches* must be made with good-quality HMA or cold mix, using a fine mix with adequate binder for workability, compaction, and durability, after cleaning, drying, and applying a tack coat in the existing potholes. The patching material could be of locally available cold mix, mix produced according to specifications, or propriety cold mix. The compatibility between aggregate and asphalt binder must be ensured for mixes. In the conventional method, a pothole is patched by placing the materials and compacting, say, with truck tires to keep at least a 3–6 mm crown on the pothole. Prior to placing the materials, the water and loose debris from the pothole must be removed. Cutting the pothole sides to nearly vertical surfaces, applying tack coat on the sides, and utilizing a small roller or plates for compaction would result in a better patching. A large variety of spray injection techniques are now becoming available for pothole patching.

18.5.1.2 Primary Preventive Maintenance Activities

Slurry seal: A slurry seal consists of a mixture of well-graded fine aggregates and mineral filler with dilute asphalt emulsion, often with additives or modifiers such as hydrated lime. The mix is generally produced by a slurry machine at the site and applied to the pavement with a squeegee. Slurry seal is effective in sealing minor cracks, reducing the potential of raveling, preventing further oxidation of the asphalt of the surface layer, and improving the friction properties of the pavement. Slurry seals with specific gradations are applied for specific pavements with different types of traffic (low, moderate, and heavy). The process requires a curing period of 2–6 h, is not effective in pavements with heavy cracking, and should be used with caution in areas with superelevation where the equipment may not have good control over the flow of the slurry. Compaction by a light rubber-tired roller may be needed.

Chip seal: A chip seal consists of an application of asphalt binder or rapid-setting emulsion followed with an application of aggregate layer (Figure 18.6). If multiple layers are used (such as double or triple seal coats), finer gradations are used in each successive layer. Precoated aggregates could also be used, and one-size aggregates are often used. Compaction of the aggregate layer is required by a steel-wheeled or rubber-tired (preferred) roller. The minimum pavement temperature for the use of a chip seal is 15°C. Chip seals can be used as only wearing layers in pavements with light traffic. If aggregates are not properly embedded, windshield damage can result. A *rubberized* chip seal contains ground tire rubber in addition to the usual components of a chip seal. Better resistance against climate- and traffic-induced stress is obtained in this seal, as well as resistance against reflective and minor fatigue cracks. It is referred to as a stress-absorbing membrane (SAM) or stress-absorbing membrane interlayer (SAMI) when used underneath an HMA overlay.

Sandwich seal: This is the application of a chip seal using two layers of aggregate with one application of asphalt in between. Enhanced frictional properties can be obtained with this method.

Cape seal: A cape seal (originally from Cape Province of South Africa) consists of a chip seal covered with a slurry seal. It prevents the occurrence of loose stones in chip seals, and could be used in pavements with high traffic volumes. It does require a curing period of 2–6 h.

Fog seal: A fog seal consists of the application of a diluted asphalt emulsion (slow or medium setting) without the application of any aggregate. It is used for preventing raveling and for sealing surfaces to resist oxidation of the surface layer. The temperature should be above 16°C for application, and the layer needs some curing time before opening to traffic. The pavement should be porous enough to absorb some of the asphalt emulsion, and until some of the emulsion of the surface is worn away by traffic, friction may be reduced. A better friction can be obtained by the use of *sand seal*, in which the sand is applied over a rapid- or medium-setting emulsion layer.

FIGURE 18.6 Chip seal. (Courtesy of Maureen Kestler, USDA Forest Services, Washington, DC.)

Microsurfacing: Microsurfacing consists of the application of a mixture of high-quality aggregates and polymer-modified emulsion binder, with additives. The components are mixed in a traveling pug mill, and areas with traffic do not require any compaction. Microsurfacing is used to seal the pavement and prevent raveling. It can also be used to resist oxidation of the surface layer, improve friction, and fill in minor ruts. This process uses relatively fast-setting emulsions and requires mix design for determination of the asphalt content, and sophisticated equipment for laydown.

Thin HMA overlay: An open-graded friction course can be used. Generally such layers are ½ to 1 in. thick, with gap-graded aggregates, and increasingly with polymer-modified asphalt binders. These layers are used to facilitate quick drainage of water and improve tire–pavement contact and hence friction. Also known *as porous friction courses* (PFC), these layers can contain ground tire rubber as well as high asphalt contents to improve adequate durability. Generally these layers do not require any compaction, but do require sealing of cracks, if any, as well as relatively low voids in the underlying layers.

Ultrathin friction course: This is a procedure in which an HMA layer made with polymer-modified asphalt and gap-graded aggregate is placed on a heavy tack coat of polymer-modified emulsion asphalt. Products such as Novachip are applied in one pass (binder, aggregates, and screeding) by a single specialized piece of equipment. It is used for sealing the surface and minor cracks, increasing friction, and reducing tire noise.

The reader is advised to view the latest NCHRP publication on emulsion-based chip seals, available at http://onlinepubs.trb.org/onlinepubs/nchrp/nchrp_rpt_680.pdf.

18.5.2 RECYCLING

The Asphalt Recycling & Reclaiming Association's (ARRA) publication *Asphalt Recycling Manual* (ARRA, 2003; and see www.arra.org) is a good source of information on asphalt pavement recycling. The common types of recycling operations include hot mix recycling, hot in-place recycling (HIR), cold in-place recycling (CIR), and full-depth reclamation (FDR). Of these, hot mix recycling is used very commonly for producing hot mix asphalt, which can be used as overlays in preventive maintenance or as thick layers in rehabilitation. Hot in-place and CIR are commonly used for preventive maintenance operations, whereas full-depth reclamation is generally used for rehabilitation work. The preventive maintenance applications are discussed first, followed by the rehabilitation methods. (The FHWA guidelines for recycling for state and local governments are available at http://www.fhwa.dot.gov/pavement/recycling/98042/.)

18.5.2.1 Hot In-Place Recycling

Hot in-place recycling has been described as an on-site, in-place method that rehabilitates deteriorated asphalt pavements and thereby minimizes the use of new materials. Basically, this process consists of four steps: (1) softening of the asphalt pavement surface with heat; (2) scarification and/or mechanical removal of the surface material; (3) mixing of the material with recycling agent, asphalt binder, or new mix; and (4) laydown and paving of the recycled mix on the pavement surface. The primary purpose of hot in-place recycling is to correct surface distresses not caused by structural inadequacy, such as raveling, cracks, ruts and holes, and shoves and bumps. It may be performed as a single-pass operation or a multiple-pass operation. In a single-pass (Figures 18.7 and 18.8) operation, the virgin materials are mixed with the restored reclaimed asphalt pavement (RAP) material in a single pass, whereas in the multistep process, a new wearing course is added after recompacting the RAP materials.

The advantages of hot in-place recycling are that elevations and overhead clearances are preserved, it is comparatively economical, and it needs less traffic control than the other rehabilitation techniques. This process can also be used to recoat stripped aggregates, reestablish crown and drainage, modify aggregate gradation and asphalt content, and improve surface frictional resistance.

FIGURE 18.7 Schematic of single-pass hot in-place recycling equipment. (Courtesy of Wirtgen, GmbH, Windhagen, Germany.)

FIGURE 18.8 Hot in-place recycling.

Hot in-place recycling is usually performed to a depth of 20–50 mm (3/4–2 in.), with 25 mm (1 in.) being a typical depth.

18.5.2.2 Cold Recycling

Cold recycling can be divided into two main parts—CIR and central plant (cold mix) recycling. CIR can further be divided into two parts—CIR (limited to the asphalt surface or/surface plus binder layer) and full-depth reclamation (recycling of asphalt layer plus part of the granular base material). Cold milling is also used for obtaining materials for hot mix recycling (discussed later in this chapter).

18.5.2.2.1 Cold Milling

The advent of cold milling (cutting) has revolutionized the recycling of asphalt pavements. Cold milling has been defined as the method of automatically controlled removal of pavement to

a desired depth with specially designed equipment, and restoration of the surface to a specified grade and slope, free of bumps, ruts, and other imperfections. The Asphalt Recycling &
Reclaiming Association guideline specifications for cold milling require that the milling machine
be power operated, self-propelled, and self-sufficient in power, traction, and stability to remove
a thickness of HMA surface to a specified depth. The alternative to cold milling is ripping and
crushing operations with earthmoving equipment, scarifiers, grid rollers, or rippers, which are not
commonly used today.

The modern cold-milling equipment has tungsten carbide teeth on drums, with variable cutting width for a variety of pavements/lanes, and excellent maneuverability for difficult milling
situations like around utility inspection covers in city streets. Figure 18.9 shows a few examples
of milling equipment. In many cases, such equipment has been integrated with a system of
spraying recycling/stabilizing materials, mixing, and compacting, in recycling equipment, as
shown in Figure 18.10.

Milling drum   Close-up view of milling drum in   Schematic of changeable
                    recycler WR 4200 showing teeth   tooth holder, lower part welded
                                                          to drum upper part exchangeable

Closed milling width, 2.80 m

Maximum milling width, 4.2 m

Schematic of variable width milling

Extension cutters   Position of foamed asphalt bar

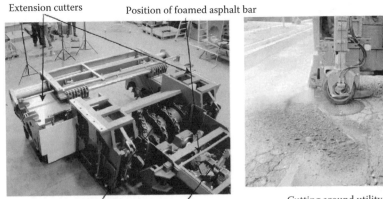

Cutter cutter   Cutting around utility on city street
          Twin-shaft pugmill

Milling (cutting) drums in WR 4200

FIGURE 18.9 Modern cold-milling equipment. (Courtesy of Mike Marshall, Wirtgen GmbH, Windhagen,
Germany.)

Integrated cutting, spraying, mixing and compaction unit with options for using multiple stabilizing materials

Variable mixing space

FIGURE 18.10 Combined milling and recycling equipment. (Courtesy of Mike Marshall, Wirtgen GmbH, Windhagen, Germany.)

18.5.2.2.2 Cold In-Place Recycling

CIR is defined as a rehabilitation technique in which the existing pavement materials are reused in place. The materials are mixed in-place without the application of heat. The RAP material is obtained by milling (cutting), planning, or crushing the existing pavement. Virgin aggregate or recycling agent or both are added to the RAP material, which is then laid and compacted. The use of CIR can restore old pavements to their desired profile, eliminate existing wheel ruts, restore the crown and cross slope, and eliminate potholes, irregularities, and rough areas. It can also eliminate transverse, reflective, and longitudinal cracks. Some of the major reasons for the increased use of CIR are the increased scarcity of materials, particularly gravel and crushed rock; the method's high production rate and potential of cost savings; minimum traffic disruption; the ability to retain the original profile; the reduction of environmental concerns; and a growing concern for depleting petroleum reserves.

The steps in CIR consist of preparation of the construction area, milling of the existing pavement, the addition of recycling agent and virgin materials, laydown, compaction, and placement of the surface course.

The addition of new aggregates may not be necessary in some projects. At present, two different methods are used for CIR: the single machine and the single-pass equipment. The single machine or single-pass equipment is capable of breaking, pulverizing, and adding recycling agents in a single pass (Figure 18.11). The single-pass equipment train consists of a series of equipment, each capable of a particular operation. The usual components are a cold-milling machine, portable crusher, travel-plant mixer, and laydown machine.

Sometimes cold central plant recycling is conducted to make use of existing RAP stockpiles and/or make recycled mixes for later use. Sophisticated equipment is available to prepare such mixes, using one or more recycling additives, as shown in Figure 18.12.

18.5.2.2.3 Full-Depth Reclamation

Full-depth reclamation has been defined as a recycling method where all of the asphalt pavement section and a predetermined amount of underlying materials are treated to produce a stabilized base course. Different types of additives, such as asphalt emulsions, and chemical agents, such as calcium chloride, Portland cement, fly ash, and lime, are added to obtain an improved base. The five main steps in this process are pulverization, introduction of additive, shaping of the mixed material, compaction, and application of a surface or a wearing course. If the in-place material is not sufficient to provide the desired depth of the treated base, new materials may be imported and included in the processing. This method of recycling is normally performed to a depth of 100–300 mm (4–12 in.). The major advantages and benefits of full-depth reclamation are as follows:

1. The structure of the pavement can be improved significantly without changing the geometry of the pavement and shoulder reconstruction.
2. It can restore old pavement to the desired profile, eliminate existing wheel ruts, restore crown and slope, and eliminate potholes, irregularities, and rough areas. Pavement-widening operations can also be accommodated in the process. A uniform pavement structure is obtained by this process.
3. It can eliminate alligator, transverse, longitudinal, and reflection cracking. Ride quality can be improved.
4. Frost susceptibility may be reduced.
5. The production cost is low, and only a thin overlay or chip seal surfacing is required on most projects.
6. Engineering costs are low.
7. Materials and energy are conserved, and air quality problems resulting from dust, fumes, and smoke are eliminated. The process is environmentally desirable, since the disposal problem is avoided.

Full-depth reclamation has been recommended for pavements with deep rutting, load-associated cracks, nonload-associated thermal cracks, reflection cracks, and maintenance patches such as spray, skin, pothole, and deep hot mix. It is particularly recommended for pavements having a base or subgrade problem. In this method, the first step is to rip, scarify, or pulverize or mill the existing pavement to a specified depth. The resulting material can be processed further for size reduction and mixed with recycling agents and new materials, if required. These include the multiple-step sequence, two-step sequence, single machine, and single-pass equipment train. Modern equipment is available for single-pass FDR, with the option of using water, emulsion, foamed asphalt, or cement slurry as the recycling additive. Figure 18.13 shows an example of an FDR process with cement slurry.

Foamed asphalt is being used increasingly in FDR. Foaming facilitates better dispersion of the asphalt into the materials to be recycled. A small amount of water is sprayed into hot asphalt as it is mixed with pulverized recycled pavement and soil. As the hot liquid and water mix, the liquid expands in a mini-explosion, creating a thin film of asphalt with about 10 times more coating potential. In modern single-pass equipment, foamed asphalt is created within the equipment in a separate foaming chamber and is directly added to the pulverized road material. An example of such equipment is shown in Figure 18.14.

Project Information

Traffic	Existing pavement structure	Distress in existing pavement	CIR method used
38,500 vehicles per day, 16% trucks	150–1200 mm HMA, 200 mm cement treated base, 300 mm aggregate subbase	Block cracking, wheel path potholes, fatigue cracking, severe aging of the surface mix	Recycle with 1.5% cement and 2.5% foamed asphalt; recycling depth: 60 mm for shoulders, 100 mm for mainline; apply fog seal with sand, with application of diluted emulsion at the rate of 0.008–0.0100 L/m^2

Cracks in existing pavement

Asphalt tanker
Wm1000 cement
Slurry mixer
WR 4200 recycler

Recycling train

WR 4200 recycler

Recycling with traffic in adjacent lane

WR 4200 fitted with integral vogele AB500 tamping and vibrating paving screed

Recycled material immediately behind the WR 4200 paving screed

Compaction with steel drum roller

Finish rolling with rubber tired roller

Recycled shoulder   Recycled no.1 lane   Mill and fill no.2 lane   Recycled shoulder

4 h after completion of recycled lane 1.
Traffic running on recycled mat with fog/sand seal

FIGURE 18.11   Example of a cold in-place recycling project. (Courtesy of Mike Marshall, Wirtgen, GmbH, Windhagen, Germany.)

Cold mixing plant-KMA 200
components

Paving 5 in. thick base layer with vogele paver and
compaction with double drum
vibrating roller and rubber-tired roller

Schematic of KMA 200 cold mixing plant

Paving the second lane: First lane opened to
traffic immediately after compaction

FIGURE 18.12 Example of a central plant cold recycling project. (Courtesy of Mike Marshall, Wirtgen, GmbH, Windhagen, Germany.)

18.5.2.3 Hot Mix Recycling

Hot mix recycling has been defined as a method by which RAP is combined with new aggregate and an asphalt cement or recycling agent to produce HMA. The RAP may be obtained by pavement milling with a rotary drum cold-milling machine or from a ripping/crushing operation.

The objective of the next step, crushing, is to reduce the RAP to the maximum acceptable particle size. One example of such a limit is that at least 95% of the RAP passes the 2 in. sieve. Cold-milling machines can crush the RAP in place, whereas in the ripping/crushing operation, front-end loaders are generally used to break up the pavement material so that it can be loaded into a truck for crushing at a central plant. The amount of aggregate degradation by cold milling is a function of the aggregate top size and gradation of the HMA pavement. For crushing in a central plant, different types of crushers are available, for example, compression crushers and impact crushers.

Variant: injection of emulsion; premixing cement and
water in the slurry mixer WM 1000 for injection as slurry

Schematic of FDR with cement slurry

FDR using WR 2500S recycler and WM 100 cement
slurry equipment

300 mm FDR with cement slurry with 5.8% cement

FIGURE 18.13 Example of a full-depth reclamation project with cement slurry. (Courtesy of Mike Marshall, Wirtgen, GmbH, Windhagen, Germany.)

FIGURE 18.14 Schematic and use of foamed asphalt recycling equipment. (Courtesy of Mike Marshall, Wirtgen, GmbH, Windhagen, Germany.)

A "RAP breaker" or "lump breaker" is also used between the bin and the belt for size reduction. Impact crushers are most widely used in recycling. To produce a homogeneous RAP product, the RAP material is first blended thoroughly with a front-end loader or a bulldozer, and then crushed to downsize the top stone size in the RAP to one smaller than the top size in the HMA being produced (e.g., 16 mm for a 19.5 mm top-size mix). This is to ensure that the asphalt aggregate blend is broken as much as possible and there is no oversize material.

RAP from different sources and containing different asphalt content and aggregates with different gradations should be stockpiled separately. RAP can be stockpiled either before or after processing, and a front-end loader or a radial stacker can be used for the purpose. The two major problems associated with stockpiling are consolidation and moisture retention. In the past, it was believed that low, horizontal stockpiles are better than high and conical stockpiles, which can result in reagglomeration of RAP. However, experience has actually proven that large, conical stockpiles are better and that RAP does not recompact in large piles. Actually, there is a tendency to form a crust over the 20–25 cm (8–10 in.) of pile depth. This crust can be easily broken by a front-end loader. Also, the crust tends to shed water and prevent the rest of the pile from recompacting. In the case of a low, horizontal stockpile, this 20–25 cm (8–10 in.) crust is also present.

Modifications are required in the batch plant to recycle RAP since attempts to introduce RAP directly with the virgin aggregates result in excessive smoke and material buildup problems in the dryer, hot elevator, and screen tower. The most widely used method for batch plant hot mix recycling is the "Maplewood method." Even though variations exist, basically a separate cold feed bin introduces the RAP into the weigh hopper or the pugmill by a chute and belt conveyor (Figure 18.15). There the RAP is joined by the virgin material coming from the cold feed bins through the dryer and the screen decks. The temperature to which the virgin aggregates are super-heated depends on the properties of the RAP material.

The RAP material cannot be processed in normal drum mix plants since excessive "blue smoke" is produced when the RAP comes in contact with the burner flame. The condition is further aggra-vated by the buildup of fine aggregates and asphalt binder on metal flights and end plates. It has been suggested that most of the smoke problem is caused by the light oils in soft grades of asphalt binder used to rejuvenate the aged asphalt in the RAP. Although the smoke problem could be solved by various processes such as lowering the HMA plant's production rate, decreasing the moisture content of the RAP, lowering the discharge temperature of the recycled mix, introducing additional combustion air, and decreasing the percentage of RAP, it was found that a more effective way to rectify the problem was to modify the drum mix plant.

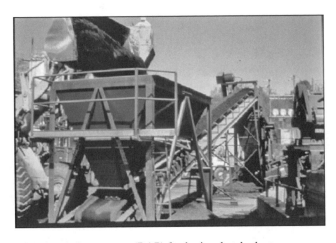

FIGURE 18.15   Reclaimed asphalt pavement (RAP) feeder in a batch plant.

FIGURE 18.16 RAP feeder in a drum plant.

Although variations exist in the process, basically the center entry method (Figure 18.16) is the most widely used method for hot mix recycling in a drum mix plant. In this method, the RAP is introduced into the drum downstream of the burner flame to mix with the superheated new aggregates. The hot virgin aggregates heat up the RAP material by conduction. The RAP is protected from coming in direct contact with the burner flame by a dense veil of aggregate added prior to the point where the RAP is added. It is very important to have the veil of virgin aggregate. Otherwise, overheating of RAP can result in blue smoke, and it may not be possible to use the design amount of RAP material. Sometimes special flight design, steel ring dams, or circular steel flame shields are utilized to force the RAP to mix with the virgin aggregates before being subjected to the high gas treatment. These techniques eliminate the "blue-smoke" problem.

Since the late 1980s, two new drum designs for more efficient heat transfer to RAP material during mixing have been developed. These are the double-barrel and triple-drum design. The double-barrel counterflow drum mix plant has more mixing space than a conventional drum mixer. The shell of the drum is used as the shaft of the coater. A 3.0–3.3-m (10–11-ft) diameter coater is created with an extremely large insulated mixing area. The virgin aggregate material is dried in the inner drum and superheated to 315°C–343°C (600°F–650°F) (when running 50% RAP). It then drops through the wall of the drum and meets with the RAP in the annular space. Approximately 1½ min of mixing time occurs in this outer shell. Since the outer shell does not rotate, easy access is available to add various other recycle components to the process as they become necessary and available. The heat of the inside barrel is transferred through the rotating shell to mixing in the annular space. The outer shell of the double barrel remains at approximately 49°C (120°F) at all times, leading to a very efficient plant. In this method, the virgin and the RAP material are not exposed to the hot gases or to the steam of the drying process, and thus the light oils are not removed from the mix. In the outer section of the double barrel, due to the moisture removed from the RAP, a steam or inert atmosphere occurs, resulting in a much lower oxidation or short-term aging of the recycled HMA mix in the mixing chamber. Another benefit derived from this type of plant is the much longer life occurring with the bags in the baghouse due to the relatively lower temperature of the exhaust gases. As dust is discharged from the baghouse through a rotary airlock on the double-barrel plant, a screw conveyor is used to transfer the mix back into the outer shell. The holes through which the virgin aggregates are directed into the outer shell are also responsible for channelizing any smoke from the inner mixing section to the outer space. The pollutants go directly to the flame, where they are burnt. This results in reduced emission and blue smoke. The counterflow dryer design also leads to higher production rates with much lower fuel consumption. The triple-drum design also uses an outer shell; however, a stainless steel cylinder is used to enclose the combustion chamber. This cylinder (without any flight or steps of a regular drum) is believed to be effective in transferring heat to the RAP material through conduction and radiation. The virgin aggregate is introduced from the opposite end of the burner flame. The RAP material is introduced in the annular space formed by the outer shell. The superheated virgin aggregates fall into the annular space and mingle with the RAP material. The factors controlling the production limit in a drum mix plant are the moisture content and ambient temperature of the RAP and new aggregate. A practical limit of 30:70 (30% RAP and 70% new aggregate) has been recommended in the available literature, although research continues to enable the use of a higher percentage of RAP.

The latest FHWA state-of-the-art report on recycling RAP is available at http://www.fhwa.dot.gov/publications/research/infrastructure/pavements/11021/11021.pdf.

18.6 MAINTENANCE AND REHABILITATION OF CONCRETE PAVEMENTS

18.6.1 JOINT AND CRACK SEALING

Cracks and joints are sealed to prevent water and noncompressible materials from entering and causing potential damage. Different types of sealants are used for this purpose:

Hot-pour liquid sealants are heated and liquefied and then poured into cracks and joints. They can be opened to traffic after they have cooled and gained enough rigidity. They have a service life of 3–5 years.

Silicone sealants are silicone polymer compounds that are poured into joints at ambient temperatures, are commonly ready for traffic loads within 30–60 min of application, and have a service life of 8–12 years.

Compression seals are preformed rubber compounds (commonly neoprene materials) that are forced into a joint under compression. They form a seal by pushing against the walls of the joint, are immediately ready for traffic, and have a service life exceeding 15 years.

18.6.2 SLAB STABILIZATION

The purpose of slab stabilization is to fill voids that have been created beneath the slab due to a combination of pumping, erosion, or consolidation of the base. If left untreated, these voids could grow in size and could potentially contribute to faulting, corner breaks, or cracking. These voids are best filled by pumping grout (cement and fly ash slurry) through predrilled holes through the slab.

18.6.3 DIAMOND GRINDING

Diamond grinding is used to restore pavement surface friction or smoothness. The process uses gang-mounted diamond saw blades to grind off a thin surface layer approximately 2–20 mm thick. The undesirable roughness may have been caused by faulting, studded tire wear, or slab curling. Although not desirable for new construction, diamond grinding can also be used in newly constructed slabs to attain initial specified smoothness.

Patches constructed in rigid pavements are used to treat localized slab problems such as spalling, popouts, scaling, joint deterioration, corner breaks, and punchouts. A partial-depth patch is used if the pavement distress is limited in depth; otherwise, a full-depth patch should be used. HMA is commonly used for emergency patches. However, fast-setting cements are available and can be used to reduce the setting time to a few hours. When constructing permanent patches in PCC pavements, Portland cement or high early strength gain cement should be used.

Spalls may be caused by the infiltration of incompressible materials into joints when they contract and open during cold weather. During warm weather the pavement expands, closing the joints. However, the presence of incompressible materials in the joints will prevent the joints from closing and produce high compressive stresses along the joint faces or walls. As these compressive stresses build up, they may cause spalling of the concrete along the joints. Force may cause spalling at both the top and bottom of the pavements. Other causes of spalling include dowel bar misalignment, corrosion of metal joint inserts, reinforcing steel that has been placed too near the surface, D-cracking, alkali–silica reactions, and lack of consolidation of concrete near the joints.

Partial-depth patches are used to restore localized areas of slab damage that are confined to the upper one-third of slab depth. This is commonly used for moderate spalling and localized areas of severe scaling that do not exceed 3 in. depth and that cover an area less than 12 ft^2.

Full-depth patches are used to restore localized areas of slab damage that extend beyond the upper one-third of slab depth or originate from the slab bottom. Full-depth patches are commonly used for repairing severe spalling, punchouts, corner breaks, severe slab cracking, and localized areas of severe scaling (ACPA, 1995).

Additional information on methods for determining the optimal timing for the application of preventive maintenance treatments in rigid pavements can be found in NCHRP Report 523 (Optimal Timing of Pavement Preventive Maintenance Treatment Applications; http://onlinepubs.trb.org/onlinepubs/nchrp/nchrp_rpt_523.pdf).

18.6.4 Load Transfer Devices

Rehabilitation of load transfer devices between two adjoining slabs involves cutting a slot perpendicular to the joint between the two slabs, inserting a dowel, and refilling the slot with a fast-setting polymer concrete. Combined with other pavement restoration methods, rehabilitation of load transfer devices has the potential to extend the service life of rigid pavements by 10–20 years, depending on the pavement condition at the time of the repair and subsequent pavement traffic loading and environmental conditions.

In 1992, the FHWA Special Project (SP-204; FHWA, 2006) was undertaken to encourage industry to develop equipment for economically constructing slots for retrofit load transfer. This equipment would make load transfer across faulted joints or working cracks a cost-effective maintenance and rehabilitation technique, and would extend the service life by providing positive load transfer for undoweled or transversely cracked JPCP or for working cracks in underreinforced JRCP.

18.6.5 Precast Panels for Repair and Rehabilitation

The use of precast concrete pavements is being evaluated for the application of rapid repair of localized failures in concrete pavements (such as full-depth repairs) and to rehabilitate long lengths of existing poorly performing asphalt and concrete pavements. Other benefits include better control over concrete batching, forming, and curing; increased performance and durability from post-tensioning; savings through reduced pavement thickness; and reduced construction time lane closures.

FHWA demonstration projects in Michigan and Colorado using precast paving for full-depth repairs of jointed concrete pavements were opened to traffic within 6–8 h of lane closure. Precast 6 × 12 ft panels were used to repair deteriorated joints and slabs along sections of the interstate highway system. The repairs involved removal of deteriorated concrete, preparation of the base support, placement of a fast-setting bedding material, installation of precast panels, and installation of retrofitted dowel bars at the transverse joints. These projects are under evaluation to determine long-term performance.

In 2002, the Texas DOT successfully completed the first pilot project utilizing precast panels for pavement construction. Numerous lessons were used in this pilot project, including the following design and construction techniques: (1) use of longitudinal post-tensioning to tie together a series of precast panels to provide a jointless slab length of 250 ft, where each precast panel was also pretensioned in the transverse direction; (2) use of a thinner precast slab compared to the need for a thicker conventional jointed slab; (3) novel post-tensioning techniques; and (4) placement of the precast panels directly over an asphalt concrete base or leveling course.

Concrete Pavement Technology Program (CPTP) Task 52 is the continuation of the FHWA's SP-205 field demonstration program. The SP-205 is developing field-tested guidance on concrete pavement rehabilitation and repair techniques as well as strategies that emphasizes the do's and don'ts, and whys and whens, for concrete pavement restoration (CPR) and preventive maintenance of concrete pavements. The rehabilitation and maintenance strategies considered are full-depth patching, partial-depth patching, subsealing, joint resealing, retrofitted load transfer, and grinding and grooving. Periodic evaluation of the field test sites is being carried out under CPTP. About 40 sites are under evaluation.

18.6.6 PORTLAND CEMENT CONCRETE OVERLAYS

PCC overlays are increasingly being used as a rehabilitation technique for both existing PCC and hot mix asphalt pavements by potentially extending service life, increasing structural capacity, reducing maintenance requirements, and lowering life cycle costs. Improvements in PCC paving technology such as the use of zero-clearance pavers, fast-track paving concepts, and high early strength PCC mixtures have greatly increased the ability of PCC overlays to serve as a viable rehabilitation alternative. The FHWA in cooperation with the American Concrete Institute (ACI) has developed a synthesis report, "PCC Overlays State of the Technology Synthesis Report." This material presents the latest information on the design, construction, and performance of PCC overlays. It describes design and construction techniques for the four types of PCC overlays that are commonly used in highway pavement applications: bonded, unbonded, conventional whitetopping, and ultrathin whitetopping. Information is also provided on the selection of PCC overlays as a possible rehabilitation alternative for existing pavements.

Whitetopping is a Portland cement concrete overlay on existing asphalt concrete pavement. It can be used as a road surface course where other paving materials and methods have failed due to rutting or general deterioration. There are three types of whitetopping: conventional (thickness greater than 8 in.), thin (thicknesses over 4 but less than 8 in.), and ultrathin (2–4 in.). Ultrathin whitetopping (UTW) is a bonded, fiber-reinforced concrete overlay. UTW is designed for low-speed traffic areas or areas with a lot of stop-and-go traffic, such as street intersections, bus stops, or toll booths. Joint spacing is critical to a good performing UTW. The use of a short joint spacing is common in both directions of the slab. In effect, a mini-block paver system is formed. The limited experience with UTW indicates that joint spacing should be no more than 12–18 in. each way per inch of whitetopping thickness. For example, a 3-in. UTW surface should be jointed into 3×3 or 4×4 ft^2. Joints are sawed early to control surface cracking.

18.7 WARRANTY PROJECTS

In many cases, pavement agencies are now requiring contractors to provide warranties for the life of the pavement. Such warranties can range from a 2 year life for a chip seal to an extended period of say 20 year life for a pavement structure. A warranty is the contractor's assurance (generally through posted bonds) that the pavement will last the number of years for which it has been designed and built. Warranted projects could cost higher initially but the life cycle cost is generally lower. The standards of performance are generally set by the agency, but the quality control job could be borne mainly by the contractor. Design-build jobs that are conducted by one or a group of contractors are becoming more common in the area of warranted projects. The use of innovative concepts is widely encouraged in such projects. Contractors generally need to be prequalified to bid for warranty projects, and the contractor is evaluated in terms of a number of factors such as quality, safety, timeliness, and contract execution for prequalification.

QUESTIONS

18.1 What are the different steps in a pavement management system?

18.2 What are the different types of mixes available for maintaining asphalt pavements?

18.3 List and briefly describe the different types of asphalt pavement recycling processes.

18.4 How does a maintenance engineer decide between partial-depth and full-depth repair in rigid pavements?

18.5 What materials should be used for partial-depth and full-depth repair in rigid pavements?

18.6 How does a designer select the appropriate technology for joint and crack sealing?

19 Airport Pavements

19.1 TYPES, IMPORTANCE, AND SPECIFICATIONS

Essentially there are three different pavement areas in an airport—runways, taxiways, and ramps or aprons. Runways (Figure 19.1) are used for takeoff and landing of aircrafts, taxiways are pavements that connect the aprons to the runways, and aprons are areas that are used for aircraft parking, loading, and unloading. Generally runways can be distinguished by white marks on them, while taxiways have yellow markings. Helipads, used for the landing and takeoff of helicopters, have a distinct H sign marked on the area.

Airport pavements must be constructed under strict guidelines and specifications to avoid damage to aircrafts as well as injuries to passengers. Loose particles of damaged airport pavement mix or materials can cause foreign object damage (FOD). Loose particles can damage jet engines by getting sucked in, can damage propellers, and may become deadly projectiles.

An example of specifications for airport pavement hot mix asphalt (HMA) is the Federal Aviation Administration's (FAA) Item P401 (Plant Mix Bituminous Pavement; http://www.faa.gov/airports/engineering/construction_standards/; FAA, n.d.). It specifies materials, composition, construction methods, materials acceptance, contractor quality control, method measurement, basis of payment, as well as relevant ASTM and AASHTO standards. Additional guidance can be obtained from engineering briefs such as the FAA's Engineering Brief EB59A (http://www.faa.gov/airports/engineering/engineering_briefs/media/EB_59a.pdf; FAA, n.d.).

The FAA provides guidance for the design, construction, and maintenance of airports through their specifications (http://www.faa.gov/airports; FAA, n.d.). AC150/5320-6E (FAA, 2009) is for airport pavement design and evaluation, whereas AC 150/5100-13B (FAA, 2011) is for nonprimary airports. The airport pavement design process has been developed from a nomograph/chart-based empirical procedure to spreadsheet-based design as well as the latest layered elastic pavement design (LEDFAA [FAA, 2004]; and FAA Rigid and Flexible Iterative Elastic Layered Design, or FAARFIELD [FAA, 2008]). AC 150-5380-6B (FAA, 2007) gives guidance on the maintenance of airport pavements. AC 150/5370-10E (FAA, 2009) gives standards for the construction of airports; it contains important updates on existing specifications. Examples of important engineering briefs are EB59A, Item P401 (Plant Mix Bituminous Pavements, or Superpave) (FAA, 2006), and EB60 (Semi Flexible Wearing Course for Apron Pavements) (FAA, 2001). The latter EB specified the use of a porous asphalt mix layer with voids filled with highly modified cement grout, which is used as an alternative to coal tar sealant for fuel resistance. Such layers (Densiphalt; see http://www.densiphalt.com; Densit, n.d.) have been used by Massport and the Port Authority of New York and New Jersey (PANYNJ) as well as in the Port of Baltimore in the United States. Densiphalt consists of open-graded asphalt mix with 25%–30% voids, limestone filler, and cellulose fiber. The mortar for the grout is composed of special cement with microsilica, superplaticizer, quartz sand, and water with a water–cement ratio of 0.27. For construction, the existing surface is removed to a depth of 2–3 in. and replaced with the open-graded material. The special grout is then mixed with water and then applied on the surface of the open-graded material, which removes the air voids; a curing compound is applied to control moisture release from the surface. EB No. 62 specifies the use

FIGURE 19.1   Runways and taxiways. (Courtesy of Massachusetts Port Authority, Boston, MA.)

of polymer composite micro-overlay (applied as a slurry) for a fuel-resistant wearing surface as an alternative to a coal tar seal coat.

19.2   STRUCTURAL DESIGN OF AIRPORT ASPHALT MIX PAVEMENTS

The FAA-specified structural design method of airport pavements has evolved from a empirical CBR-based "spreadsheet" design method, which was based on the concept of equivalent load and departure to a layered elastic (for flexible pavements) and finite element (for rigid pavements) based methods. The former method was specified in AC 150/5320-6D, which was cancelled in 2009 and replaced by AC 150/5320-6E, that contains the new method (discussed in Section 19.5). However, in this book, both procedures are presented.

The design of airport pavements can be done with airfield pavement design software following the Advisory Circular AC 150/5320-6D (Airfield Pavement Design and Evaluation). The design process started with the basic shear failure–based California Bearing Ratio (CBR) method developed and modified by the U.S. Army Corps of Engineers in the 1960s and 1970s. The main drawback with this design method is its inaccurate representation of specific aircraft loads/gears and limited outputs with a lack of ability to conduct sensitivity analysis.

The *formula* for relating thickness of pavement to CBR is as follows:

$$t = \alpha(A_c)^{0.5}\left[-0.0481 - 1.1562\left(\log\frac{CBR}{P}\right) - 0.6414\left(\log\frac{CBR}{P}\right)^2 0.473\left(\log\frac{CBR}{P}\right)^3\right]$$

where
   α is the load repetition factor
   A_c is the tire contact area, in.^{2}
   CBR is the CBR of the layer being considered
   P is the tire pressure (psi) at depth t used in calculating the ESWL

The basis for this formula is equal deflection between the ESWL and multiple gear, assuming equal contact area.

Currently, a spreadsheet-based design (based on the CBR method, with gear configurations related through empirical data and theoretical concepts) is used, which has been made user-friendly with the use of Visual Basic macros. The design software is available from http://www.faa.gov/airports_airtraffic/airports/construction/design_software/(FAA, n.d.).

This spreadsheet-based design allows for inputs for subgrade compaction requirements, layer conversion, multiple subbase layers, and charts on thickness versus CBR and thickness versus annual departures. For the aircraft mix, the design is based on "critical aircraft," and allows consideration of frost in design. Note that the design curves on the basis of which this spreadsheet software has been developed are provided in Advisory Circular AC 150/5320-6D in Figures 3.2 through 3.15. This design method is intended to result in pavements that would last for 20 years without needing any major maintenance if no major changes in forecasted traffic are encountered.

For asphalt pavements, there are 14 design curves that give the total pavement thickness as well as the thickness of the HMA surface for different types of aircrafts over a particular subgrade. These thicknesses are provided for both critical areas (with concentrated/departing traffic) and noncritical areas (with arriving traffic). The thickness of the HMA surface courses for noncritical areas is provided on the charts, whereas those for the base and subbase courses for noncritical areas can be considered as 0.9 times the thickness of the corresponding layers in the critical areas.

The software provides two options—the main module for designing airports with aircrafts heavier than 30,000 lb, and a submodule one for designing airports with aircrafts weighing less than 30,000 lb. The design for the main module is based on a pavement structure consisting of a pavement surface layer (HMA), a stabilized or unstabilized base layer, a stabilized or unstabilized subbase layer(s), and a subgrade.

There are 10 basic steps in the main module. For the proper completion of the design process, it is important that the steps are completed in the proper numerical order. (Please use Circular AC 150/5320-6D to refer to the figures and tables mentioned in the following steps.)

Step 1. Enter airport name and data: this step involves the input of the airport name, city and state, Airport Improvement Program (AIP) project number, as well as designer's name and comments, if any.

Step 2. Enter subgrade CBR and frost code: it has been suggested that the design CBR value should be equal to or less than 85% of all the subgrade CBR values, which corresponds to a design value of one standard deviation below the mean CBR value. Knowing the type of the soil (such as gravelly soil), the percentage finer than 0.02 mm by weight (say, 5%), and the soil classification (say, GW), one can determine the frost code (FG-1) from Table 2.4 in Advisory Circular AC 150/5320-6D. The frost code is used along with the frost penetration depth information to decide on a specific frost protection system for the pavement.

Note that for specific aircraft types and loads, requirements regarding the compaction of subgrade of cohesive and noncohesive soils are provided in Table 3.2 in Advisory Circular AC 150/5320-6D.

Step 3. Enter subbase information. For the subbase, the number of subbase layers, CBR, and frost code are required.

Step 4. Select default aggregate base. For FAA design (P209), crushed aggregate base is the default which can be substituted by either aggregate case (P208) only if the gross weight of the design aircraft is less than 60,000 lb and the thickness of the surface HMA layer is increased by 1 in. or with lime rock base (P211) material.

Step 5. Calculate frost penetration depth. This step is optional, to compare against the required frost protection depth. In this step, the required inputs are the Air Freezing Index (degree days °F) and the unit weight of the soil (in pcf). The frost penetration depth is calculated from the interpolation of the following data, developed on the assumption of a 12-in. (300-mm) thick rigid pavement or a 20-in. (510-mm) thick flexible pavement.

Frost Penetration Depth (in.)

Degree Days, °F	Soil Unit Weight (pcf)			
	100	115	125	150
200	20.5	21.5	23.8	25.5
400	27.5	30.5	35	38.5
600	34	38	44.5	49
800	40	44.5	54	59
1000	45	51	62	69
2000	69.5	79	102	113
3000	92	105	140	156
4000	115	130	177	205
4500	125	145	197	225

Note that the Air Freezing Index is a measure of the combined duration and magnitude of below-freezing temperatures occurring during any given freezing season. The average daily temperature is used in the calculation of the Freezing Index. For example, if the average daily temperature is 12° below freezing for 10 days, the Freezing Index would be $= 12° \times 10$ days $= 120$ degree days. It is recommended that the design Air Freezing Index be based on the average of the three coldest winters in a 30 year period, if available, or the coldest winter observed in a 10 year period.

For economic reasons, the maximum depth of frost protection suggested is 72 in. There are different ways of providing frost protection to the pavement, depending on the frost code of the soil. Seasonal frosts can cause nonuniform heave and loss of soil strength during melting, resulting in possible loss of density, development of pavement roughness, restriction of drainage, and cracking and deterioration of the pavement surface. Whether such effects are possible or not depends on whether the subgrade soil is frost susceptible, there is free moisture to form ice lenses, and the freezing front is deep enough to penetrate the subgrade soil.

The frost susceptibility of the soil is affected primarily by the size and distribution of voids (where ice lenses can form), and is determined through empirical correlations from the soil classification (Unified Soil Classification System) and the percentage finer than 0.02 mm. The table relating these parameters to the frost codes is given as Table 2.4 in Advisory Circular AC 150/5320-6D. It can be assumed that sufficient water for creating ice lenses is present if the moisture content of the soil is at or greater than 70% of its saturation. It is advised that sufficient moisture for detrimental frost action is assumed to be present all the time.

The protection from frost action can be provided either by making sure that deformations from frost actions are limited or avoided, or by making sure that the pavement has adequate structural capacity during the frost melting (and ignoring the heave due to freezing). The most costly option is complete frost protection, in which the difference between the pavement depth designed from load-carrying consideration only and the depth of frost penetration (assuming the latter is greater than the former) is replaced with nonfrost-susceptible materials to protect the subgrade soil from frost. This could involve replacement of a significant amount of the subgrade soil. The complete frost protection method is recommended only for pavements on FG-3 and FG-4 soils, which are extremely variable in horizontal extent.

The limited subgrade frost protection is meant for limiting the effects of frost, and allowing a tolerable (based on experience) level of heave. In this case, 65% of the depth of frost penetration should be made up of nonfrost-susceptible materials. This design method is recommended for FG-4 soils except where the conditions require complete protection, as well as for FG-1, FG-2, and FG-3 soils when the functional requirements of the pavement permit a minor amount of frost heave. It is recommended that consideration be given to using transition sections where horizontal variability of frost heave potential permits.

In the third method, the reduced subgrade strength method, the CBR of the subgrade is reduced (on the basis of the frost group/code) so as to design a thicker pavement and hence have adequate strength during the thawing or melting of ice lenses. The reduced CBRs versus the frost group data are given in Table 3.1 in Advisory Circular AC 150/5320-6D. This method is recommended for pavements on FG-I, FG-2, and FG-3 subgrades which are uniform in horizontal extent or where the functional requirements of the pavement will permit some degree of frost heave. The method is also permitted for variable FG-1 through FG-3 subgrades for less sensitive pavements which are subject to slow-speed traffic and for which heave can be tolerated.

For airport pavements in permafrost regions, if the complete protection system is not found practical, the method of reduced subgrade strength can be used. The system of providing insulation panels to protect the permafrost can also be used (with a case-by-case approval from the FAA, as no design standard exists), with careful construction and possible restriction of aircraft loads to prevent failure of the insulation panels.

Steps 6–8. Enter aircraft mix. There are two directives regarding the thickness of the pavement over and above those that are determined through design. First, for different types of aircrafts (and different loads for each of those), minimum base course thickness has been specified in Table 3.4. Second, for number of departures in excess of 25,000, the total pavement thickness should be increased according to Table 3.5.

The design method is based on the determination of pavement thickness for each of the aircrafts in the list, with consideration of both load and number of departures, and the selection of the most critical aircraft, which requires the thickest pavement. If accepted by the designer, the thickness of the pavement required for the critical aircraft is considered as the design thickness from this point onward in the design process.

In the design process, it is assumed that 95% of the gross weight is carried by the main landing gears and 5% is carried by the nose gear of aircrafts. For the weights of civil aircrafts, FAA Advisory Circular AC 150/5300-13A can be consulted. The designs are based on the maximum anticipated takeoff weights with the assumption that this conservative approach would offset any unforeseen changes in operation and traffic, and also the ignoring of the arriving traffic.

The design curves on which the software is based provide thickness for four different sets of aircrafts—single-gear, dual-gear, dual-tandem-gear, and wide-body aircrafts. Once the critical or design aircraft is determined, each of the other aircrafts is converted to an equivalent design aircraft, first by converting into the equivalent gear, and then by converting to the equivalent annual departures, of the design aircraft. Each wide-body aircraft is treated as a 300,000-lb (136,100 kg) dual-tandem aircraft when computing equivalent annual departures. Finally, all the equivalent annual departures are summed up, and the pavement is designed for the design aircraft with the total equivalent annual departures. However, the heaviest aircraft must be considered to provide for adequate thickness of the surface layer, depth of compaction, and drainage structures.

Note that the lateral distribution of aircraft traffic along the path of travel (in the form of a normal distribution) or aircraft wander is taken into consideration in the design process with the help of the "departure-to-coverage ratio." The input is in terms of annual departures. Departures are converted to "coverages." Coverage is a measure of the number of maximum stress applications that occur on the surface of the pavement due to the applied traffic. One coverage occurs when all points on the pavement surface within the traffic lane have been subjected to one application of maximum stress, assuming the stress is equal under the full tire print. Assuming normal distribution of aircraft traffic along the travel path, each departure (also called *pass*) of an aircraft can be converted to coverages using a single pass-to-coverage ratio. These ratios are given in Table 5 in Appendix 2 of Advisory Circular AC 150/5320-6D. Annual departures are converted to coverages by multiplying by 20 and dividing that product by the pass-to-coverage ratio, which is tabulated with respect to 12 air-craft/gear types. There is also a figure in Appendix 2 of Advisory Circular AC 150/5320-6D (Figure 3), which shows plots of "load repetition factor" versus coverages. After the pavement is designed for coverages, the thickness is multiplied by the appropriate load repletion factor to determine the final thickness for each specific type of aircraft (this step is done prior to the selection of the critical aircraft).

Step 9. Compute thickness for stabilized layers. This step requires the selection of a stabilized layer and input of the equivalency factor, for a base and/or the top layer of the subbase layer(s). Guidance regarding the selection of proper equivalency factors is provided in the software.

Step 10. Go to the design summary. In this step, the design calculations done in Step 8 are repeated with the information provided in Step 9, and a summary showing the different input parameters, designed parameters, as well as recommended guidelines is shown, along with options for printing, on the screen. Options are also available to see plots of total pavement thickness versus number of departures for the design aircraft and versus CBR of the subgrade.

A step-by-step example is provided next.

Example 19.1

Consider airport pavements to be built on an area with subgrade soil which has the following properties: percentage passing No. 200 sieve = 85%; liquid limit on minus #40 material = 35%; plastic limit on minus #40 material = 7%; percentage finer than 0.02 mm = 12%; and results from three CBR tests conducted in the laboratory = 6, 8, and 7. The unit weight of the soil is 115 pcf.

From a weather station very close to the airport site, the average of the three coldest winters in a 30 year period shows that the design Air Freezing Index is 1100 degree days.

The forecasted air traffic for the airport is as follows:

Aircraft	Gear Type	Annual Departures	Average Maximum Takeoff Weight (lb)
B727-200	Dual	2000	190,500
B737-200	Dual	5000	128,600
B747-200	Double dual tandem	1200	833,000
ABA300-B2	Dual tandem	1000	304,000

1. From the plasticity chart in Figure 2.2, the soil is classified as ML; from Table 2.3, the soil is described as sandy silt.
2. Design CBR = Average – 1* standard deviation = 7 – 1 = 6.
3. From Table 2.4, the frost group/code = FG-4.
4. Consider sufficient water to be present in the subgrade to cause detrimental frost action.
5. Using Figure 2.7, the depth of frost penetration is 47 in.
6. Determine the required thickness of the pavement for each aircraft.

Aircraft	Gear Type	Maximum Takeoff Weight (lb)	Annual Departures	Required Thickness (in.)
B727-200	Dual	190,500	2000	34 in., Figure 3.3
B737-200	Dual	128,600	5000	29 in., Figure 3.3
B747-200	Double dual tandem	833,000	1200	45 in., Figure 3.7
ABA300-B2	Dual tandem	304,000	1000	34 in., Figure 3.5

7. Determine the design aircraft. From this table, it can be seen that the B747-200 requires the thickest pavement, and hence it should be considered as the critical or design aircraft. Since the B747-200 has double-dual-tandem gear, all traffic should be grouped into the double-dual-tandem configuration.

8. Group forecast traffic into the landing gear of the design aircraft.

Aircraft	Gear Type	Annual Departures	Multiplied by Factor (Table Provided in 305a in AC150/5320-6D)	Equivalent Double-Dual-Tandem Gear Departure
B727-200	Dual	2000	0.6	1200
B737-200	Dual	5000	0.6	3000
B747-200[a]	Double dual tandem	1200	1.0	1200
ABA300-B2	Dual tandem	1000	1.0	1000

[a] Each wide-body aircraft is treated as a 300,000-lb dual-tandem aircraft when computing equivalent annual departures.

9. Convert aircraft to the equivalent annual departures of the design aircraft.

Aircraft	Gear Type	Equivalent Double-Dual-Tandem Gear Departure (A)	Wheel Load[a] (lb) (B)	Wheel Load of Design Aircraft (lb) (C)	Equivalent Annual Departure of the Design Aircraft $(10^{[(\log A) * (B/C)^{1/2}]})$
B727-200	Dual	1200	45,240	35,625	2950
B737-200	Dual	3000	30,543	35,625	1658
B747-200[a]	Double dual tandem	1200	35,625[b]	35,625	1200
ABA300-B2	Dual tandem	1000	36,100	35,625	1047
				Total	6855

[a] Main gears carry 95% of the gross load; divide the load carried by each gear by the number of wheels.

[b] Each wide-body aircraft is treated as a 300,000-lb dual-tandem aircraft when computing equivalent annual departures.

10. Final design. Final design should be made for 6855 annual departures for a B747-200 weighing 833,000 lb. See Figure 3.7 (this figure is reproduced as Figure 19.2 in this book) regarding design curves for a B747-200; designed total thickness is 47.5 in. Next, consider Figure 3.7 (here, Figure 19.2) again, with a subbase CBR of 20. The combined (base + surface) thickness is obtained as 18 in. From Figure 3.7, the thickness of the HMA surface required for critical areas is 5 in.

Therefore, the thickness of the base course is $(18 - 5) = 13$ in. (if a stabilized course with a 1.4 layer equivalency number is used, the thickness is reduced to $13/1.4 = 9.5$ in.).

The thickness of the subbase is therefore $(47.5 - 18) = 29.5$ in. (if a stabilized course with a 1.6 layer equivalency number is used, the thickness is reduced to $29.5/1.6 = 18.5$ in.).

Since the stabilized base and subbase are required for new pavements designed to accommodate aircraft weighing 100,000 lb or more, this should be done. For the base course, consider a plant mix bituminous concrete layer with a layer equivalency of 1.4 (compared to a P209 crushed aggregate base). For the subbase course, consider the use of a P209 crushed aggregate base course with a layer equivalency of 1.6 (compared to a P-154 subbase course). Note that the frost penetration depth is 51 in. Therefore, to provide complete frost penetration, the thickness of the pavement section should be increased from $(5 + 9.5 + 18.5) = 33-51$ in. This means a layer of $(51 - 33) = 18$ in. of nonfrost-susceptible material must be placed below the designed pavement section.

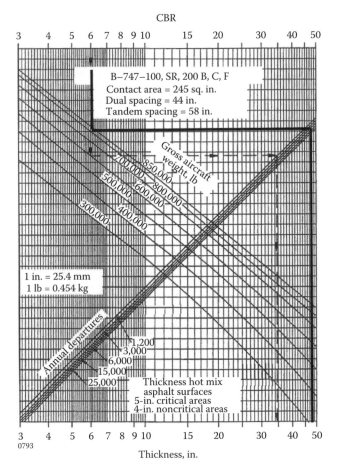

FIGURE 19.2 Design curves for B747 family of aircraft. (From Federal Aviation Administration (FAA), Airport pavement design and evaluation, AC 150/5320-6D, U.S. Department of Transportation, Washington, DC, 1995.)

The final design is as follows:

Layers	Thickness Required (in.)		
	Critical Areas	**Noncritical Areas**	**Edge**
HMA surface (P401)	5	4 (from Chart 3–7)	3.5 (0.7 * T, thickness in critical areas; minimum is 0.7T; Figure 3.1)
HMA base (P401)	9.5	9 (0.9T)	7 (0.7T)
P209 crushed aggregate subbase	18.5	17	17 (0.7T does not apply for the subbase)
To provide complete frost protection: nonfrost-susceptible material		18 in. below pavement section	
To provide limited frost protection: 65% of frost penetration depth must have nonfrost-susceptible material		0 in., none required	

The summary information for this design, as obtained from the spreadsheet program, is shown in Figure 19.3. The plots of thickness versus CBR of the subgrade and annual departure are shown in Figure 19.4.

Flexible pavement design for		10/31/2005
Trial-1		AC method
City-1		
Engineer-1 Airport design engineer	AIP No.	1
Example problem		

47.5″   Total thickness required (in.)
No thickness adjustments required

Stabilized base/subbase are required

Initial pavement cross section		Stabilized or modified cross section		Factors
5″	Pavement surface layer (P-401)	5″	P-401 Plant mix bituminous pavements	
13″	Base layer (P-209)	9.5″	P-401, Plant mix bituminous pavements	1.4
29.5″	Subbase #1 (P-154) CBR= 20	18.5″	P-209, Crushed aggregate base course	1.6
0″	Subbase #2 CBR= 0	0″	Material as defined by user	
0″	Subbase #3 CBR= 0	0″	Material as defined by user	

Frost considerations

Frost considerations			
115 1b/cf	Dry unit weight of soil		
1000	Degree days °F		
51″	Frost penetration depth		
6	Original CBR value of subgrade soli		
6	CBR value used for the subgrade soli	Non-frost code for subgrade soil	
20	CBR value used for subgrade #1	Non-frost code for subbase #1	
0	CBR value used for subgrade #2	Non-frost code for subbase #2	
0	CBR value used for subgrade #3	Non-frost selection for subbase #3	

Design aircraft information		
The design aircraft is a BOEING747 - 780,000 lb – ()		
833,000 lb	Gross weight	20 Design life (years)
6,857	** Equivalent annual departures of a 300,00 lb dual tandem gear — see para. 305 AC 150/5320-6D	

Subgrade compaction requirements for design aircraft

Non-cohesive soils		Cohesive soils	
Compaction	Depth required	Compaction	Depth required
100%	0–23″	95%	0–9″
95%	23–41″	90%	9–18″
90%	41–59″	85%	18–27″
85%	59–76″	80%	27–36″

See Appendix 5 to AC 150/5320-6D, Airport Design and Evaluation for application, for application of this software

FIGURE 19.3   Summary output.

19.2.1   DESIGN OF FLEXIBLE PAVEMENT OVER EXISTING PAVEMENT

For flexible pavements, two options are allowed—HMA overlay on existing flexible pavement and HMA overlay over existing rigid (PCC) pavement. Note that the FAA does not allow overlay pavement sections containing granular separation courses between the existing and new layers (called *sandwich pavement*) because of the possibility of the granular layers getting saturated with water, and the trapped water having a detrimental effect on the performance of the pavement.

19.2.1.1   HMA Overlay on Existing Flexible Pavement

For this option, the method used is known as the *thickness deficiency approach*, and it consists of the following steps.

1. Using the design curves or software presented earlier, determine the thickness required for a new pavement for the load and annual departures. Determine the thickness of all of the pavement layers.

(a)

(b)

FIGURE 19.4 Change in thickness with (a) CBR and (b) departures. Note that the CBR of the different layers can be determined by evaluation tests, such as those using nondestructive testing (NDT). Also, rounding off thicknesses of equivalent layers (derived using layer equivalencies) is not recommended.

2. Compare the required thickness of the new pavement with the existing pavement, and determine the thickness of the overlay. In this step, the existing base may be considered as the subbase and the existing surface may be considered as the base. A higher quality material layer may be converted to a lower quality material layer, using layer equivalencies provided in Tables 3.6 through 3.8 of Advisory Circular AC 150/5320-6D. While using the layer equivalencies, proper consideration should be given to the condition of the existing pavement—since defects such as surface cracking, a high degree of oxidation, and evidence of low stability would reduce the equivalency factors.

Example 19.2

Consider an existing pavement with subgrade CBR = 6; the HMA surface course is 5 in. thick, the base course is 8 in. thick, and the subbase is 12 in. thick. The subbase CBR is 15 in. Frost action is negligible. Assume an existing pavement is to be strengthened to accommodate a dual-wheel aircraft weighing 150,000 lb and with an annual departure level of 6000.

HMA surface: 4 in.; base: 14 in.; subbase: 15 in.; total thickness: 33 in.

The total thickness required to protect the subgrade is 33 in.; the combined thickness of the surface and the base should be 18 in. to protect the subbase.

The existing pavement is 8 in. deficient in total pavement thickness. Assume that the HMA surface of the existing pavement can be substituted as a base layer with a layer equivalency of 1.5. If 4 in. of the existing HMA surface layer is considered as the base in this manner, the thickness of the new base becomes $(8 + 4 * 1.5) = 14$ in. This satisfies the base course requirement, leaving 1 in. of HMA surface. A 3-in. HMA overlay on top of this would satisfy the HMA surface thickness requirement.

19.2.1.2 HMA Overlay on Existing Rigid Pavement with or without Existing HMA Overlay

To design HMA overlay on existing rigid pavements, the thickness deficiency approach, considering the requirement for a new rigid pavement and the existing rigid pavement, is used. The following formula is used:

$$t = 2.5(Fh_d - C_b h_e)$$

where the different parameters are explained as follows:

t is the required HMA overlay thickness, in.

F is the factor controlling degree of cracking in the base rigid pavement

h_d is the exact (not rounded off) thickness of new rigid pavement required, in., considering the modulus of subgrade reaction, k, of the existing subgrade and the flexural strength of the existing concrete

C_b is the condition factor indicating the structural integrity of the existing rigid pavement, with values ranging from 1 for pavements with slabs having nominal structural cracking to 0.75 for slabs with structural cracking

These two conditions are explained in Figures 4.4 and 4.5 in Advisory Circular AC 150/5320-6D. Note that the overlaying of slabs with severe structural cracking and hence C_b less than 0.75 is not recommended, as this could lead to severe reflective cracking. In such situations, it is advised that such slabs be replaced and load transfer along inadequate joints be restored (and increase C_b as a result), and then an overlay be applied. NDT can be used to determine an appropriate value of C_b, a single value is recommended to be used for an entire area, and the value should not be varied along a pavement feature and h_e is the thickness of existing rigid pavement, in.

The factor F is used to control or limit the amount of cracking expected in the concrete underneath the applied HMA overlay. The failure mechanism is assumed to be like this: the underlying rigid pavement cracks progressively with traffic load repetitions until the average size of the slab pieces reaches a critical value, beyond which shear failure occurs with a significant increase in deflection under traffic. F is considered to be a function of subgrade strength and traffic volume, and is recommended as 1.0; it can also be determined from Figure 4.3 in Advisory Circular AC 150/5320-6D.

To design a HMA overlay over an existing HMA layer over an existing rigid pavement, the same formula given earlier is to be used, assuming that there is no existing HMA layer. Then the new HMA overlay thickness should be determined by subtracting from the required overlay thickness the thickness of the existing HMA layer or part thereof (if the existing HMA layer is not in such a condition as to be considered in whole).

Example 19.3

Consider an existing 12-in. thick rigid pavement is to be strengthened to provide adequate design for 6000 departures of a dual-wheel aircraft weighing 150,000 lb. The flexural strength of the existing concrete is 700 psi, and the modulus of subgrade reaction is 300 pci. $C_b = 0.95$.

Using Figure 3.18 (from AC150/5320-6D), the required single-slab thickness is found to be 13.1 in. The F factor is considered to be 1.

$$t = 2.5(Fh_d - C_b h_e)$$

$$t = 2.5(1 * 13.1 - 0.95 * 12)$$

$$t = 4.25 \, in.$$

Now suppose that in the existing pavement there was a 2-in. thick HMA overlay on top of the rigid pavement. The calculation for the required new HMA overlay would have proceeded as earlier, ignoring the presence of the old overlay. Now that it is known that a 4.25-in. HMA overlay is needed, the thickness of the existing overlay must be subtracted from 4.25 in. to determine the required new overlay thickness. Based on NDT and/or engineering judgment, it is now seen that, because of its existing condition, the existing 2-in. HMA can at best be considered as an HMA layer of effective thickness of 1.25 in.; the new overlay that should be used is $(4.25 - 1.25) = 3$ in.

19.3   DESIGN OF CONCRETE PAVEMENTS

AC 150/5320-6D specifies the design procedure for rigid pavements using charts. A subbase of at least 4-in. thickness is required unless certain conditions regarding soils and drainage are satisfied (given in Table 3.10 in AC/5320-6D). Stabilized subbases are required for all pavements carrying aircrafts weighing more than 100,000 lb. For the subgrade, the compaction requirements are less strict compared to the requirements for flexible pavements. Guidance is provided for compaction of cohesive and noncohesive soils (such as 90% of maximum density for cohesive soils in the fill sections). The use of a nonrepetitive static plate load test (AASHTO T-222) is recommended for the determination of the modulus of subgrade reaction, k. Recommended values for subgrades for different thicknesses of subbase are provided (Figure 19.5).

The thickness of concrete slabs is determined using charts for a variety of aircrafts. The inputs required for using these charts are concrete flexural strength, subgrade modulus, gross weight of the design aircraft, and annual departure for the design aircraft. Figure 19.6 shows an example of using the design charts. The dashed line shows the sequence of input values. Note that the given charts are for jointed edge loading, and based on the consideration of loading, either tangent to or perpendicular to the joint.

FIGURE 19.5 Effect of stabilized subbase on subgrade modulus. (From Federal Aviation Administration (FAA), Airport pavement design and evaluation, AC 150/5320-6D, U.S. Department of Transportation, Washington, DC, 1995.)

19.4 DESIGN FOR AIRPORT PAVEMENTS WITH LIGHT AIRCRAFTS

Aircrafts weighing less than 30,000 lb are characterized as light aircrafts by the FAA, and airport pavements handling such aircrafts are designed on the basis of methods presented in Chapter 5 of AC150/5320-6D. The design of flexible pavements is conducted with the use of the gross aircraft weight and the CBR of the subgrade (Figure 19.7). Once the total thickness of the pavement is determined, the thickness obtained by considering a CBR of 20 is subtracted from it to get the thickness of the subbase. The thickness corresponding to a CBR of 20 is made up of the surfacing and the base. The minimum thickness of HMA over granular base is 2 in. Note that for airports handling aircrafts weighing less than 12,500 lb, highway HMA mixes used by the state department of transportation could be utilized. For rigid pavements, two thicknesses are specified: 5 in. for aircrafts weighing less than or equal to 12,500 lb, and 6 in. for aircrafts weighing more than 12,500 lb.

19.5 ADVANCED DESIGN METHODS

The FAA has developed a suite of design methods/software (FAARFIELD; www.airporttech.tc.faa. gov/pavement/3dfem.asp; FAA, 2008) for both flexible and rigid pavements. These design methods have been evaluated and calibrated on the basis of full-scale testing at the National Airport Pavement Test Facility (NAPTF; http://www.airporttech.tc.faa.gov/naptf/; FAA, 2006). The flexible pavement design method is the same as in the original layered elastic analysis program, LEDFAA (with LEAF as the main computational program), and utilizes subgrade strain and asphalt layer tensile strain as the two primary responses for design. For rigid pavement, the design method utilizes a 3D-finite element method (FEM) model (NIKE-3D is the computational program) of an edge-loaded slab on multiple layers for design. Currently the software is available for download at http://www.airporttech.tc.faa.gov/naptf/download/index1.asp.

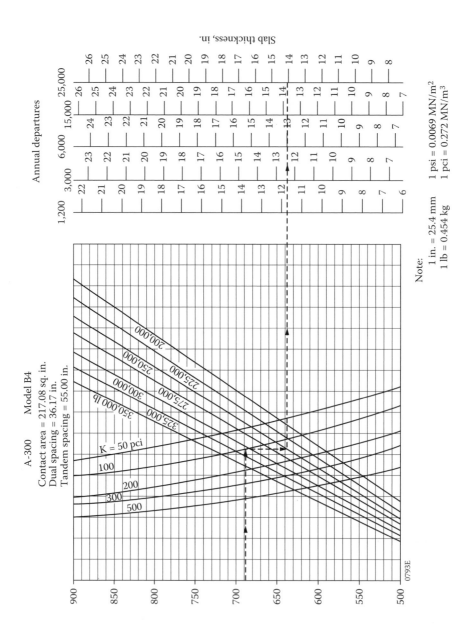

FIGURE 19.6 Example of design chart for rigid pavement. (From Federal Aviation Administration (FAA), Airport pavement design and evaluation, AC 150/5320-6D, U.S. Department of Transportation, Washington, DC, 1995.)

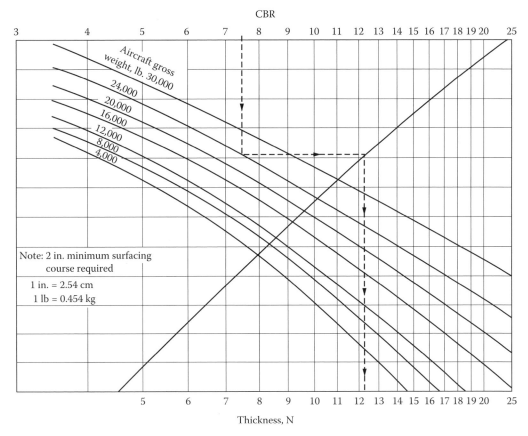

FIGURE 19.7   Design curves for flexible pavement, light aircraft. (From Federal Aviation Administration (FAA), Airport pavement design and evaluation, AC 150/5320-6D, U.S. Department of Transportation, Washington, DC, 1995; Courtesy of David Morrow, Dynatest Consulting Inc., Westland, MI.)

The salient features of the new layered elastic/FE based methods are as follows:

1. A fixed 20-year design life is considered as the FAA standard.
2. The new method analyzes the effect (damage due to) of each aircraft separately and determines a thickness that is required to resist the cumulative damage, rather than converting the air traffic mix to a single design aircraft and all annual departures to an equivalent annual departure of the design aircraft.
3. The design for fatigue failure in the new method is based on the concept of cumulative damage factor (CDF) rather than on the "design aircraft." For a single aircraft and constant annual departure, FAA has defined CDF as

$$CDF = \frac{\text{Number of applied load repetitions}}{\text{Number of allowable repetitions to failure}}$$

or

$$CDF = \frac{\text{Annual departures} * \text{life in year}}{\dfrac{\text{Pass}}{\text{Coverage ratio}} * \text{coverages to failure}}$$

or

$$\text{CDF} = \frac{\text{Applied coverages}}{\text{Coverages to failure}}$$

4. One coverage occurs when a unit area of the pavement experiences the maximum strain for flexible pavement and stress for rigid pavement by a given aircraft. For flexible pavements and rigid pavements, the number of coverages means the measure of the number of repetitions of the maximum strain occurring at the top of the subgrade, and for a rigid pavement, it means the measure of repetitions of the maximum stress occurring at the bottom of the PCC layer. The concept of effective tire width is used in FAARFIELD for calculating the pass-to-coverage ratio (P/C). The effective tire width in a flexible pavement, for shear failure of subgrade, is considered at the top of the subgrade by drawing response lines at 1:2 slope from the edge of the contact patches on the surface to the top of the subgrade (either separate or combined tires). In the case of rigid pavements, the effective tire width is considered as the nominal tire contact patch width at the surface of the pavement.

19.5.1 ASPHALT PAVEMENTS

For asphalt pavements, the primary reason for moving to a more sophisticated design from the CBR-based spreadsheet program is the necessity to design pavements for aircrafts with significantly higher loads and more complex gear configurations, compared to the aircrafts on which the original design curves were developed, as well as compared to the "wide-body aircrafts" for which adjustments had already been made to the original design curves/methods. Two notable examples of such aircraft are the Boeing B777 and the Airbus A380. The load and gear configurations of these aircrafts are shown in Figure 5.7 in Chapter 5. LEDFAA, being a M-E design, should be capable of conducting a more rational analysis and hence a more accurate design of pavements required by aircrafts such as the two mentioned earlier.

The failure criteria are based on two responses: compressive strain on top of the subgrade, to prevent rutting; and the tensile strain at the bottom of the HMA layer, to prevent bottom-up fatigue cracking. The material properties, for which default values are provided in the software, are stiffness modulus, E, Poisson's ratio, μ, and layer thickness, t. The main distinctive feature of this design process is the way of considering traffic and layer characteristics. There is no equivalent departure or design aircraft—the damage for each aircraft is calculated and added. The entire aircraft fleet mix must be input in the design.

The cumulative damage factor (CDF) is calculated as follows:

$$\text{CDF} = \sum \frac{n_i}{N_i}$$

where
    n_i are the actual passes of the individual aircraft
    N_i are the allowable passes of the individual aircraft

With the CDF, 1 means the entire life of the pavement is used up.

The gear location and lateral distribution of the aircraft along the path of travel for each aircraft are considered, and the CDF is calculated for each 10-in-wide strip over a total 820-in. width. Miner's rule is used to sum the damage for each layer. Note that the vertical strain on top of the subgrade is computed at multiple points, with all of the wheels in the main gear contributing to the computed strain (e.g., 16 wheels for the B747 and 20 wheels for the A380). The models are as follows.

19.5.1.1 For Vertical Strain ε_v on Top of the Subgrade

$$\text{Number or coverages to failure, } C = \left(\frac{0.004}{\varepsilon_v}\right)^{8.1}$$

where $C \leq 12{,}100$.

$$C = \left(\frac{0.002428}{\varepsilon_v}\right)^{14.21}$$

where $C > 12{,}100$.

19.5.1.2 For Horizontal Strain e_h at the Bottom of the Surface Layer

$$\log_{10} C = 2.68 - 5 * \log_{10} \varepsilon_h - 2.665 * \log_{10} E_A$$

where E_A is the modulus of the asphalt mix surface layer.

FAARFIELD/FEDFAA allows the user to input a fleet of aircraft, for which an inbuilt library contains all relevant information such as gear layout, loads, and tire pressure. The pavement structure can be checked for the aircrafts for a design period, or, for a specific design period, the thickness required to be added can be determined. Note that the structural property of the layers used in this method is elastic modulus (E), and no longer CBR. The following expression can be used to convert one form to the other: E, psi $= 1500$ (CBR).

19.5.2 RIGID PAVEMENTS

For rigid pavements, the CDF is calculated using the horizontal edge stress at the bottom of the Portland cement concrete (PCC) layer. The edge and the interior stresses are determined using 3D finite element models. The steps in design consist of the following: (1) compute interior stress, (2) compute 75% of the free edge stress with gear oriented parallel to the slab edge, (3) compute 75% of the free edge stress with gear oriented perpendicular to the slab edge, and (4) select the highest of the three cases as the working stress for design.

The failure model is of the following form:

$$\frac{DF}{F_{cal}} = \left[\frac{F_s'bd}{(1-\alpha)(d-b)+F_s'b}\right] * \log C + \left[\frac{(1-\alpha)(ad-bc)+F_s'bc}{(1-\alpha)(d-b)+F_s'b}\right]$$

where

DF is the design factor, defined as R/σ, R is the flexural strength of concrete and σ is the concrete tensile strength

F_{cal} is the stress calibration factor, 1.13

F_s' is the stabilized base compensation factor, $F_s' = 0.3$ for stabilized base, 1 for nonstabilized base

$$\alpha = 0.8, \quad a = 0.5878, \quad b = 0.2523, \quad c = 0.7409, \quad d = 0.2465$$

Failure is defined as the number of coverages for $SC = 80$ at any given value of R/σ.

Figure 19.8 shows the steps that are used for the design of flexible and rigid pavement design with FAARFIELD. A step by step guide, with an example, is provided in AC 150/5320-6E which can be downloaded from http://www.faa.gov/documentLibrary/media/Advisory_Circular/150_5320_6e.pdf.

PECASE (Pavement Transportation Computer Assisted Structural Engineering) is a software that has been developed by the U.S. Triservices (Army, Navy and Air Force) for the design and evaluation of roads and airfield, and is available for download at: https://transportation.wes.army.mil/pcase/. The design methodology, which includes those based on CBR as well as layered elastic

Startup window

Structure window—conventional flexible

FIGURE 19.8 Steps in design of flexible pavement with FAARFIELD. (Courtesy of David Morrow, Dynatest Consulting Inc., Westland, MI.)

analysis, is presented in the document "UFC 3-260-02," which is available from http://www.wbdg.org/ccb/DOD/UFC/ufc_3_260_02.pdf. The basic flexible pavement design approach is similar to that of the FAA method, and allows the evaluation of fatigue under stabilized layers as well as for different seasons. For rigid pavement, the pavement thickness requirement is calculated using a mechanistic fatigue analysis, with stresses that are calculated under design aircraft with the Westargaard edge-loaded model. The calculated edge stresses are used in field (full-scale accelerated loading) derived equations relating concrete flexural strength and repetitions of aircraft loadings.

Right click copies structure picture; arrows identify iteration layer

Help file

FIGURE 19.8 (continued)   Steps in design of flexible pavement with FAARFIELD. (Courtesy of David Morrow, Dynatest Consulting Inc., Westland, MI.)

(continued)

19.6 NONDESTRUCTIVE TESTING AND REHABILITATION OF AIRFIELD PAVEMENTS

The purpose of using nondestructive testing and evaluation is to identify failure mechanisms in pavements and hence to develop the most cost-effective treatments/remedies. The process of evaluation should include a condition survey, a review of records, the development of a test plan, testing and evaluation, and the preparation of a report.

The records should provide information on user knowledge, as built drawings, aerial photos, maintenance information, and results from prior evaluations. The network to be evaluated can be

Airplane window

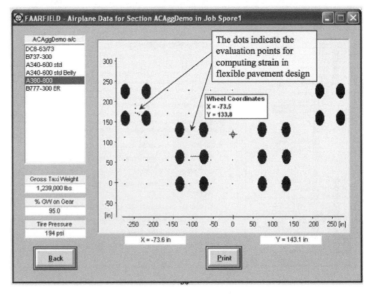

View gear window

FIGURE 19.8 (continued)   Steps in design of flexible pavement with FAARFIELD. (Courtesy of David Morrow, Dynatest Consulting Inc., Westland, MI.)

defined on the basis of type of material (such as HMA or PCC) and its use (ramp, taxiway, or runway), thickness, age, and/or subsurface layers.

The condition survey is very important for checking surface conditions and for avoiding foreign object debris. It consists of working to identify distress—its quantity and severity—and combining the different data into a pavement condition index (PCI), which is related to the condition of the

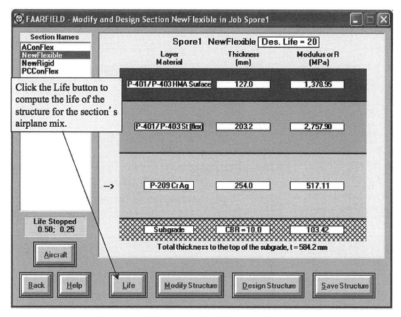

Structure window—standard FAA flexible pavement

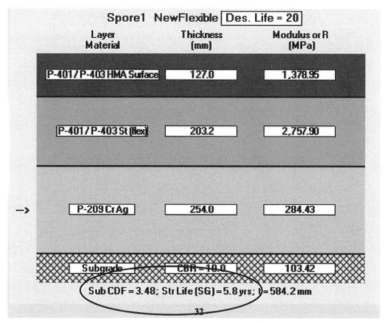

Structure life results

FIGURE 19.8 (continued) Steps in design of flexible pavement with FAARFIELD. (Courtesy of David Morrow, Dynatest Consulting Inc., Westland, MI.)

(*continued*)

pavement. The higher the PCI, the better is the condition. The data can be used to conduct condition analysis and develop pavement condition versus time models using software such as MicroPaver. This also helps in identifying pavements whose PCIs are close to or beyond critical PCI. Functional characteristics such as smoothness and surface friction also require testing.

When developing the testing plan for pavements, one can plan for destructive, semidestructive, or nondestructive testing. Destructive testing includes digging test pits; semidestructive testing can be coring or conducting a dynamic cone penetrometer (DCP); nondestructive testing can be tests such as falling weight deflectometer (FWD) or ground-penetrating radar (GPR).

Scrolling down airplane window to see CDF for each aircraft

Click the design structure button to compute the minimum base
thickness for a subbase of CBR=20, and the required subbase thickness
for a 20-year life

FIGURE 19.8 (continued) Steps in design of flexible pavement with FAARFIELD. (Courtesy of David Morrow, Dynatest Consulting Inc., Westland, MI.)

Generally, pavement coring is done to obtain the thickness of the layers (to be used for analysis with GPR or FWD data) as well as to test HMA cores for their properties. The coring and removal of the upper layers can also provide access for DCP testing or for obtaining subbase, base, or subgrade soil samples.

Soil boring can be done with a drill rig, and can be done with split spoon sampling and the standard penetration test (SPT). For NDT, the DCP testing can be done to check the stiffness of the

Structure design results

Uncheck enable automatic base design in options window if needed

FIGURE 19.8 (continued) Steps in design of flexible pavement with FAARFIELD. (Courtesy of David Morrow, Dynatest Consulting Inc., Westland, MI.)

(*continued*)

layers below the HMA layers, and the results can be converted to CBR (and to modulus, if required). The GPR can be used to determine the thickness of the different layers in the pavement, and detect buried structures or hidden conditions such as voids or trapped moisture.

The FWD is the most commonly used NDT for determination of the structural condition of the pavement. It applies an impulse load with a load pulse duration of approximately 25 ms. It represents a moving load more accurately than any other existing type of deflection-based NDT.

For example, round base thickness to 250 mm and rerun design

Start structure for new rigid pavement design

FIGURE 19.8 (continued)   Steps in design of flexible pavement with FAARFIELD. (Courtesy of David Morrow, Dynatest Consulting Inc., Westland, MI.)

Example: First LEAF phase: PCC thickness changed from 300 mm to 393.5 mm

Example: Second FEM Phase: Thickness changed from 393.5 mm to 388.2 mm

FIGURE 19.8 (continued) Steps in design of flexible pavement with FAARFIELD. (Courtesy of David Morrow, Dynatest Consulting Inc., Westland, MI.)

The Road Rater, a vibratory loading apparatus, is also used for testing. It can operate at 5–60 cycles per second, with loads of 1000–8000 lb.

The heavy weight deflectometer (HWD) is also used instead of the FWD to apply higher loads. However, note that if the layers are not sensitive to the load magnitude (force amplitude), the FWD can be used in place of the HWD.

The FWD/HWD applies a load and simultaneously measures the deflection at different locations. There are three relevant ASTM standards for deflection-measuring equipment. D4695-96 (Standard Guide for General Pavement Deflection Measurements) specifies the use of static, vibratory, and impulse load equipment; D4602-93 (Standard Guide for Nondestructive Testing of Pavements Using Cyclic Loading Dynamic Deflection Equipment; 2002) specifies the use of vibratory testing equipment; and D4694-96 (Standard Test Method for Deflections with a Falling Weight Type Impulse Load Device) specifies the use of impulse loading equipment.

The data from FWD testing can be used for both qualitative and quantitative evaluation of the pavement. The qualitative evaluation helps in understanding the variability of structural conditions within the network, whereas the quantitative evaluation provides the data needed as inputs for design (modulus, E). It can also provide information regarding voids in PCC and load transfer across joints in PCC pavements.

The impulse or dynamic stiffness modulus (ISM or IDM) provides an indication of the stiffness of the overall pavement. It is defined as follows:

$$I(D)SM = \frac{L}{d_0}$$

where
    L is the applied load
    d_0 is the maximum deflection

The I(D)SM can be plotted along the length of a runway, for example, to evaluate the variability of the structural condition in the initial stage of analysis. The more detailed analysis involves backcalculation of layer moduli, which can be used with a layered elastic design program. The backcalculation program BACKFAA can be downloaded from the FAA website (http://www.airporttech.tc.faa.gov/naptf/download/index 1 asp; FAA, 2008).

For PCC pavements, closed-form solutions are used for the determination of stiffness from FWD test results. Test locations are at the center of the slab for layer properties, transverse joints for load transfer, slab corners for voids, and longitudinal joints for load transfer. Load transfer efficiency as well as voids can be evaluated.

The results of backcalculation analysis can be used in both the spreadsheet-based CBR design and LEDFAA. The backcalculated modulus of the subgrade can be used directly in LEDFAA, while it can be used to estimate CBR (through correlation, E = 1500 CBR) to be used in the spreadsheet design. In the pavement evaluation part, the primary inputs are traffic, pavement thickness and composition, pavement condition, layer properties, and subgrade strength.

Traffic, in the form of aircraft type, frequency, and weights, can be obtained from the airport master plan (forecast) as well as from the airport manager's office. The pavement thickness and composition data can be obtained from cores, test pits, as-built plans, and pavement management system reports.

The pavement condition inputs come from analysis of pavement condition data, which includes description and quantity of distresses and the PCI at different points along the feature. The layer properties include moduli values backcalculated from FWD, DCP, thicknesses from cores or GPR, and laboratory testing (split tensile test for PCC, and CBR on soils from boring).

For analysis of NDT data for estimation of layer properties for use in spreadsheet design, one can evaluate deflection basins, select analytical sections, and determine subgrade modulus—and then estimate CBR from correlation. The conventional FAA design method is based on CBR for flexible pavements and the Westergaard method for rigid pavements. For mechanistic methods, the layered elastic method is used for flexible pavements (which can be assumed as semi-infinite layers) and FEM is used for rigid pavements (which have joints).

The report of output of the pavement evaluation process includes allowable loads and the remaining life of a pavement feature (such as a runway). For the design aircraft, the allowable loads for different sections of the pavement are determined, and the lowest allowable load in that pavement is reported. Similarly, for the different sections the allowable number of load repetitions and hence the remaining lives are determined.

19.7 ACN-PCN

All airports in countries belonging to the International Civil Aviation Organization (ICAO) are obligated to report the strength of pavements (with bearing strengths greater than 12,500 lb) in terms of a standardized system, known as the ACN-PCN system. Details of this method and relevant calculations are provided in Advisory Circular AC 150/5335-5B (FAA, 2011).

ACN is defined as a number that expresses the relative effect of an aircraft at a given configuration on a pavement structure for a specified standard subgrade strength. PCN is a number that expresses the load-carrying capacity of a pavement for unrestricted operations.

The ACN-PCN system is structured so that a pavement with a particular PCN value can support, without weight restrictions, an airplane that has an ACN value equal to or less than the pavement's PCN value. This is possible because ACN and PCN values are computed using the same technical basis. Note that the ACN-PCN system is only intended as a method of reporting relative pavement strength so airport operators can evaluate the acceptable operations of airplanes, and is not meant for pavement design or evaluation.

The ACN-PCN system is based on four standard levels of subgrade strengths (modulus of subgrade reaction for PCC pavements, and CBR for flexible pavements). The ACN for each aircraft is provided by the aircraft manufacturer, and can be determined using the FAA's COMFAA software (www.airporttech.tc.faa.gov/naptf/download/index 1.asp; FAA, 2008).

The ACN for rigid pavements is derived on the basis of the Westergaard solution for a loaded elastic plate on a Winkler foundation (interior load case), assuming a concrete working stress of 399 psi. For flexible pavements, the CBR method based on Boussinesq's solution for stresses and displacements in a homogeneous, isotropic elastic half-space is used.

To standardize the ACN calculation and to remove operational frequency from the relative rating scale, the ACN-PCN method specifies that ACN values be determined at a frequency of 10,000 coverages.

A single wheel load is calculated to define the landing gear–pavement interaction for each type of pavement section. The derived single wheel load implies equal stress to the pavement structure and eliminates the need to specify pavement thickness for comparative purposes. This is achieved by equating the thickness derived for a given airplane's landing gear to the thickness derived for a single wheel load at a standard tire pressure of 181 psi (1.25 MPa). The ACN is defined as two times the derived single wheel load (expressed in thousands of kilograms).

The determination of a pavement rating in terms of PCN is a process of determining the ACN for the selected critical (i.e., most demanding) airplane and reporting the ACN value as the PCN for the pavement structure.

The PCN can be determined in two ways—the airplane method (simpler) or the technical evaluation method (requires more time and resources).

The airplane method is useful when information on traffic and runway characteristics is limited and engineering analysis is neither possible nor desired. In the airplane method, the ACN for all airplanes currently permitted to use the pavement facility is determined and the largest ACN value is reported as the PCN. The assumption in using this method is that the pavement structure has the structural capacity to accommodate all airplanes in the traffic mixture and that each airplane is capable of operating on the pavement structure without restriction. Note that the pavement capacity can be significantly overestimated if an excessively damaging airplane, which uses the pavement on a very infrequent basis, is used to determine the PCN.

Example 19.4

Consider a flexible pavement runway with a subgrade strength (CBR) of 10 and traffic having the operating gross weights and ACNs shown next.

Aircraft	Operating Weight (lb)	Tire Pressure (psi)	% Gross Weight on Main Gear for ACN	ACN (from COMFAA)	Annual Departures
B737-300	130,000	195	90.86	32	5000
A300-B4	370,000	205	94.00	57	2000
B767-300ER	370,000	190	92.40	50	1500
B777-200	600,000	215	95.42	52	500

The PCN should be expressed as follows:

PCN (highest number from the aforementioned table)/F or R (depending on flexible or rigid)/ code for CBR (given in Table 2.2)/code for tire pressure (given in Table 4.2)/U or T (depending on whether the airplane method or the technical evaluation method is used)

Therefore, in this example, the PCN is

PCN57/F/B/X/U

FAA discourages the use of the aircraft method on a long-term basis because of the potential inaccuracies. The technical evaluation method of calculating PCN is explained in Appendix 2 (with examples in Appendix 3) of AC 150/530-5335-5B. COMFAA (http://www.faa.gov/airports/engineering/design_software/) can be utilized to calculate the PCN using this method. The basic steps, as mentioned in AC 150/530-5335-5B, are as follows: (1) determine the traffic volume in terms of type of aircraft and number of annual departures/traffic cycles of each aircraft that the pavement will experience over its life; (2) determine the appropriate reference section to use based on the number of wheels on main gears; (3) determine pavement characteristics, including the subgrade CBR and equivalent pavement thickness; (4) calculate the maximum allowable gross weight for each aircraft on that pavement at the equivalent annual departure level; (5) calculate the ACN of each aircraft at its maximum allowable gross weight; and (6) select the PCN from the ACN data provided by all aircraft. Pass, coverage, traffic cycle ratio are defined in AC 150/530-5335-5B.

19.8   REHABILITATION OF PCC AIRPORT PAVEMENTS

The final step is to decide on the rehabilitation option, depending on whether the problem is functional or structural. In deciding on a specific method, the considerations should include those of economy, in terms of initial construction cost as well as life cycle cost (in terms of present worth), operational disruption, local capabilities, and reliability. Alternatives include mill and fill, perpetual pavement, overlay, break and seat, rubblization, recycling, or reconstruction.

The condition of the pavement section is measured in terms of serviceability, expressed as a PCI. The serviceability of a pavement goes down with time after construction, and there is a critical PCI (65 for primary airports and 55 for small airports—details given in the Government Accountability Office report GAO/RCED-98-226; U.S. General Accounting Office, 1998) below which rehabilitation must be done to improve the condition of the pavement. If the condition of a pavement warrants rehabilitation, as mentioned earlier, an investigation must be conducted to determine the cause of the failure and select the best rehabilitation alternative. The investigation can be done on the basis of existing information, a visual condition survey (as part of the pavement management system), nondestructive testing, pavement coring and boring, traffic and pavement analysis, as well as drainage analysis.

There are two available options for rehabilitating a PCC airport pavement using HMA. The first is an HMA overlay over the existing PCC pavement, with saw cut and seal on top of the joints or providing a crack relief layer. The second is the application of an HMA overlay over a fractured or rubblized PCC pavement.

The sequence of work in the first option with the crack relief layer includes repairing the existing concrete slabs, installing edge drains, placing a crack relief layer, providing a HMA leveling course, and finally applying a HMA surface course. The work with the crack/break and seat approach includes removing the HMA overlay, if any; correcting drainage problems; cracking/breaking concrete; seating cracked pieces; removing and patching soft areas; sweeping pavement surfaces; applying a tack coat; and placing an HMA leveling and surface course. The modulus of cracked/broken and seated concrete can be between 500 and 800 ksi. The rubblization process includes removing HMA overlay, if any; correcting drainage problems; rubblizing PCC pavement (generally, the top half or two-thirds is rubblized); rolling rubblized concrete; applying a prime coat; and placing HMA leveling and overlay courses. The modulus of rubblized concrete can range from 200 to 500 ksi.

The break and seat process is considered to be more costly and more effective compared to the rubblization process.

19.9   CONSTRUCTION QUALITY CONTROL AND ACCEPTANCE TESTING

The FAA's Advisory Circular 150/5370-10A presents standards for specifying the construction of airports. The main features of the quality control and quality assurance (QC/QA) of HMA (which is specified under P401 specifications) are that the contractor has the full responsibility for all QC testing, and the engineer has the full responsibility for all QA testing, with the provision that in both QC and QA, the testing laboratory (or laboratories) meets the requirements of ASTM D-3666 (Standard Specification for Minimum Requirement of Agencies Testing and Inspecting Bituminous Materials). The contractor must employ technicians who are certified by the appropriate agency, and should submit the QC plan to the engineer.

The main scope of work for QC includes the following:

1. Asphalt content: a minimum of two tests per lot
2. Aggregate gradation: a minimum of three per lot
3. Moisture content of aggregate: a minimum of one per lot
4. Moisture content of mixtures: a minimum of one per lot
5. Temperature: a minimum of four times per lot to determine temperatures of the dryer, asphalt in the storage tank, mixture at the plant, and mixture at the site
6. In-place density: required to ensure that specified density is achieved
7. Sampling: at the discretion of the engineer
8. Control charts: for the production process
9. Documentation: includes daily reports

The main scope of work for QA involves the following:

1. Tests for stability, flow, and air voids: for each sublot and evaluation of acceptance on the basis of the percent within limits (PWL) specification; air voids are calculated on the basis of the maximum specific gravity measured in each sublot.
2. Mat and joint density: evaluation of acceptance for each lot on the basis of PWL; cores from mats are taken at a minimum of 1 ft distance away from the joint; location of method of determination of joint density (such as with cores taken directly at the joint) varies; evaluation of acceptance of joint density on the basis of PWL; one core is taken from each of four sublots in a lot.
3. Thickness: measured from cores taken for density measurement.

4. Smoothness: measured with a 12-ft straight edge perpendicular and parallel to the center-line at a minimum distance interval of 50 ft; compacted surface shall not vary more than 3/8 in. for base courses and ¼ in. for surface courses; remedial actions are required when more than 15% of all measurements within a 2000 yd² lot exceed tolerance.
5. Grade: measured by running levels at intervals of 50 ft or less; compacted mat shall not vary from the gradeline elevations and cross sections shown in the plans by more than ½ in.; remedial actions are necessary, when more than 15% of all the measurements within a lot are outside the specified tolerance.

Example 19.5

For plant-produced mix, variability of the following can be found:

1. Stability (lb) = ±270
2. Flow (0.01 in.) = ±1.5
3. Air voids (%) = ±0.65

For in-place mixes (results from tests with cores, which are compared to results from tests on laboratory-compacted samples) a few of the important items are as follows:

1. For mat density, contractor should target 98.5% (mat density variability is 1.3%).
2. For joint density, contractor should target 96.5% (joint density variability is 2.0%–2.1%).
3. For air voids, mixes designed at 3.5% have the most tolerance for acceptance.

A few other important items are as follows:

1. During production, check stability and flow on the job mix formula.
2. Obtain thickness from cores.
3. Resampling is done for mat density only.
4. Test section is required, whose dimensions are provided, along with specific instructions regarding how it should be tested ("300 ft long, 20–30 ft wide with a longitudinal joint included; treated as a lot with three sublots for acceptance and pay").

19.10 CONSTRUCTING, CHECKING, AND IMPROVING FRICTION/ SKID RESISTANCE DRAINAGE OF RUNWAYS

Checking the friction numbers of runways and improving them if necessary are critical for the safe operation of airports. Good friction is important for traction and good braking performance, and to avoid skidding and loss of control of aircrafts. Good friction is provided as surface finish, as well as through regular checking and maintenance operations. FAA Advisory Circular AC 150/5320-12C provides guidance on the measurement, construction, and maintenance of skid-resistant airport pavement surfaces.

For the design and selection of construction materials of pavement structures, drainage factors (such as slope) and micro- and macrotexture of the surface should be taken into consideration. Microtexture, contributed by fine-particle surfaces, provides friction at low speeds and when making contact between tire and residual rainwater; while macrotexture, provided by the roughness of the pavement surface as a whole, is more important at high speeds, and for providing drainage during rainstorms. Macrotexture can be improved by selecting suitable coarse aggregates (those with resistance to polish, a rough texture, and an angular shape). Skid resistance for asphalt mix pavements can be improved with the use of a 0.75–1.5-in. thick layer of permeable friction course (PFC), which has coarse gradation and large interconnected voids for the quick drainage of water. Chip seals and aggregate slurry seals can also be used for temporary enhancement of skid resistance.

For PCC pavement, brush, broom, or burlap drag or wire combing/wire tining can be used to texture the surface while it is still in plastic state.

For both asphalt mix and PCC pavements, grooving can be done by the transverse saw cut method. For new asphalt mix pavements, grooving is recommended after 30 days of paving; while for PCC pavements, a vibrating ribbed plate or a ribbed roller can be used to cut grooves while it is in plastic state, or a saw cut can be used if it is in a hardened state. The FAA specifies the depth and width of grooves as 1.4 in.× 1.4 in., at ½ in. center-to-center spacing, for the entire runway, perpendicular to the direction of aircraft movement. Note that the transverse grooving is done actually to facilitate forced escape of water under fast moving aircraft traffic (during rainstorm) and hence improve contact between tire and the pavement surface and reduce the chance of hydroplaning.

Over time, the skid resistance of surfaces deteriorates due to the polishing action of the aggregates as well as the accumulation of materials such as rubber from tires (especially in touchdown areas) in the void spaces of PFCs or surface grooves. The problem is made worse by the spilling of jet fuels and presence of ice and snow on the pavement surface. Therefore, regular friction surveys at specified intervals (depending on the number of jet aircrafts landing daily per runway end, specified by the FAA) should be conducted, using continuous friction-measuring equipment (CFME), preceded by visual inspection. In addition, periodic inspection of groove depth should be made, and corrective actions should be taken if 40% of the grooves are found to be equal to or less than 1/8 in. in depth and/or width for a distance of 1500 ft.

The friction surveys should be conducted 10 and 20 ft to the right of the runway centerline at speeds of 40 and 60 mph, using water from the CFME equipment to wet surfaces in front of the friction-measuring tires to provide water of at least 0.04 in. depth. Periodic inspections of water accumulation in depressed areas during rainstorms must be made, and corrective actions should be taken for water depths exceeding 1/8 in. The FAA provides friction numbers (Mu) for minimum, maintenance/planning, and new constructions, depending on the type of equipment used, for 40 and 60 mph measurements. For example, the three numbers for the Dynatest Runway Friction tester (Figure 19.9) are 0.5, 0.6, and 0.82, respectively, for 40 mph, and 0.41, 0.54, and 0.72 for 60 mph testing. The NASA grease smear method is used for evaluating the macrotexture of the pavement surface.

(a)   (b)

FIGURE 19.9   (a) Friction tester and (b) closeup of water nozzle friction testing. (Courtesy of David Morrow, Dynatest Consulting Inc., Westland, FL.)

The rubber removal techniques may affect the structural integrity of the pavement, and hence the use of PFCs in airports with aircraft traffic has not been recommended by the FAA. Generally, contaminants such as rubber are removed by using high-pressure water, approved chemicals, high-velocity impacts, or mechanical grindings, under dry conditions, until the friction number of the contaminated area has been raised to within 10% of that of the uncontaminated areas. Also, to check whether the techniques result in structural damage, a test section can be used for demonstration.

Painted areas in airport pavements can provide less friction, and the skid-resistance properties of such areas can be improved by adding silica sand or glass beads to paints.

Guidelines regarding airport pavement roughness criteria (profiling equipment and procedures, data reduction and simulations, and profile index) are provided by the FAA at http://www.airporttech.tc.faa.gov/pavement/25rough.asp. Guidelines for measurement, quantification and application of runway roughness are also provided by the Boeing company at http://www.airporttech.tc.faa.gov/NAPTF/Download/Roughness/Boeing%20D6-81746%20Roughness.pdf. An application of the criteria for temporary ramping during surface overlay construction is also provided. A software, ProFAA for computing pavement elevation profile roughness index, is also available at the FAA website (http://www.airporttech.tc.faa.gov/pavement/25rough.asp).

19.11   ASPHALT MIXES FOR AIRPORT PAVEMENT

The mix design procedures discussed in Chapter 14 are applicable for airport pavements. In many cases (such as FAA), the Marshall design procedure is being used, although a change to the Superpave (as discussed in Chapter 14) system is expected in the near future. There are some specific issues regarding the mix design and construction of asphalt mix pavements for airports, which are discussed next.

19.11.1   FUEL-RESISTANT MIX

Jet fuel from parked aircrafts can damage asphalt mix on the surface, through raveling and loss of materials, rutting and shoving. Coal tar–based sealants can be effective in resisting this damage but their use is not preferred due to their significantly different coefficient of thermal expansion (as compared to HMA) and hence potential of cracking, and health hazards through exposure to fumes and direct contact. More and more, specially formulated polymer modified asphalts have been used to prepare fuel resistant HMA. Mass loss (e.g., maximum allowable 1%) of an asphalt mix when immersed in jet fuel for 24 h (plain immersion or immersion and application of brushing) has been used to check the suitability of a material. Coal tar sealers have been applied in the past, although currently specialty products (tar free), such as Sealoflex and StellaFlex FR, are being used. Fuel-resistant asphalts (often modified with polymer) have been used in many airports such as those in Boston and New York City (La Guardia), the United States; Cairo, Egypt; Saint Maarten, the Netherlands; and Kuala Lumpur, Malaysia.

19.11.2   CONSTRUCTION AND MAINTENANCE OF LONGITUDINAL JOINTS

Damaged longitudinal joints are of very serious concern in airfield pavements. Loose materials from such areas can cause FOD to aircrafts, leading to loss of life and equipment. The different good techniques discussed in Chapter 14 can be adopted for airfield pavement construction for obtaining good joints. Wherever possible, suitable-size aggregates (finer aggregates produce better joints) and echelon paving should be used. Typical distresses such as cracking and raveling in longitudinal joints happen in about 4–5 years after construction. Initially a crack occurs along the joint, which leads to secondary cracks. Such cracks should be sealed promptly before they become a problem. Hot poured rubberized asphalt is usually used for crack sealing.

19.11.3 TIME OF CONSTRUCTION

In many cases, airport pavement work is conducted at night under a tight schedule (to avoid busy hours of aircraft traffic). Proper management of construction work is crucial for a successful operation, and contractors with proven experience in handling airport jobs are preferred. Mixes and materials must be obtained from plants that are close by, and if required, mixes that require lower temperatures, and hence can be stored on site (such as warm mix asphalt), can be considered.

19.11.4 DE-ICING AND ANTI-ICING

Ethylene glycol is commonly used for anti-icing and de-icing of runway. Typically this fluid is made up of 90% of ethylene glycol and 10% water. For deicing of aircrafts, Type I and IV fluids are commonly used. Type I fluid, which consists of primarily propylene glycol or ethylene glycol, is used for deicing and cleaning of aircrafts on the apron during de-icing operations at the gate. Type IV fluid is used to improve hold over times once the aircraft is already de-iced and is designed to slide off the aircraft prior to lift off. This fluid is generally made up of about 65% glycol and a polymer thickener to create a gel.

19.12 MAINTENANCE OF AIRPORT PAVEMENTS

A variety of maintenance techniques are available for asphalt and concrete airport pavements, as summarized in Table 19.1. ACRP Synthesis 22 (http://onlinepubs.trb.org/onlinepubs/acrp/acrp_syn_022.pdf) gives a complete synthesis of different maintenance and rehabilitation (preservation) techniques and also provides a catalog (in the form of fact sheets) for the different techniques.

TABLE 19.1
Maintenance Techniques for Airport Pavement

For HMA Pavements	For Concrete Pavements	Applicable for Both Types of Pavements
1. Sealing and filling of cracks (with hot- or cold-applied sealants)	1. Joint and crack sealing (with bituminous, silicone, or compression sealants)	1. Texturization using controlled shot blasting
2. Small area patching (using hot mix, cold mix, or proprietary material)	2. Partial-depth repairs (using AC, PCC, or proprietary materials)	2. Diamond grinding
3. Spray patching (manual chip seal or mechanized spray patching)	3. Full-depth repairs (using AC, PCC, or proprietary materials)	3. Microsurfacing
4. Machine patching with AC material	4. Machine patching using hot mix	
5. Rejuvenators and seals	5. Slab stabilization and slabjacking	
6. Texturization using fine milling	6. Load transfer	
7. Surface treatment (chip seal, chip seal coat)	7. Crack and joint stitching	
8. Slurry seal	8. Hot-mix overlays	
9. Hot-mix overlay (includes milling of AC pavements)	9. Bonded PCC overlay	
10. Hot in-place recycling		
11. Cold in-place recycling		
12. Ultra-thin whitetopping		

Source: Hajek, J. et al., Common airport pavement maintenance practices. A synthesis of Airport Practice. ACRP Synthesis 22, Transportation Research Board, Washington, DC, 2011.

The reader is also advised to download the asphalt and concrete airport pavement related reports from the Airport Asphalt Pavement Technology Program (AAPTP) and from

http://www.aaptp.us/reports.html
http://resources.acpa.org/index.php?option=com_docman&task=cat_view&gid=1&Itemid=2
http://www.iprf.org/products/JP007P%20-%20Airport%20Best%20Practices%20Manual.pdf

QUESTIONS

19.1 What are the different types of pavements in an airport?

19.2 Design an asphalt pavement runway to be built on an area with subgrade soil which has the following properties. Use the spreadsheet method.

Percentage passing No. 200 sieve = 80%; liquid limit on minus #40 material = 32%; plastic limit on minus #40 material = 5%; percentage finer than 0.02 mm = 16%; results from three CBR tests conducted in the laboratory = 5, 6, 5. The unit weight of the soil is 112 pcf.

From a weather station very close to the airport site, the average of the three coldest winters in a 30 year period shows that the design Air Freezing Index is 1500° days.

The forecasted air traffic for the airport is as follows:

Aircraft	Gear Type	Average Annual Departures	Maximum Takeoff Weight (lb)
B737-200	Dual	5000	128,600
ABA300-B2	Dual tandem	1000	304,000

19.3 Design the aforementioned pavement using FAARFIELD/LEDFAA 2.

19.4 Name a very important form of maintenance work for airport pavement, and describe briefly how it is conducted.

20 Nondestructive Tests

Nondestructive tests used in pavement engineering can be broadly divided into several categories: nuclear equipment, deflection based, electromagnetic, and seismic equipment. Examples of such equipment are discussed in the following sections.

20.1 NUCLEAR GAGE

The basic operational principle of a nuclear gage is that a source (such as Cesium-137) of nuclear particles (such as photons) is inserted into (e.g., 6 in.) or placed on the surface of the layer whose density is to be measured. As these particles are released into the layer, a detector on the equipment on the surface of the layer detects and counts the number of particles coming back to that point. As the density increases, the number of detected particles decreases. Working on this principle, the gage is calibrated using a block of known density, and then it is used for determining the density by using the calibration factor.

The properties that can be measured include the density of hot mix asphalt (HMA) or soil as well as the moisture content of soil. The mode of transmission of the particles from the source inside the layer to the detectors can be either direct (when the source is inserted into the layer) or backscatter (in which case the particles are reflected off layers, when the source is placed on the surface). A set of Geiger–Mueller tubes is used as detectors (two are used for a thin-layer density gage) for measuring density, whereas an He 3 tube is used as a detector (of neutrons) for moisture content. The available nuclear gages have sufficient capabilities of storing data that can be collected from multiple tests, and are generally powered by rechargeable NiCd batteries.

The relevant American Society for Testing and Materials (ASTM) standard is D2922. The equipment needs to be verified for calibration by measuring the density of a block of known density, ensuring that the reading is within ±2 pcf of the actual density of the material at each depth, and comparing the densities measured at any time to those measured earlier (say, an average of the last four). Factors that can influence this "standard count" include temperature, the time elapsed between measurements, the presence of any affecting material (such as a wall, other gage, or person) close to the gage, and operator error. In the backscatter technique, the measurements are very sensitive to surface roughness, and special care must be taken to make the surface level with water or fine sand (note that this leveling would increase the density by 1%–1.5%).

The ratio of the particle count obtained on a test material to that obtained from a standard block is known as the *count ratio*. The gage is calibrated by measuring the count ratio for blocks of materials such as aluminum, magnesium, granite, and limestone of known densities. Procedures are available for ensuring the short- and long-term electronic stabilities of the system, and correcting measurement anomalies due to factors such as a natural decrease in intensity of particles due to decay of the source material, test location close to a wall, varying material composition or surface features, and absorption of rays because of the elemental composition of the material. Typical equipment is shown in Figure 20.1.

20.2 FALLING WEIGHT DEFLECTOMETER

An falling weight deflectometer (FWD) is extensively used for estimation of pavement layer moduli and for determination of the structural condition of pavements. The information obtained from FWD testing can be used in structural analysis to determine capacity, estimate expected

FIGURE 20.1   Schematic of backscatter operation and nuclear gage equipment gage used on HMA.

performance life, and design a rehabilitation plan for pavements. Deflections prior to and after pavement rehabilitations are done to evaluate the effectiveness of specific rehabilitation methods. An FWD can also be used to test load transfer efficiency (LTE) of joints within concrete pavements. American Society for Testing and Materials standards are available for the use of an FWD for pavement deflection-based testing (see *ASTM 4694, 4695-96*). Standard calibration procedures for load cells and deflection sensors are also available.

In this testing, an impulse load is generated by dropping a mass ranging from 6.7 to 120 kN through different heights onto a base plate through a set of rubber buffers (Figure 20.2). There are two sets of sensors—a load cell to measure the applied load and an array of geophones (which are located on the surface of the pavement) to measure the resulting deflection. FWD tests can be conducted at different load levels, although some protocols would require the deflections to be normalized to a 9000-lb load and "corrected" for temperature (if different from the standard 77°F). For example,

$$\text{Adjusted Center Deflection} = D_o \left(1.598837 - 0.009211683 * T^{0.96} \right) \qquad (20.1)$$

where T is the temperature of the asphalt pavement surface (°F).

An example of FWD equipment is the Dynatest Model 8000 FWD, which has a load range of 7–120 kN. It is a fast (up to 60 test points per hour) and reliable tool that can be operated by a single person. This is self-sufficient equipment, with sensors for the measurement of air and surface temperature, and includes Windows-based ELMOD 4 software for data analysis.

20.2.1   DIRECT USE OF DEFLECTIONS

Using the deflections noted at the different sensors, a series of expressions can be used to determine relevant properties of the pavement, such as the following.

The *area parameter* (which can be used together with the center deflection) represents the normalized area of a slice taken through the deflection basin from 0 to 0.9 m (0 to 3 ft):

$$\text{Area parameter} = \frac{6(D_0 + 2D_{12} + 2D_{24} + D_{36})}{D_0}$$

where D_0, D_{12}, D_{24}, and D_{36} are the deflection readings (in inches) from sensors located 0, 12, 24, and 36 in. from the center of the FWD loading plate, respectively:

(a)   FWD unit on trailer   (b)

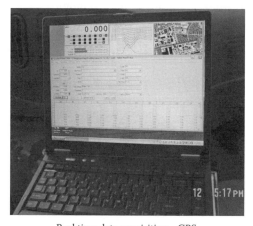

(c)   Real time data acquisition—GPS
coordinates, load and deflections

FIGURE 20.2 Falling weight deflectometer. (a) FWD unit on trailer. (b) Array of geophones on the pavement surface. (c) Real time data acquisition—GPS coordinates, load, and deflections.

$$\text{Area correction factor} = 0.7865 + \left(1.4578 * 10^{-4} * T^{1.68}\right)$$

where T is the temperature of the asphalt pavement surface (°F).

The surface curvature index (SCI) and shape factors are indicators of the relative stiffness of the upper pavement layers:

$$SCI = D_0 - D_2$$

$$\text{Shape factor} = \frac{(D_0 - D_2)}{D_1}$$

where
D_0 is the center deflection
D_1 and D_2 are the deflections at the first and second sensors, respectively (located at 8 and 12 in. from the load)

Subgrade modulus is calculated as follows:

$$\text{Subgrade modulus} = -466 + \left(9000 * \frac{0.00762}{\left(\dfrac{D_{36}}{1000} \right)} \right)$$

where the subgrade modulus is given in psi, and D_{36} is the deflection reading (in inches) from the sensor located 36 in. from the center of the FWD loading plate.

A software program for calculating the aforesaid parameters (FWD) can be downloaded free from the Washington State Department of Transportation (DOT) website at http://www.wsdot.wa.gov/Business/MaterialsLab/Pavements/PavementDesign.htm#PavementDesignRequirements.

20.2.1.1 Relationship between Deflection Bowl Parameters and Stresses and Strains at Various Locations in the Pavement

Tensile strain at the bottom of the asphalt layer is calculated as follows:

$$\text{Log} \, \varepsilon_{r1,0} = -1.06755 + 0.56178 \log h_1 + 0.03233 \log d_{1800} + 0.47462 \log \text{SCI}_{300}$$

$$+ 1.15612 \log \text{BDI} - 0.68266 \log \text{BCI}$$

$\varepsilon_{r1,0}$ is the maximum horizontal strain at the bottom of the asphalt layer, $\mu m/m$
h_1 is the thickness of the asphalt mix layer, mm
d_r is the deflection at distance r of the load center, μm
$\text{SCI}_{300} = d_0 - d_{300}$, μm
$\text{BDI} = \text{base damage index} = d_{300} - d_{600}$, μm
$\text{BCI} = \text{base curvature index} = d_{600} - d_{900}$, μm

Tensile strain at pavement surface is calculated as follows:

$$\varepsilon_{r1,b} = 194.895 - 20.7769 \, \text{SCI}_{300}^{0.5}$$

where $\varepsilon_{r1,b}$ is the tensile strain at pavement surface, μm.

Compressive vertical strain at the top of the unbound base is calculated as follows:

$$\text{Log} \, \varepsilon_{vb} = 1.5615 + 0.3743 \log \text{SCI}_{300} + 1.0067 \log \text{BDI} + 0.8378 \log d_0$$

$$-1.9949 \log d_{1800} + 0.6288 \log d_{300}$$

The equation is valid only for the following conditions:

1. The pavement should not be an inverted pavement so $E_1 > E_2 > E_3 > E_4$
2. The stiffness of the upper layer should not exceed four times the stiffness of the underlying layer
3. Applicable only for weak bases (e.g., $E_2 < 1000$ MPa)

Tensile strain at the bottom of the bound base is calculated as follows:

$$\text{Log} \, \varepsilon_{r,20} = 0.0931 + 0.4011 \log d_0 + 0.3243 \log d_{1800} + 0.4504 \log d_{300}$$

$$-0.9958 \log d_{900} + 0.8367 \log \text{BDI}$$

Compressive vertical strain at the top of the subbase and subgrade

1. Subbase is stiffer than subgrade

$$\text{Log } \varepsilon_{v3} = 2.48589 + 0.34582 \log SCI_{300} + 0.16638 \log d_{1800}$$
$$-0.68746 \log(h_1 + h_2) + 0.47432 \log BDI$$

2. Subbase is less stiff than subgrade

$$\text{Log } \varepsilon_{v3s} = 1.52887 + 0.39502 \log SCI_{300} - 0.84168 \log d_{1800}$$
$$-0.60888 \log(h_1 + h_2) + 0.43195 \log BDI - 0.78407 \log BCI + 1.73707 \log d_{600}$$

3. Subgrade

$$\text{Log } \varepsilon_{v4} = 2.48589 + 0.34582 \log SCI_{300} + 0.16638 \log d_{1800}$$
$$-0.68746 \log(h_1 + h_2 + h_3) + 0.47432 \log BDI$$

where
  ε_{v3} is the vertical compressive strain at the top of the subbase, μm/m
  ε_{v3s} is the vertical compressive strain at the top of the subbase when this layer has a lower
    stiffness than the subgrade, μm/m
  ε_{v4} is the vertical compressive strain at the top of the subgrade, μm/m

20.2.2 BACK-CALCULATION

Another approach to using the FWD data is to "back-calculate" moduli of the different layers of the pavement, using mechanistic principles. This process involves using layer thickness and Poisson's ratio (which must be known, estimated, or assumed with reasonable accuracy), assuming moduli of the different layers ("seed" moduli), and determining the different deflections at the locations of the FWD sensors. The process is repeated until the calculated deflection basin is found to be approximately identical to the measured deflection basin (within tolerances). An example of a criterion to stop the back-calculation process is that the summation of the "errors" (the difference between calculated and actual deflections) is less than 5%. Note that the modulus of the subgrade can be determined from the FWD deflections at the outer sensors without back-calculation.

Back-calculation can be done by different software, some of which is freely available. The Ever-calc software is available free as part of the Everseries suite of software from the Washington State DOT website at http://www.wsdot.wa.gov/Business/MaterialsLab/Pavements/ PavementDesign.htm#PavementDesignRequirements (2007). This is a mechanistic-based pavement analysis program that uses an iterative process of matching the measured surface deflections with the theoretical surface deflections calculated from assumed moduli. The program is capable of handling up to five layers and can be run with or without a rigid or stiff base. User-provided seed moduli are not required. The seed moduli can be estimated using the internal regression equations that are based on relationships between layer moduli, load, and various deflection basin parameters. Provisions are available for nonlinear material behavior. The main steps are indicated later. Note that other software (such as MODULUS, available from the Texas Transportation Institute at http://tti.tamu.edu; n.d.) have different features but similar steps.

20.2.2.1 Steps in Performing Back-Calculation with Evercalc Program

1. The Evercalc program is opened.
2. "File" is selected, and "General File" is chosen.
3. The file is opened, it is named, and the units are changed to U.S. units.
4. The appropriate number of layers is selected.
5. The number of sensors is chosen to be 9, and the plate radius is input as 5.9 in.
6. "Stiff Layer" and "Temp. Correction" are both unselected, and the "Sensor Weigh Factor" is chosen to be uniform (as an example).
7. The radial offsets for sensors 1 through 9 are selected to be 0, 8, 12, 18, 24, 36, 48, 60, and 72, respectively.
8. The layers are then filled:
    Select the appropriate layer IDs.
    For HMA, select Poisson's ratio as 0.35. For soil layers, select 0.4.
    Input the values of "Initial Modulus," "Min. Modulus," and "Max. Modulus."
9. The max iterations are chosen to be 5, and the "RMS Total%" and "Modulus Tot.%" are also chosen to be 5.
10. "Stress and Strain Location" is chosen, the general file is saved, and the main menu is displayed.
11. Next, the Deflection file is opened.
12. The appropriate pavement temperature is selected.
13. The appropriate layer thicknesses are selected.
14. The appropriate number of drops is selected.
15. The deflection data from the FWD are put into the section with the drops.
16. The file is saved (should be the same prefix as the general file), and the main menu is displayed.
17. The back-calculation is performed by choosing "File." "Perform Backcalculation" is pressed, and then "Interactive Mode" is chosen. The name of the general file is chosen, the "OK" button is pressed, and then the name of the corresponding deflection file is chosen. Next the output file is chosen, and then the summary file. This prompts the output, and "OK" is pressed at the bottom and a graph is displayed. This graph shows how closely the input into the program matches the calculated values.
18. The window is then closed, and "Print" is chosen in the top tool bar. "Print/View Output" is chosen, and the file is selected to view the summary.
19. The "Average RMS(%)" is viewed.
20. This whole process is repeated, with the modulus values of the layers changed, until the error is near 6% or less.
21. Tips: if the bottom of the calculated deflection basin deviates from the actual data, the modulus of the top layer needs to be changed; changing the middle two layers controls the middle of the deflection basin, and the top part of the deflection basin is controlled by changing the modulus of the bottom layer.

Numerous studies have been conducted on the accuracy of load and deflection measurements as well as on the refinement of the back-calculation procedures, resulting in continuous improvement of both equipment as well as analysis procedures. As a result, the FWD has become the principal nondestructive testing tool for pavement engineers. The use of back-calculated moduli using an FWD has been recommended for rehabilitation of pavements by the newly developed NCHRP-1-37A pavement design guide. Such use is absolutely necessary for determination of moduli of subsurface layers as well as for determination of moduli and consistency of new and innovative pavement materials. Examples of such use include the testing of full depth reclaimed pavements with different types of additives, such as foamed asphalt. Furthermore, the FWD is being used by numerous researchers in the determination of pavement layer properties in full-scale test sections and for the evaluation of subgrades under unusual conditions as well as the effect of environment and new materials on pavement properties.

One key requirement for the successful application of an FWD is the use of accurate layer thickness and condition data. Without the use of any nondestructive tool, the only solution is to take cores, in sufficient numbers, such that accurate estimates of layer thickness and conditions can be made. However, taking cores defeats the whole idea of nondestructive testing and hence is not an attractive option. Nondestructive instruments such as ground-penetrating radar (GPR) have now become very valuable tools for pavement engineers for providing the data necessary for using an FWD.

20.2.2.1.1 Modeling Thin Layers

Back-calculation will not be able to give a separate modulus for a thin surface course. If the layer is too thin (insensitive), it must be combined with the layer of similar material that lies below it. If the modulus of a layer does not influence the surface deflections, one cannot use deflections to determine the layer modulus. The term "sensitivity" is used to describe the effect of a layer's modulus on the surface deflections. An insensitive layer can have almost any modulus and it will have little effect on the deflection. The best approach to use with an insensitive layer is to give it a fixed modulus or combine it with an adjacent (similar) layer.

20.2.2.1.2 Modeling Subgrades

The upper portion of a subgrade is affected by weather. It goes through seasonal cycles of freeze–thaw and/or wet–dry, and thus its modulus changes over time. Even if the depth is arbitrary, it is better to model the upper subgrade as a separate layer

20.2.2.1.3 Modeling Bedrock

Shallow bedrock (25 ft deep or less) must be included in the pavement model. As discussed earlier, use the a/r method to check for shallow bedrock.

20.2.2.1.4 Stress-Dependent Materials

Pavement materials are often modeled as a linear elastic material, for which E = constant. However, most pavement materials are stress dependent. They can be stress-hardening (clean gravels) or stress-softening (wet, fine-grained materials).

20.2.2.1.5 Spatial Variations

Layer thicknesses, depth to water table, and depth to bedrock all vary along the road. Materials will also vary along the road, and loads and traffic may vary along the road. To take these variations into considerations, follow these steps:

1. Back-calculate each individual test point
2. Conduct pavement analysis or design at all test points
3. Put the results in rank order
4. Then take the 85th percentile result for overlay design or remaining life calculation

20.2.2.1.6 Seasonal Variations

Temperature and moisture conditions in the pavement vary over time. Also, these conditions can vary within a single day. To take seasonal variations into considerations, consider the following:

1. Back-calculated moduli are valid only for the date and conditions of test.
2. For important projects, test the pavement at two or more times in the year.
3. Use Miner's hypothesis to account for seasonal variability.

20.2.2.2   Detection of Voids in PCC Pavements

FWD data are used to detect voids where pavement layers have no support. To fill voids under a PCC pavement, injection holes are drilled into the pavement and a grout of cement, fly ash, and water is pumped through the holes. This procedure is referred to as "undersealing." Before and after drilling holes, voids in the pavement are detected using FWD data. The FWD loading plate is placed as close as possible to the slab corner, and the LTE to the adjacent joint is measured. If the measured deflections fall out of a range determined by the state engineer as acceptable, undersealing procedures begin.

Because voids under PCC-bridge approach slabs contribute to premature cracking, early detection of these voids is crucial to avoid costly replacement and rehabilitation measures. The Missouri DOT recommends that FWD should be used to determine voids under PCC slabs. This recommendation assumes that the FWD and operator are available, undersealing is being considered as a preventive maintenance treatment, and one or more of the following conditions are met:

- Long-lane closures for proof-roll testing are not desirable (e.g., at bridge approaches with reduced shoulder widths and high-volume routes).
- Fewer personnel than required with proof rolling are available for testing.
- The pavement shoulder is unstable for accurate proof-rolling measurements.
- More clear and quantifiable indications of undersealing improvements than proof rolling can provide are desired (i.e., AASHTO rapid void detection procedure).

20.2.2.3   Detection of Nonresilient Pavement Layer Behavior

Because the FWD has replaced the Benkelman beam as the primary pavement analysis and design device, measured layer moduli now include plastic deformations as well as recoverable deformations. Mechanistic design practices assume that all layers behave resiliently. In the past, these additional plastic deformations were assumed negligible; however, nonresilient behavior may be observed given a load of significant magnitude. The practice of "16 (FWD weight) drops at four load levels with four replicates at each drop height or load level" may result in nonresilient behavior. Such behavior can be detected by statistical tests. Two statistical methods of nonresilient behavior detection were tested using FWD tests at Cornell University. Tests were performed from February until May 2003. No trends were observed through ANOVA (analysis of variance) tests but chi-squared variance tests on the center sensor data revealed nonresiliency during the spring–thaw season. ANOVA tests will detect systematic variations; however, if the deflections are not always generally increasing or decreasing for a given load level, the test does not detect when nonresilient behavior is occurring.

20.2.2.4   Evaluation of Experimental Paving Materials

Crushed aggregate, a popular base course for pavements, have become progressively more expensive. To save money on base courses, Florida DOT has sponsored recycling concrete aggregate (RCA) research. FWD data were used to test various RCA mixes and the results show RCA to be a viable base course for roadway pavements. Ultra-thin whitetopping (UTW) was evaluated using FWD data in Minnesota. FWD data were collected 1 year after an experimental UTW pavement test section was constructed at the Minnesota Road Research test facility. PCC thickness varied from section to section; the study's intent is to determine an ideal PCC thickness. Strain data captured by the FWD showed a good bonding condition between the lower bituminous surface and the new PCC wearing course. Although an optimal UTW overlay design is not yet determined, "the dynamic strain measurements indicate that there is a better bond between the asphalt and the overlay in the thinner sections." It was also observed that the magnitude of the strains in the thinner sections were more dependent on the stiffness of the asphalt than the number of equivalent single axle loads accumulated.

20.2.2.5 Determination of Load Transfer Efficiency for Jointed PCC Pavements

Adjacent JPC slabs should move together when a load is applied to one of them; faulting can result if they do not. The degree to which adjacent slabs move together is defined as load transfer efficiency, LTE. LTE is calculated from FWD deflections by placing the load cell on one PCC slab and then placing a sensor on an adjacent unloaded slab. When the weight is dropped, the measured deflections are used to calculate LTE with the following equation:

$$\text{LTE} = 100\% * \frac{D_{\text{unloaded}}}{D_{\text{loaded}}}$$

where
  D_{unloaded} represents the deflection of the unloaded PCC slab
  D_{loaded} represents the deflection of the loaded PCC slab

20.3 PORTABLE FALLING WEIGHT DEFLECTOMETER

The portable FWD (Figure 20.3) is a single-person-use portable device that uses an accelerometer to determine the deflection due to a falling weight. The results are given as maximum deflection, and the bearing capacity modulus is calculated from it. This device can be used to estimate the stiffness of the upper layer of the pavement or to compare the stiffness and hence determine the effectiveness of compaction. The weight is of 10 kg falling through a height of 80 cm onto a plate of diameter ranging from 140 to 200 mm. The peak force is 25 kN, with impulse duration of 25–30 ms.

20.4 ROLLING WHEEL DEFLECTOMETER

The rolling wheel deflectometer (RWD) is a truck-based system that can be used to measure the deflection under moving dual tires of an 18-kip single-axle load for primarily flexible pavements. The equipment (Figure 20.4) consists of a semi-trailer with four laser units in the undercarriage and close to the pavement surface. Of the four laser units, three (located forward and outside of the deflection basis) are used for measuring the unloaded pavement profile, while the fourth one (located between and under the dual tires) is used for determination of the deflected profile. The deflection is measured as the difference between the deflected and the undeflected profile. A reading is taken at every 0.1 in., and the data are averaged to reduce the random error of measurement. The tractor unit contains all of the required data acquisition system, and the equipment can also record digital images and videos of the test section during testing as well as GPS coordinates. The test data can

FIGURE 20.3 Portable falling weight deflectometer. (Courtesy of Maureen Kestler/Michael Santi.)

(a)

Laser units on beam

(b)   (c)

FIGURE 20.4   Rolling wheel deflectometer. (a) RWD semi tractor trailer. (b) Laser units on beam. (c) Laser units between and under wheels.

be used for network-level analysis in pavement management as well as for delineation of segments for rehabilitation, and to establish limits of deflection for new pavements and load restrictions for spring–thaw periods. Since it can measure the deflection while moving at highway speeds, there is no need for lane closures and traffic control during its operation, and as much as 200–300 lane miles can be tested in 1 day.

20.5   GEOGAUGE (SOIL STIFFNESS GAUGE) FOR SOIL COMPACTION

The GeoGauge (Soil Stiffness Gauge; Figure 20.5) can be used for measuring soil layer stiffness as part of the soil compaction control process. This handheld equipment works by measuring the impedance at the surface of the soil. It measures the stress imparted to the surface and the resulting surface velocity as a function of time. The GeoGauge imparts very small displacements to the soil ($<1.27 \times 10^{-6}$ m or <0.00005 in.) at 25 steady-state frequencies between 100 and 196 Hz. Stiffness is measured as a ratio of force to displacement, and can be used to determine the modulus. The stiffness is determined at each frequency, and the average of the 25 is displayed. The GeoGauge is supported on the soil by a ring-shaped foot through isolators. Shakers attached to the foot measure force and displacement. Moduli can be calculated with the use of Poisson's ratio. The equipment is powered by batteries and can store multiple readings, and the data can be downloaded to a computer. The equipment is calibrated by measuring force and displacement of a reference mass.

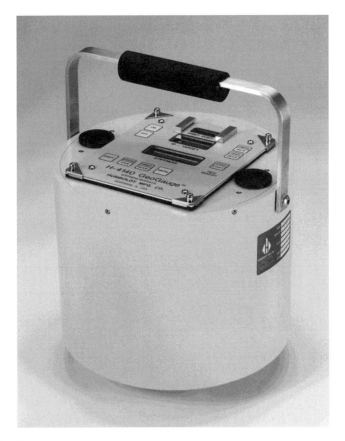

FIGURE 20.5 GeoGauge equipment.

20.6 GROUND-PENETRATING RADAR

The GPR works on the principles of radiating short electromagnetic pulses from an antenna and analyzing the amplitude and velocity of the reflected waves from pavement layers (Figure 20.6). The pavement thickness and density can be computed from these amplitudes and arrival times. The GPR can be used as a handheld device for use on short stretches of pavement, but for pavement thickness measurement of long stretches of test pavements generally a vehicle-mounted 1 GHz horn antenna GPR system is used (Figure 20.6). Data are collected at normal driving speeds, and passes can be made out in the wheelpaths and centerline of each lane in each direction. The equipment consists of an electronic box and an antenna, together with a tether and a laptop computer with the software required for data acquisition and analyses. The data can then be analyzed to develop continuous thickness profiles, which are very useful for determination of pavement condition (along with other devices such as the FWD) for rehabilitation or estimation of remaining life, as well as for segmentation of a project for proper consideration of mix design and construction methods for rehabilitation or recycling operations.

20.7 PORTABLE SEISMIC PAVEMENT ANALYZER

The portable seismic pavement analyzer (PSPA; Figure 20.7) is a rapid nondestructive testing device that provides the modulus of the top pavement layer in real time. It is a portable unit that consists of a "source," two "receivers," and an electronics box, and comes with a tether and a laptop computer with the software that controls both data acquisition as well as analysis. The analysis procedure uses

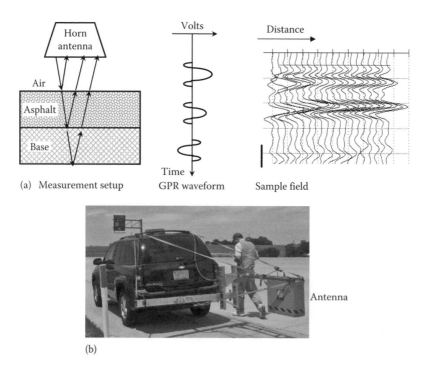

(a) Measurement setup   GPR waveform   Sample field

(b)

FIGURE 20.6   (a) Schematic of operation and (b) ground penetrating radar equipment GPR. (Courtesy of Kenneth Maser, Infrasense, Inc., Arlington, MA.)

the surface wave energy to determine the variation in modulus with wavelength (strictly speaking, surface wave velocity with wavelength). The analysis method implemented in the PSPA is called the ultrasonic surface waves (USW) method, which is a simplified version of the spectral-analysis-of-surface-waves (SASW) method. Briefly, this method utilizes the surface wave energy to determine the variation in modulus with wavelength. For simplicity, the surface wave velocity is converted to modulus using mass density and Poisson's ratio.

Up to a wavelength equal to the thickness of the top pavement layer, the moduli from the dispersion curve are equal to the actual moduli of the layer. As such, the modulus of the topmost layer can be directly estimated without a need for back-calculation.

20.8   FREE FREE RESONANT COLUMN TEST

The basic principle of this test is that the resonant frequency that is developed within a specimen of a certain material when it is subjected to an impulse load is dependent on its dimension and stiffness. Hence in this test, a specimen of known dimension is subjected to a tapping by hammer (Figure 20.8) fitted with a load cell, and an accelerometer is used to record the resulting signal at the other end of the specimen. The amplitude–frequency plot, displayed by a data acquisition system, which is connected to both load cell and the accelerometer, is used to determine the resonant frequency. The resonant frequency is then used to calculate the modulus, using the mass density (ASTM C-215).

20.9   ULTRASONIC TEST

In this test (Figure 20.9), a transducer-generated electric pulse is used to cause a mechanical vibration in one side of a specimen, and a transducer is used to sense the propagating wave at the other end of the specimen. The time of travel of the wave through the specimen is measured, and the modulus of the material is calculated using the bulk density and Poisson's ratio.

(a)   (b)

(c)

FIGURE 20.7   Portable seismic pavement analyzer (PSPA). (a) PSPA with data acquisition system. (b) Source (foreground) and two receivers in PSPA unit. (c) Laptop computer shows results of analysis within seconds.

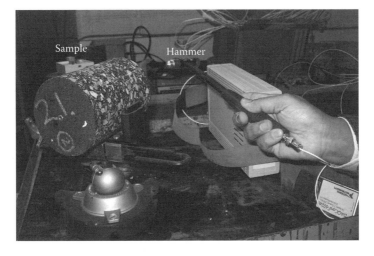

FIGURE 20.8   Free Free resonant column test.

FIGURE 20.9   Ultrasonic test.

FIGURE 20.10   MIT scan equipment on rails on concrete slab. (Courtesy of FHWA, Atlanta, GA.)

20.10   MAGNETIC INDUCTION TOMOGRAPHY

The magnetic induction tomography (MIT)-Scan equipment (Figure 20.10) uses magnetic tomography to detect and evaluate metal dowels placed in concrete pavements. The equipment consists of five sensors that emit an electromagnetic pulse and detects the induced magnetic field in the dowels. It is towed on rails across a concrete pavement slab, and both numerical and graphical data are output quickly from dedicated software in the attached computer. The device utilizes a number of sensors to capture redundant data, as well as filtering techniques to detect both the position and orientation of dowels inside the concrete.

QUESTION

20.1   Use any back-calculation software to estimate the moduli of the different layers in a pavement in a rocky fill area in Maine, which has the following results, from an FWD test. Make reasonable assumptions, as required.

Layer 1: HMA surface and binder (consider as one layer) thickness = 4 in.; layer 2: granular
   base thickness = 12 in.; layer 3: subgrade
FWD test details and data:
Plate radius (in.): 5.9; number of sensors: 7
Sensor offsets (in.): .0, 8.0, 12.0, 18.0, 24.0, 36.0, and 60.0
Load (lbf): 8953.0
Measured deflections (mils): 7.250, 5.420, 4.690, 4.000, 2.960, 2.230, and 1.690

21 Economic Analysis and Cost-Saving Concepts

21.1 ECONOMIC ANALYSIS

A proper economic analysis should be conducted for every pavement construction project. In order to conduct economic analysis, it is necessary to understand some key concepts and the techniques that help us to compare alternatives on the basis of economics and have reasonable estimates of different types of costs associated with pavement projects.

21.1.1 ENGINEERING ECONOMY

Economic analysis can be conducted at two levels—first, at the network level to determine the feasibility and scheduling of a project and, second, at the project level to achieve maximum economy within a specific selected project by comparing different alternatives. For the second case, all of the alternatives must be able to satisfy the project requirements and should be considered over the same time period. The economic evaluation of the different alternatives must consider accurate (as far as possible) estimates of economic variables and costs and benefits associated with the pavement. Life cycle cost (LCC) analysis provides a rational method of conducting such economic analysis.

21.1.2 CONCEPT OF LIFE CYCLE COST

The life cycle cost means the total cost that is incurred during the complete life cycle of the pavement. This includes construction, maintenance, and rehabilitation costs and considers return from salvage, if any. The utility of computing LCC is that it may be more prudent to spend more money to build a better pavement initially and spend comparatively less money in maintenance and rehabilitation than it is to save money by building a lower quality pavement, which will require more and/ or more frequent maintenance and rehabilitation and therefore will require spending significantly more money in the long run.

The techniques used for conducting economic analysis to compute LCC vary, and they depend on many factors such as the level at which the analysis is being conducted (network or project), the applicability of the techniques to the specific type of project (e.g., public or privately owned), and the consideration of level of acceptance from the decision maker and the public (which, e.g., may be more sensitive to initial construction costs). However, all applicable techniques must consider future costs and benefits and the differences between the different types of alternative pavement types for the specific project.

21.1.3 TECHNIQUES

The different steps involved in LCC analysis are establishing alternatives, determining timings of activities, estimating costs, and computing the costs, respectively. Two of the more commonly used economic indicators are the net present value (NPV) or net present worth (NPW) and equivalent uniform annual cost (EUAC). NPV is the discounted monetary value of expected net benefits (i.e., benefits minus costs). In this method, monetary values are assigned to benefits and costs, which are discounted using a discount rate to determine their present values (PV), and then the PV of the costs is subtracted from the PV of the benefits. NPV is calculated using the following formula:

$$NPV = \text{Initial cost} + \sum_{k=1}^{N} \text{Rehabilitation cost}_k \left[\frac{1}{(1+i)^{n_k}} \right]$$

where
   i is the discount rate, the percentage figure representing the rate of interest that money can be assumed to earn over the period of time under analysis
   n is the year of expenditure

Note that the discounting (using a discount rate, i) reflects the time value of money, and it expresses the benefits and costs at different times in terms of a common unit of measurement. The real discount rate of 3%–5% has been suggested for use with real dollar cost estimates.

   In the EUAC method, the NPV of all discounted costs and benefits is expressed as uniform annual payments throughout the analysis period and is useful for situations where budgets are established on an annual basis. EUAC is calculated as follows:

$$EUAC = NPV \left[\frac{i(1+i)^n}{(1+i)^n - 1} \right]$$

where
   NPV is the net present value
   i is the discount rate
   n is the year of expenditure

The concepts of NPV and EUAC are illustrated in Figure 21.1.

Example 21.1

For an analysis period of 20 years, there are two alternative pavement designs with equal salvage values. The first alternative is to reconstruct the pavement when it reaches failure, while the second alternative is to improve the condition of the pavement at the 8th, 16th, and 24th years.
   The costs are as follows:
   Alternative 1: initial cost: $18 million; cost to reconstruct after 20 years: $9 million.
   Alternative 2: initial cost: $15 million; cost for improvement at 8th, 16th, and 24th years: $5 million each occurrence
Which alternative should be selected?

Solution

Use a discount rate of 4%.

$$NPV = \text{Initial cost} + \sum_{k=1}^{N} \text{Rehabilitation cost}_k \left[\frac{1}{(1+i)^{n_k}} \right]$$

$$NPV_{\text{alternative1}} = \$18\,\text{million} + \$9\,\text{million} \frac{1}{(1+0.04)^{20}} = \$22.1\,\text{million}$$

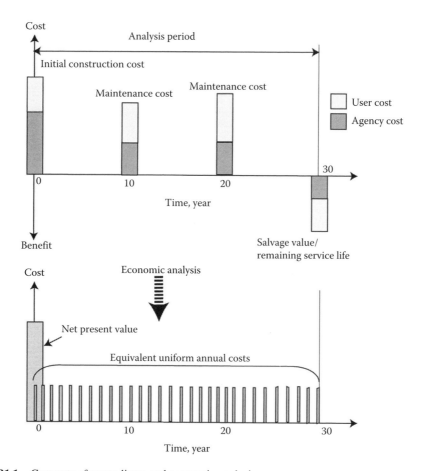

FIGURE 21.1   Concepts of expenditure and economic analysis.

$$\text{NPV}_{\text{alternative2}} = \$15\,\text{million} + \$5\,\text{million}\frac{1}{(1+0.04)^8} = \$5\,\text{million}\frac{1}{(1+0.04)^{16}}$$

$$+ \$5\,\text{million}\frac{1}{(1+0.04)^{24}} = \$23.3\,\text{million}$$

Therefore, alternative 1 has the lower LCC and should be considered for selection.

21.1.4   Costs in Life Cycle Cost Analysis

The costs include those incurred by the pavement-managing entity (such as a department of trans-portation) as well as the users, because of the construction or maintenance/rehabilitation work. Note that costs that are common to alternatives can be excluded from the LCC analysis. The same concepts are applicable for benefits as well.

The agency costs include those associated with construction, maintenance, and rehabilita-tion; engineering and administrative costs; as well as traffic control costs, and should take into account the salvage value at the end of the analysis period. In case the design life of one alternative exceeds the other, then the remaining service life (RSL) must be taken into consideration. The user costs, which are relatively difficult to estimate, include those associated with vehicle operation, travel time, and accidents. Guidance on evaluating travel time costs can be obtained from a U.S.

Department of Transportation (USDOT) memorandum from http://www.dot.gov/sites/dot.dev/files/docs/vot_guidance_092811c_0.pdf.

21.1.5 Probabilistic versus Deterministic Approach

In the earlier discussions, the values of the outputs (such as NPV) were obtained by using equations, assuming unique values of the input, and this resulted in specific values of the output. This is the *deterministic* approach, in which the uncertainty associated with the inputs, such as future costs, discount rate, or year of rehabilitation, is not considered, and the values are considered to be constants. However, in real life, the input values are not constant—they are variable. In the more sophisticated *probabilistic* method, this variability of input values (such as initial costs and discount rate) is considered and the output is determined as a (normal) distribution (rather than a single value). Using a cumulative distribution, one can estimate the different costs (corresponding to different percentiles or probability) and also determine which alternative is more variable. The method uses Monte Carlo simulation, which picks random numbers from the distribution of the input variables, outputs the results also as a random variable, and then sums up all the results to provide a distribution of the output parameter. The use of this probabilistic approach enables the user to quantify the risk (risk analysis) associated with each option (due to the variability of input parameters) and hence make a more informed decision on the basis of risk the agency is willing to tolerate.

21.1.6 Information on Life Cycle Cost Analysis for Pavements

Since the early 1990s, the U.S. Federal Highway Administration (FHWA) has been actively sponsoring workshops on LCC and encouraging users to employ LCC analysis to evaluate alternatives for different projects. The use of LCC analysis has been made compulsory for all projects in excess of $25 million that have federal involvement in the United States. Simultaneously, the FHWA Office of Asset Management has been offering a very useful collection of tools for conducting LCC analysis—these tools include detailed and summary documents and spreadsheet-based software (RealCost), complete with manuals. The materials can be downloaded free from the following website: http://www.fhwa.dot.gov/infrastructure/asstmgmt/lcca.cfm.

Information on discount rates can be obtained from the U.S. Office of Management and Budget (OMB) through its Circular No. A-94 at http://www.whitehouse.gov/omb/circulars/a094/a094.html#1.

LCC analysis software can also be obtained from the Asphalt Pavement Alliance (APA) website at http://asphaltroads.org/LifeCycleCostAnalysis.

21.1.7 Software for Running Life Cycle Cost Analysis

Different software are available for conducting LCC analysis. The process is illustrated with the use of two such software—RealCost (an LCC analysis according to the FHWA best practice methods) and AirCost (a life cycle analysis tool for airport pavements, available from http://www.aaptp.us/aircost.html).

For learning the RealCost spreadsheet-based software, the reader is advised to view the manual that is downloadable from the FHWA website and the step-by-step instructions given in the Caltrans Interim RealCost Manual that is unzipped from the RealCost package, once the software is installed. The RealCost software opens up with a "switchboard" on a spreadsheet that allows the user to input data into the spreadsheet. The various steps and input details are summarized in Figure 21.2. A comprehensive list of design alternatives for pavement projects is provided in the Caltrans manual.

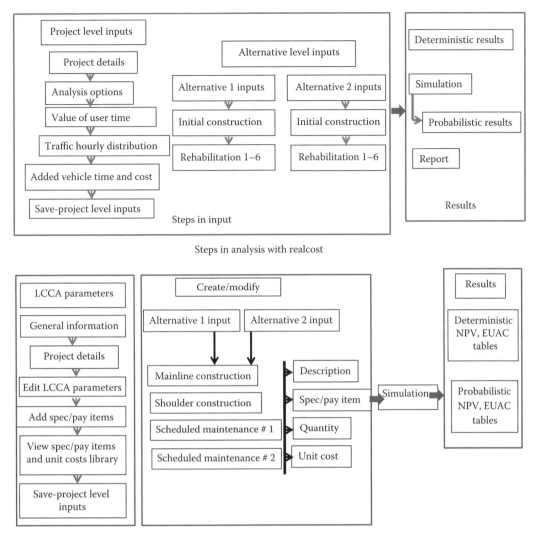

FIGURE 21.2 Steps in the use of RealCost and AirCost.

The analysis period needs to be selected on the basis of guidelines that are provided—it depends on the alternatives and their design lives. For the selected analysis period, maintenance or rehabilitation treatments need to be decided after the initial construction. Typical discount rates vary from 3 to 5. For each alternative, the Maintenance and Rehabilitation (M&R) schedule must be established. Typical M&R schedules are provided in the Caltrans manual and are decided on the basis of existing/new pavement type, proposed final surface type, pavement design life, climatic region, and maintenance service level. A guide to the selection of the proper M&R table is provided in a flow chart in the manual. Since the software allows up to 6 rehabilitation activities, interim maintenance projects are converted into annualized cost and entered separately in RealCost.

The costs that are estimated include agency costs (initial, maintenance, rehabilitation cost, and RSL value) and user costs (travel time costs and other than routine maintenance vehicle operating costs that is incurred by the user). Guidance for estimating project support costs is also given. Maintenance costs include those required for routine, preventive, and corrective maintenance, while rehabilitation cost is the cost for future rehabilitation that is required for a specific alternative. A step-by-step procedure for the calculation of rehabilitation cost is given in the guide.

The reduction of the capacity of the facility and traffic flow leads to the user costs (typical values are given). The RSL is defined as the difference between the service life and the analysis period and is calculated on the basis of total cost of the last scheduled rehabilitation activity and the percentage of service life remaining at the end of the analysis period. The LCC is determined with the help of the concept of discounting (to compare the money at a common point of time) and, specifically, the PV approach.

AirCost is also a spreadsheet-based software that consists of four components—life cycle analysis parameter specification, scenario alternative analysis, LCC analysis simulation, and viewing simulation results. It also provides a "switchboard" interface to input data and run the analysis. The general steps are shown in Figure 21.2. Because this software has been developed specifically for airport pavements, certain items in areas such as project details contain airport specific inputs, for example, information on lighting, striping and grooved pavement area. Other important features include the total airport daily revenue and its growth rate for the LCCA parameters. For the analysis variables, option is provided for selecting deterministic or probabilistic—normal, uniform, and triangular—approach. A pay item/cost library is inbuilt in the spreadsheet. The data sheet for the mainline and other construction/rehabilitation provides the user the option of including the salvage value analysis for the project. The final results/output can be saved as an Excel file.

21.2   COST-SAVING CONCEPTS

Over the years, cost-saving concepts in pavement design and construction have been developed on the bases of economic analysis as well as necessity. Two such concepts are perpetual pavement and the recycling of pavement materials. The concept of perpetual pavement is based on economic analysis—it provides a better pavement now with higher initial cost, to prevent excessive costs in the future. The concept of recycling has grown partly on the basis of necessity and partly on that of engineering economics. Periodic shortages of asphalt (as a result of shortage of gasoline) and ever increasing costs of asphalt and construction processes have made pavement agencies recycle more and more pavement materials over the years. In fact, recycling is a routine activity for most pavement agencies, and the last few decades have seen tremendous improvements in recycling equipment and processes all over the world.

21.2.1   PRINCIPLES OF PERPETUAL PAVEMENTS

Perpetual pavements refer to full-depth asphalt pavements that are engineered to resist typical distresses and survive for long periods of time, without requiring any major rehabilitation. The basic requirements of such a pavement are as follows:

1. The bottom must have high asphalt content, and the overall thickness should be large enough to prevent fatigue cracking (by having a large number of allowable repetitions to failure as well as by having a relatively small tensile strain).
2. The intermediate layer should be durable and stable—requiring strong aggregate structure as well as sufficient resistance against moisture damage.
3. The wearing course must possess sufficient resistance against rutting and wearing failures and be impermeable, although, if needed, an open-graded friction course should be used to improve driving conditions.

The concept of design of perpetual pavement is shown in Figure 21.3.

Over time, a perpetual pavement is expected to have distresses on the surface or within 2–3 in. of the surface, which can be repaired relatively easily and economically by replacing or improving the top layer only, for example, through resurfacing or recycling. This is a significantly

Surface	Zone of high quality HMA or OFGC; 1.5–3 in.
Intermediate	Zone of high compression; high modulus rut-resistant material; 4–7 in.
Base	Zone of maximum tensile strain; flexible fatigue-resistant material; 3–4 in.
Subgrade	Pavement foundation

FIGURE 21.3 Concept of perpetual pavement. (Adapted from Newcomb, D.E. et al., Concepts of perpetual pavements in perpetual bituminous pavements, Transportation Research Circular, No. 503, Transportation Research Board, National Research Council, Washington, DC, December 2001.)

better option compared to a conventional pavement, in which fatigue cracks coming up from the bottom part would necessitate the replacement of the entire pavement structure that is up to the depth of the crack initiation. The time interval of resurfacing of perpetual pavements can be in the range of 12 years, giving them a relatively longer life of 25 years or more (with two resurfacings).

The structural and mix design of the asphalt-rich bottom layer (the base, which is in the zone of the maximum tensile strain and hence the zone of fatigue crack initiation) can be based on the "maximum limiting" strain of 60 $\mu\varepsilon$, whereas that for preventing subgrade rutting can be based on the "maximum limiting" strain of 200 $\mu\varepsilon$.

The materials used for constructing the different layers must be selected carefully, with proper consideration of the type(s) of distresses that can be expected in each specific layer. The foundation (subgrade) must be well compacted and have adequate resistance against swelling or frost heaves and should be capable of supporting necessary construction equipment. The base should have a relatively high content (to give a high density of the layer) of an asphalt whose high-temperature grade should be the same as that of the surface layer, while the low-temperature grade could be the same as that of the layer above this layer (the intermediate layer). Performance tests should include those for rutting, fatigue, as well as moisture damage. The intermediate layer should contain rut-resistant aggregate structure and an optimum content of asphalt of the same high-temperature grade as that of the surface layer and low-temperature grade one grade lower than that of the surface layer. For aggregates, either large-size stones or smaller stones with stone-on-stone contact can be used. Performance tests should include those for rutting and moisture damage. The wearing (surface) layer should be made up of polish-resistant aggregates with stone-on-stone structure (gradation) and an optimum amount of asphalt with appropriate (or one grade higher) high- and low-temperature grade. This asphalt may contain polymers and/or fibers, as required for the mix, to achieve high density, impermeability, and rutting resistance. At a minimum, performance testing for rutting should be conducted.

Construction of a perpetual pavement should be done using proper equipment and quality control/assurance techniques. Key factors are the use of volumetric properties to check the quality of hot mix asphalt (HMA), and sophisticated tools such as laser for detecting (and hence preventing) the segregation of HMA, a nuclear gage for assuring proper density, ground-penetrating radar for ensuring adequate and uniform thickness, and profilometers for smoothness. Continuous access to an adequately staffed and equipped quality control laboratory is critical.

Once the perpetual pavement is constructed, it is necessary to monitor it for distresses, such that adequate steps can be taken when the critical type and level of distress are observed. Annual surveys of distress and ride quality are necessary. In general, the resurfacing can consist of the milling of 2–4 in. of the surface and replacing it with a similar or slightly better mix. Structural evaluation can be triggered by the observation of critical distresses on the surface and can be conducted with the use of a falling weight deflectometer and ground-penetrating radar (and cores). If necessary, a slight increase in thickness can be made during resurfacing. Proper tack coat must be used to ensure bonding between the existing pavement and the new wearing layer.

Detailed information on perpetual pavement can be obtained from the APA webpage at http://asphaltroads.org/PerpetualPavement.

21.2.2 ECONOMIC BENEFITS OF RECYCLING

When properly selected, all the different types of recycling methods are usually cheaper than the conventional rehabilitation methods, even though the relative savings will depend on the kind of recycling technique used. The primary savings in hot and cold mix asphalt pavement recycling come from savings in the cost of virgin asphalt cement, whereas the savings in hot in-place recycling come by the elimination of transportation costs and use of very little amount of virgin material. The major savings in the case of cold in-place recycling come by eliminating the need for a fuel or emission control system, since the process is done at ambient temperature; the elimination of transportation costs; and the addition of only a small percentage of virgin asphalt binder. Savings of up to 40%, 50%, 55%, and 67% can be achieved by using hot mix recycling, hot in-place recycling, cold in-place recycling, and full-depth reclamation, respectively. These savings are achieved when one of the recycling methods is used in place of a conventional method or some other recycling method. For concrete pavements, existing materials can be rubblized and the layer can be compacted and used as a base for an asphalt pavement, or the existing concrete can be recycled with new aggregates for a new concrete pavement.

In addition to the material and construction cost savings, a significant amount of cost savings (in terms of user costs) can be realized by the reduced interruptions in traffic flow when compared with conventional rehabilitation techniques. Recycling can be used to rejuvenate a pavement or correct a mix deficiency and conserve material and energy—options not available with the conventional paving techniques. A conventional overlay may require upgrading shoulders to maintain a profile, raising guard rails to maintain the minimum safety standard, and restricting overlays below bridges to maintain underpass height. On the other hand, recycling can effectively be used to maintain the highway geometry and thus result in substantial overall savings as well.

Information on recycling pavements can be obtained from the FHWA's website at http://www.fhwa.dot.gov/pavement/recycling/.

QUESTIONS

21.1 What is life cycle cost analysis?

21.2 What are the different types of costs associated with a pavement?

21.3 ADetermine which of the alternatives presented next has the lower life cycle cost. Assume that both have the same life of 30 years and the same salvage value. Assume a discount rate of 4%.

Alternative A: initial cost: $15 million; rehabilitation at the 10th and 20th years: each cost $4 million.

Alternative B: initial cost: $17 million; rehabilitation at the 15th year: $4.5 million.

21.4 What is the basic concept of perpetual pavement?

21.5 Consider two alternatives with equal lives for constructing an asphalt pavement: option 1 involves milling and removing the existing 2-in. layer and replacing it with 4 in. of HMA; option 2 uses cold in-place recycling to recycle the top 1.5 in. of the existing pavement and then overlays it with 2 in. of new HMA. Obtain the costs of milling, recycling, and construction of new HMA from your local HMA industry (e.g., an asphalt association such as the National Asphalt Pavement Association [NAPA] and/or the Asphalt Recycling and Reclaiming Association [ARRA], or the local or state department of transportation), and select the best option.

22 Instrumentation in Asphalt and Concrete Pavement

Analysis of data from properly instrumented pavement test sections can provide invaluable information for the proper design and rehabilitation of pavements. Such data consist of environmental as well as response data from different layers, the more important ones of which are discussed in the following sections.

22.1 TEMPERATURE

A thermocouple is generally used for determination of temperature. Its principle of operation is based on the fact that the voltage drop across dissimilar metals, which are placed in contact, is a function of temperature. There can be different types of thermocouples. For example, in a T type, a twisted-stranded-shielded-soldered pair of thermocouple wire (constantan and copper) is used (Figure 22.1). The exposed end of the thermocouple is surrounded by copper tubing, which is attached to the cable insulation by heat-shrinkable Teflon tubing.

22.2 SOIL MOISTURE CONTENT

Soil moisture content is measured using the time domain reflectometry (TDR) technique, which is based on the change in dielectric constant of a material with a change in moisture content. The dielectric constant of dry soil can range from 3 to 8 (whereas that of free water is 81). A TDR probe is shown in Figure 22.2. This probe (CS615; Campbell Scientific) has two parallel conducting rods held on one end by an epoxy head, containing a bistable multivibrator. A suitable oscillation frequency of the vibrator is used for measurement of dielectric constant and to determine moisture content with the help of a data acquisition system.

22.3 FROST DEPTH

The depth of frost in soil is determined using the concept of measurement of soil resistivity, which differs widely from an unfrozen state (20,000–50,000 Ω normally) to a frozen state (from 500,000 up to several million ohms). Electrical resistance is measured between the conductors mounted along a cylindrical probe. The depth of frost penetration is measured from the resistance data collected from adjacent pairs of electrodes down the resistivity probe, noting the depth of change in resistivity. Copper rings in PVC rods (Figure 22.3) are generally used for measuring resistivity.

FIGURE 22.1   Thermocouples. (Courtesy of Lauren Swett.)

FIGURE 22.2   Time domain reflectometry (TDR) probe. (Courtesy of Lauren Swett.)

FIGURE 22.3   Frost depth measurement instrument. (Courtesy of Lauren Swett.)

22.4   STRAIN IN ASPHALT OR CONCRETE PAVEMENT AND SOIL LAYERS

The horizontal strain in an asphalt or concrete pavement layer is determined on the basis of change in resistance of a body due to its elongation or shortening. The elongation or shortening is caused by stresses coming from the traffic load. Two features are important for the proper functioning of the strain gages. First, the strain gage must be embedded in material with high flexibility and low stiffness and have sufficient durability against stresses and temperature fluctuations. Second, the strain gage must be securely anchored in the pavement to behave as "part" of the pavement. A Dynatest PAST II strain gage is shown in Figure 22.4 as an example.

FIGURE 22.4 Asphalt or concrete strain gage. (Courtesy of Lauren Swett.)

FIGURE 22.5 Soil strain gage. (Courtesy of Lauren Swett.)

It has a resistance of 120 Ω and a gage factor of 2.0 and can endure a temperature range of $-30°C$ to $+150°C$. The maximum strain it can record is 1500 microstrain. For soils, for measuring the vertical strain, generally linear variable deformation transducers (LVDT) such as those with a ± 5 mm range are used. Figure 22.5 shows a Dynatest Soil and Stress Deformation Transducer (FTC-1).

22.5 STRESS IN SOIL LAYERS

Soil pressure cells are used for the measurement of vertical stress in different types of soils. The critical features include sensitivity and resolution of the pressure and resistance to the damaging effects of loading and the environment. An example of a pressure cell (Figure 22.6) is the SOPT gage from Dynatest, which consists of a liquid-filled cell with an integrated pressure transducer within a thin membrane, all consisting of pure titanium. Measurable pressures range from 1.5 to 120 psi for different size aggregates.

FIGURE 22.6   Soil pressure cell. (Courtesy of Lauren Swett.)

22.6 DEFLECTION IN LAYERS

Multidepth deflectometer (MDD) (Figure 22.7) is used to determine the deflection of the different layers in the pavement under traffic loading. The instrument consists of a series of LVDTs at different depths stringed on a column that is inserted through a hole in the pavement. The data from the device can be acquired and viewed in real time, using sophisticated data acquisition system.

22.7 DATA ACQUISITION SYSTEMS

Data are obtained from in-place instruments through the use of appropriate data acquisition systems. Such data acquisitions systems are controlled by a suitable software (such as LAB VIEW) through a computer system, and include appropriate modules or devices to interface between the instruments and the computer. In many cases, data loggers are used to collect and store data, which are then periodically transferred to a computer by using a suitable software and connection cable. As required the software can be used to trigger data acquisition on the basis of some other signal, such as data coming from a weigh-in-motion (WIM) signal (as a truck passes by the instrumented section). The relevant information about the instruments (such as gage factor or calibration constant) is input into the software, which then makes all the relevant computations on the basis of the acquired data and stores the data in ASCII or spreadsheet format or even displays data in graphical format. Various types of hardware and software are available—an example is shown in Figure 22.8.

FIGURE 22.7 Multi-depth deflectometer (MDD) (a) schematic, (b) installation, and (c) data acquisition. (Courtesy of CTL Group, Skokie, IL.)

(a)   (b)

FIGURE 22.8 Data acquisition systems. (a) Data acquisition system with module in computer. (b) Close-up view of the module showing connection from a strain gage.

QUESTIONS

22.1 List the various instruments used in pavements for environmental and load response–related data.

22.2 Conduct a literature search on Transportation Research Information System (TRIS) on the Internet to obtain at least one report on full-scale instrumentation in pavements. Prepare a review of the report with particular emphasis on the type of instruments, installation, and experience of the researchers.

23 Specialty Applications

23.1 ASPHALT MIXTURES

Over the years different types of asphalt mixtures have been developed in different parts of the world for specialty applications. Some examples of such mixes are presented in the following sections.

23.1.1 POLYMER-MODIFIED ASPHALT

Polymers are used in asphalt to increase the range of temperature in which the asphalt mix can be used without any distress and to improve bonding of asphalt binder and aggregates. Elastomers and plastomers are the two major types of polymers used in asphalt. Block copolymers, random polymer, and natural and synthetic latex are examples of elastomers, which are used primarily for decreasing temperature susceptibility, increasing tensile strength at high strain, and improving the cohesion and adhesion of asphalt binder. For example, styrene butadiene rubber (SBR) is used for reducing the rutting potential of asphalt mixes at high temperatures. Plastomers consist of polyethylene and ethylene copolymers, and are used primarily to decrease temperature susceptibility, increase modulus, or increase tensile strength. For example, ethyl vinyl acetate (EVA) is used for modifying the strength, stiffness, and workability of the asphalt mix. Polymers can be either preblended with the liquid asphalt binders or blended with the mix at the hot mix asphalt (HMA) plant. A growing use of polymer-modified asphalt is in chip seal coats and other thin-layer maintenance applications, as an interlayer for resisting reflective cracking, as well as in other specialty mixes such as the open-graded friction course (OGFC) and stone matrix asphalt (SMA). The use of polymers has become almost routine in many applications, and this has necessitated the development of special characterization tests (such as the direct tension test and the multiple stress creep and recovery test or MSCR) and conditioning tests (such as the modified rotating German flask or MGRF) for polymer-modified asphalt binders. Two issues that are specifically important for the good performance of polymer-modified asphalt mixes are storage stability and degradation. Storage stability involves the compatibility of the polymer and asphalt binder when stored for long time periods at elevated temperatures, and degradation refers to the deterioration of the properties of the polymer due to the presence of oxygen at high temperatures.

Polymer-modified binders can be produced by physical blending or cross-linking (with a reagent) and the important factors are the reagent, dosage of asphalt binder and polymer, use of high shear mixer, and proper storage. For the mixing process, critical parameters are time, temperature, speed, and shear of the agitation system. Depending on the amount of polymer, three types of polymer-binder blends can be produced—asphalt continuous phase (<3% polymer content), two co-continuous phase (3%–7% polymer content), and polymer phase matrix (polymer content >7%). For a specific type of polymer, the compatibility depends on the polymer content and the properties of the base asphalt. Cross-linked blends are favored over physical blends because of superior performance and also homogeneity, which leads to storage stability, reduced storage cost (no stirring), and reduced problems if there is delay and better control quality. In Europe, the use of polymer-modified binder is controlled by EN 14023 specification, which is based on consistency (Pen and Ring and Ball), cohesion (FD, DTT, and Vialit), durability (RTFOT) and flash point, brittleness (Fraass breaking point), strain recovery (elastic recovery), as well as additional tests for plasticity range, storage stability, and ER after RTFOT; in the United States, Superpave with MSCR test is utilized. The time required for putting a structural overlay for rehabilitation can be significantly prolonged by using polymer-modified binder in the surface and base layers.

For information on modified asphalts, the reader is advised to see the Association of Modified Asphalt Producers (AMAP) website (www.modifiedasphalt.org).

23.1.2 Asphalt Rubber Mixes

Crumb rubber, obtained from recycling vehicle tires, has emerged as a commonly used additive in asphalt binder in the United States. Started primarily as a way of recycling tires and avoiding filling up landfill space, the use of crumb rubber has grown steadily since the early 1990s and continues to grow as different advantages of its use are being investigated. Generally, crumb rubber is used at about 15%–25% of the total mass of the binder. It is crushed to a maximum size (generally 1 mm) before being added to asphalt binders at an elevated temperature (wet process) or to aggregates at an HMA plant (dry process). The reported benefits of using crumb rubber in asphalt mixes include greater flexibility, rougher surface texture, greater resistance to aging, and a decrease in draindown potential of asphalt binder. These qualities translate to greater fatigue life, better surface friction, improved elasticity, and ability to add a higher amount of asphalt binder to increase film thickness in mixes such as OGFC. The use of rubberized asphalt (or what is known as asphalt rubber in the United States) continues to grow in chip seals and other maintenance mixes, as an interlayer, as well as in structural HMA, in the United States as well as in Europe (e.g., in Spain). The extra cost of the crumb rubber can be offset by the reduced thickness of layers and/or the enhanced life of the pavement. In many cases, successful use of crumb rubber–modified binder (CRMB) has been made to increase the durability of OGFC and porous friction course (PFC) mixes. Generally, CRMB contains a processing agent for achieving better particle distribution (trans-polyoctenamer reactive polymer, Vestenamer). In many cases, CRMB is accepted in lieu of styrene butadiene styrene (SBS) or styrene butadiene (SB)-modified binder. Proper specification regarding feed control, processing agent, particle size and mixing must be followed to prevent phase separation. A wealth of information is available from the Rubber Pavement Association website (www.rubberpavements.org).

23.1.3 Stone Matrix Asphalt

Stone matrix asphalt (SMA) is a binder-rich, aggregate, interlocked, rut-resistant, and durable asphalt mix that can be used for high-volume, heavy, as well as slow or standing traffic pavement, such as that in industrial loading areas, bus lanes, or container terminal areas. The key feature of SMA is its unique gradation, which consists of 20%–30% passing the 4.75 mm sieve material, and hence it is of relatively coarse nature. Based on available guidelines, the gradation must be determined as part of the mix design process, through the use of the concept of "stone-on-stone contact." In principle, it refers to a situation in which the fine aggregates have been reduced to a point where the coarse aggregate particles are in contact with each other, resulting in a strong load-bearing coarse aggregate skeleton.

The mix consists of a relatively high amount of mineral filler (percentage passing the 0.075 mm sieve)—around 10%, for example, compared to around 5% for a typical dense graded mix. The high percentage of mineral filler acts as a stabilizer for the asphalt binder, which is also needed at a relatively high content ($\geq$6% compared to around 5% for a dense graded mix). Fibers are also added to the binder to reduce the tendency of draindown of asphalt binder from the mix.

Because SMA derives its strength primarily from the coarse aggregate skeleton, aggregates from only strong parent rock types such as granite, basalt, or quartzite and with fractured faces are generally used. Also, the design gradation should be strictly followed (or adjusted appropriately) during production. Attention to quality control, especially aggregate gradation, is critical

---

TABLE 23.1

Steps in Mix Design of SMA

1. Selection of materials—aggregates, asphalt binder, additive, if any, filler material, and fiber
2. Making sure that the materials meet the required specifications
3. Blending of aggregates to meet target gradation for specific nominal maximum aggregate size, and selection of trial gradation
4. Selecting percentage of fiber and additives in asphalt
5. Determination of dry rodded voids in coarse aggregates, VCA_{DRC}, using AASHTO T19
6. Compaction of SMA with trial gradation and 6% asphalt content
7. Determination of voids in coarse aggregate of the mix, VCA_{mix}
8. Checking whether the VCA_{mix} is less than or equal to VCA_{DRC} (if not, a different gradation should be tried), and whether the voids in mineral aggregate (VMA) is equal to or greater than 17%
9. Selecting the design aggregate gradation
10. Compaction of SMA samples with selected aggregate gradation and trial asphalt contents
11. Checking whether the voids (Voids in Total Mix, VTM) are 3%–4%
12. Selecting the design asphalt content

---

for the successful production of SMA. Similarly, the temperature of the mix should be carefully adjusted to allow good mixing, prevent draindown, and provide adequate temperature for laydown and compaction.

Design air voids are around 3%, while those in compacted mats should not exceed 5%. Fibers are generally added at 0.3%–1.5% by mass of the total mix as stabilizers to prevent draindown of the mix. The fibers can be of mineral or cellulose type, and vary in length from 1 to 6 mm. For example, cellulose fibers are about 1 mm long and 0.05 mm thick. Steps in the mix design of SMA are shown in Table 23.1.

The time of mixing in the plant should be increased by 5–15 s, compared to that for dense graded mixes. Mixing and compaction temperatures are dependent on many factors such as type of asphalt, but generally the compaction process should be completed before it cools down below 130°C. SMA layer thicknesses vary from 2.5 to 5 times the maximum aggregate size and are generally put down in 38–50 mm thick layers. If the relatively high amount of mineral filler is not available as aggregates, lime or fly ash could be used through mineral filler feeder, lime feeder, or cold feed bins. Fat spots or puddles of asphalt binder can show up on the surface if more than adequate binder has been used. In the field, compaction should start immediately after placement, using a >10-ton static steel-wheeled roller. A vibratory roller can be used only sparingly, and a rubber-tired roller should be used only for finishing. Sand can be spread on finished surfaces to reduce skid hazards, if any.

Once the surface asphalt layer wears off under traffic, the surface texture of SMA provides good friction, and also helps in noise reduction, which can be as much as 2.5 dB. SMA costs 20%–30% more than dense graded mixes because of factors such as higher quality of materials required, larger mixing time, and more intensive quality control.

23.1.4 POROUS FRICTION COURSE

Porous friction courses (PFC; sometimes referred to as *open-graded friction courses*) are layers made of asphalt mix with a high amount of interconnected air voids that drain out water quickly during a rainstorm, reducing glare, splash, and spray and the potential of hydroplaning; improving friction; and hence reducing the potential of accidents significantly. PFCs also contribute to the reduction of tire noise on pavements.

PFCs consist of very coarse "open/gap" gradation, in which materials passing certain sieve sizes are missing. The aggregates should be durable, rough textured, and resistant to abrasion and polish. Typically the gradation consists of 5%–15% passing the 2.36 mm sieve. The asphalt content can range from 5% to 6.5%, with 5% to 7% polymer by weight of the binder and 0.15% to 0.5% fibers by weight of the mix. It has a relatively high amount of asphalt binder to provide adequate resistance against raveling and premature aging, often modified with polymer. The mix may also consist of a suitable stabilizer, such as fiber. Design as well as in-place voids in PFCs can range from 10% to 20%.

Fibers in the PFC can be added to the pugmill in packaged or loose form in a batch plant or by blowing into a drum in a drum plant. Pelletized form fiber can be added to the pugmill or drum also. The mixing time needs to be increased, and storage times in silos should be limited to very short times to prevent draindown.

Haul distances should be limited, and trucks carrying PFCs should have a thorough coating of asphalt-release agent on their beds to prevent bonding of mix on the bed as well as tarping to prevent cooling down and formation of crust on the surface of the mix.

To prevent trapping of water, PFC should be placed on impermeable asphalt pavement surfaces only. It should preferably not be used on a brand-new dense graded pavement where the in-place voids are still relatively high. Before the placement of the PFC, the underlying pavement can also be sealed using an appropriate amount of slow-setting emulsion tack coat (that can penetrate surface voids). The PFC layer should be open (daylighted) on the edges of the pavement to allow the free flow of water permeating through it.

Use of a materials transfer device is recommended, and the presence of a hot screed helps in preventing pulling of the mat under the paver. Generally a few passes of a static steel-wheeled roller is sufficient for properly seating the PFC (there is no need for compaction). Specific areas such as intersections or turning areas, ramps, and curbed areas may not be suitable for PFCs since they can ravel and shove or prevent draining.

Since they are critical in draining water quickly, the voids in the PFCs should be prevented from getting filled up with materials such as sand from snow treatments. Also, since they are placed in thin lifts, such as those with 1/2- to 5/8-in. thickness, snow plows may not be used to prevent lifting and dislodging of the mix. Use of anti-icing chemicals is a better option, and pores clogged with debris can be cleaned with high-pressure cleaners.

23.1.5 WARM MIX ASPHALT

Warm mix asphalt (WMA) is a concept that involves the use of an additive and/or the modification of the asphalt mix process to lower the viscosity of asphalt binder and hence enables the production of asphalt mixes at a relatively lower temperature (generally, 10°C–75°C, depending on the technology used). There are four widely used WMA production processes. The WAM-Foam® technology involves the use of a two-component binder system called WAM-Foam (warm asphalt mix foam). In this process, a soft-foamed binder and hard-foamed binder are added at different stages during plant production. In the Aspha-Min® process, a synthetic zeolite called Aspha-Min is used during mixing at the plant to create a foaming effect in the binder. Sasobit, Fischer–Tropsch paraffin wax, and Asphaltan B®, a low-molecular-weight esterified wax, are mixed directly with the asphalt binder, whereas Evotherm is a nonproprietary product that includes additives to improve coating and workability.

In the WAM-Foam process, the mixing is done in two stages. In the first stage, a soft binder is used to provide effective coating of aggregates at a lower temperature (typically 212°F–250°F, or 100°C–120°C). In the second phase, a harder binder, in the form of a powder, foam, or emulsion, is added to bring the grade of the total binder to the level of a binder used for a conventional HMA. Aspha-Min is a manufactured synthetic sodium aluminum silicate, better known as zeolite. These zeolites contain about 21% water by mass. When the zeolites are added to the aggregates at the same time as the binder (around 185°F–360°F, or 85°C–182°C), the water is released, resulting in the creation of foamed asphalt, which in turn improves the coating of the aggregates and workability of the mix. Zeolites (commonly used amount as 0.3% by mass of the mix) can be added directly to the pugmill of a batch plant or through a recycled asphalt pavement (RAP) feed or pneumatic feeder in a drum plant. Sasobit is a synthetic paraffin wax that melts around 210°F (99°C) and reduces the viscosity of the asphalt binder. This enables the production temperatures to be reduced by 18°F–54°F (10°C–30°C). Sasobit is added directly to the asphalt binder at the plant, generally at the rate of 1.5%–3.0% by mass of the asphalt binder. Asphaltan B is a mix of wax constituents and higher molecular weight hydrocarbons, which also reduces the viscosity of the asphalt. It is generally used at 2%–4% by weight of the asphalt binder and can be added directly to the binder or during mixing.

Evotherm is a chemical that is delivered in an emulsion with high asphalt residue. The water from this emulsion is given out when it is mixed with hot aggregates. The product is supposed to enable the production of HMA at a significantly lower temperature, and enhance workability, adhesion, and coating of the asphalt mixes.

The benefits of WMA over conventional HMA are many. Because of the lowering of the production temperature, less oxidative aging of the asphalt binder and hence the production of a superior asphalt mix can be expected. A lowering of temperature also means savings in fuel (burner) costs, as well as lowering of emissions, and hence improvement in air quality. It also means that longer haul distances could be used (WMA with lower temperature cools slower than HMA with relatively high temperature), and the construction season in many parts of the world that have colder climates in winter can be extended. The lower temperature of WMA has proven to be beneficial in paving overlays with pavements with sealed cracks and joints—unlike the high temperature in HMA, the relatively low temperature did not cause formation of steam in joints or liquefaction of the sealant material and resulting bumps. Contractors also find it easier to compact WMA—for example, four roller passes instead of 20 (used for HMA) could be sufficient. Compaction of joints is also found to be easier with WMA. Workers are reported to be in a better state at the end of the day, while paving with WMA, since they are exposed to less heat. The use of WMA is also beneficial for mixes that require higher than conventional temperatures, greater asphalt content, polymer-modified binders and are more sensitive to mix/ambient temperatures, such as open-graded friction courses (OGFC) and asphalt rubber mixes. Finally, it offers the potential of recycling larger amounts of RAP materials.

As the use of WMA grows in different parts of the world and in the United States, research and field trials keep on generating useful information on this mix, and guidelines for successful use of this product are being developed. Important considerations that are specific to this kind of mix are as follows: the use of the proper grade of asphalt binder, the need to ensure that aggregates are free of moisture (since they are heated to a lower temperature), the effect of low-viscosity binder in early stages of life of the pavement, the use of appropriate lower temperatures for unconventional mixes, the effect of lower temperature on compaction, and the effect, if any, of the WMA additives on the low-temperature properties of the asphalt mixes. Although WMA may need a lower amount of asphalt binder to compact (during mix design in the laboratory), current guidelines encourage the use of asphalt content as would be used for a conventional HMA. The WMA homepage (warmmixasphalt.com) contains a growing list of publications related to WMA.

23.1.6 Ultrathin Wearing Course

Ultrathin wearing courses are provided on pavements that have adequate structural capacity but exhibit surface distress such as loss of skid resistance and surface cracks. Such a wearing course is generally laid down with specialized equipment, with materials consisting of polymer-modified asphalt emulsion and a coarse-graded asphalt mix, in a lift thickness ranging from 10 to 20 mm. The layer is applied in one pass within a relatively short period of time (20–40 m/min), enabling

TABLE 23.2
Typical Mix Design for an Ultrathin Wearing Course

Sieve (mm)	% Passing
1/2 in. (12.5 mm)	100
3/8 in. (9.5 mm)	95
4 (4.75 mm)	35
8 (2.36 mm)	24
16 (1.18 mm)	18
30 (0.6 mm)	12
50 (0.3 mm)	9
100 (0.15 mm)	7
200 (0.075 mm)	5.2
% Asphalt content, AC 20	5.2

Mixing temperatures: 143°C–168°C.
Plant: Drum mixing plant, capacity 90–135 mg/h.
Tack coat: CRS-2P SBR latex–modified asphalt emulsion; application
    rate: 1 ± 0.2 L/m^2.

minimum disturbance to traffic. The use of a specially formulated polymer-modified emulsion binder allows a good bond of the layer with the existing surface and hence helps in avoiding delamination that is common in conventional thin lift HMA layers. The thin layer facilitates rapid curing and hence quick opening to traffic. The use of gap graded mix can also result in lowering of traffic noise and reduction of splash and spray during rainstorm. Prior to the application of this layer, it is important to seal existing cracks, patch distress areas, and also diamond grind concrete surface, if necessary. The specialized paver (e.g., Novapaver, manufactured by Midland Machinery Company for Novachip, a specialized mix) has a combination tamping bar-vibratory screed that helps in reduction of roller effort in compacting the mix. Double steel drum rollers are used specifically to seat the aggregates in the polymer emulsion. Table 23.2 shows mix design information of a project. This type of mix has been found to be suitable for high traffic volume roads and can be considered as an alternative to chip seals, microsurfacing, and OGFC.

23.2 CONCRETE WHITETOPPING

Whitetopping refers to the placement of a layer of concrete over existing HMA pavements as a rehabilitation operation. The thickness of ultrathin whitetopping ranges from 2 to 4 in., and fiber-reinforced concrete is used in such cases. Before the placement of whitetopping, the existing HMA surface is textured for providing good bond between the new and old materials. The behavior of a pavement with bonded whitetopping is considered to be different from that of other pavements. The composite section results in much lower stress in the concrete section and hence allows the use of thinner section. Many successful applications have been noted, and research continues on the better design concepts of whitetopping layers (NCHRP, 2004). For more information, please see the ACPA (www.acpa.com) and FHWA websites (FHWA, 2005).

QUESTIONS

23.1 What are the different types of polymers used in asphalt mixes?
23.2 How does asphalt rubber improve HMA pavements?

23.3 How is an SMA mix designed?

23.4 What are the main characteristics of porous friction courses?

23.5 What are the different processes of producing warm mix asphalt?

23.6 Using current cost data for fuel for dryer drums in HMA plants (from local industry), estimate the savings in dollars per ton if a warm mix asphalt is produced at 125°C instead of a hot mix asphalt at 150°C.

24 Sustainable Pavement Engineering

24.1 NEED FOR PAVEMENTS

The need for safe, durable, and smooth riding pavements on highways and in airports cannot be contended. Construction and maintenance of pavements cost money—which are mostly provided through taxes that are collected from the citizens. The money is spent on employing workers, quarrying and manufacturing usable products, transporting them, and laying down and compacting the materials. In this whole process, which gets repeated to different extents during the maintenance and rehabilitation cycles, man-made material is laid down to cover virgin (most likely arable) ground (making it un-arable from that point onward), a significant amount of natural resources is utilized, (nonrenewable) energy is consumed, landfills are created, and different types of chemicals (in the form of gases) are emitted to the environment (all of these processes are irreversible). Thus, pavement engineering is a high impact activity. So what is the solution?

24.2 FIRST CONSIDERATION

First and foremost, the pavement infrastructure should be viewed as part of a larger "system," and not viewed alone. Buildings, pavements, and natural ground/water bodies (nature) form the built environment in which we live and breathe. Hence, pavements need to be compatible and if possible, complementary to the other two components—buildings and nature. This means construction and maintenance of pavements should not be taxing the natural resources, energy utilization should be minimized, byproducts in the form of emissions should not be harming the environment, and the layout of the pavement should not be significantly impacting the natural environment. Broadly speaking, this includes layout and type, design and construction of pavement. It is only through this process we can ensure that our future generations will have no/less problem of continuing the work of building or maintaining/rehabilitating the pavement infrastructure. And, this is the very essence of sustainability. True sustainability consists of three aspects—social, economic, and environmental sustainability. The discussions in the following paragraphs refer to either one or more of these three different types of sustainability.

24.3 DESIGN OF LAYOUT OF PAVEMENTS

Some of the more important issues with the layout/geometric design of a pavement can be summarized as follows:

1. Protection of forests: forests provide a home to many species of plants and animals and at the same time serve as a source of economic products such as timber and enjoyment for tourists. Roads should be preferably designed away from forests, or in a way that creates minimum impact on them.
2. Protection of wild life: considerations must be made to provide features that would minimize the effect of road construction on animals and subsequent vehicle–animal collisions (such as protected crossovers).

3. Consideration of unique geological conditions, such as steep slopes with vegetation or caves/landforms should be made. Precaution should be taken to avoid such areas where a slight disturbance of the landscape would lead to a significant irreversible disturbance of the local environment.

4. Context-sensitive design should be adopted: the layout of the road should conform to the values of the communities through which it is passing, by making sure that is its integrated with the local character—such as flora and fauna in rural areas.

5. Access to critical areas such as hospitals and police stations should be provided, such that the road helps the local communities and in return the people value the road.

6. Consideration of the impact of diversion routes on the local environment, during the construction of new pavement, should be made. Proper forecasting methods and mitigation procedures should be adopted.

7. Consideration of traffic growth and providing adequate room for further growth should be made, such that traffic saturation does not lead to a drastic change in the character of the community and significant construction activity could be avoided for a longer period of time.

The issues listed earlier can be represented in terms of problems such as deforestation, noise pollution, crossing of habitat areas, flooding/storm water management, creation of wastage, urban heat island effect, drinking water pollution, air pollution, traffic jams, soil erosion, and use of natural resources and energy.

These problems could be solved with the use of one or more of a variety of methods, which include habitat connection across roadways through underpasses or overpasses, fencing along the road for animal protection, amphibian rescue fences, storm water pollution prevention plan, pedestrian, bicycle, and transit access, noise barriers, buffer zones, or specially planted slopes/embankments to reduce noise pollution, intersections with interchanges, use of pervious surfaces in pavement or shoulders, runoff treatment practices to clean runoff water, avoiding sensitive aquifers, making the locations of service connections easily accessible, providing easy access to emergency services, and defining an implementation strategy to be followed during construction.

Some of these desirable features are shown in Figure 24.1. An example of sustainable pavement construction is maintaining the established wildlife pattern during the construction of a highway. In this case, a deer crossing was maintained with the use of a high-profile arch structure that followed a serpentine alignment conforming to the existing footprint of a creek in the construction area (Figure 24.2).

24.4 CONSTRUCTION OF PAVEMENTS

For construction of asphalt pavements, the key factors that need to be considered are using local materials (to reduce transportation-related energy expenses), lowering the need for new materials, reducing production and construction temperatures, using more environment-friendly products (such as bio-asphalt), and environmentally friendly layers/mixes (such as those that would allow the percolation of water into the soil layers and hence facilitate recharging of aquifer, and those that have reduced noise levels), utilizing mixes that would allow recycling or reuse of discarded materials, and longer lasting pavements, and conducting a proper life cycle analysis.

The favorable impact of using sustainable pavements is far reaching. For example, if a recycling procedure is used to rehabilitate an existing pavement, then the following benefits can be ensured:

1. Less use of new natural resources such as mineral aggregates
2. Significant reduction in transportation cost of new materials
3. Less amount of energy use in obtaining and processing new materials (aggregates)

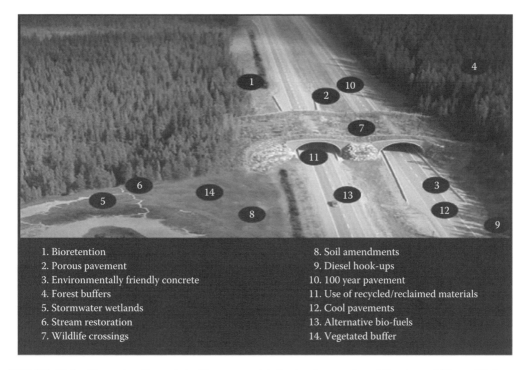

FIGURE 24.1 Elements of a sustainable pavement infrastructure system. (Courtesy of Green Highways Partnership, US EPA Region 3.)

1. Bioretention
2. Porous pavement
3. Environmentally friendly concrete
4. Forest buffers
5. Stormwater wetlands
6. Stream restoration
7. Wildlife crossings
8. Soil amendments
9. Diesel hook-ups
10. 100 year pavement
11. Use of recycled/reclaimed materials
12. Cool pavements
13. Alternative bio-fuels
14. Vegetated buffer

FIGURE 24.2 Deer crossing. (Courtesy of CONTECH Construction Products Inc., Olde West Chester, OH, reprinted from sitesolutionsmag.com.)

4. Less amount of energy use in manufacturing new materials (asphalt binder, cement)
5. Less emission and pollution of the environment
6. Reduction or avoiding of filling up precious landfill space
7. Significant reduction in overall transportation energy, specifically if an in-place recycling process is utilized

TABLE 24.1
Different Types of Recycling Methods

Type of Pavement	Primary Distress	Recycling Method
Asphalt	Shallow rutting, raveling, bleeding, low skid resistance, corrugations, longitudinal cracking, slippage, longitudinal joint distress	Hot in-place, hot mix
	Deep rutting, fatigue cracks, pavement edge and slippage, block cracking, transverse and reflection cracking	Cold in-place, hot mix
	Deep rutting and cracking, problems in base course	Full-depth reclamation
Concrete	Cracks	Rubblization

There is a variety of recycling techniques that are available today, each having its applicability under specific conditions, as shown in Table 24.1.

24.5   USE OF WASTE AND BYPRODUCTS IN PAVEMENTS

There is a variety of industrial waste and byproduct materials that have the potential of use in asphalt and concrete pavement layers. Table 24.2 shows a list of such products that have been identified by the FHWA. The important things that need to be considered are appropriate methods to evaluate the suitability of a waste or a byproduct material for a specific use, costs of use of such materials—both initial and life cycle and the environmental impacts of the use of such materials. Detail user guidelines of these aspects are available at http://www.fhwa.dot.gov/publications/research/infrastructure/pavements/97148/index.cfm.

A pervious layer in roads would allow the percolation of rain water to the lower layers, and offer the following benefits, that are related to both groundwater and storm water/flood control:

1. Recharge of ground water, and hence a reduction in the potential of lowering of groundwater and the related problems.
2. Detention of rain water and hence a lowering of the storm water volume that ends up in road sewer system.
3. Filtering of storm water that is contaminated by different methods on the road; studies show that a significant amount of suspended solids, phosphorous, zinc, and hydrocarbons are removed.

For asphalt pavements, a typical pervious pavement is obtained by using asphalt-treated permeable base over a layer of pervious stone layer over the subgrade (with filter material) as shown in Figure 24.3. For concrete pervious pavements, there is the option of using interlocking/segmented concrete pavers. Even if not used in mainline, storm water tree trenches can allow the detention of a significant amount of storm water for sufficient time before releasing it into the sewer system (Figure 24.4).

Low-energy mixes, particularly WMA, are becoming more and more common these days, because of their inherent advantage of lower energy cost and lower emission potential. Another key factor in considering WMA is the fact that because of lower temperature, the asphalt binder undergoes less aging, and hence a better pavement mix obtained during construction—which translates to longer life, and hence less frequent rehabilitation cycles. There are different types of WMA techniques available, such as zeolites, wax, emulsion, and foam—with newer products and techniques being developed, and one or more suitable methods can easily be adopted for regular use by the industry. The selection of the specific method should be made on the basis of the availability of the material, expertise, and experience with that specific technology.

TABLE 24.2

Waste and Byproducts and Their Applications in Pavements

Application	Material
Aggregates in hot mix asphalt	Blast furnace slag, coal bottom ash, coal boiler slag, foundry sand, mineral processing wastes, municipal solid waste combustor ash, nonferrous slags, reclaimed asphalt pavement, roofing shingle scrap, scrap tires, steel slag, waste glass
Aggregates in cold mix asphalt	Coal bottom ash, reclaimed asphalt pavement
Aggregates in seal coat or surface treatment asphalt mix	Blast furnace slag, coal boiler slag, steel slag
Asphalt concrete—mineral filler	Baghouse dust, sludge ash, cement kiln dust, lime kiln dust, coal fly ash
Asphalt binder modifier	Roofing shingle scrap, scrap tires
Aggregates in Portland Cement Concrete (PCC)	Reclaimed concrete
Supplementary cementitious materials in PCC	Coal fly ash, blast furnace slag
Granular base	Blast furnace slag, coal bottom ash, coal boiler slag, mineral processing wastes, municipal solid waste combustor ash, nonferrous slags, reclaimed asphalt pavement, reclaimed concrete, steel slag, waste glass
Embankment or fill	Coal fly ash, mineral processing wastes, nonferrous slags, reclaimed asphalt pavement, reclaimed concrete, scrap tires
Aggregates in stabilized base	Coal bottom ash, coal boiler slag
Cementitious materials in stabilized base (pozzolan, pozzolan activator, or self-cementing material)	Coal fly ash, cement kiln dust, lime kiln dust, sulfate wastes
Aggregates in flowable fill	Coal fly ash, foundry sand, quarry fines
Cementitious material in flowable fill (pozzolan, pozzolan activator, or self-cementing material)	Coal fly ash, cement kiln dust, lime kiln dust

Source: Adapted from User Guidelines for Waste and Byproduct Materials in Pavement Construction, Publication Number: FHWA-RD-97-148, http://www.fhwa.dot.gov/publications/research/infrastructure/pavements/97148/004.cfm.

FIGURE 24.3 Pervious pavement cross section. (Courtesy of Richard Bradbury, Maine Department of Transportation, Augusta, ME.)

FIGURE 24.4 Stormwater tree trench concept applied in Green Street project in Philadelphia. (Courtesy of Glen J. Abrams, AICP, Watersheds Planning Manager, City of Philadelphia, Philadelphia, PA.)

The use of binders manufactured from bio-resources is being researched and conducted in different parts of the world. These binders are essentially derived from biomass—agricultural and forest residue—with the help of biochemical or thermochemical processes. One such example of bio-binder that is fully derived from vegetable is Vegecol from COLAS. Such binders can be utilized for modifying asphalt (<10% replacement), extending asphalt binder (>25% replacement), or totally replacing asphalt. Different grades of such binder could also be used for cleaning road equipment or partially replacing fluxing agents. The advantages of using such binders are significant reduction in pollution—no SOx and major reduction in NOx emissions, and manufacturing from 100% renewable agricultural products. Furthermore, the temperature of mix production could be reduced significantly.

One of the most exciting technologies that are being developed in pavement engineering is the combined use of the recycling and warm mix concepts to produce good performing and long-lasting pavements. Binder from renewable source could also be used in this process. Using sophisticated materials and plant/process control technology, it is now possible to effectively recycle close to 100% reclaimed asphalt pavement (RAP) material, at a reduced temperature (e.g., below 130°C as opposed to 150°C) to meet emission requirements, and still obtain a pavement that has excellent performance under both high and low temperatures, and that can be opened to traffic quickly after construction, even at a temperature that is higher than that at which conventional pavements could be opened. Overall, this translates to savings in money and time, materials and natural resources and energy, and significant reduction in emission. An example of such a project in New York City is shown in Figure 24.5. In this case, a high RAP content recycled HMA was designed with the combined use of appropriate rejuvenator and plant technology. The properties of the rejuvenator are shown in Table 24.3. The steps in the production process are shown in Figure 24.6.

Recycling of discarded tires in the form of crumb rubber in asphalt pavement mixes is now an accepted practice in many parts of the world. Apart from allowing the user to avoid the filling up of precious landfill space with hazardous waste, this process allows the user to

1. Incorporate more asphalt binder (thicker film) in mixes to make them longer lasting and durable
2. Use a coarser gradation for specific uses such as drainage of water
3. Reduce noise levels from pavements significantly

The key to successful use of rubber-modified asphalt binder is successful incorporation of the rubber into the binder/mix. As is the case for many technologies, this is heavily dependent on available plant and construction equipment. For example, appropriate technology must exist to uniformly blend the crumb rubber mechanically or chemically with the binder, ensure that the binder remains in that state during transportation, and for proper quality control test methods to test and prevent the separation of the rubber and the binder prior to laydown and compaction. In terms of recycling of other materials, artificial aggregates such as blast furnace slag, generated from steel and iron production, could be successfully incorporated in paving mixes.

The production of Portland cement, the key constituent in concrete pavements, and clinker, required for manufacturing of cement, are a major source of CO_2 emission in the modern world. The use of limestone and supplemental cementitious materials (SCMs, such as fly ash from thermal power plants using coal, and rice husk ash) provides attractive options of reducing the use of cement and thereby cutting down its CO_2 footprint. Blended cement, consisting of Portland cement and SCMs, is becoming popular. For maintenance of concrete pavements, modern techniques such as dowel bar retrofits, cross-stitching, partial-depth repairs, joint and crack resealing, slab stabilization, and diamond grinding could be used to improve the conditions of concrete roads and at the same time avoid or reduce the need for energy intensive major repairs/rehabilitation work.

The concept of perpetual asphalt pavement has been recently proposed as a sustainable technology. The objective of this technique is to provide thick and appropriately constructed (with specific mixes for different layers) such that the critical responses (such as tensile strain) in the pavement are kept very low, and as a result no major structural failure could be expected for a long period of time. The only deterioration that could be expected will be in the form of surface distresses, which could be relatively easily and economically addressed by less involved rehabilitation procedure, such as in-place surface recycling. Obviously these pavements would cost more than conventional pavements initially, but if constructed properly, they allow the users and the highway agencies to enjoy a better quality and durable pavement with significantly reduced cycle

FIGURE 24.5 Example of 100 % recycling; (a) milling, (b) removal of excess fines, (c) portable plant that can be set up near the road being recycled, (d) material is heated to laydown temperature, (e) ½% to 1% rejuvenator is added at discharge, (f) conventional procedure is used for laydown and compaction. (Courtesy of Robert Frank, RAP Technologies, LLC, Linwood, NJ.)

of rehabilitation (the pavements need to be maintained properly), and hence significant reduction in activities that could negatively impact the environment.

In terms of longevity, proponents of the concrete pavement industry argue that the longer life of concrete pavements translates to reduced needs for new natural resources, energy and waste disposal. Another important factor is that because of lower deflection in concrete pavements and hence lower rolling resistance, the mileage is increased by about 3.85% in concrete pavements over asphalt pavements (with similar roughness), which translates to lower amount of fuel spent and hence lower amount of greenhouse gas emission. Furthermore, because the surface of a concrete pavement reflects a higher amount of solar energy than an asphalt pavement, concrete pavement helps in reducing the indirect contribution of pavements toward the urban heat island effect.

TABLE 24.3

Properties of Rejuvenator (Renoil 1736)

Viscosity SUS @ 100°F, 494   Pour point °F (°C), −10 (−23)

Viscosity SUS @ 210°F, 51   Flash point °F (°C), COC420 min (216)

Viscosity cSt @ 40°C, 92.2   Sulfur wt%, 4.3

Viscosity cSt @ 100°C, 7.4   Aniline point °F (°C), 55 (13)

Specific gravity @ 60°F, 0.966   Clay gel analysis

Asphaltenes%, 0; polars%, 7; aromatics%, 65; saturates%, 28

Density lb/gal @ 60°F, 8.04   Carbon analysis%

Aromatic%, 30; naphthenic, %15; paraffinic%, 55; refractive
index, 1.5218

Source: Courtesy of Renkert Oil, Morgantown, PA.

Production steps

1. RAP millings delivered by truck:
2. Mechanical sizing: Stockpiled millings are run through a diesel powered Powerscreen Chieftain 800 screening plant; Fed by a rubber tired loader into the Hopper; Conveyed to a double deck screen to make a fine and sized aggregate; Oversize material is conveyed to an electric powered impact crusher; Debris and tramp material is scalped from the crusher discharge by a small in line scalping screen; Debris is stacked for offsite disposal.
3. Reduced material is returned to the main conveyor for screening. Fines and sized material are stockpiled in yard for use in manufacturing
4. Manufacturing
   a. Cold feed: Fines and sized materials are feed by loader to feed hoppers and metered into a rotary shell dryer for heating to 300°F.
   b. Rotary dryer: Dryer is fired with #2 fuel oil; Produces 150–200 tons finished product per hour.
   c. Material is heated then discharged to a drag slat conveyor for surge storage.
   d. A rejuvenator is sprayed onto the material as it leaves the dryer.
   e. Dryer is maintained at a slight negative pressure to vent combustion gases and fugitive emissions to the air pollution control device.
   f. Surge storage: Finish product is stored in an insulated silo for load out into conventional dump trucks. Silo is equipped with weigh hopper rather than truck scale to provide certified tickets for legal sale.
5. Air pollution control baghouse: Removes airborne particulate and aerosol mist
   a. Cyclonic separator is use to drop out coarse fines that are manually removed and mixed with screened fines to be reintroduced
   b. Water spray provides adiabatic cooling to 160°F
   c. Disposable fiberglass pocket filters remove micron size particulate
   d. Fiberbed filters remove aerosol mist
   e. Scrubber liquor is pumped periodically and stored in a hold and haul tank
   f. Exhaust gases comply with 0.04 grains per SCF and 10% opacity at 40 CFR 60; Air flow approximately 30,000 ACFM at 30% moisture

FIGURE 24.6 Steps and components in 100% recycling.

And at nighttime, the lighter color of the concrete pavement surface helps in a significant reduction in the amount of energy that is needed for illumination.

Life cycle analysis (LCA) in pavement engineering is an established concept, and is utilized in many, if not most, high-cost projects. However, traditionally, the cost is evaluated in terms of materials/construction/salvage cost only. For sustainable pavement engineering, the environmental and social cost must also be evaluated. Furthermore, the cost of extending the life of a pavement (and thus lowering the life cycle cost) must be evaluated in the context of the amount of natural resources and energy that are spent on producing the premium materials that are used to extend its life. The impact of the pavement construction/maintenance and rehabilitation should be evaluated in terms of all three components of sustainability—social, economic, and environmental. Examples of such framework and components of framework are the Sustainability Life Cycle Assessment (SLCA), the IRF Greenhouse Gas Calculator, and PaLATE (Pavement Life cycle Assessment Tool for Environmental and Economic Effects).

24.6 WORKERS

Hundreds of thousands of human beings—men, women, and in many cases even children—work in pavement-related projects all over the world. In many cases, this work goes on under extreme conditions of weather, in areas that are remote from basic facilities such as school or hospitals, and under conditions of emissions/pollution from construction conditions (smoke/gas/particulate emission). This makes the entire realm of pavement engineering socially unsustainable. Conditions must be improved such that environmental and economic justice are prevalent among the pavement workers. While there are many things that need to be done, first and foremost it is the responsibility of pavement engineers to select materials and methods such that working conditions are improved. For example, switching to WMA instead of hot mix asphalt will reduce harmful emissions significantly (Figure 24.7). Secondly, engineers and authorities must enforce strict safety regulations and provide for the basic necessities such as protective body and eye wears.

24.7 PAVEMENT–BUILDING–NATURE SYMBIOSIS

As pointed out earlier, pavements need to be considered as part of the built environment—going one step forward, in the future, pavements need to be considered as part of a symbiotic system that also includes buildings and the nature. Imagine a day when we will have smart pavements that will be able to self-control temperatures and adapt to changing environmental conditions to sustain loads without deforming or cracking, will be able to harvest heat energy from the solar radiation falling on them, and supply this to buildings for various uses, and in return, the buildings will be keeping the pavements below a critical temperature, thus extending their lives, and also reducing the amount of heat that is radiated back into the environment, and hence indirectly reducing the urban heat island effect. Multidisciplinary research, as indicated in Figure 24.8, needs to be conducted to make this vision a reality.

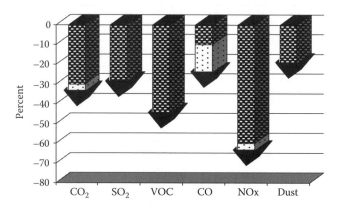

FIGURE 24.7   Range of reduction in emission through the use of WMA.

Utilize the network of existing paved areas to provide energy to buildings and reduce urban heat island effect of paved areas

Systems dynamics

Consideration of infrastructure/paved area and buildings as symbiotic unit

Architectural engineering

Solar energy highway

Temperature of pavement drops and radiated heat is reduced; pavement life is extended, and energy comsumptions of adjacent buildings are reduced and air quality is improved

High conductivity pavement material the surface transmitting heat to pipes

Civil/materials/mechanical engineering

Solar pump pumps cold fluid through pipes in the asphalt pavement

Hot fluid coming out of pipe

Cold fluid returning to pipe

Energy capture

Environmental/ mechanical engineering

Heat exchanger increases fluid temperature as required

Adequately hot fluid comes out for use as hot fluid or for heating other fluid in turbine for generating electricity or to be used in absorptive chillers for air conditioning

Energy generation

Energy storage

Energy transmission

FIGURE 24.8 Multidisciplinary approach to consider pavement infrastructure as part of the built environment.

24.8 REGULATORY BODIES AND IMPETUS FOR SUSTAINABILITY

Adopting sustainable practices can reduce many of the environmental impacts of pavement construction. However, both impetus and regulations ("carrots and stick") need to be in place to accelerate the adoption of sustainable practices. Impetus in the form of prestigious certification, first voluntary, and then gradually, mandatory, should be introduced. An example of such certification is the U.S. Green Building Council's Leadership in Energy and Environmental Design (LEED) in the United States, which awards silver, gold, and platinum certifications for projects in increasing order of contribution toward "improving performance across all the metrics that matter most: energy savings, water efficiency, CO_2 emissions reduction, improved indoor environmental quality, and stewardship of resources and sensitivity to their impacts." Regulatory bodies such as the Environmental Protection Agency (EPA) in the United States enforce strict emission standards to prevent the emission of harmful gases and particulates into the atmosphere.

24.9   HUMAN FACTOR

Pavements are engineered systems that need to be planned, designed, constructed, maintained, and rehabilitated by knowledgeable and skilled people. However, for achieving sustainable pavement engineering, we need socially responsible engineers, with broad vision and compassion for humanity. Therefore, the preparation for sustainability starts today in our colleges and universities with the students who will be joining the profession tomorrow. Professors, administrators, and the tax-paying general public need to be educated also on the impacts of pavement engineering and the different ways to minimize such impacts. This can be achieved by establishing multidisciplinary centers of excellence in sustainable pavement engineering in major colleges and universities in conjunction with the industry and the government in the different parts of the country. While education on existing and new concepts of sustainability can be transmitted by this method, the centers should also focus on developing, through research, new sustainable concepts and practices. And in that research it is very important that engineers look at the future—innovative, sophisticated materials and techniques, as well as in the past—to determine what has survived over centuries without harmfully affecting our environment.

QUESTIONS

24.1 What are the different ways in which the environment is affected by pavement construction?

24.2 Name a few ways in which roadways can be designed to have minimum impact.

24.3 What is a type of asphalt mix that can be used for lowering energy use? How does it work?

24.4 What are some of the features of environment-friendly concrete materials?

24.5 How can the development of low or minimum impact pavement design and construction be promoted?

25 Environmental Mitigation in Transportation Projects*

Engineers, planners, and public decision-makers recognize the numerous and significant benefits arising from transportation improvement or expansion projects. However, transportation projects often lead to adverse impacts on the environment. In an ideal sense, all negative impacts to both the natural and built environments would be avoided with creative engineering, excellent design, and exceptional construction techniques. Developing these improved practices are often the pursuits of graduate study and academic research. In practical applications, mitigation activities are implemented to offset any unfavorable consequences.

Good mitigation design considers a holistic evaluation of short- and long-term impacts. There is not currently one standard method for this process, so this chapter utilizes various U.S. Federal and State regulatory requirements to build a framework for identifying, measuring, and monitoring impacts to be addressed by mitigation activities.

25.1 HOW TRANSPORTATION IMPACTS THE ENVIRONMENT

Transportation projects are often associated with other development projects that expand or improve the built environment. The effects may be immediate and within the project site (replacing a natural surface with pavement) or longer term and carried a significant distance away from the project location (air quality impacts from vehicle emissions). The U.S. Code of Federal Regulations (CFR) Title 40 "Protection of the Environment" Section 1508 defines direct, indirect, and cumulative impacts as follows:

- Direct effects—those "which are caused by the action and occur at the same time and place."
- Indirect (secondary) effects—those "which are caused by the action and are later in time or farther removed in distance, but are still reasonably foreseeable. Indirect effects may include growth-inducing effects and other effects related to induced changes in the pattern of land use, population density or growth rate, and related effects on air and water and other natural systems, including ecosystems."
- Cumulative impact "is the impact on the environment which results from the incremental impact of the [proposed] action when added to other past, present, and reasonably foreseeable future actions regardless of what agency (Federal or non-Federal) or person undertakes such other actions. Cumulative impacts can result from individually minor but collectively significant actions taking place over a period of time." (40 CFR 1508.7)

Table 25.1 assumes these definitions and lists the types of adverse impacts that might result from transportation projects. This table is not an exhaustive list of all possible outcomes, but is intended to illustrate potential issues that may need to be addressed. Depending upon the locations and magnitude of the project, mitigation of some of these impacts may be required by national or local laws and regulations. For example, flood zones, wetlands, coastal areas, historic districts, habitat areas for endangered or threatened species, and other designated places are likely to have more stringent rules for mitigation. In many cases, mitigation of the broader indirect and cumulative impacts may be considered beyond the scope of the transportation improvement project. However, there may be

---

*This chapter was written by Suzanne LePage, AICP Adjunct Professor, Worcester Polytechnic Institute (WPI) Worcester, MA.

TABLE 25.1

Potential Adverse Transportation Project Impacts

	Direct Effects	Indirect Effects	Cumulative Impacts
Land	• Habitat destruction and/or disruption on or near project site • Conversion of impervious surface to pervious surface • Changes to slope and grading	• Conversion of natural environments or low-intensity land uses to developed parcels as a result of improved access and mobility	• Increased percentage of impervious surfaces within a watershed • Fragmentation of habitats and/or migratory paths • Degradation of habitat quality
Water	• On-site hydrologic changes resulting from land impacts listed above • Contaminated runoff (especially sediment) during construction process • Wetland impacts (filling and/or fragmentation) • Stream relocation	• Anthropogenic impacts resulting from improved access to or near waterbodies (e.g., motor boats) • Increases in volume and peak flow rates of stormwater runoff (neighborhood scale) • Increased contaminants in stormwater runoff generated from the newly paved surface • Changes in stream flows, water levels, tidal flows, flood patterns, etc.	• Increased contaminants in stormwater runoff generated from the changes in land use that may result from improved access and mobility • Increased groundwater withdrawals and utilization of surface water bodies to support growth and development • Degradation of habitat quality
Air	• Localized air quality impacts during construction	• Increased emissions from vehicle use (consider both traffic volumes and cold starts for projects at trip ends, such as residential neighborhoods, parking facilities, etc.)	• Changes in regional conformity status re: U.S. Clean Air Act (attainment, non-attainment, and/or maintenance)
Human[a]	• Resident or business displacement (takings) • Mobility disruptions during construction • Cultural resource impacts (archeological, historic, architectural, scenic views, etc.)	• Demographic changes associated with growth and development (can be positive or negative) • Noise, traffic, and other neighborhood impacts of increased nearby development	• Increased tax rate resulting from additional cost of community services (public safety, school expansion needs, infrastructure, etc.)

[a] Important note regarding human impacts: In the United States, special consideration must be taken to ensure that federal projects are not disproportionately impacting minority and low-income populations. See Section 25.6 regarding Environmental Justice.

some mitigation techniques or design strategies that can address more than one impact. As such, this holistic view early in the project scoping process can be beneficial.

25.2 MODEL FOR ASSESSING IMPACTS AND DEVELOPING MITIGATION MEASURES

The U.S. National Environmental Policy Act (NEPA) provides an effective model for scoping, assessing, and mitigating transportation projects and their impacts. In fact, there are numerous laws in other nations that are similar in NEPA's intent and implementation. These include, but are not limited to, Japan (Law No. 81 of 1997); China (Order of the President No. 77, 2003); European Union (Council Directive 85/337/EEC); and Germany (Federal Law Gazette I p. 2081, 1997).

The primary purpose of the U.S. NEPA Public Law 91-190 (1969) was to establish a review process for federal projects and policies that could impact the environment and/or irreversibly use

http://environment.transportation.org/environmental_issues/nepa_process/#bookmarksubEnviron
mentalImpactStatements

FIGURE 25.1 Steps in the review process.

natural resources. All federal agencies must comply, including the Federal Highway Administration (FHWA) and the Federal Transit Administration (FTA). AASHTO's Center for Environmental Excellence provides an overview of how the NEPA process impacts most transportation projects (http://environment.transportation.org/environmental_issues/nepa_process/#bookmarksubCategor icalExclusions). As illustrated in Figure 25.1, once an action is proposed, the first step is to coordinate with other agencies and analyze the action to determine if the impacts are significant. To accomplish this, there are essentially three levels of analysis. A categorical exclusion (CE) is an action that has been evaluated and defined by a federal agency as not having a significant impact on the environment. For example, administrative activities, bicycle/pedestrian facilities, noise barriers, projects included in a State's Highway Safety Plan, and emergency repairs are considered CEs. If the proposed activity is not a CE, and if coordination and initial screening are not sufficient to determine the level of significance of the impacts, then an environmental analysis (EA) is conducted. The EA may result in a finding of no significant impact (FONSI), although mitigation may still be required. Or, the EA may determine that the third level of analysis, an environmental impact statement (EIS), is needed to evaluate alternatives and mitigation options. More information on NEPA can be found at http://www.epa.gov/oecaerth/basics/nepa.html

Many states have laws that are similar to the NEPA (State Environmental Policy Acts, SEPAs). These include Arkansas, California, Connecticut, Florida, Indiana, Massachusetts, Michigan, North Carolina, Minnesota, New York, Virginia, Washington, and Wisconsin. Many of these have similar goals, but may contain requirements that are different from that of NEPA. For example, NEPA and many states define applicability on the project proponent and/or funder. NEPA specifies that federal agencies must comply with its requirements, and many SEPAs designate publicly funded projects as those that fall under their jurisdiction. Massachusetts, however, requires any project that exceeds minimum thresholds to comply with its rules, regardless of funding source. Table 25.2 compares a number of these acts and lists websites where additional information can be obtained.

Using NEPA and these state laws as models, the following general steps of a process to address mitigation can be defined:

1. *Project conception*: Define the project need; propose an action; coordinate with stakeholders
2. *Impact assessment*: Screen projects for potential impacts and assess their significance

TABLE 25.2
Environmental Review Processes at the State Level

State	Name	Applicability	Thresholds	Required Documents
California	California Environmental Quality Act (CEQA) http://ceres.ca.gov/ceqa	All new "projects" undertaken or approved by state and local public agencies	Possibility that project may have "significant effect on the environment" likely due to "direct or indirect physical changes" to the environment (some exemptions exist)	Notice of Preparation (NOP) Draft Environmental Impact Report (DEIR) Final Environmental Impact Report (FEIR) Notice of Determination (NOD)
Connecticut	Connecticut Environmental Policy Act (CEPA) http://www.ct.gov/dep/cwp/view.asp?A=2709&Q=324144	All projects implemented or funded by the State (exceptions made for emergencies and administrative actions)	Any action that might create a significant impact on the state's land, water, air, or other environmental resources	Environmental Classification Document (ECD) Scoping notice (published in Environmental Monitor) Environmental impact evaluation (EIE) Record of decision (ROD)
Massachusetts	Massachusetts Environmental Policy Act (MEPA) http://www.env.state.ma.us/mepa/	Any project that exceeds the review thresholds	Many defined: Land disturbances, endangered/threatened species, wetlands, waterways, and tidelands, water consumptions, wastewater, transportation, energy Air, solid, and hazardous waste, historical and archaeological resources, areas of critical environmental concern (ACEC)	Environmental Notification Form (ENF) Draft Environmental Impact Report (DEIR) Supplementary Environmental Impact Report (SEIR) Final Environmental Impact Report (FEIR) Record of Decision (ROD)
Montana	Montana Environmental Policy Act (MEPA) http://leg.mt.gov/css/services%20division/lepo/mepa/default.asp	All projects implemented or funded by the State (exceptions made for administrative and ministerial actions)		Environmental Assessment (EA) Environmental Impact Statement (EIS) Notice of Decision (NOD)

State	Applicability	Thresholds	Document types
New York	New York State Environmental Quality Review Act (SEQR) http://www.dec.ny.gov/permits/357.html	Projects and planning activities undertaken, funded, or approved by state agency or unit of local government	Type I and Type II actions (Type I more likely to need EIS)— Statewide and agency thresholds are established Above "minimum criteria," which are set by each state agency	Environmental Assessment Form (EAF) Draft Environmental Impact Statement (DEIS) Final Environmental Impact Statement (FEIS) SEQR findings form
North Carolina	North Carolina Environmental Policy Act of 1971 http://www.ncga.state.nc.us/enactedlegislation/statutes/html/bychapter/chapter_113a.html	All projects undertaken, approved by, or funded by state agencies or (if privately funded) involve state-owned lands		Environmental Assessment (EA) Finding of No Significant Impact (FONSI) Environmental Impact Statement (EIS) Record of Decision (ROD)
Washington	State Environmental Policy Act (SEPA) http://www.ecy.wa.gov/programs/sea/sepa/e-review.html	Any government "action" which is non-exempt—exemptions can be defined on a regional or local basis	Thresholds can be set locally; Checklist provided covers impacts to earth, air, water, plants, animals, energy and natural resources, environmental health, noise, land and shoreline use, housing, aesthetics, light and glare, recreation, historic and cultural preservation, transportation, public services, and utilities	Environmental Checklist Draft Environmental Impact Statement (DEIS) Final Environmental Impact Statement (FEIS) Determination of Significance (DS) Determination of Nonsignificance (DNS) Notice of Action Taken (NAT)
Wisconsin	Wisconsin Environmental Policy Act (WEPA) http://dnr.wi.gov/topic/CompAssist/WEPA.html	All projects undertaken, approved by, or funded by state agencies or (if privately funded) involve state-owned lands	Type I–IV actions	Environmental Assessment (EA) Environmental Impact Statement (EIS)

3. *Alternatives analysis*: Develop and compare alternatives and determine appropriate mitigation as needed
4. *Public review*: While public involvement is inherent among all steps of the process, a formal review of the draft proposals should always be conducted, and revisions undertaken as needed

The next few sections discuss these steps in more detail by elaborating on regulatory requirements and offering additional methods that are commonly utilized. The Federal Highway Administration describes a similar process in "Community Impacts Assessment: A Quick Reference for Transportation," available in print or online at http://www.ciatrans.net/CIA_Quick_Reference/Purpose.html

25.3   PROJECT CONCEPTION

The need for a transportation improvement is often identified when there is a failure in the transportation system, such as traffic congestion, pavement disrepair, or safety deficiencies. Transportation projects are also commonly conceived in relation to other development activities, such as land subdivisions, major retail or industrial development, or natural resource extraction (e.g., logging and mining). The desire for new or expanded roadways, airports, railroads, and other facilities may also result from broader planning activities with the goals of economic development, accommodating population growth, or directing future development activities to the most appropriate locations. In all of these cases, there are numerous stakeholders that have an interest in how the project proceeds, and many of them focus on minimizing negative impacts to the natural environment, other infrastructure, under-represented populations, and various cultural resources. Identifying and coordinating with these stakeholders at the project conception stage will ultimately allow for a more efficient project design and implementation.

At the U.S. Federal level, the Council on Environmental Quality (CEQ) was created by NEPA to draft regulations, oversee the process, and publish annual reports on the national's environmental conditions and processes. The CEQ directs the federal agencies to

1. Integrate the NEPA process with other planning at the earliest possible time to ensure that planning and decisions reflect environmental values, to avoid delays later in the process, and to head off potential conflicts
2. Rigorously explore and objectively evaluate all reasonable alternatives
3. Ensure that environmental information is available to public officials and citizens before decisions are made
4. Encourage and facilitate public involvement in decisions that affect the quality of the human environment (Section 1500.2)
5. Emphasize interagency cooperation before the environmental impact statement is prepared

Within this coordinated process, the project proponent(s) can identify applicable regulations, and determine the level of analysis that will be needed to ensure that impacts are avoided, minimized, or appropriately mitigated. For federal agencies, NEPA stipulates three possibilities. The coordination efforts may easily determine that the proposed action is either considered a categorical exclusion or that a more robust analysis in the form of an environmental impact statement (EIS) is needed to address impacts and mitigation. If it is unclear whether or not an EIS is warranted, NEPA directs the proponent to prepare and environmental assessment (EA) to determine the level of significance of the likely impacts.

25.4   IMPACT ASSESSMENT

Assessing a transportation project's impacts is an essential step in developing mitigation methods, and preparing numerous alternatives for consideration. The NEPA regulations (40 CFR Part 1508) define an "environmental assessment" as "a concise public document for which a Federal Agency is responsible that serves to

1. Briefly provide sufficient evidence and analysis for determining whether to prepare an environmental impacts statement or a finding of no significant impact
2. Aid an Agency's compliance with the Act when no environmental impact statement is necessary
3. Facilitate preparation of a statement when one is necessary"

The regulations also stipulate that contents of an EA shall describe the need for the proposed action, alternatives, impacts of the proposed action and its alternatives, as well as a listing of agencies and persons consulted in the development of the EA. Alternatives should always include a "no build" scenario, which helps to establish the baseline for impact assessments. Note that the EA does not need to develop or select mitigation measures at this stage. It is intended to assess the extent and severity of the impacts. If significant, then an EIS will be required, and that document will address mitigation techniques and comparisons between alternative options.

Examples of Environmental Assessments may be found online—often by perusing Federal Agency or State DOT websites. A sample from Colorado for major improvements to State Highway 7 resulted in a FONSI and can be found at http://www.coloradodot.info/projects/SH7EA

Many states similarly require some kind of impact assessment as a screening measure to determine whether a full alternatives analysis will be necessary. The Massachusetts Environmental Policy Act (MEPA) stipulates that all projects must be compared to a series of quantifiable review thresholds, which will determine what level of MEPA review is required, if at all. The comparison to these thresholds may result in finding that no MEPA review is required, that an environmental notification form (ENF) and some MEPA review is required, or that an ENF and a mandatory environmental impact report (EIR) is necessary. For example, if a proposed project is expected to alter designated significant habitat or more than 2 acres of priority habitat, then an ENF and MEPA review will be required. Other thresholds define limits for direct alteration of land, creation of impervious surfaces, water withdrawals, energy use, and many others. The review thresholds for MEPA can be found at http://www.env.state.ma.us/mepa/regs/11-03.aspx

Other general methods for determining the extent of environmental impacts range from simple checklists to more comprehensive activities utilizing Geographic Information Systems (GIS). The Tahoe Regional Planning Agency in Nevada provides a comprehensive checklist for initial determination of environmental impact. The 25 page checklist covers issues for land, air, water, vegetation, wildlife, noise, light and glare, land use, natural resources, risk of upset, population, housing, transportation/circulation, public services, energy, utilities, human health, scenic resources/community design, recreation, and archeological/historic resources. It can be accessed at http://www.trpa.org/default.aspx?tabid=221. FHWA provides a number of case studies in the use of GIS in transportation applications, some of which are for environmental impact assessments. See http://www.gis.fhwa.dot.gov/reports.asp for more information.

25.5   ALTERNATIVES ANALYSIS

The best way to avoid, minimize, or mitigate the impacts of transportation projects is to consider a variety of alternative approaches and designs that can meet the identified need. The impacts and potential mitigation measures for each alternative can be developed, and a comparative analysis between them can be conducted. In this manner, the least impact-intensive alternative can be selected. Other factors that may be considered in the final selection might include cost-efficiency, practicality of implementation, public support, and the overall benefit attained from the proposed transportation project.

NEPA requires the submission of an "Environment Impact Statement" (EIS) for all federally funded projects. The EIS is a detailed statement that describes the impacts of the proposed activities and its alternatives on the environment. It should contain the following:

1. Environmental impact
2. Unavoidable adverse impact of the proposed activity, if any
3. Alternatives
4. Relationship between local short-term use of the environment and the maintenance and enhancement of long-term productivity
5. Irreversible and irretrievable commitments of resources in the proposed activity, if any

The EIS process involves several steps, such as submitting a notice of intent (NOI), conducting the alternatives analysis (including no-build scenario) with assessment of impacts and mitigation, publishing a Draft EIS, public review and comment, and submitting a Final EIS. Details on submitting EIS are available at http://www.epa.gov/compliance/nepa/submiteis.

The EIS formally begins with the NOI being published in the Federal Register. The NOI not only informs the public that an EIS will be prepared for a particular project, it also should outline the proposed action and any alternatives; describe the scoping process, including information about scoping meetings; and inform the public about who should be contacted with questions. As defined by 40 CFR Section 1508.25, the project scope should include the range of actions, alternatives, and impacts to be considered in the EIS. Scoping is based on a start and end time of analysis, the extent of the physical area for the evaluation of impacts, the level of action definition, meaningfulness to decision-making, availability of information, and the regulatory focus of the EIS.

Major sections of the Draft EIS include purpose and need; alternatives; affected environment; environmental consequences; comments and coordination; and a list of preparers. The Federal Highway Administration published a Technical Advisory (T6640.8A) to provide guidance on the preparation of an EIS. The Advisory provides specification on both format and content. It is available on the Internet at http://environment.fhwa.dot.gov/projdev/impta6640.asp. Some of the prominent items are summarized here.

The *Purpose and Need* Statement is a critical piece of writing in the EIS. The problem that the action is intending to address needs to be clearly and effectively described. This will drive the development of alternatives as well as provide a basis for discussing the no-build scenario. FHWA encourages the use of visuals (charts, maps, photographs, tables, etc.) to make this description effective and easy to read by members of the public.

The discussion of *Alternatives* needs to describe the preferred alternative, the no-build scenario, and a range of other options. FHWA recommends consideration of various build alternatives, transportation systems management (TSM) options as well as mass transit in urban areas.

The *Affected Environment* describes the area of impact of the proposed project. Data and information that provide a broad context should be included, including adopted land use, transportation, or other plans of local jurisdictions.

The *Environmental Consequences* section will detail the various impacts of the proposed alternatives within the affected environment as well as describe appropriate mitigation measures. FHWA suggest two options for organizing this section: (1) discuss impacts and mitigation separately for each alternative or (2) discuss the impacts in turn and describe which alternatives are generating these impacts and how they may be mitigated. Trend analysis is typically conducted to estimate cause-and-effect relationships between the impact and the ecosystems. For any project, the area's potential for growth and forces of change are also considered.

Mitigation can be defined as actions that avoid, minimize, rectify, reduce, or compensate for the undesirable impacts of development activities. Five actions that could be taken are location modification, design modification, construction measures, operational conditions, and right of way measures and replacement land. Examples of such activities for the construction of highway would involve one or more of the following: changing the location of a bridge or road to avoid or reduce the effect on a wetland, elevating a roadway on piers, limiting construction period to avoid breeding season of wildlife, prohibiting trucks or heavy vehicles from the use of a road, purchasing or creating wetlands to compensate for those affected or destroyed by the highway construction. The EIS will describe all practicable mitigation measures that have been adopted and the means to monitor them. The mitigation techniques can be identified from the stipulations of the NEPA. The requirements of mitigation can also arise from an agreement between the transportation agency and a private group of citizens who are affected by the project. The requirements could be dictated by the federal or state government laws or agreements between transportation agencies and environmental groups. The specific mitigation requirements will depend on the specific impact such as those on historic site or air or water quality.

The Mid-Atlantic region of EPA maintains a database of Environmental Impact Statements at http://www.epa.gov/reg3esd1/nepa/eis.htm. Other EIS samples can be obtained from the Northwestern University Transportation Library and on microfiche at the EPA Headquarters Repository—accessible through many libraries' interlibrary loan program.

Most state environmental review processes require similar contents in the EIS and EIR documents (see Table 25.2).

An example from China can also be reviewed: Environmental Impact Assessment Xi'an Urban Transport Project, http://www.adb.org/documents/environment/prc/xian_urban.pdf

25.6 PUBLIC INVOLVEMENT AND REVIEW

The NEPA process stipulates specific timelines and procedures for public review and comment and assumes that appropriate measures are taken to include public input into the development of the various NEPA documents. Public comment is formally accepted in writing or as oral testimony at public hearings. Minimum review-and-comment periods are specified as

- Thirty days beginning with the issuance of the NOI
- Forty-five days beginning with the issuance of the Draft EIS
- Thirty days beginning with the issuance of the Final EIS

There are other, more robust models for public involvement in transportation projects, however. Many of these are discussed by FHWA at http://www.fhwa.dot.gov/environment/pubinv2.htm. This site includes a guide for residents on the project development process and how the public can be effectively involved; best practices on involving people with limited English proficiency; assistance with utilizing visualization to make public participation more meaningful; and other public involvement techniques and case studies.

In 1994, the U.S. president, Bill Clinton, signed the EO 12898, *Federal Actions to Address Environmental Justice in Minority Populations and Low Income Populations* (59 FR 7629) to identify and address, as required, *disproportionately* high adverse human health or environmental effects of each agency's programs policies, and activities on minority populations and low-income populations. The resulting USDOT directive was the order titled, Actions to Address *Environmental Justice In Minority Populations and Low-Income Populations* dated (1995). The primary issues in enforcing this directive need to be addressed through an intensive public involvement process, such as those during project development meetings. Steps may consist of the development of a study area and a reference population, determination of affected people, and impact evaluation. The methods that are utilized consist of the use of census data, survey data, and statistical analysis of trends. Table 25.3 shows an example of the results of such a study.

25.7   ENFORCEMENT AND POST-PROJECT MONITORING

Once identified, the mitigation efforts or commitments are indicated in construction specification and project plans. Enforcement may be made through a simple memorandum of understanding between the overseeing agency and the transportation construction agency or it may require withholding of funds from grantees, or could be facilitated by the existence of performance bonds that are required for many road construction projects. If administrative follow-ups are not successful, citations and formal litigation proceedings may be the last resort for rectifying failed enforcement. In the United States, the relevant enforcement authorities are the EPA and the Army Corps of Engineers. EPA, established in 1970, has the authority to implement and enforce the various laws. It reviews and comments on EISs prepared by other federal agencies. It also charges the COR to review the EIS, especially for projects that could impact water bodies.

In some cases, the approval of highway construction projects in a specific state would require the state to show that the new construction would not lead to violation of any environmental (such as air) quality standard and nor will it make difficult to achieve compliance with the mitigation system in order to get approved by the federal agency.

Improvement in enforcement of mitigation measures can be made by administrative mechanism or judicially, with the help of different provisions and acts, as appropriate for the country/state. Examples of such provisions in the United States are those of the NEPA, Section 4(f) of the DOT Act, National Historic Preservation Act, Clean Water Act and the Clean Air Act. Enforcement through the court could also be made with the help of enforcement of agreements by parties to the agreement, third-party beneficiary law, or under the Nuisance law.

Note that the mitigation measures need to be tracked and monitored for a certain number of years (say 3–5 years for wetland replacement) to evaluate the effectiveness of the mitigation procedure. An example of a law governing mitigation monitoring is the Clear Water Act. Section 404 of this act states that if a high construction activity leads to the filling of natural wetlands, compensation must be provided by the creation of an artificial wetland. The U.S. Army Corps of Engineers, as directed by the EPA, is in charge of ensuring the compliance with the Clean Water Act for any project that is undertaken by a DOT.

The subject of post-project monitoring and analysis for improving environmental impact mitigation has been researched extensively, and a number of methods have been suggested. For example, the United Nations has suggested the use of Post-Project Analysis (PPA) as a tool for improving the assessment of environmental impact, in order to ensure or facilitate the enforcement of mitigation activities and also to learn from a specific project for future applications.

In the PPA method, first, the importance of environmental concern and its relationship to economic development have been recognized. The use of EIA has been presented as the starting point for the evaluation of effects of an economic activity before the development of that activity, through the investigation of the effects, identification of mitigation activities, and overall environmental management. To be effective, EIAs must be able to predict the impact of a project accurately and convincingly, and allow the formulation of a mitigation plan even in the case of lack of certainty regarding those impacts. The PPA is presented as a cost-effective tool to improve the EIA process. The PPAs are defined as environmental studies that are conducted during all phases of a project implementation—before construction, during construction or operation, and at the time of abandonment. The process consists of examining, documenting, and analyzing impacts and the effectiveness of impacts of the projects, in terms of scientific and technical aspects and the management system, during their implementation. The PPS process helps in identifying cases

TABLE 25.3
Example of Environmental Justice Study from North Carolina

Issue	Adverse Effect[a]	Mitigation[b]	Mitigation Benefiters
Relocations	*Center alternative*: 20 Residential (34% minority and 14% low-income); 8 businesses (0% minority) *Southern alternative*: 49 residences (19% minority and 13% low-income); 13 businesses (0% minority)	Displaced residents will be provided relocation assistance by DOT Relocation personnel. Every reasonable effort will be made to relocate all displaced residences to areas within the general community of Wrightsboro	The benefiters of relocation assistance would be those who are displaced, (Please see Adverse Effects column.)
Communities	*Center alternative*: Possible relocation of minority elder of Rockhill neighborhood. Possible encroachment on the Moore Family Cemetery *Southern Alternative*: None	Please see mitigation for relocations mentioned earlier. Additionally, the configuration of the proposed interchange at US 117/NC 133 (Castle Hayne Road) was revised to avoid the Moore Family Cemetery	The congregation of the St. James A.M.E. Church, residents of the Rockhill Road neighborhood, and Moore family members would be the only benefiters of these measures. All are predominantly minority
Community services	*Center alternative*: Possible elimination of existing Chair Road access to the St. James A.M.E. Church parking area caused by the widening of US 117/NC 133 in front of this church *Southern alternative*: None	Chair Road will be realigned to eliminate its skew and to maintain the current access to the St. James A.M.E. Church Additionally, any current or future widening of Castle Hayne Road will be done to the east, and away from the church	The congregation of the St. James A.M.E. Church would be the only benefiters of this mitigation. As noted earlier, it is predominantly minority
Safety and traffic	No adverse effect	Because no adverse effects anticipated, no mitigation measures are necessary. However, the elimination of the skew at Chair Road and the redesign of the interchange at Castle Hayne will benefit roadway travel in this vicinity	The predominant benefiters comprise the congregation of the St. James A.M.E. Church and residents of the greater Rockhill Road neighborhood. Both are predominantly minority
Noise	*Center alternative*: 18 persons (estimated; 75.6% minority; 7.8% low-income) *Southern alternative*: 32 persons (estimated: 12.5% minority; 19.0% low-income)	*Center alternative*: With one noise barrier, 8 persons (estimated: 75.6% minority; 7.8% low-income) *Southern alternative*: No noise barrier justified under preliminary evaluation: 32 persons (estimated: 12.5% minority; 19.0% low-income)	The benefiters of the noise barrier are the residents of the predominantly minority neighborhood along Granny Road. Noise wall lengths will be reinvestigated during final design. Results will be coordinated with the Community Liaison

(continued)

TABLE 25.3 (continued)

Example of Environmental Justice Study from North Carolina

Issue	Adverse Effect[a]	Mitigation[b]	Mitigation Benefiters
Air quality	No adverse effect	None	Not applicable
Land use and value	No adverse effect	None	Not applicable
Visual impacts	*Center alternative*: The highest density residential area in the vicinity of a visually sensitive area is the Northchase Subdivision, predominantly non-minority. *Southern alternative*: The highest density residential area in the vicinity of a visually sensitive area is the Oakley Loop neighborhood, predominantly non-minority	Landscaping along visually sensitive areas	Benefiters of the landscaping will include travelers of the Bypass (racially and socioeconomically representative of the area) and the residents adjacent to these areas. The more densely populated of these residential areas are non-minority neighborhoods, (Please see Adverse Effects column.)

Source: Lane, L.B., Hoffeld, S. and Griffin, D., *Journal of the Transportation Research Board*, 1626, 131, 1998.

Notes: The expected mean percentage of minority persons within the adapted study area is 25.5%; the expected mean percentage of persons having low income within the adapted study area is 19.6%.

[a]   This table summarizes the adverse effects from the proposed alternatives on issues identified by the communities of concern. This summary does not list all effects but is limited in scope to summarizing the adverse effects that may have potential disproportionate impacts to minority or low-income populations.

[b]   Mitigation includes mitigative measures to combat associated adverse effects and (for the case of safety and traffic) ameliorative measures that will be implemented to enhance benefits of the project.

where the mitigations measures are actually necessary, and improve the EIA process by providing feedback about its effectiveness. The 22 clauses of PPA, as put forth by the UN in their 1990 publication (*Post-project analysis in environment impact assessment, 1990*), are shown in Table 25.4. The entire publication is available at http://www.unece.org/env/documents/1990/ece.envwa.11.e.pdf.

25.8 TRANSPORTATION PLANNING AND REGIONAL MITIGATION APPROACHES

Regional and statewide transportation planning, such as that which is mandated in urban areas in the United States, offers the opportunity to address many of the long-term cumulative impacts of transportation projects as well as provide a foundation for assessing and minimizing the short-term direct impacts. Beginning in 1962, the U.S. Federal Highway Act has required the development of long-range transportation plans in urbanized areas (as defined by the U.S. Census) with populations of greater than 50,000 people. These requirements are implemented by Metropolitan Planning Organizations (MPOs), which are policy boards that make decisions on transportation funding and priorities within the region. Over the years, the Act has been reauthorized under different names, and some of the more current Acts have included specific provisions for dealing with mitigation of highway improvement projects.

TABLE 25.4
Post-Project Analysis

The task force concluded that post-project analyses are a very effective and necessary means of continuing the EIA process into the implementation phase because of their uses for the following purposes:

a. To monitor compliance with the agreed conditions set out in construction permits and operating licenses

b. To review predicted environmental impacts for proper management of risks and uncertainties

c. To modify the activity or develop mitigation measures in case of unpredicted harmful effects on the environment

d. To determine the accuracy of past impact predictions and the effectiveness of mitigation measures in order to transfer this experience to future activities of the same type

e. To review the effectiveness of environmental management for the activity

With a view to promoting the use of effective and efficient PPAs as well as to strengthening EIA processes in the project implementation phase, *it is recommended that*

1. PPA should be used to complete the EIA process by providing the necessary feedback in the project implementation phase both for proper and cost-effective management and for EIA process development.

2. ECE Governments should apply the specific recommendations set out hereafter to suitable projects and should report back to the Senior Advisers in three years' time on the results of those PPAs and experience gained in their implementation.

Relationships between environmental impact assessment and post-project analysis

3. A preliminary plan for the PPA should be prepared during the environmental review of a project; the PPA framework should be fully developed when the EIA decision on the project is made.

4. The PPA should focus on important impacts about which there is insufficient information; identification of these impacts and their priorities is undertaken during the environmental review process.

5. The authority to undertake a PPA should be linked to the EIA process so that the concerns identified for inclusion in the PPA during the environmental review can be properly addressed.

6. The conditions of approval for a project should be such that the environmental management for that project will take into account the findings of the PPA.

7. PPAs should be done for all major projects with potentially significant impacts. In addition, for other projects, focused PPAs may be suitable either for environmental management of the project or to learn from the project.

Content of PPA

8. The development of hypotheses to test should be a part of PPAs. The hypotheses will depend greatly on the nature of the PPA and may involve comparisons of impacts with predictions or with standards, or they may relate to how well the environmental management system worked.

9. In order to undertake PPAs effectively, baseline data relevant to the hypotheses should be collected and be as complete as possible.

10. Monitoring and evaluation of the data collected in the monitoring process should be an essential part of PPA. These steps are needed in order to test the hypothesis.

11. Documentation of the project and its impacts should be encouraged in order to improve PPAs.

PPA development and design

12. The first and most crucial step in developing a PPA should be to define its purpose. This would include the development of a specific purpose and focus for each component of the PPA.

13. Once the purpose of the PPA is known and its conceptual content identified (from the environmental review), it is essential to define the roles and responsibilities of the various participants in the PPA—the proponent, the various government agencies, scientific and technical advisers, and the public.

14. Management and participant responses required in the light of PPA findings should be, as far as possible, specifically addressed.

15. The need to deal with environmental surprises must be built into PPAs. Monitoring should be done in such a way that unexpected results have a good chance of being detected and those responsible for the PPA should have the power to respond appropriately to unexpected results.

(continued)

TABLE 25.4 (continued)
Post-Project Analysis

16. The use of independent experts to help design the PPA should be encouraged as it leads to a better and more credible PPA.

17. The detailed development of the PPA should consider features such as the different phases of the project (preconstruction, construction, operation, and abandonment), the need for integration of different aspects being studied, and the need to relate the effects being monitored to the project (separating out confounding effects of other activities).

PPA management

18. As a tool for managing PPAs, advisory boards consisting of industry, government, contractors, independent experts, and public representatives, should be used. Such boards with well-defined terms of reference increase the credibility and quality of the PPA.

19. Public participation in the PPA should be encouraged.

20. PPA reports should be made public.

21. The use should be encouraged of independent researchers to do those parts of a PPA that are particularly sensitive and for which work done by the proponent (or possibly even by a government agency) may not be regarded as credible.

22. PPAs should be managed adaptively with opportunities to refine them, depending on the results obtained. More effort should be put into examining those effects that are observed and important, and less effort should be expended on those effects which the PPA indicates are not resulting in significant impacts.

Source: United Nations. Economic Commission for Europe, *Post-Project Analysis in Environmental Impact Assessment*, Volume 3 of Environmental Series, United Nations, Geneva, Switzerland, 1990.

Notes: (1) An analysis of the information, which led to the conclusion and recommendations, is contained in Chapter III B. (2) The United Kingdom fully concurs with the principle that the actual effects of a development project on the environment should be evaluated, after development consent has been given, both during the construction phase and subsequently during the project's operation. Such evaluation needs to be continuous and to be accompanied by powers to enforce environmental standards and conditions of operation, to require action to remedy adverse environmental effects and to secure improvements, for example by continually upgrading plant to the standard of the best available technology not entailing excessive cost, or by the tightening of pollution controls in the light of new evidence of risk. The United Kingdom has long-standing provisions of this kind, applying to all relevant installations, not only those that have been subject to environmental impact assessment. These provisions will be refined and expanded by planned new legislation on the introduction of an integrated system of pollution control for industry. In the United Kingdom's view, the elaborate post-project procedures recommended by the task force would not improve on these arrangements. (In accordance with document ECE/ENVWA/9, paragraph 42.)

The *Intermodal Surface Transportation Efficiency Act of 1991* (ISTEA) provided a framework for making transportation project-related decisions more compliant with environmental needs and regulations, such as the Clean Air Act. Demonstrating that transportation improvements will not worsen air quality is still a requirement for states and MPOs today. ISTEA also required the completion of the major investment study (MIS) as a standardized way to demonstrate compliance with planning requirements and considerations of environmental impacts. One of the goals was to consider mobility and travel needs at a corridor level, rather than at a localized roadway level, in order to conduct an effective alternatives analysis. The MIS process is no longer formally required, but many agencies still use the principles of the MIS in the development of transportation improvement projects.

The *Safe, Accountable, Flexible, Efficient Transportation Equity Act: A Legacy for Users* (SAFETEA-LU) (Pub. L. 109-59, August 10, 2005), required long-range statewide transportation plans to include a discussion of potential environmental mitigation activities and potential areas to carry out these activities, including activities that may have the greatest potential to restore and maintain the environmental functions affected by the long-range statewide transportation plan.

The requirement is intended to encourage consideration of environmental impacts at the planning level—well in advance of project scoping and design. In addition, SAFETEA-LU required that states and MPOs demonstrate consultation efforts with various stakeholders, including groups and agencies whose missions include environmental protection. The intent was to ensure consistency between various planning efforts and, again, to identify mitigation opportunities early in the planning process.

These requirements recognize that the impacts of transportation projects are felt beyond the geographic limits of the project and in some cases, beyond the lifecycle of the project. The U.S. Federal Highway Administration published an assessment of various state and regional efforts regarding regional mitigation in 2009. The report, entitled "Environmental Mitigation in Transportation Planning: Case Studies in Meeting SAFETEA-LU Section 6001 Requirements," is available at http://www.environment.fhwa.dot.gov/integ/pubcase_6001.asp. Two of the case studies included in this report are summarized briefly here.

The Minnesota Department of Transportation began implementing a wetlands banking program in the 1980s. By 2005, the established system offered the state DOT an easy way to demonstrate compliance with SAFETA-LU's new requirements. Essentially, a "no net loss of wetlands goal" was established, and an ecosystem approach was taken to identify the highest value and/or most at-risk wetlands at regional and watershed levels. These are purchased, preserved, and used as credits to offset wetland impacts of future transportation improvement projects. The DOT estimates that they have enough wetlands "banked" to address wetlands mitigation for another 10–15 years of transportation projects. This approach is considered superior to mitigating at the project-specific level, and the State has since implemented a similar approach for historic bridges. In addition, they are broadening the program to address habitat connectivity, not just wetlands.

The San Diego Association of Governments (SANDAG) developed a Regional Comprehensive Plan in 2004, which addressed a broad range of issues, such as housing, transportation, and the environment. It is used as a companion to the long-range transportation plan and includes an inventory of baseline natural resource data that can be used to monitor environmental health as well as impacts to transportation-related construction. This approach supports the State's 1991 Natural Community Conservation Planning Act, which bases conservation efforts regarding endangered species on ecosystem approach rather than focusing on a single species. SANDAG also raises funding through a ½-cent sales tax, the proceeds of which support acquisition of land that will be preserved or used to restore habitat that is slated to be disturbed by transportation projects. The benefit of conducting land conservation at the regional planning level is that acquisitions can be made more efficiently and at lower costs than they would be if they were carried out at the project scoping and design level. In addition, the value of an entire ecosystem can be preserved upfront, rather than addressing site-specific impacts after transportation design decisions are made.

In summary, the development of mitigation strategies for transportation improvement projects should never be limited to the site-specific level. In many areas, there are likely regional or state-wide planning efforts, in place or underway, that will assist the transportation project engineer in identifying high-priority impacts to be avoided or addressed; offering policy guidance on efficient mitigation techniques; providing natural resource data (often in a GIS-based platform) to be used in impact analyses; and/or supporting communication with various stakeholders that may be impacted by the transportation project. Working within these frameworks can also ease the burden of regulatory compliance and ultimately address indirect and cumulative impacts of transportation projects.

QUESTIONS

25.1 Describe the difference between a direct and an indirect effect of a transportation project.

25.2 List five potential impacts of a roadway improvement project that adds an additional travel lane in each direction of a 6 mile roadway segment, paves and widens the shoulders, and constructs sidewalks with granite curbing on one side of the roadway.

25.3 Identify a state that has passed an Environmental Policy Act, similar to the National Environmental Policy Act (NEPA), to regulate environmental impacts of development projects. Cite the name and regulation number for the Act. To which entities or under what circumstances does the legislation apply?

25.4 Why is a "no build" scenario an important component of an alternatives analysis?

25.5 Find a recently published Draft Environmental Impact Statement (DEIS) for a transportation improvement project. Imagine you are representing the community in which the project is being constructed. Identify three to five items that you would like to see addressed differently than proposed in the final EIS.

25.6 Discuss the implications of Environmental Justice in transportation improvement projects.

25.7 Provide an example of a successful strategy for addressing cumulative impacts of transportation projects in a watershed of a major public drinking water supply.

Conversion Factors

U.S. Units to SI Units

From	To	Multiply by:
Length		
Inches, in.	Millimeters, mm	25.4
Feet, ft	Meters, m	0.305
Yards, yd	Meters, m	0.914
Miles, mi	Kilometers, km	1.61
Area		
Square inches, in.^{2}	Square millimeters, mm^2	645.2
Square feet, ft^2	Square meters, m^2	0.093
Square yard, yd^2	Square meters, m^2	0.836
Acres, ac	Hectares, ha	0.405
Square miles, mi^2	Square kilometers, km^2	2.59
Volumes		
Fluid ounces, fl oz	Milliliters, mL	29.57
Gallons, gal	Liters, L	3.785
Cubic feet, ft^3	Cubic meters, m^3	0.028
Cubic yards, yd^3	Cubic meters, m^3	0.765
Mass		
Ounces, oz	Grams, g	28.35
Pounds, lb	Kilograms, kg	0.454
Short tons (2000 lb), T	Megagrams (metric ton), Mg (t)	0.907
Temperature		
Fahrenheit, °F	Celsius, °C	(F − 32)/1.8
Illumination		
Foot-candles, fc	Lux, lx	10.76
Foot-Lamberts	Candela/m^2	3.426
Force, Pressure/Stress		
Poundforce, lbf	Newtons, N	4.45
Poundforce per square inch, psi	Kilopascals, kPa	6.89

(*continued*)

SI Units to U.S. Units

From	To	Multiply by:
Length		
Millimeters, mm	Inches, in.	0.039
Meters, m	Feet, ft	3.28
Meters, m	Yards, yd	1.09
Kilometers, km	Miles, mi	0.621
Area		
Square millimeters, mm^2	Square inches, $in.^2$	0.0016
Square meters, m^2	Square feet, ft^2	10.764
Square meters, m^2	Square yard, yd^2	1.195
Hectares, ha	Acres, ac	2.47
Square kilometers, km^2	Square miles, mi^2	0.386
Volume		
Milliliters, mL	Fluid ounces, fl oz	0.034
Liters, L	Gallons, gal	0.264
Cubic meters, m^3	Cubic feet, ft^3	35.314
Cubic meters, m^3	Cubic yards, yd^3	1.307
Mass		
Grams, g	Ounces, oz	0.035
Kilograms, kg	Pounds, lb	2.202
Megagrams (metric ton), Mg (t)	Short tons (2000 lb), T	1.103
Temperature		
Celsius, °C	Fahrenheit, °F	1.8C + 32
Illumination		
Lux, lx	Foot-candles, fc	0.0929
Candela/m^2	Foot-Lamberts	0.2919
Force, Pressure/Stress		
Newtons, N	Poundforce, lbf	0.225
Kilopascals, kPa	Poundforce per square inch, psi	0.145

References

Abou-Ayyash, A. 1974. Mechanistic behavior of continuously reinforced concrete pavement. PhD thesis, University of Texas at Austin, Austin, TX.

Abrams, D. A. 1918. Design of concrete mixtures. Lewis Institute, Structural Materials Research Laboratory, Bulletin No. 1, PCA LS001.

ACI Committee 211. 1991. Standard practice for selecting proportions for normal, heavyweight and mass concrete, ACI 211, pp. 1–91. Farmington Hills, MI: American Concrete Institute.

ACI Committee 318. 2002. Building code requirements for structural concrete, ACI 318-02, and Commentary, ACI 318R-02. Farmington Hills, MI: American Concrete Institute.

Alavi, S., J. F. LeCates, and M. P. Tavares. 2008. Falling weight deflectometer usage, Synthesis of Highway Practice, NCHRP Synthesis 381. Washington, DC: Transportation Research Board.

Allen, D. L. and R. C. Deen. 1986. A computerized analysis of rutting behavior of flexible pavements, Transportation Research Record No. 1095. Washington, DC: Transportation Research Board.

American Association of State Highway and Transportation Officials (AASHTO). 1986. *Guide for Design of Pavement Structures*, Vol. 2. Washington, DC: AASHTO.

American Association of State Highway and Transportation Officials (AASHTO). 1998. *Supplement to the AASHTO Guide for Design of Pavement Structures Part II-Rigid Pavement Design and Rigid Pavement Joint Design*. Washington, DC: AASHTO.

American Association of State Highway and Transportation Officials (AASHTO). 2000. *MDM-SI-2, Model Drainage Manual*, 2000 Metric Edition. Washington, DC: AASHTO.

American Association of State Highway and Transportation Officials (AASHTO). 2001. *A Policy on Geometric Design of Highways and Streets*. Washington, DC: AASHTO.

American Association of State Highway and Transportation Officials (AASHTO). 2007. *Standard Specifications for Transportation Materials and Methods of Sampling and Testing*, 27th edn., M 145–91 (2004). Washington, DC: AASHTO.

American Association of State Highway and Transportation Officials (AASHTO). N.d. [Home page]. www.transportation.org

American Concrete Institute (ACI) Committee 325. 2002. *Guide for Design of Jointed Concrete Pavements for Streets and Local Roads*. Farmington Hills, MI: ACI 325, 12R-02.

American Concrete Pavement Association (ACPA). 1990. Guidelines for unbonded concrete overlays, Technical Bulletin TB-005.0 D. Skokie, IL: ACPA.

American Concrete Pavement Association (ACPA). 1995. Construction of portland cement concrete pavements, National Highway Institute Course No. 13133, ASHTO/FHWA/Industry Joint Training. Washington, DC: Federal Highway Administration, Department of Transportation.

American Concrete Pavement Association (ACPA). 1999. Survey of states' concrete pavement design and construction practices. Skokie, IL: ACPA.

American Concrete Pavement Association (ACPA). 2002. Portland cement concrete overlays: State of the technology synthesis, Product Code SP045P. Skokie, IL: ACPA.

American Concrete Pavement Association (ACPA). N.d. Whitetopping: State of the Practice, Product code EB210P. Skokie, IL: ACPA.

American Society for Testing and Materials (ASTM). 1989. *Annual Books of ASTM Standards, Road and Paving Materials: Traveled Surface Characteristics*, Vol. 04.03. Philadelphia, PA: American Society for Testing and Materials.

American Society for Testing and Materials (ASTM). N.d. ASTM International, www.astm.org.

American Society for Testing and Materials (ASTM). Various years. *Annual Book of ASTM Standards, Soil and Rock*, Vol. 04.08. Philadelphia, PA: American Society for Testing and Materials.

American Society of Civil Engineers (ASCE). 2004. *Introduction to Mechanistic-Empirical Pavement Design, Reference Manual*, Seminar. Baltimore, MD: ASCE.

Anderson, D. A. and T. W. Kennedy. 1993. Development of SHRP binder specifications. *Journal of AAPT* 62: 508.

Asphalt Institute. 1983. *Computer Program DAMA: User's Manual*. Lexington, KY: AI.

Asphalt Institute. 1984. *Mix Design Methods for Asphalt Concrete and Other Hot Mix Types*, Manual Series No. 2 (MS-2), May. Lexington, KY: AI.

Asphalt Institute. 1991. *Thickness Design: Asphalt Pavements for Highways and Streets*, Manual Series No. 1 (MS-1). Lexington, KY: AI.

Asphalt Recycling & Reclaiming Association (ARRA). 2003. *Basic Asphalt Recycling Manual*. Annapolis, MD: ARRA.

Ayres, M., Jr. 1997. Development of a rational probabilistic approach for flexible pavement analysis. PhD dissertation, University of Maryland at College Park, College Park, MD.

Bahia, H. U. and D. A. Anderson. 1995. The SHRP binder rheological parameters: Why are they required and how do they compare to conventional properties? Preprint Paper No. 950793, Presented at the *Transportation Research Board Meeting*. Washington, DC: Transportation Research Board.

Baladi, G. 1989. Fatigue life and permanent deformation characteristic of asphalt concrete mixes, Transportation Research Record No. 1227. Washington, DC: Transportation Research Board.

Barber, E. S. and C. L. Sawyer. 1952. Highway subdrainage. *Proceedings, Highway Research Board*, pp. 643–666. Washington, DC: Highway Research Board.

Barksdale, R. D., Ed. 1991. *The Aggregate Handbook*. Washington, DC: National Stone Association.

Boltzmann, L. 1874. Zur Theorie der elastichen Nachwirkung, Sitz. Kgl. Akad. Wiss. Wien (Math-Naturwiss Klasse) 70: 275–306.

Bradbury, R. D. 1938. *Reinforced Concrete Pavements*. Washington, DC: Wire Reinforcement Institute.

Brademeyer, B. 1988. VESYS modification, Final report, FHWA DTFH61-87-P-00441. Washington, DC: U.S. Department of Transportation.

Burati, J. L., R. M. Weed, C. S. Hughes, and H. S. Hill. 2003. Optimal procedures for quality assurance specifications, FHWA-RD-02-095. Washington, DC: Federal Highway Administration.

Bureau of Reclamation. 1977. *Method for Determining the Quantity of Soluble Sulfate in Solid (Soil and Rock) and Water Samples*. Denver, CO: Bureau of Reclamation.

Byrum, C. R. 1999. Development of a high sped profiler based slab curvature index for jointed concrete pavements. PhD dissertation, University of Michigan, Ann Arbor, MI.

Byrum, C. R., W. Hansen, and S. D. Kohn. 1997. The effect of pcc strength and other parameters on the performance of PCC pavements. *Sixth International Purdue Conference on Concrete Pavement Design and Materials for High Performance*, pp. 373–393. Indianapolis, IN.

Carey, W. N. and P. E. Irick. 1960. The pavement serviceability-performance concept, Highway Research Bulletin 250. Washington, DC: Highway Research Board.

Casagrande, A. and W. L. Shannon. 1952. Base course drainage for airport pavements. *Proceedings of the American Society of Civil Engineers* 77: 792–814.

Chamberlain, E. J. 1987. A freeze thaw test to determine the frost susceptibility of soils, Special Report 87-1. Hanover, NH: USA Cold Regions Research and Engineering Laboratory.

Chang, H. S., R. L. Lytton, and H. S. Carpenter. 1976. Prediction of thermal reflection cracks in West Texas, Report No. TTI-2-8-73-18-3. College Station, TX: Texas Transportation Institute, Texas A&M University.

Chou, Y. T. 1981. Structural analysis computer programs for rigid multicomponent pavement structures with discontinuities—Wesliqid and Weslayer. Report 2, Manual for the Wesliqid finite element program, Tech. Report GL-81-6, Vicksburg, MS: U.S. Army Waterways Experiment Station.

Choubane, B. and M. Tia. 1992. Nonlinear temperature gradient effect on maximum warping stresses in rigid pavements. Transportation Research Record No. 1370. Washington, DC: Transportation Research Board, National Research Council, pp. 11–19.

Christopher, R. A. and M. L. Hines, March 1999. Enforcement of environmental mitigation commitments in transportation projects: A survey of federal and state practice. *NCHRP, Legal Research Digest* 42.

Coduto, D. P. 1999. *Geotechnical Engineering: Principles and Practices*. Upper Saddle River, NJ: Prentice Hall.

CRCP Manual 2012; Rasmussen, R. O., Rogers, R., Ferragut, T., 2012. *Continuously Reinforced Concrete Pavement Design and Construction Guidelines*, Federal Highway Administration and Concrete Reinforcing Steel Institute.

D'Arcy, H. 1856. *Les Fountaines Publiques de la Ville de Dijon*. Paris, France: Dalmont.

Darter, M. I. 1988. A comparison between Corps of Engineers and ERES Consultants, Inc. rigid pavement design procedures. Technical Report. Savoy, IL: United States Air Force SAC Command.

Darter, M. I. and E. J. Barenberg. 1976. Zero-maintenance pavements: Results of field studies on the performance requirements and capabilities of conventional pavement systems Federal Highway Administration, FHWA-RD-76-105. Washington, DC: Federal Highway Administration.

Darter, M. I. and E. J. Barenberg. 1977. Design of zero-maintenance plain jointed concrete pavement, volume 1—Development of design procedures, FHWA-RD-77-111. Washington, DC: Federal Highway Administration.

Daugherty, R. L. and A. C. Ingersoll. 1954. *Fluid Mechanics with Engineering Applications*. New York: McGraw-Hill.

Davids, W. G., Turkiyyah, G. M., and J. Mahoney. 1998. EverFE—A new rigid pavement finite element analysis tool. Transportation Research Record No. Washington, DC: National Research Council, pp. 69–78.

Delatte, N. J., Safarjalani, M., and N. B. Zinger. 2000. Concrete pavement performance in the southeastern United States. Report No 99247, Tuscaloosa, AL: The University Transportation Center for Alabama. http://utca.eng.ua.edu/projects/final_reprts/99247report.pdf

Dellate, N. J. 2008. *Concrete Pavement Design, Construction and Performance*. New York: Taylor & Francis.

Federal Aviation Administration (FAA). 1989. Airport design, AC 150/5300-13. http://www.faa.gov/documentLibrary/media/Advisory_Circular/150_5300_13a.pdf, accessed December 12, 2012.

Federal Aviation Administration (FAA). 1995a. Airport pavement design and evaluation, AC 150/5320-6D. Washington, DC: U.S. Department of Transportation.

Federal Aviation Administration (FAA). 1995b. *LEDFAA User's Manual*, AC 150/5320-16, October. Washington, DC: U.S. Department of Transportation.

Federal Aviation Administration (FAA). 2001. Engineering brief # 60. http://www.faa.gov/airports/engineering/engineering_briefs/media/EB_60.pdf, accessed December 12, 2012.

Federal Aviation Administration (FAA). 2004. LEDFAA: Layered elastic design. www.airporttech.tc.faa.gov/pavement/26ledfaa.asp, accessed December 12, 2012.

Federal Aviation Administration (FAA). 2006a. Engineering brief # 59A. http://www.faa.gov/airports/engineering/engineering_briefs/media/EB_60.pdf, accessed December 12, 2012.

Federal Aviation Administration (FAA). 2006b. Item P-401 plant mix Bituminous pavements, Engineering brief EB59A, May 12. Washington, DC: FAA.

Federal Aviation Administration (FAA). 2006c. National airport pavement test facility. www.airporttech.tc.faa.gov/naptf/, accessed December 12, 2012.

Federal Aviation Administration (FAA). 2006d. Standardized method of reporting airport pavement strength: PCN, AC 150/5335-5a. http://rgl.faa.gov/Regulatory_and_Guidance_Library/rgAdvisoryCircular.nsf/0/f4b2780536535ff7862571f7006848c2!OpenDocument&Click=, accessed December 12, 2012.

Federal Aviation Administration (FAA). 2007a. Guidelines and procedures for maintenance of airport pavements, AC No: 150/5380-6B. http://www.faa.gov/documentLibrary/media/advisory_circular/150-5380-6B/150 _5380_6b.pdf, accessed December 12, 2012.

Federal Aviation Administration (FAA). 2008a. Advanced airport pavement design procedures: FAARFIELD—3D finite element based design procedure. www.airporttech.tc.faa.gov/pavement/3dfem.asp, accessed December 12, 2012.

Federal Aviation Administration (FAA). 2008b. [Software information page]. http://www.airporttech.tc.faa.gov/natptf/download/index1.asp#soft, accessed December 12, 2012.

Federal Aviation Administration (FAA). 2009. Standards for specifying construction of airports, AC No. 150/5370-10E. http://www.faa.gov/documentLibrary/media/advisory_circular/150-5370-10E/150_5370_10e.pdf, accessed December 12, 2012.

Federal Aviation Administration (FAA). 2011. Development of state standards for nonprimary airports. http://www.faa.gov/documentLibrary/media/Advisory_Circular/150_5100_13b.pdf, accessed December 12, 2012.

Federal Highway Administration (FHWA). 1961. Design charts for open-channel flow, Hydraulic Design Series No. 3 (HDS 3). Washington, DC: U.S. Government Printing Office.

Federal Highway Administration (FHWA). 1989. Concrete pavement drainage rehabilitation, FHWA Report, Experimental Project No. 12. Washington, DC: U.S. Department of Transportation.

Federal Highway Administration (FHWA). 1990a. Concrete pavement joints, Technical Advisory 5040.30. Washington, DC: Federal Highway Administration.

Federal Highway Administration (FHWA). 1990b. Highway subdrainage design by microcomputer: (DAMP), FHWA-IP-90-012. Washington, DC: U.S. Department of Transportation.

Federal Highway Administration (FHWA). 1992. Drainage pavement system participant notebook, FHWA-SA-92-008, Demonstration Project No. 87. Washington, DC: U.S. Department of Transportation.

Federal Highway Administration (FHWA). 1998a. Geosynthetic design and construction guidelines, FHWA-HI-95-038, NHI Course No. 13213, April. Washington, DC: U.S. Department of Transportation.

Federal Highway Administration (FHWA). 1998b. FHWA, Life Cycle Cost Analysis in Pavement Design Demonstration Project No. 115, FHWA-SA-98-040. Washington, DC: U.S. Department of Transportation.

Federal Highway Administration (FHWA). 1999. Pavement subsurface drainage design, FHWA-HI-99-028, NHI Course No. 131026, April. Washington, DC: U.S. Department of Transportation.

Federal Highway Administration (FHWA). 2001. Drainage requirements in pavements (DRIP), Developed by Applied Research Associates for the Federal Highway Administration, Contract No. DTFH61-00-F-00199. Washington, DC: U.S. Department of Transportation.

Federal Highway Administration (FHWA). 2002. Office of asset management, life-cycle cost analysis primer, August. Washington, DC: U.S. Department of Transportation.

Federal Highway Administration (FHWA). 2003. Distress identification manual for the long term pavement performance program, FHWA-RD-03-031, June. Washington, DC: U.S. Department of Transportation.

Federal Highway Administration (FHWA). 2005a. Computer based guidelines for concrete pavements, Vol. 2: Design and construction guidelines and HIPERPAVEII user's manual, Publication No. FHWA-HRT-04-122. Washington, DC: U.S. Department of Transportation.

Federal Highway Administration (FHWA). 2005b. Ultra-thin white topping (UTW) project. http://www.fhwa.dot.gov/pavement/utwweb, accessed December 12, 2012.

Federal Highway Administration (FHWA). 2006. Retrofit load transfer: Sharing the load. SP-204. www.fhwa.dot.gov/Pavement/concrete/sp204.cfm, accessed December 12, 2012.

Fick, G. J. 2008. *Testing Guide for Implementing Concrete Paving Quality Control Procedures*. National Concrete Pavement Technology Center/Center for Transportation Research and Education Iowa State University, ISU/FHWA Cooperative Agreement No. DTFH 61-06-H-00011. Ames, IA: Iowa State University.

Finn, F. N., K. Nair, and C. Monismith. 1973. Minimizing premature cracking of asphalt concrete pavements, NCHRP Report 195, June. Washington, DC: National Cooperative Highway Research Program and National Research Council.

Finn, F., C. L. Saraf, and R. Kulkarni. 1986. PDMAP, Development of pavement structural subsystems, NCHRP Report 291. Washington, DC: Transportation Research Board.

Foley, G., S. Cropley, and G. Giummarra. 1996. Road dust control techniques—Evaluation of chemical dust suppressants' performance, Special Report 54. Vermont South, Victoria, Australia: ARRB Transport Research.

Fonseca, O. A. and M. W. Witczak. 1996. A prediction methodology for the dynamic modulus of in-place aged asphalt mixtures. *Proceedings, Association of Asphalt Paving Technologists* 65: 532–572.

Fredlund, D. G. and A. Xing. 1994. Equations for the soil-water characteristic curve. *Canadian Geotechnical Journal* 31: 521–532.

Friberg, B. F. 1938. Design of dowels in transverse joints in concrete pavements. *ASCE Transactions* 64(2): 1809–1828.

Fuller, W. B. and S. E. Thompson. 1907. The laws of proportioning concrete. *ASCE Transactions* 59: 67–143.

Garber, N. J. and L. A. Hoel. 2002. *Traffic and Highway Engineering*, 3rd edn. Clifton Park, NY: Thomson Learning.

Gisi, A. J. and S. S. Bandy. 1980. Swell prediction of natural soils in kansas, Memorandum. Topeka, KS: Kansas Department of Transportation.

Goldbeck, A. T. 1919. Thickness of concrete slabs. Public Roads.

Guidance on the Use of the MSCR Test with the AASHTO M320 Specification, Asphalt Institute Technical Advisory Committee 2 December, 2010. downloaded from: http://amap.ctcandassociates.com/wp/wp-content/uploads/Guidance-on-Using-MSCR-with-AASHTO-M320_Final.pdf

Haas, R., F. Meyer, G. Assaf, and H. Lee. 1987. A comprehensive study of cold climate airfield pavement cracking. *Journal of Asphalt Paving Technologists Association* 56: 198–245.

Hajek, J. J. and R. C. G. Haas. 1972. Predicting low-temperature cracking frequency of asphalt concrete pavement, Transportation Research Record No. 407. Washington, DC: Transportation Research Board.

Halladay, M. 1998. The strategic highway research program: An investment that has paid off. *Public Roads* 61 (5, March–April 1998). http://www.tfhrc.gov/pubrds/marapr98/shrp.htm, accessed December 12, 2012.

Hanna, A. N. 1994. SHRP-LTPP Specific pavement Studies: Five-Year Report, Strategic Highway Research Program, National Research Council, Washington, DC.

Hansen, W., Jensen, E., Mohr, P., Jensen, K., Pane, I., and A. Mohamed. 2001. *The Effects of Higher Strength and Associated Concrete Properties on Pavement Performance*. Washington, DC: Federal Highway Administration, FHWA-RD-00-161.

Harichandran, R. S., M. S. Yeh, and G. Y. Balidi. 1989. MICH-PAVE user's manual, Final Report, FHWA-MI-RD-89-032. East Lansing, MI: Department of Civil and Environmental Engineering, Michigan State University.

Harwood, D. W., D. J. Torbie, K. R. Richard, W. D. Glauz, and L. Elefteriadou. 2003. Review of truck characterization as factors in roadway design, NCHRP Report 505. Washington, DC: Transportation Research Board.

Heinrichs, K. W., M. J. Liu, M. I. Darter, S. H. Carpenter, and A. M. Ioannides. 1989. Rigid pavement analysis and design, FHWA-RD-88-068, June. Washington, DC: Federal Highway Administration, Research, Development, and Technology.

Heukelom, W. 1973. An improved method of characterizing asphaltic Bitumens with the aid of their mechanical properties. *Proceedings, Association of Asphalt Paving Technologists* 42: 67–98.

Holtz, R. D. and W. D. Kovacs. 1981. *An Introduction to Geotechnical Engineering.* Upper Saddle River, NJ: Prentice Hall.

Huang, Y. H. 1993. *Pavement Analysis and Design.* Englewood Cliffs, NJ: Prentice Hall.

Huang, Y. H. 2004. *Pavement Analysis and Design*, 2nd edn., p. 60. Upper Saddle River, NJ: Pearson Education, Inc.

Huber, G. A. 1994. Weather database for the superpave mix design system, Strategic Highway Research Program Report SHRP 648A. Washington, DC: Transportation Research Board, National Research Council.

Ioannides, A. M., M. R. Thompson, and E. J. Barenberg. 1985. Westergaard solutions reconsidered, Transportation Research Board No. 1043. Washington, DC: Transportation Research Board.

Janssen, D. J. and M. B. Snyder. 1994. Resistance of concrete to freezing and thawing, SHRP-C-391. Washington, DC: Strategic Highway Research Program. http://onlinepubs.trb.org/onlinepubs/shrp/SHRP-C-391.pdf

Jones, D. 1999. Holistic approach to research into dust and dust control on unsealed roads. *Proceedings of the Seventh International Conference on Low-Volume Roads*, TRR No. 1652, Vol. 2. Washington, DC: Transportation Research Board.

Jones, G. M., M. I. Darter, and G. Littlefield. 1968. Thermal expansion-contraction of asphaltic concrete. *AAPT Proceedings* 37: 56–100.

Kallas, B. F. and V. P. Puzinauskas. 1972. *Flexural Fatigue Tests on Asphalt Paving Mixtures: Fatigue of Compacted Bituminous Aggregate Mixtures. STP 508.* Philadelphia, PA: American Society for Testing and Materials.

Kaloush, K. E. and M. W. Witczak. 2000. Development of a permanent to elastic strain ratio model for asphalt mixtures: Development of the 2002 guide for the design of new and rehabilitated pavement structures, NCJRP 1-37A Inter Team Technical Report, September. College Park, MD: University of Maryland.

Khazanovich, L. and A. M. Loannides. 1994. Structural analysis of unbonded concrete overlays under wheel and environmental loading. Transportation Research Record No. 1449. Washington, DC: Transportation Research Board.

Koerner, R. M. and B.-L. Hwu. 1991. Prefabricated highway edge drains, Transportation Research Record No. 1329. Washington, DC: Transportation Research Board.

Kok, A. W. M. 1990. A PC program for the analysis of rectangular pavements structures. *Proceedings, Second International Workshop on the Design and Rehabilitation of Concrete Pavements*, pp. 113–120. Sigüenza, Spain.

Lane, L. B., S. Hoffeld, and D. Griffin. 1998. Environmental justice evaluation: Wilmington bypass, Wilmington, North Carolina. Transportation research record, *Journal of the Transportation Research Board*, 1626: 131–139.

Larson, G. and B. J. Dempsey. 1997. Enhanced integrated climatic model, Version 2.0, Final Report, Contract DTFA MN/DOT 72114. Urbana, IL: Department of Civil Engineering, University of Illinois.

Leahy, R. B. 1989. Permanent deformation characteristics of asphalt concrete. PhD dissertation, University of Maryland at College Park, College Park, MD.

Lee, Y. H. and M. I. Darter. 1995. Development of performance prediction models for Illinois continuously reinforced concrete pavements, TRR 1505. Washington, DC: Transportation Research Board.

Lytton, R. L., D. E. Pufahl, C. H. Michalak, H. S. Liang, and B. J. Dempsey. 1990. An integrated model of the climatic effects on pavements, Report No. FHWA-RD-90-033, Texas Transportation Institute, Texas A&M University. McLean, VA: Federal Highway Administration.

Lytton, R., J. Uzan, E. Fernando, R. Roque, D. Hiltunen, and S. Stoffels. 1993. Development and validation of performance prediction models and specifications for asphalt binders and mixes, strategic highway research program, Report No. SHRP-A-357. Washington, DC: U.S. Department of Transportation.

Majidzadeh, K. 1981. Implementation of a pavement design system, Final Report, Research Project EES 579. Columbus, OH: Ohio State University.

Majidzadeh, K. E. M., D. Kauffmann, V. Ramsamooj, and A. T. Chan. 1970. Analysis of fatigue and fracture of bituminous paving mixtures, Report No. 2546. Washington, DC: U.S. Bureau of Public Roads, Research and Development.

Maupin, G. W., Jr. and J. R. Freeman, Jr. 1976. Sample procedure for fatigue characterization of bituminous concrete, FHWA-RD-76-102, June. Washington, DC: U.S. Department of Transportation.

May, R. W. and M. W. Witczak. 1992. An automated asphalt concrete mix analysis system. *Proceedings of the AAPT*, Vol. 61, pp. 154–187. Charleston, SC.

McCullough, B. F. and M. L. Cawley. 1981. CRCP design based on theoretical and field performance, *Proceedings of the Second International Conference on Concrete Pavement Design*, Purdue University, West Lafayette, IN.

McCullough, B. F. and G. E. Elkins. 1979. CRC *Pavement Design Manual*. Austin, TX: Austin Research Engineers.

McGennis, R. B., S. Shuler, and H. U. Bahia. 1994. Background of superpave asphalt binder test methods, FHWA-SA-94-069. Washington, DC: U.S. Department of Transportation.

McLeod, N. W. 1970. Influence of hardness of asphalt cement on low temperature transverse pavement cracking. *Sixth World Highway Conference*, Canadian Good Roads Association, Montreal, Quebec, Canada.

McLeod, N. W. 1976. Asphalt cements: Pen-vis number and its application to Moduli of stiffness. *American Society of Testing and Materials Journal of Testing and Evaluation* 4(4).

Mindess, S. and J. F. Young. 1981. *Concrete*. Englewood Cliffs, NJ: Prentice-Hall.

Miner, M. A. 1945. Cumulative damage in fatigue. *ASME Transactions* 67: A159–A164.

Moavenzadeh, F., J. E. Soussou, H. K. Findakly, and B. Brademeyer. 1974. Synthesis for rational design of flexible pavements part 3: Operating instructions and program documentation, Reports for FH 11-776. Washington, DC: U.S. Department of Transportation.

Molenaar, A. A. A. 1983. Structural performance and design of flexible road constructions and asphalt concrete overlay. PhD dissertation, Delft University of Technology, Delft, the Netherlands.

Monismith, C. L., J. A. Epps, D. A. Kasianchuk, and D. B. McLean. 1972. Asphalt mixture behavior in repeated flexural, Report No. TE 70-5. Berkeley, CA: University of California, Institute of Transportation and Traffic Engineering.

Monismith, C. L., D. A. Kasianchuk, and J. A. Epps. 1967. Asphalt mixture behavior in repeated flexural: A study on in-service pavement near Morro Bay, California, IER Report TE67-4. Berkeley, CA: University of California.

Mooney, M. A., R. V. Rinehart, N. W. Facas, O. M. Musimbi, D. J. White, and P. K. R. Vennapusa. 2010. Intelligent soil compaction systems, NCHRP Report 676. Washington, DC: Transportation Research Board.

Moulton, L. K. 1979. Design of subsurface drainage systems for control of groundwater. Transportation Research Record No. 733. Washington, DC: Transportation Research Board.

Moulton, L. K. 1980. Highway subsurface design, FHWA-TS-80-224. Washington, DC: U.S. Department of Transportation.

National Asphalt Pavement Association (NAPA). 2003. Constructing quality hot mix asphalt pavements: A troubleshooting guide and construction reference, QIP 112. Lanham, MD: NAPA.

National Cooperative Highway Research Program (NCHRP). 1994. Long-term performance of geosynthetics in drainage applications, NCHRP Report 367. Washington, DC: Transportation Research Board, National Research Council, National Cooperative Highway Research Program.

National Cooperative Highway Research Program (NCHRP). 2004. Thin and ultra-thin whitetopping, NCHRP Synthesis 338. Washington, DC: Transportation Research Board, National Research Council, National Cooperative Highway Research Program.

Neville, A. M. 1981. *Properties of Concrete*, 3rd edn. London, U.K.: Pitman.

Newcomb, D. E., M. Buncher, and I. J. Huddleston. 2001. Concepts of perpetual pavements in perpetual bituminous pavements, Transportation Research Circular, No. 503, December. Washington, DC: Transportation Research Board, National Research Council.

Office of Management and Budget. 1992. (Transmittal Memo No. 64). Circular No. A-194, revised, October 29. www.whitehouse.gov/omb/circulars/a094/a094.html#l

Older, C. 1924. Highway research in Illinois. *Transactions of ASCE*, 87.

Packard, R. G. and S. D. Tayabji. 1985. New PCA thickness design procedure for concrete highway and street pavements. *Proceedings of the Third International on Concrete Pavement Design and Rehabilitation*, pp. 225–236. West Lafayette, IN: Purdue University.

Paris, P. C., M. P. Gomez, and W. P. Anderson. 1961. A rational theory of fatigue. *Trend in Engineering* 13(7): 9.

PCA. 1984. Thickness design for concrete highway and street pavements. Skokie, IL: Portland Cement Association.

PCA, 1991. Design and construction of joins for concrete highways. Portland, OR: Concrete Paving Technology, Portland Cement Association.

Peter, C. T., Steven, H. K., Gerald, F. V. et al. 2007. *Integrated Materials and Construction Practices for Concrete Pavement: A State-of-the-Practice Manual*. Washington, DC: Federal Highway Administration, FHWA HIF-07-004-2007.

PIARC. 1987. Combating concrete pavement slab pumping. PIARC Technical Committee on Concrete Roads.

Poole, T. S. 2005. *Guide for Curing of Portland Cement Concrete Pavements.* Washington, DC: Federal Highway Administration, FHWA-RD-02-099.

Portland Cement Association (PCA). 1984. *Thickness Design for Concrete Highway and Street Pavements.* Skokie, IL: Portland Cement Association.

Portland Cement Association (PCA). 2002. *Design and Control of Concrete Mixtures*, 14th edn., Engineering Bulletin 001. Skokie, IL: PCA.

Powell, W. D., J. F. Potter, H. C. Mayhew, and M. E. Nunn. 1984. The structural design of bituminous pavements, TRRL Laboratory Report 1132. Wokingham, U.K.: Transportation and Road Research Laboratory.

Powers, T. C. 1932. Studies of workability of concrete. *Journal of the American Concrete Institute*, 28: 419.

Qinwu, Xu., Ruiz, M. J., Chang, G. K., Dick, J. C., Garber, S. I., and R. O. Rasmussen. 2009. *Computer Based Guidelines for Concrete Pavements: HIPERPAV® III User Manual.* Washington, DC: Federal Highway Administration, FHWA DTFH61 99C 00081.

Raad, L. and J. L. Figueroa. 1980. Load response of transportation support systems. *Journal of Transportation Engineering, ASCE* 106(1): 111–128.

Rasmussen, R. O., Rogers, R., and Ferragut, T. 2012. *Continuously Reinforced Concrete Pavement (CRCP) Design and Construction Guidelines.* Federal Highway Administration and Concrete Reinforcing Steel Institute.

Read, J. and D. Whiteoak. 2003. Shell bitumen. *The Shell Bitumen Handbook*, 5th edn. London, U.K.: Thomas Telford.

Ridgeway, H. H. 1976. Infiltration of water through the pavement surface, Transportation Research Record No. 616, pp. 98–100. Washington, DC: Transportation Research Board.

Roberts, F. L., P. Kandhal, E. R. Brown, D.-Y. Lee, and T. W. Kennedy. 1996. *Hot Mix Asphalt Materials, Mixture Design, and Construction.* Lanham, MD: National Asphalt Pavement Association (NAPA) Education Foundation.

Rose, G. and D. Bennett. 1994. Benefits from research investment: Case of Australian accelerated loading facility pavement research program, Transportation Research Record No. 1455. Washington, DC: Transportation Research Board.

Ruiz, J. M., Kim, P. J., Schindler, A. K., and R. O. Rasmussen. 2001. Validation of HIPERPAV for prediction of early age jointed concrete pavement behavior. Transportation Research Record 1778. Washington, DC, pp. 17–25.

Salsilli, R. A., E. J. Barenberg, and M. I. Darter. 1993. Calibrated mechanistic design procedure to prevent transverse cracking of jointed plain concrete pavements. *Proceedings of the 5th International Conference on Concrete Pavement Design and Rehabilitation.* West Lafayette, IN: Purdue University.

Schapery, R. A. 1986. Time-dependent fracture: Continuation aspects of crack growth. *Encyclopedia of Materials Science and Engineering*, ed. M. B. Bever. Elmsford, NY: Pergamon Press.

Shell. 1978. *Shell Pavement Design Manual: Asphalt Pavements and Overlays for Road Traffic.* London, U.K.: Shell International Petroleum.

Smith, R. 2004. Forensic investigation of pavement failures. Darling Heights, Queensland, Australia: University of Southern Queensland.

Solaimanian, M. and P. Bolzan. 1993. Strategic highway research program report SHRP-A-637: Analysis of the integrated model of climate effects on pavements. Washington, DC: Transportation Research Board, National Research Council.

Solaimanian, M. and T. W. Kennedy. 1993. Predicting maximum pavement surface temperature using maximum air temperature and hourly solar radiation, Transportation Research Record No. 1417. Washington, DC: Transportation Research Board, National Research Council.

Spangler, M. G. 1942. *Stresses in the Corner Region of Concrete Pavements Bulletin 157.* Ames, IA: Iowa State College.

Tabatabie, A. M. and E. J. Barenberg. 1980. Structural analysis of concrete pavement systems. *ASCE Transportation Engineering Journal* 106(5): 493–506.

Tangella, S., C. Rao, J. Craus, J. A. Deacon, and C. L. Monismith. 1990. Summary report on fatigue response of asphalt mixtures, Interim Report No. TM-UCB-A-003A-89-3, Project A-0030A Strategic Highway Research Program. Berkeley, CA: University of California, Institute of Transportation Studies.

Tayabji, S. D. and B. E. Colley. 1983. Improved pavement joints. Transportation Research Record 930. Washington, DC: Transportation Research Board, National Research Council, pp. 69–78.

Tayabji, S. D. and B. E. Colley. 1986. Analysis of jointed concrete pavements, FHWA/RD-86/041. Washington, DC: U.S. Department of Transportation.

Taylor, H. F. W. 1997. *Cement Chemistry*. London, U.K.: Thomas Telford.

Thompson, M. R. 1987. ILLI-PAVE based full-depth asphalt concrete pavement design procedure. *Proceedings of the Sixth International Conference on Structural Design of Asphalt Pavements*. Ann Arbor, MI: University of Michigan.

Timoshenko, S. and J. N. Goodier. 1951. *Theory of Elasticity*, 2nd edn. New York: McGraw-Hill. (Originally published in 1934.)

Tseng, K. H. and R. L. Lytton. 1986. Prediction of permanent deformation in flexible pavement materials. *ASTM Symposium, Implication of Aggregates in the Design, Construction and Performance of Flexible Pavements*, New Orleans, LA.

Ullidtz, P. 1980. *Engineering Manual*. College Station, TX: Military Construction University.

Ullidtz, P. 1987. *Pavement Analysis*. New York: Elsevier.

United Nations. Economic Commission for Europe. 1990. *Post-Project Analysis in Environmental Impact Assessment*, Volume 3 of Environmental Series. Geneva, Switzerland: United Nations.

U.S. General Accounting Office. 1998. Airfield pavement: Keeping nation's runways in good condition could require substantially higher spending, GAO/RCED-98-226. Washington, DC: U.S. GAO.

Van der Poel, C. 1954. A general system describing the viscoelastic properties of bitumens and its relation to routine test data. *Journal of Applied Chemistry* 4(May): 221–236.

Verstraeten, J., J. E. Romain, and V. Veverka. 1977. The Belgian road research center's overall approach to asphalt pavement structural design. *Proceedings of the Fourth International Conference on Structural Design of Asphalt Pavements*, Vol. 1, pp. 298–324. Ann Arbor, MI.

Vesic, A. S. and S. K. Saxena. 1969. Analysis of structural behavior of road test rigid pavements, Highway Research Board No. 291. Washington, DC: Highway Research Board.

Von Quintus, H. L., M. I. Darter, and J. Mallela. 2007. *Manual of Practice-Interim Mechanistic-Emperical Pavement Design Guide*. NCHRP 1-40B. Washington, DC: Transportation Research Board.

Von Quintas, H. L., J. A. Sherocman, C. S. Hughes, and T. W. Kennedy. 1991. Asphalt-aggregate mixture analysis system: AAMAS, NCHRP Report No. 338, March. Washington, DC: National Cooperative Highway Research Program, National Research Council.

Washington State Department of Transportation. 1995. Everstress: Version 5. Seattle, WA: Washington State Department of Transportation.

Washington State Department of Transportation, Software Programs, 2012. http://www.wsdot.wa.gov/Business/MaterialsLab/Pavements/PavementDesign.htm

Westergaard, H. M. 1926a. Stresses in concrete pavements computed by theoretical analysis. *Public Roads* 7: 25–35.

Westergaard, H. M. 1926b. Analysis of stresses in concrete pavement due to variations of temperature. *Proceedings, Highway Research Board* 6: 201–215.

Westergaard, H. M. 1927. Theory of concrete pavement design. *Proceedings, Highway Research Board, Part 1*, 7: 175–181.

Westergaard, H. M. 1939. Stresses in concrete runways of airports. *Proceedings of the 19th Annual Meeting of the Highway Research Board*, pp. 90–112. Washington, DC: National Research Council.

Westergaard, H. M. 1948. *New Formulas for Stresses in Concrete Pavements in Airfields*, Vol. 113. Washington, DC: American Society of Civil Engineers Transactions.

Witczak, M. W., H. Von Quintus, K. Kaloush, and T. Pellinen. 2000. Simple performance test: Test results and recommendations. Superpave Support and Performance Models Management. NCHRP 9-19. Task C Report, November. Tempe, AZ: Arizona State University.

Wood, S. L. 1991. Evaluation of long-term properties of concrete. *ACI Materials Journal* 88(6, November–December): 630–642.

World Bank. 1995. *HDM III: The Highway Design and Maintenance Standards Model*. Washington, DC: World Bank.

Yao, Z. 1990. Design theory and procedure of concrete pavements in China. *2nd International Workshop on the Theoretical Design of Concrete Pavements*, Siguenza, Spain.

Yoder, E. J. and M. W. Witczak. 1975. *Principles of Pavement Design*, 2nd edn. New York: John Wiley.

Yu, H. T., M. I. Darter, K. D. Smith, J. Jiang, and L. Khazanovich. 1998. Performance of concrete pavements, Vol. III—Improving concrete performance, Report No. FHWA-RD-95-111. Washington, DC: Federal Highway Administration.

Bibliography

ACI 318. 2002. Building code requirements for structural concrete. Farmington Hills, MI: American Concrete Institute.

ACI Committee 214. 1977. Recommended practice for evaluation of strength test results for concrete, ACI 214–77. Reapproved 1997. Farmington Hills, MI: American Concrete Institute.

ACI Committee 302. 2002. Concrete floor and slab construction. Farmington Hills, MI: American Concrete Institute.

ACI. N.d. [Home page]. http://www.concrete.org/general/home.asp, American Concrete Institute.

ACPA 1993, 1996a, 1998.

Acum, W. E. A. and L. Fox. 1951. Computation of load stresses in a three layer elastic system. *Geotechnique* 2(4): 293–300.

Ahlborn, G. 1972. *ELSYM5: Computer Program for Determining Stresses and Deformations in Five Layer Elastic System.* Berkeley, CA: University of California.

Ahlvin, R. C. and H. H. Ulery. 1962. Tabulated values for determining complete pattern of stresses, strains and deflections beneath a uniform circular load on a homogeneous half space. *Highway Research Board Bulletin* 342: 1–3.

Ahmed, N. V., R. L. Lytton, J. P. Mahoney, and O. T. Phillips. 1978. Texas rehabilitation and maintenance district optimization systems, Research Report No. 207–3. College Station, TX: Texas Transportation Institute, Texas A&M University.

American Association of State Highway and Transportation Officials (AASHTO). 1993. *AASHTO Guide for Design of Pavement Structures.* Washington, DC: AASHTO.

Asphalt Emulsions Manufacturers Association (AEMA). 1997. *A Basic Asphalt Emulsion Manual*, 3rd edn., Manual Series No. 19. Annapolis, MD: AEMA.

Asphalt Pavement Alliance. N.d. Life cycle cost analysis software, http://asphaltroads.org/economics/apa-releases-new-life-cycle-cost-software.html

Asphalt Pavement Alliance. N.d. Perpetual pavements, http://asphaltroads.org/PerpetualPavement

Association of Modified Asphalt Producers. N.d. [Home page]. http://modifiedasphalt.org/

Bahia, H. U. and D. A. Anderson. 1994. *The Pressure Aging Vessel (PAV): A Test to Simulate Rheological Changes due to Field Aging*, ASTM Special Technical Publication 1241. Philadelphia, PA: American Society for Testing and Materials.

Barenberg, E. J. and M. R. Thompson. 1992. *Calibrated Mechanistic Structural Analysis Procedures for Pavements,* Phase 2 of NCHPR Project 1-26. Washington, DC: National Cooperative Highway Research Program and Transportation Research Board.

Bhatti, M. A., J. A. Barlow, and J. W. Stoner. 1996. Modeling damage to rigid pavements caused by subgrade pumping. *Journal of Transportation Engineering* 122(1, January–February): 12–21.

Bolander, P. and A. Yamada. 1999. Project leader dust palliative selection and application guide, Online document: The forest service. Washington, DC: United States Department of Agriculture (USDA), http://www.fs.fed.us/eng/pubs/html/99771207/99771207.html

Boussinesq, J. 1885. *Application des Potentiels a L'Etude de L'Equilibre et du movement des Solides Elastiques.* Paris, France: Gauthier-Villars.

British Standards Institution. N.d. [Home page]. http://www.bsigroup.com/

Brown, E. R. N.d. Class notes for CE 589, Pavement construction. Auburn, AL: Civil Engineering Department, Auburn University.

Brown, E. R. N.d. Class notes from CE 688, Asphalt mix design. Auburn, AL: Civil Engineering Department, Auburn University.

Brown, E. R. and L. A. Cooley, Jr. 1999. *Developing Stone Matrix Asphalt Mixtures for Rut-Resistant Pavements*, Project D-09-08, NCHRP Report 425. Washington, DC: Transportation Research Board, National Center for Asphalt Technology, National Research Council.

Burmister, D. M. 1943. The theory of stresses and displacements in layered systems and application to the design of airport runways. *Proceedings of the Highway Research Board.* Washington, DC: Highway Research Board.

Burmister, D. M. 1962. Applications of Layered system concepts and principals to interpretations and evaluations of asphalt pavement performances and to design and construction. *Proceedings, International Conference on the Structural Design of Asphalt Pavements*, pp. 441–453, Ann Arbor, MI: University of Michigan.

Butos, M., H. E. DeSolminihac, M. I. Darter, A. Caroca, and J. P. Covarrubias. 1998. Calibration of jointed plain concrete pavements using long-term pavement performance, Transportation Research Record No. 1629. Washington, DC: Transportation Research Board.

Cedergren, H. 1994. America's pavements: World's longest bathtubs. *Civil Engineering* (September): 56–58.

Cedergren, H. R., J. A. Arman, and K. H. O'Brien. 1973. Development of guidelines for the design of subsurface drainage systems for highway pavement structural sections, Report No. FHWA-RD-73-14. Washington, DC: U.S. Department of Transportation.

CERL. 2006. MicroPaver. *Drainage of Highway and Airfield Pavements*. Champaign, IL: U.S. Army Construction Engineering Research Laboratory.

Choubane, B., G. A. Sholar, J. A. Musselman, and G. C. Page. 1999. A ten-year performance evaluation of asphalt-rubber surface mixes, Preprint Paper No. 00926. Presented at the *Transportation Research Board Meeting*, Washington, DC.

Christensen, D. W. and D. A. Anderson. 1992. Dynamic mechanical test data for paving grade asphalt cement. *Journal of AAPT* 61: 117.

Christory, J. P. 1990. Assessment of PIARC recommendations on the combating of pumping in concrete pavements. *Sixth International Symposium on Concrete Roads*, Madrid, Spain.

Concrete Pavement Technology Program (CPTP). N.d. CPTP Status Report: Task 65 engineering ETG review copy. Washington, DC: U.S. Department of Transportation.

Council on Environmental Quality, Executive Office of the President. 1978. Regulations for Implementing the Procedural Provisions of the National Environmental Policy Act, Code of Federal Regulations Title 40, Parts 1500–1508.

Craik, N. 2008. *The International Law of Environmental Impact Assessment: Process, Substance and Integration*. New York: Cambridge University Press.

Crovetti, J. A. and B. J. Dempsey. 1993. Hydraulic requirements of permeable bases, Transportation Research Record No. 1425. Washington, DC: Transportation Research Board.

Daniels, T. and D. Katherine. 2003. *The Environmental Planning Handbook for Sustainable Communities and Regions*. Chicago, IL: Planners Press, American Planning Association.

Darter, M. I., K. T. Hall, and C.-M. Kuo. 1994. Appendices, support under concrete pavements. Washington, DC: National Cooperative Highway Research Program.

Darter, M. I., K. T. Hall, and C. Kuo. 1995. Support under Portland cement concrete pavements, NCHRP Report 372. Washington, DC: National Cooperative Highway Research Program.

Darter, M. I., S. A. LaCourseiere, and S. A. Smiley. 1979. Performance of continuously reinforced concrete pavement in Illinois, Transportation Research Record No. 715. Washington, DC: Transportation Research Board.

DeLong, D. L., M. G. F. Peutz, and A. R. Korswagen. 1973. Computer Program BISAR, Layered systems under normal and tangential surface loads, External Report AMSR.0006.73. Amsterdam, the Netherlands: Koninklijke/Shell Laboratorium Amsterdam.

Dempsey, B. J. 1982. Laboratory and field studies of channeling and pumping, Transportation Research Record No. 849, 1–12. Washington, DC: Transportation Research Board.

Dempsey, B. J. 1988. Core flow capacity requirements of geocomposite fin-drain materials used in pavement subdrainage, Transportation Research Record No. 1159. Washington, DC: Transportation Research Board.

Dempsey, B. J. 1993. Performance of prefabricated geocomposite subdrainage system in an airport runway. Federal Aviation Administration Report DOT/FAA/RD-93/23. Washington, DC: FAA.

Densit. N.d. [Home page]. http://www.densit.com/

Downing, D. and R. B. Noland. 2000. Environmental consequences of reducing the federal role in transportation: The legal framework. Washington, DC: Transportation Research Board.

Elliott, R. P. and M. R. Thompson. 1985. Mechanistic design concepts for conventional flexible pavements, Transportation Engineering Series No. 42. Urbana, IL: University of Illinois.

Emerson, D. and G. Benz. 1998. Major investment studies and environmental documentation: Clearing up the confusion. Washington, DC: Transportation Research Board.

Fassman, E. A. and S. L. Yu. 2000. Estimating evapotranspiration from small mitigated wetland sites. Washington, DC: Transportation Research Board.

Federal Aviation Administration (FAA). N.d. U.S. Department of Transportation, [Home page]. www.faa.gov

Federal Aviation Administration (FAA). N.d. Airports. www.faa.gov/airports_airtraffic/airports/

Federal Aviation Administration (FAA). N.d. Design software. www.faa.gov/airports_airtraffic/airports/construction/design_software/

Federal Aviation Administration (FAA). N.d. Engineering briefs. http://www.faa.gov/airports/engineering/engineering_briefs/index.cfm

Federal Aviation Administration (FAA). N.d. Pavement design and construction. http://www.faa.gov/airports/engineering/pavement_design/

Federal Highway Administration (FHWA). 1994a. Pavement deflection analysis. NHI Course No. 13127, Participant Workbook, Publication No. FHWA-HI-94-021. Washington, DC: Federal Highway Administration.

Federal Highway Administration (FHWA). 1994b. Pavement deflection analysis, NHI Course No. 13127, Publication No. FHWA-HI-94-021. Washington, DC: U.S. Department of Transportation.

Federal Highway Administration (FHWA). 1995. Geotextile engineering manual, course text, Publication No. FHWA-HI-89-050. Washington, DC: U.S. Department of Transportation.

Federal Highway Administration (FHWA). 2004a. Concrete pavement technology update. http://www.fhwa.dot.gov/pavement/concrete/cptu201.pdf, accessed December 12, 2012.

Federal Highway Administration (FHWA). 2004b. Framework for LTPP forensic investigations—Final, April 2004. Washington, DC: Federal Highway Administration.

Federal Highway Administration (FHWA). 2008. Recycling: Current projects and activities. www.fhwa.dot.gov/pavement/recycling, accessed December 12, 2012.

Federal Highway Administration (FHWA). N.d. Asset management: Life-cycle cost analysis. www.fhwa.dot.gov/infrastructure/asstmgmt/lcca.cfm, accessed December 12, 2012.

Federal Highway Administration (FHWA). N.d. [Home page]. U.S. Department of Transportation, http://www.fhwa.dot.gov/, accessed December 12, 2012.

Federal Highway Administration (FHWA). N.d. Pavements: LTPPBind. www.fhwa.dot.gov/pavement/ltpp/ltppbind.cfm, accessed December 12, 2012.

Federal Highway Administration (FHWA). N.d. Preservation. www.fhwa.dot.gov/preservation, accessed December 12, 2012.

Fehr, R. 1999. Environmental health impact assessment: Evaluation of a ten-step model. *Epidemiology* 10: 618–625.

Foster, C. R. and R. G. Ahlvin. 1954. Stresses and deflections induced by a circular load. *Proceedings of Highway Research Board* 33: 467–470.

Foundation for Pavement Preservation. N.d. Welcome to the foundation for pavement preservation. www.fp2.org

Fox, L. 1948. Computation of traffic stresses in a simple road structure, Road Research Technical Paper No. 9. Crowthorne, U.K.: Department of Scientific and Industrial Research, Road Research Laboratory.

Hajek, J., J. W. Hall, and D. K. Hein. 2011. Common airport pavement maintenance practices, ACRP Synthesis 22, A Synthesis of Airport Practice. Washington, DC: Transportation Research Board.

Halstead, W. J. 1983. Criteria for use of asphalt friction surfaces, NCHRP Synthesis 104. Washington, DC: Transportation Research Board.

Hayhoe, G. F. 2004. LEAF—A new layered elastic computational program for FAA pavement design and evaluation procedures. Paper presented at the *FAA Administration Technology Transfer Conference*, April 2002, http://www.airporttech.tc.faa.gov/naptf/Downloads/p-26.pdf, accessed December 12, 2012.

Heitzman, M. S. 1992. State of the practice: Design and construction of asphalt paving materials with crumb rubber modifier, FHWA SA-92-022. Washington, DC: U.S. Department of Transportation.

Highway Community Exchange, Federal Highway Administration (FHWA). N.d. Highway Community Exchange Online Forum. https://www.transportationresearch.gov/dot/fhwa/hcx/default.aspx

Highway Community Exchange, Federal Highway Administration (FHWA). N.d. NCHRP 1-37A (Mechanistic-Empirical) Pavement Design Guide. https://www.transportationresearch.gov/dot/fhwa/hcx/default.aspx

Hines, M. L. 1993. *Asphalt Cement Properties Improved by Styrelf: Laboratory and Field Data.* Wichita, KS: Koch Materials Company.

Hoerner, T. E., M. I. Darter, L. Khazanovich, L. Titus-Glover, and K. L. Smith. 2000. Improved prediction models for pcc pavement performance-related specifications, Vol. 1, Final Report. Washington, DC: Federal Highway Administration.

Horak, E. 2008. Benchmarking the structural condition of flexible pavements with deflection bowl parameters. *Journal of the South African Institution of Civil Engineering* 50(2): 2–9.

Ioannides, A. M., C. M. Davis, and C. M. Weber. 1999. Westergaard curling solution reconsidered, Transportation Research Record No. 1684. Washington, DC: Transportation Research Board.

Ioannides, A. M. and M. I. Hammons. 1996. Westergaard type solution for edge load transfer, Transportation Research Record No. 1525. Washington, DC: Transportation Research Board.

Jones, A. 1962. Tables of stresses in three layer elastic systems: Stress distribution in earth masses, Bulletin 342. Washington, DC: Highway Research Board.

Kandhal, P. S. and L. Lockett. 1997. Construction and performance of ultrathin asphalt friction course, NCAT Report No. 97-5, September. Auburn, AL: National Center for Asphalt Technology.

Kerali, H. G. R. 2000. *Highway Development and Management (HDM-4), Vol. 1: Overview of HDM-4.* Paris and Washington, DC: PIARC (World Road Association) and the World Bank.

King, G. N., H. W. Muncy, and J. B. Prudhomme. 1986. Polymer modification: Binder effect on mix properties. *Journal of Asphalt Paving Technologists Association,* 55: 519–540.

Kosmatka, S. H., B. Kekhoff, and W. Panarese. 2002. *Design and Control of Concrete Mixtures,* 14th edn. Skokie, IL: Portland Concrete Association.

Lane, L. B., S. Hoffeld, and D. Griffin. 2000. Determining disproportionate impacts in environmental justice evaluations Wilmington bypass. Wilmington, NC: TRB.

Likos, W. J. and N. Lu. 2001. Automated measurement of total suction characteristics in the high suction range: Application to the assessment of swelling potential, Transportation Research Record No. 1755. Washington, DC: Transportation Research Board.

Little, D. N., M. R. Thompson, R. L. Terrel, J. A. Epps, and E. J. Barenberg. 1987. Soil stabilization for roadways and airfield, Final Report. Panama City, FL: Engineering and Services Laboratory, Air Force Engineering and Services Center, Tyndall Air Force Base.

Lohani, B., J. W. Evans, H. Ludwig, R. R. Everitt, R. A. Carpenter, and S. L. Tu. 1997. *Environmental Impact Assessment for Developing Countries in Asia. Volume 1—Overview,* 356pp. Manila, Philippines: Asian Development Bank.

Lytton, R. L., R. L. Boggess, and J. W. Spotts. 1976. Characteristics of expansive clay roughness of pavements, Transportation Research Record No. 568, 9–23. Washington, DC: Transportation Research Board.

Lytton, R. L., D. T. Phillips, and C. V. Shanmugham. 1982. The Texas rehabilitation and maintenance system. Paper presented at the *Fifth International Conference on the Structural Design of Asphalt Pavements,* p. 613. Delft, the Netherlands, August 23–26.

Majidzadeh, K. 1988. A mechanistic approach to rigid pavement design. In *Concrete Pavements.* London, U.K.: Elsevier.

Masad, E. 2005. Aggregate imaging system (aims): Basics and applications, Report No. FHWA/TX-05/5-1707-01-1. Washington, DC: FHWA.

McCabe, M. and B. Sadler (eds.). 2003. Studies of EIA practice in developing countries. Nairobi, Kenya: United Nations Environment Programme.

McCall, J. T. 1958. Probability of fatigue failure of plain concrete. *Proceedings of the American Concrete Institute* 55(13): 233–244.

McCullough, B. F., J. C. M. Ma, and C. S. Noble. 1980. Limiting criteria for the design of continuously reinforced concrete pavements, Transportation Research Report No. 756. Washington, DC: Transportation Research Board.

Mega Concrete LLC. N.d. [Home page, in Polish], http://www.megaconcretellc.com/

Minnesota Department of Transportation. 2000–2006. PaveCool: Asphalt pavement cooling tool. http://www.dot.state.mn.us/app/pavecool/

Mitchell, J. K. 1993. *Fundamentals of Soil Behavior,* 2nd edn. New York: Wiley International Sciences.

Najjar, Y., I. Basheer, H. Ali, and R. McReynolds. 2000. Swelling potential of Kansas soils: Modeling and validation using Ann reliability approach. Paper presented at the *Seventy-Ninth Annual Meeting of the Transportation Research Board,* January 9–13. Washington, DC: Transportation Research Board.

National Asphalt Pavement Association (NAPA). 1994. Guidelines for materials, production, and placement of stone matrix asphalt (SMA), IS 118. Lanham, MD: NAPA.

National Center for Pavement Preservation (NCPP). N.d. NCPP reference library. www.pavementpreservation.org/reference/

National Climatic Data Center (NCDC). N.d. Heavy rainfall frequencies for the U.S. http://ncdc.noaa.gov/oa/documentlibrary/rainfall.html

National Cooperative Highway Research Program (NCHRP). 1990. Calibrated mechanistic structural analysis procedures for pavements, Vol. 1, Final Report. Washington, DC: National Cooperative Highway Research Program.

National Cooperative Highway Research Program (NCHRP). N.d. Design guide: Mechanistic-empirical design of new & rehabilitated pavement structures, www.trb.org/mepdg

National Lime Association. 1991. Lime stabilization construction manual, Bulletin 326. Arlington, VA: National Lime Association.

National Research Council. 1994. Sensitivity analyses for selected pavement distress, SHRP-P-393. Washington, DC: National Research Council.

National Slag Association. N.d. [Home page]. www.nationalslag.org

National Stone, Sand & Gravel Association (NSSGA). N.d. [Home page]. www.nssga.org

Occupational Safety and Health Administration (OSHA). N.d. [Home page]. U.S. Department of Labor, Occupational Safety & Health Administration, http://www.osha.gov/

Occupational Safety and Health Administration (OSHA). N.d. OSHA Hazard Communication Standards, 29 CFR 1910.1200. www.ilpi.com/msds/osha/1910_1200.html

Page J. and L. Parkins. 1998. Cumulative impact assessment and its application to a transportation project. Washington, DC: TRB.

Paris, P. C. and F. Erdogan. 1963. A critical analysis of crack propagation laws. *Transactions of the American Society of Mechanical Engineers, Journal of Basic Engineering, Series D* 85(3): 538.

Parker, F. N.d. Class notes for CE 584, Soil stabilization. Auburn, AL: Civil Engineering Department, Auburn University.

Peattie, K. R. 1962. Stress and strain factors for three-layer elastic systems: Stress distributions in earth masses, Bulletin No. 342. Washington, DC: Highway Research Board.

Peattie, K. R. and A. Jones. 1962. Surface deflection of road structures. *Proceedings: Symposium on Road Tests for Pavement Design*, pp. 1–30, viii. Lisbon, Portugal.

Peutz, M. G. F., H. P. M. Van Kempen, and A. Jones. 1968. Layered systems under normal surface loads. *Highway Research Record* 228: 34–45.

Pfeiffer, J. P. and P. M. Van Doormaal. 1936. The rheological properties of asphaltic bitumens. *Journal of Institute of Petroleum Technologists* 22: P414.

Phillips, D. T., F. Ghasemi-Tari, and R. L. Lytton. 1980a. Rehabilitation and maintenance system: State optimal fund allocation—Program I, Research Report No. 239-4. College Station, TX: Texas Transportation Institute, Texas A&M University.

Phillips, D. T., C. V. Shanmugham, S. Sathaye, and R. L. Lytton. 1980b. Rehabilitation and maintenance system: District time optimization, Research Report No. 239-3. College Station, TX: Texas Transportation Institute, Texas A&M University.

PIARC (World Road Association) Technical Committee on Flexible Roads and on Surface Characteristics. 1993. Porous asphalt, Publication No. 08.01B. Paris, France: PIARC.

Portland Cement Association (PCA). 1992. *Soil-Cement Laboratory Handbook*. Skokie, IL: PCA.

Prithvi, S. K., F. Parker, and R. B. Mallick. 1999. Aggregate tests for hot mix asphalt: State of the practice, Transportation Research Circular No. 479. Washington, DC: Transportation Research Board, National Research Council.

Raad, L. 1982. Pumping mechanisms of foundation soils under rigid pavements, Transportation Research Record No. 849, pp. 29–37. Washington, DC: Transportation Research Board.

Rauhut, J. B., R. L. Lytton, and M. I. Darter. 1984a. Pavement damage functions and load equivalence factors, FHWA-RD-84-01. Washington, DC: U.S. Department of Transportation.

Rauhut, J. B., R. L. Lytton, and M. I. Darter. 1984b. Pavement damage functions for cost allocations, Vol. 2, FHWA-RD-84-019, June. Washington, DC: U.S. Department of Transportation.

van Rooijen, R. C., A. H. de Bondt, and R. L. Corun. 2004. Performance evaluation of jet fuel resistant polymer-modified asphalt for airport pavements. Paper presented at the *FAA Worldwide Airport Technology Transfer Conference*. Atlantic City, NJ, http://www.airporttech.tc.faa.gov/naptf/att07/2004%20Track%20P.pdf/P04038.pdf

Rubber Pavements Association. N.d. [Home page]. www.rubberpavements.org

Saarenketo, T., P. Kolisoja, K. Luiro, P. Maijala, and N. Vuorimies. 2000. Percostation for real time monitoring moisture variations, frost depth and spring thaw weakening, Transportation Research Report No. 00833. Washington, DC: Transportation Research Board.

Sayers, M. W., T. D. Gillespie, and W. D. O. Paterson. 1986a. Guidelines for conducting and calibrating road roughness measurements, Technical Paper 46. Washington, DC: World Bank.

Schwartz, D. R. 1987. D-cracking of concrete pavements, NCHRP Synthesis of Highway Practice No. 134. Washington, DC: National Cooperative Highway Research Program.

Seed, H. B. 1959. A modern approach to soil compaction. *Proceedings of the Eleventh California Street and Highway Conference*, Reprint No. 69, p. 93. Berkeley, CA: Institute of Transportation and Traffic Engineering, University of California.

Seiler, W. J. Expedient stress analysis of jointed concrete pavement loaded by aircraft with multiwheel gear, Transportation Research Report No. 1370, pp. 29–38. Washington, DC: Transportation Research Board.

Shahin, M. Y. N.d. MicroPaver. Washington, DC: U.S. Army Corps of Engineers.

Shell. 1951. *Shell Pavement Design Manual: Asphalt Pavements and Overlays for Road Traffic*. London, U.K.: Shell.

Sivaneswaran, N., L. M. Pierce, and J. P. Mahoney. 2001. *Evercalc® Pavement Backcalculation Program*, Version 5.20, March. Seattle, WA: Materials Laboratory, Washington State Department of Transportation.

Smith, K. L., K. D. Smith, L. D. Evans, T. E. Hoerner, and M. I. Darter. 1997. Smoothness specifications for pavements, Final Report, NCHRP 1-31. Washington, DC: Transportation Research Board.

Smith, K. D., T. P. Wilson, M. I. Darter, and P.A. Okamoto. 1992. Analysis of concrete pavements subjected to early loadings, Transportation Research Report No. 1370. Washington, DC: Transportation Research Board.

Spangler, E. B. and W. J. Kelly. 1964. GMR road profilometer—A method for measuring road profiles, Research Publication GMR-452. Detroit, MI: General Motor Corporation.

Strategic Highway Research Program (SHRP). 1993. Distress identification manual for long term pavement performance project, SHRP-P-338. Washington, DC: Strategic Highway Research Program.

Strickland, D. 2000. *Shell Pavement Design Software for Windows*. Wythenshawe, U.K.: Shell Bitumen.

Tayabji, S. D., P. J. Stephanos, and D. G. Zollinger. 1995. Nationwide field investigation of continuously reinforced concrete pavements, Transportation Research Report No. 1482. Washington, DC: Transportation Research Board.

Texas Transportation Institute. 2001. MODULUS. http://tti.tamu.edu

Tolman, F. and F. Gorkum. 1996. Mechanical durability of porous asphalt. Vol. 2. Paper presented at Eurasphalt and Eurobitume Congress, Strasbourg, May 7–10, 1996.

Transportation Research Board. 2004. Guide for mechanistic-empirical design of new and rehabilitated pavement structures, Final Report, Part 1: Introduction, Part 2: Design inputs, National Cooperative Highway Research Program, Copy No. 88. Washington, DC: Transportation Research Board, National Research Council.

Transportation Research Board. 2010a. http://onlinepubs.trb.org/onlinepubs/nchrp/nchrp_rpt_646.pdf

Transportation Research Board. 2010b. http://onlinepubs.trb.org/onlinepubs/nchrp/nchrp_w162.pdf

U.S. Army Corps of Engineers, Construction Engineering Research Laboratory. N.d. MicroPaver. www.cecer. army.mil/paver/Downloads.htm

U.S. Department of Transportation (USDOT). 1992. Traffic monitoring guide, FHWA-PL-92-017, October. Washington, DC: U.S. Department of Transportation.

U.S. Department of Transportation (USDOT). 2003. Memorandum: Revised departmental guidance. http://www.dot.gov/sites/dot.dev/files/docs/vot_guidance_092811c_0.pdf, 2.

Uzan, J., S. Frydman, and G. Wiseman. 1984. Roughness of airfield pavement on expansive clay. *Proceedings, Fifth International Conference on Expansive Soils*, pp. 286–291. Adelaide, South Australia, Australia, May 21–23.

Van Wijk, A. J., J. Larralde, C. W. Lovell, and W. F. Chen. 1989. Pumping prediction model for highway concrete pavements. *ASCE, Journal of Transportation Engineering* 115(2): 161–175.

Velasco, M. O. and R. L. Lytton. 1981. Pavement roughness on expansive clay, Transportation Research Record No. 790, pp. 78–87. Washington, DC: U.S. Department of Transportation.

Wallace, D. and J. S. Shalkowski. 2000. Post NEPA monitoring of environmental impacts and mitigation commitments. Washington, DC: TRB.

Wardle, L. J. 1976. Program CIRCLY: A computer program for the analysis of multiple complex circular loads on layered anisotropic media: User's Manual Geomechanics Computer Program No. 2. Melbourne, Victoria, Australia: CSIRO Division of Applied Geomechanics.

Warren, H. and W. L. Dieckmann. 1963. Numerical computation of stresses and strains in a multiple-layer asphalt pavement system. Unpublished report. Richmond, CA: Chevron Research Corporation.

Washington State Department of Transportation, 2012. http://www.wsdot.wa.gov/Business/MaterialsLab/Pavements/PavementDesign.htm

Washington State Department of Transportation, Falling Weight Deflectometer. 2012b. http://www.wsdot. wa.gov/Business/MaterialsLab/Pavements/PavementDesign.htm

Winkler, E. 1867. Die Lehre von der Elastizitat und Festigkeit [The theory of elasticity and stiffness]. Prague, Czech Republic: H. Dominicus.

Wu, C. L., J. W. Mack, P. A. Okamoto, and R. G. Packard. 1993. Prediction of faulting of joints in concrete pavements. *Proceedings, Fifth International Conference on Concrete Pavement Design and Rehabilitation*, Vol. 2. West Lafayette, IN: Purdue University, April.

Index